Skeletal Tissue Mechanics

Matthew R. Allen • Joseph Wallace

Skeletal Tissue Mechanics

Third Edition 2025

Springer

Matthew R. Allen
Department of Anatomy
Cell Biology & Physiology
Indiana University School
of Medicine
Indianapolis, IN, USA

Joseph Wallace
Weldon School of Biomedical
Engineering
Purdue University
Indianapolis, IN, USA

ISBN 978-3-031-90608-4 ISBN 978-3-031-90609-1 (eBook)
https://doi.org/10.1007/978-3-031-90609-1

This Springer imprint is published by the registered company Springer Nature Switzerland AG
The registered company address is: Gewerbestrasse 11, 6330 Cham, Switzerland

This book, as with previous editions, is dedicated to R. Bruce Martin. This book was based on his many years of lecturing and thinking about the skeleton and how it works both from an engineering perspective and from a biological one. Many of the ideas expressed in this edition were original with Bruce. He thought of them; he developed them; he applied them.

In the same spirit, we extend the dedication to David Burr, Neil Sharkey, and David Fyhrie, who contributed significantly to develop this book, which has become essential in the field of skeletal mechanics. Their faith in us to take this work into this third edition is humbling.

As with all things, we are both grateful to our families for their support. In the first edition of this book, Bruce and David creatively dedicated it to their kids by plotting their ages referenced to Basic Multicellular Unit(s) (BMU's), one of the central premises of bone remodeling and this book. We echo their words, in saying that our children have "positively remodeled our lives for the better." In the same way, our wives, Sarah and Kristine, have supported us in ways no words of thanks or love can convey.

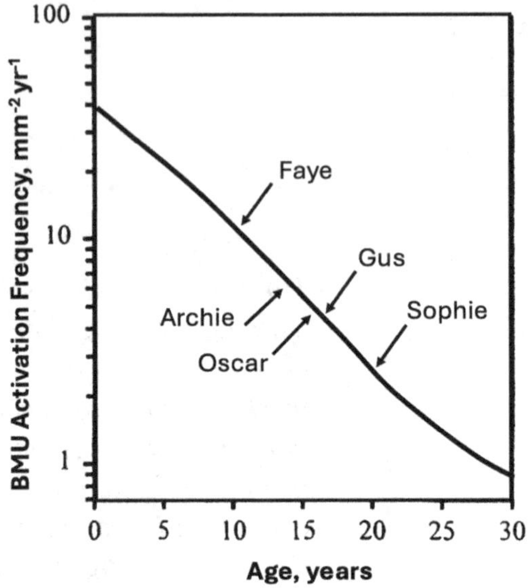

Preface

Skeletal Tissue Mechanics, first published in 1998, was intended to review the biology of skeletal tissues and their mechanical properties and biomechanical functions. It was purposefully interdisciplinary—bringing together anatomy/physiology with structural/material mechanics, with the goal of being unique among publications. The text filled a niche for skeletal researchers at boundaries of orthopedics, biomechanics, skeletal biology, and tissue engineering. The second edition of the text, published in 2015, integrated new literature and expanded the content to other musculoskeletal tissues (cartilage, ligament, tendon).

In this third edition, we have focused on overall updates to the content. This includes updating areas, such as histomorphometry, where technical advances have made some of the described approaches less relevant. We have shifted some of the content among chapters, added learning objectives, and consolidated study questions into an appendix at the end. Most notably, we have added a "Historical" chapter. This text existed in the predecessor to this text, written by Martin and Burr, which is out of print and challenging to find. This chapter provides insight into the earliest work of bone, its remodeling and its mechanical properties. Such history plays an important role in our field, and bringing this back into print will hopefully provide the next generation with a bridge to the past.

Indianapolis, IN, USA

Matthew R. Allen
Joseph Wallace

Preface

Preface to Second Edition

The first edition of *Skeletal Tissue Mechanics*, published in 1998, grew out of Bruce Martin's lecture notes for a course by the same name, which was part of the biomedical engineering curriculum at the University of California at Davis. Given this history, it is not surprising that the book is intended to be used as a textbook, not a reference book. But it also is intended to comprehensively review the biology of all the skeletal tissues, and their mechanical properties and biomechanical functions. The integration of anatomy and physiology with structural and material behavior was one of the book's principal objectives, and this remains the goal of the second edition as well. The interdisciplinary nature of the book is strengthened by the complementary blend of perspectives brought by the authors, who have backgrounds in Functional Anatomy, Anthropology, Physiology, Biomedical and Mechanical Engineering, and Orthopedics.

The first edition was primarily written for graduate and advanced undergraduate students in biomedical engineering. It fit a niche between textbooks focused on the biomechanics of locomotion and sport, and those aimed more specifically at orthopedic biomechanics, which often do not concentrate on the mechanical behaviors of skeletal tissues or the biologic processes that underlie them. In practice, the book was used more broadly, by residents in orthopedic surgery, and students at various levels who were studying anthropology, kinesiology, veterinary medicine, and skeletal physiology.

An explosion of new information—both experimental and computational—has improved our understanding of the influences on, and adaptability of, skeletal tissues. Our approaches to understand both the dynamic and the static properties of skeletal tissues have become more sophisticated as well. Because of this, the first edition of *Skeletal Tissue Mechanics* became woefully dated, and its continued use as a text mandated that we update it, and expand upon certain topics.

Besides updating and refreshing the content, and correcting some errors present in the first edition, the second edition of this book differs from the first in a number of important ways. The Biology of all skeletal tissues (bone, cartilage, ligament, and tendon) is now incorporated into a single chapter, and described using a hierarchical approach, focusing on the mechanics of the different biological levels of organization in each. The Mechanical aspects of each tissue are moved to subsequent chapters, with mechanics related to bone now at the end of the book. Having used the book ourselves, we have attempted to re-organize the chapters into a sequence that makes more sense from a didactic standpoint.

We have added new Historical Notes, set off from the main text in boxes, which we hope will increase student interest but also broaden their perspective. Additional boxes with Technical Notes will help students understand how to collect, analyze, and use information related to the textual material. We have updated the supplementary readings at the end of each chapter. We revised the exercise set we developed for students that we hope will help them to consolidate and use the information found in the chapters. Few of these exercises are simply practice at solving problems demonstrated in the book; rather, most are designed to stimulate the student to synthesize concepts based on material discussed in the text, and to develop analytical problem-solving skills that will serve them in the future. Finally, figures have been updated, re-drawn in some cases to improve content, or make them more easily understandable, and color figures are included where we felt it would enhance learning.

We appreciate both our students and our colleagues who have contributed directly and indirectly to this book. Feedback and critical thought from students was particularly helpful in identifying gaps, or confusing presentations. Discussions with colleagues are always enlightening, and informed some of the changes. Beyond that, the co-authors themselves engaged in many entertaining discussions, some philosophical and others scientific, which expanded the boundaries of our own thinking. Finally, we would like to thank Drew Brown for his help in drawing (and re-drawing, often several times) some of the new figures.

Davis, CA, USA R. Bruce Martin
Indianapolis, IN, USA David B. Burr
University Park, PA, USA Neil A. Sharkey
Davis, CA, USA David P. Fyhrie

Competing Interests The authors have no competing interests to declare that are relevant to the content of this manuscript.

Contents

Chapter 1
The Historical Development of Concepts Related to the Structure, Function, and Dynamic Nature of Bone

Abstract This chapter provides an overview of aspects of bone "history," but it is neither required reading nor will there be a test on dates (two things that make many people dislike history). The goal of this chapter is twofold. First, it aims to help the readers understand where many fundamental concepts originated (and by whom) and how, throughout the centuries, there was great debate on such topics, some of which continues today. Second, it serves an attempt to preserve history. Nine years before the first edition of this book, Bruce Martin and David Burr wrote "Structure, Function, and Adaptation of Compact Bone" (Martin and Burr, Structure, function, and adaptation of compact bone. Raven Press, 1989). It served as a foundation for the current book but is now out of print and very challenging to obtain. The first chapter "Structure, Function, and Adaptation of Compact Bone" was a rich historical account of concepts that are now foundational aspects of bone physiology. It had been left out of the first two editions of the current text. We have brought it back, slightly edited, in hopes that readers can benefit from the historical research done by Martin and Burr over 30 years ago (Martin and Burr, Structure, function, and adaptation of compact bone. Raven Press, 1989).

What You Will Learn

1. Historical discoveries and controversies regarding bone microstructure
2. Historical perspectives on bone remodeling and its role in altering bone structure/function
3. The evolution of thought on how bone structure adapts to mechanical loads
4. Historical views about the functional role of the osteon in bone mechanics

M. R. Allen, J. Wallace, *Skeletal Tissue Mechanics*,
https://doi.org/10.1007/978-3-031-90609-1_1

1.1 History of Concepts Related to Microscopic Bone Structure [1]

Soon after the invention of the compound microscope in the early seventeenth century, people began to observe and interpret the microscopic structure of bone. These early descriptions led to much speculation and discourse about whether the microscopic structure of bone was largely an artifact (or, as Bostock described it in 1825, "self-deception" [2]), and if not, what the function of the microscopic structure was. The poor quality of magnifying lenses and limited discussion about human bone led initial descriptions to focus primarily on the canal system and the "laminar" structure of bone. Interestingly, the man whose name is often associated with the "Haversian system" never actually discussed the structural unit, although he did make several other noteworthy observations that shaped the direction of early work.

1.1.1 Clopton Havers (1655–1702)

Clopton Havers is generally credited with the first description, in 1691, of the porous nature of bone. Technically, however, van Leeuwenhoek had reported his observation of an extensive canal system in bone to the Royal Society of London 13 years before this, in 1678 [3]. van Leeuwenhoek called the canals "pipes" and pointed out that they ran both longitudinally and transversely in bone. Even prior to this he had observed the channels in bone but described the pores as lobules of tissue, rather than voids, within the bone [4].

Although not first, Havers provided a more extensive description of the canal system in bone, linking it to his ideas of the lamellar nature of bone tissue [3]. Based on his studies of bone from a variety of different species, including humans, he described the canals as forming parallel strings (interlamellar substance) and plates (lamellae, which he called laminae or Lamells), contributing to our present confusion over these terms (Fig. 1.1). He noticed that the strings and plates existed around the medullary cavity in a tubular fashion and were more or less continuous from one end of the bone to the other. He also commented on deviations from this general structure in which lamellae were not tubular and did not run the length of the bone, but he conveyed that he thought these deviations were pathologic. At the ends of the bones, he described how the plates splayed out from one another to form cancellous bone. To satisfy the geometric difficulty that these plates must therefore be of different lengths, he theorized that the outermost strings were longer than the innermost. Consequently, the innermost strings could bend to form the cancellous latticework at the ends of the bone. This accounted for the gradually diminished thickness of cortical bone in regions where cancellous bone was more extensive.

Havers wrote that the pores found between the plates were irregularly spaced and consisted of two kinds—one longitudinally oriented parallel to the strings and one

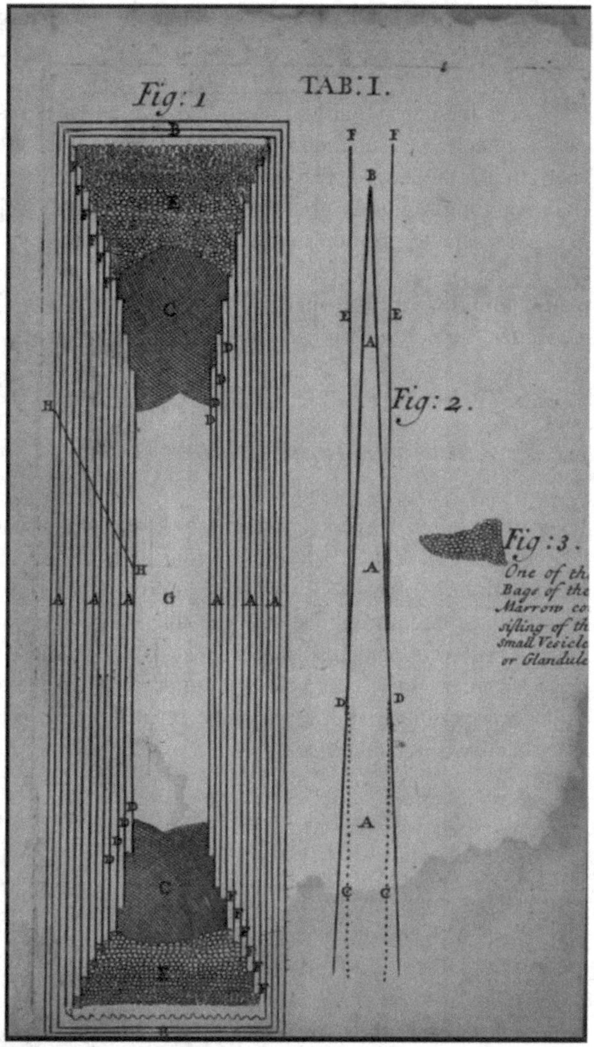

Fig. 1.1 Haver's conception of the structure of bone in the longitudinal section [3]

oriented in the transverse direction. Havers denied the presence of vessels in these pores, but did recognize that vessels passed through bone on their way to the marrow.

Havers was adamant that the pores were for the passage of the "medullary oils," which diffused between laminae to keep the bone from drying out [3]. If vessels were also found in the pores, they would be separate from the movement of the medullary oil. How Havers thought blood reached the marrow space, if not through the pores, is unclear. He distinguished different kinds of spaces in bone by the terms "Pores" (the "Haversian" canals) and "Foramina" (nutrient fora-mina), some of which were for the passage of blood vessels and some for the transfer of fluids/oils to and from the marrow. It is possible that Havers did not always distinguish between the function of blood vessels and the function of the marrow in either fortifying bone nutritionally or keeping it viable.

Havers indicated that the transversely oriented pores coming from the medullary cavity enter the endosteal bone obliquely so as not to compromise the mechanical strength of the bone. They then divide within the substance of bone, sending channels longitudinally up and down between the laminae. He understood that this was necessary to provide nutrition to as much bone as possible. Havers speculated that as the strings and plates divided at the bone ends, the vessels between them would mingle with the marrow. Note that Albinus (1697–1770) was the first to recognize the pores within cortical bone as having vascular channels, in 1754 [5].

1.1.2 Anton van Leeuwenhoek (1632–1723)

Two years after Havers's descriptions, van Leeuwenhoek (1693) classified the "Tubes and Cilinders'' in bone into four types based primarily on size [6]. Although his description of these four types is somewhat obscure, interpretation from his descriptions and drawings suggest that the three largest tubuli were osteocyte lacunae, osteons, and resorption spaces; the smallest tubuli were likely canaliculi (Fig. 1.2). Certainly, later scientists interpreted the smallest pores as canaliculi, and by the early nineteenth century, the term "minute pores" had clearly come to be associated with canaliculae. Van Leeuwenhoek therefore was almost certainly the

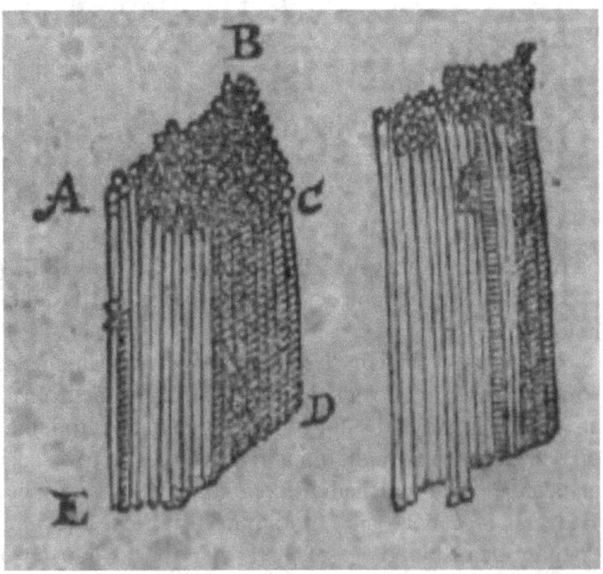

Fig. 1.2 Drawings of bone microstructure from Anthony Leeuwenhoek in 1693 [6]

first to recognize what later came to be known as the "Haversian System," and today is often referred to as simply an osteon:

> I found them placed in such an Order, as convinced me, that the Ring of these Tubuli was the Augmentation of the Bone, as I had formerly discovered it to be in Wood, especially when I saw, at a little distance, another Circle or Ring of Tubuli [6].

Although the analogy with wood leads us to believe at first that van Leeuwenhoek was discussing circumferential lamellae, reference to his figure suggests otherwise (Fig. 1.2). The figure shows a series of pores placed throughout the bone cortex. Assuming the scale of van Leeuwenhoek's drawings is correct, the diameter of the canals presented is 35–65 μm, very close to the average Haversian canal diameter of 50 μm. His figure gives no indication that he was able to visualize the lamellae around the canals. Nevertheless, his suggestions that the circles he found in bone were each composed of a series of rings ("Circle or Ring of Tubuli") and that these rings were bone rather than pores ("The Rings of these Tubuli was the Augmentation of the Bone") makes it tempting to suggest that he had observed the entire "Haversian" system.

Van Leeuwenhoek also observed radially running "Tubes" of two sizes. The smaller tubes were equivalent to the smallest longitudinal tubes, and both appeared to connect the longitudinal tubes. These tubes came to be known for many years as 'Volkman canals' although today are often simply referred to as transverse channels.

1.1.3 The Eighteenth Century

Van Leeuwenhoek's descriptions were soundly ignored, possibly because of their obscurity and brevity, and little progress was made in understanding the microscopic structure of bone over the next several decades. By the middle of the eighteenth century, descriptions of the lamellar and fibrous structure of bone were reminiscent of Havers' descriptions 40–80 years before [7, 8]. The canal system in bone was well accepted by this time and had acquired the eponym "straight pores of Havers" (Haversii pori recti) in reference to Clopton Havers [5].

The major controversy during the eighteenth century was over the presence of lamellae in bone. Between 1739 and 1743, Du Hamel (1700–1782) demonstrated the lamellar nature of bone using pulse labels of madder root, which contains alizarin and stains bone red [9]. This was the first instance of vital staining of bone, a hallmark of human and animal bone histologic study today. When the bone was viewed microscopically, it appeared as a series of rings, alternating red and white. He concluded that bone is formed of concentric laminae, deposited from the periosteum.

These experiments were refuted by the best-known anatomists and physiologists of the day (e.g., Haller, 1708–1777; Rutherford, 1695–1779; Cheselden, 1688–1752; Scarpa, 1752–1832). Haller and Rutherford [2] disputed the findings on the curious grounds that the madder had an affinity for the inorganic (earthy) part of bone and that it did not attach to the organic (animal) part of bone, as Hunter had proposed [10]. This line of reasoning demonstrates the poor understanding at that time about the deposition and subsequent mineralization of bone. Although Meckel (1837) later claimed that anatomists of the day doubted the fibrous structure of bone, it was actually the arrangement of these fibers into plates of concentric lamellae that they doubted [11]. For example, Bichat (1771–1802) felt that the laminar arrangement of bone was entirely artifactual and did not occur in nature [12]. At the same time, Bichat acknowledged that canals ran between layers of bone, although said that they were infrequent. The confusion to which all these ideas led is amply demonstrated by Bostock's statements as late as 1825:

> These concentric layers certainly do not exist, but I think it equally certain that the membrane of bone is composed of plates, ... and it is probable that the earthy matter is inserted between these plates and thus is likewise disposed to assume the laminated structure [2].

This line of reasoning, which plates of bone exist but are not organized concentrically, is a direct repetition of Havers's descriptions more than 130 years before.

In an environment in which the very lamellar structure of bone was questioned, it is not difficult to understand why van Leeuwenhoek's observations of osteons in bone were not examined more closely. Parts of this dispute may have been fueled by the feeling that if bone were disposed of in plates, then these plates must be laid down in some regular fashion, one on another, similar to the growth rings of a tree. However, observations indicated that there was no regularity to these rings, because of what we now know today to be interstitial lamellae (remodeled portions of old osteons that run at various angles to circumferential lamellae and to each other). These discrepant observations were pointed out implicitly by Lieutaud (1703–1780) in 1742 when he claimed that bone was composed of laminae and very compact fibers, highly unified, but without any particular regularity or distinct arrangement [13]. This description was mimicked by numerous biologists for the next 100 years. Meckel (1781–1833) claimed that the irregularity of the laminae caused anatomists to doubt the fibrous structure of bone. This was not strictly true. Those who felt that the laminae were artifactual never disputed the idea that bone was a fibrous structure, only that these fibers were arranged in well-defined plates.

Amid this eighteenth-century discourse, the systems of concentric lamellae constituting osteons were mostly ignored. Todd and Bowman (1845) apparently believed that Gagliardi (1660–1725) was referring to the Haversian system when he stated that the laminae in bone were held together by small nails or pins:

> ... we need not smile at the descriptions of Gagliardi, who, with imperfect means of observation, appears to have been at least faithful in his attempts to delineate nature. The periosteal and medullary layers are true plates of bone, and the Haversian systems are true fibres or pins, all connected with one another by direct continuity of tissue [14].

The idea that bone was held together with pins was discussed for years, with some simply repeating the observation nearly verbatim [12], and others actively denying it [2, 8]. Although somewhat obscure and difficult to interpret, Gagliardi's descriptions do not seem to refer very accurately to osteonal structure in bone. For instance, he calls pores "transverse Epiphyses" of the bony fibers composing the laminae and says they appear to arise within each lamina. He goes on to say that the pores generally run outward, except near the surface of the bone, where they run inward, and that they are found in great numbers even in trabecular bone. The fact that osteons do not run transversely, do not arise within each bony layer, do not run in different directions toward the periosteal surface, and are found in much fewer numbers (or in the case of secondary osteons, are seldom found) in trabecular bone are all arguments against this description applying to osteons. Gagliardi's description may therefore be in reference to transverse/Volkmann's canals as opposed to osteons.

Although Gagliardi's observations probably did not refer to Haversian systems, it is likely that the observations of a French scientist of this era did:

> It is very certain that one can see within the sheets of the sieve-like compact part [of bone] a great quantity of small bundles of different diameter [15].

This is all that De Lasone (1717–1788) says about these bundles, and it is possible that he is referring only to Haversian canals [15]. However, he also went on to discuss the channels in bone as something other than these bundles, suggesting that he made a distinction between the two morphological entities.

1.1.4 The Nineteenth Century

By the mid-nineteenth century, Haversian system observations and discussions were more common in scientific literature. Bruns (1812–1883) and Cruveilhier (1791–1874) nearly simultaneously reconnected with the topic, although neither provided any hint that these had ever been observed before. Bruns states:

> Thus I have a cross section of the middle part of a human thigh bone treated with hydrochloric acid in front of me, the cortical substance [Rindensubstanz] of which, when observed under the microscope, consists of very tightly de-posited marrow canals with their lamellae, without me being able to find anywhere between them a trace of the other lamellae that run concentrically around the whole bone, except very near the surface right under the periosteum [Beinhaut, literally, bone-skin], while with a similar specimen of bovine bone I find the latter [i.e., circumferential lamellae] in the majority and only few canals among them [16].

And Cruveilhier wrote that bone was composed of:

> exceedingly fine lamellae laid on one another. These lamellae run parallel with the surface; they also surround, concentrically, the small cavities of the compact substance [17].

The year following the observations of Cruveilhier, R. B. Todd (1809–1860) and William Bowman (1816–1892) described and defined in greater detail (with exquisite drawings) (Fig. 1.3) systems of concentric lamellae surrounding small cavities in bone, naming them "Haversian systems of lamellae" [10]. It is interesting that, after years of intense debate about the microscopic structure of bone, with little mention of its cellular nature and the interconnections between these cells, Todd and Bowman's highly accurate descriptions of secondary bone seem to emerge full blown and were not debated in detail. Perhaps this was due both to improved histological preparation techniques and advancements in microscopy.

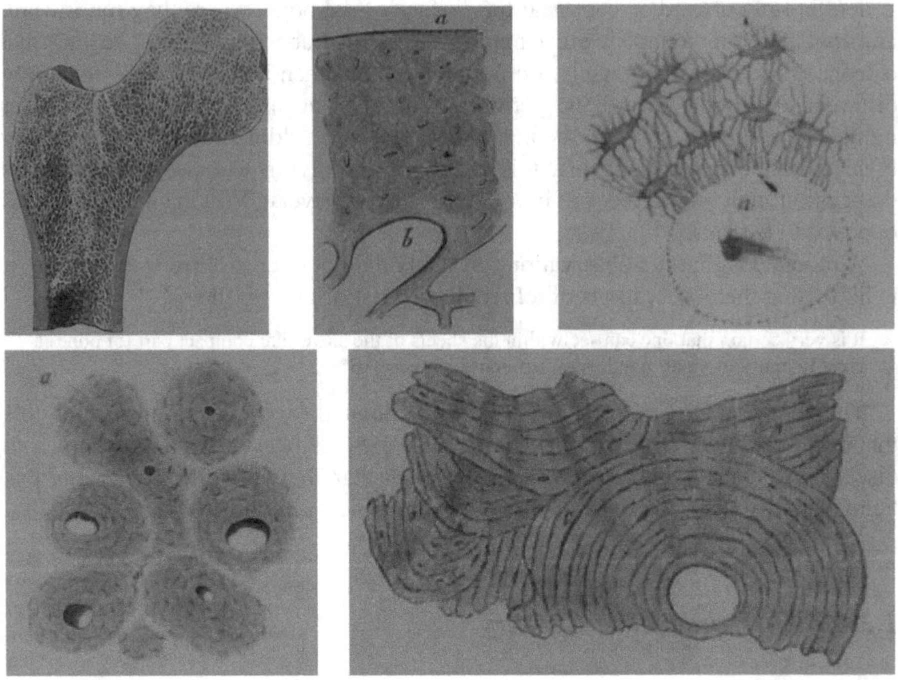

Fig. 1.3 Detailed microstructure drawings of proximal femur trabecular, cortical, and lacunar–canalicular microstructures [10]

Todd and Bowman presented a fairly accurate and rather complete summary of the orientation and disposition of lacunae and canaliculi. They also may have been the first to describe the boundary between the secondary osteon and interstitial matrix as something different than the mineralized matrix. This is the underpinning of thought that eventually led to the feeling that cement lines were hypercalcified regions. This is reflected in their figures, one of which shows the cement line as a region of increased density [10] (Fig. 1.3c). Todd and Bowman felt that the cement line was a vestige of the primary ossification center and its nonporous nature made it denser than normal.

> … the primary osseous network, formed in the intercellular substance of the temporary cartilage, must come to constitute the substance intervening between the Haversian rods, the non-porosity of which is thus satisfactorily accounted for, as well as the facility with which the rods themselves may be made to separate from one another [10].

Their description and figures are somewhat at variance. They call the cement line a "transparent interval" in which "the pores and lacunae [do not exist] to intercept the passage of the light [10]. Quekett repeats these observations and seems to mirror their feelings [18]. Interestingly, the passage quoted above may suggest that Todd and Bowman believed that bone was likely to crack at the cement line. They note this earlier, "Bone is very apt to crack in the interval between the rods" [10]. This supported the idea that they believed that the cement line was not artifactual.

By 1849, Todd and Bowman's descriptions and illustrations of Haversian bone, including the concept of the cement line, were incorporated into the general anatomy texts of the day, often with diagrams that directly copied their work [19, 20].

1.2 History of Concepts Related to Bone Remodeling

Part of the early confusion over the lamellar nature of bone tissue stemmed from a lack of general understanding about the changing nature of bone tissue throughout life. As the structure of bone became clearer, questions surrounding bone physiology and function, at the microscopic level, began to be addressed.

A pathologist, Alexander Monro (1697–1767), had originally described the concept of bone remodeling on surfaces and had understood that the resorption and formation of bone occurred throughout life [7, 21]. He also suggested that the balance between formation and resorption somehow changed with age and that this could explain both growth and the loss of bone in old age:

> there is a perpetual waste and renewal of the particles which compose the solid fibres of bones; … the addition from the fluids exceeding the waste during the growth of the bones; the renewal and waste keeping pretty near par in adult middle age; and the waste exceeding the supply from the liquors in old age; as is demonstrable from their weight. The bones of old people are thinner and firmer in their sides, and have larger cavities than those of young persons [7].

Monro appeared to recognize that the loss of bone with age occurred not only through expansion of the marrow cavity and subsequent cortical thinning, but also through loss of bone within the cortex. It makes sense that a pathologist should be the first to adequately describe intracortical bone resorption, as this is a feature that is most prominent in metabolic diseases.

Howship (1781–1841) observed in 1816 that interstitial bone could be removed by "absorption" and that the effect of this was the enlargement of the longitudinal channels within the bone cortex [22]. He also described the corrugated texture of the resorption front, which now bears his name (Howship's lacunae). He appears to have realized that the vascular system contained within the Haversian canals was an important part of the removal process. Yet his writings suggest he was not certain whether there were special vessels for this purpose or whether some vessels simply contained materials that caused bone "absorption." There is no indication that Howship understood that these "materials" were cells or that he understood the role cells play in the resorption process.

The development of primary bone from cartilage and the initial ossification of bone had been described by the early eighteenth century but the replacement throughout life of primary bone with secondary bone had not been understood. Todd and Bowman briefly touched on the topic of remodeling [10]. They discussed the formation of new Haversian systems among existing systems by invagination of the canal wall to form a new Haversian canal. In 1853 John Tomes (1815–1895) and Campbell De Morgan (1811–1876) provided the first detailed theories about the remodeling functions of bone, alongside yet more exquisite drawings [23] (Fig. 1.4).

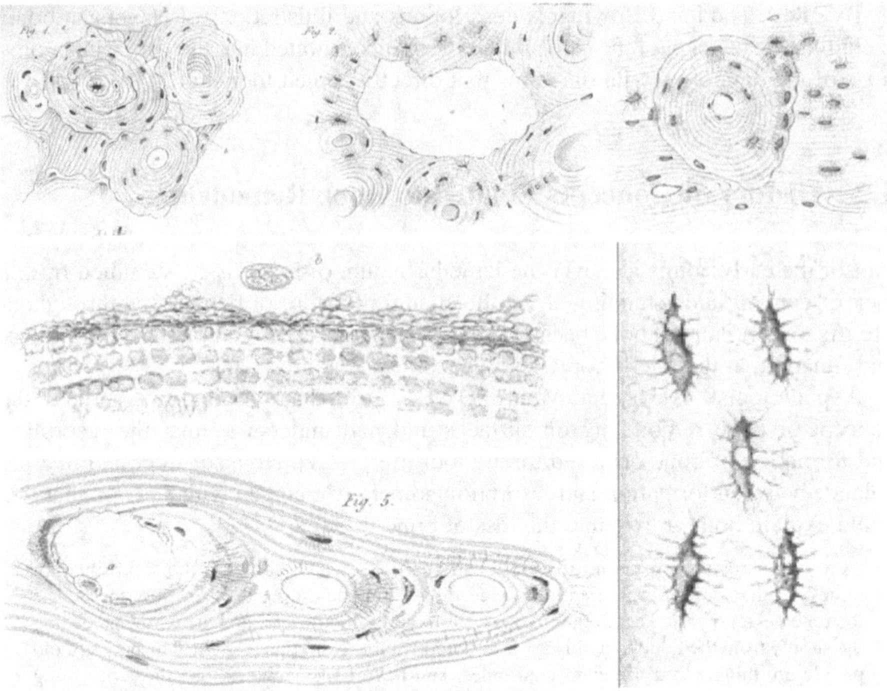

Fig. 1.4 Tomes and De Morgan, in the mid-1800s, provided exquisitely detailed drawings of bone and its cells/structure [23]

Although van Leeuwenhoek may have been the first to observe the presence of large resorption spaces, their relationship to the Haversian system does not seem to have been recognized prior to Tomes and De Morgan. Howship had discussed the presence of resorption spaces in bone disease but gave no indication that the process was an ongoing, naturally occurring one [22]. Tomes and De Morgan, however, noticed that the irregularity of these spaces corresponded to the peripheral outline of other Haversian systems (i.e., the cement line), and theorized that these "Haversian spaces" must precede the formation of a new osteon. Moreover, they correctly identified the interstitial lamellae as remnants of older Haversian systems that had been partially removed at the time of reformation (Fig. 1.4).

Because of these observations, Tomes and De Mogan correctly characterized the Haversian system as replacement bone, suggesting the term "secondary bone" to distinguish it from primary bone derived from cartilage. In this context, they touched on the differences between primary and secondary Haversian systems as well. They noted that in recently formed Haversian systems the canaliculi do not cross the "layer of transparent tissue" (cement line), whereas many more "anastomizes" are observed in older Haversian systems (it is not clear how they determined the age of a given Haversian system). This observation was also made by Todd and Bowman who noted that "newer Haversian systems" were almost certainly areas of remodeled bone. Beyond observations, however, Tomes and De Morgan attributed no significance to the differences between primary and secondary microscopic structures; indeed, they were not even routinely recognized as separate entities.

The mechanism for bone remodeling was not clear at this time, and one can imagine that Tomes and De Morgan's observations may have been criticized on the grounds that, because bone is so hard, cells could not be capable of resorbing it. Indeed, they raise this objection themselves. This criticism is remarkable in view of Howship's well-documented observations of resorption fronts nearly 40 years before. According to Howship, Cruik-shanks, Hooper, and Hatchett had suggested even earlier that the first step in bone resorption might be the breakdown and disaggregation of the solid parts from the "gelatinous matter" [22].

Tomes and De Morgan eventually reached the same conclusion that because bone gradually mineralizes after deposition, it could equally well be slowly demineralized prior to its resorption [23]. Instead of implicating cells in this process, they suggested that both mineralization and demineralization were caused by the organic matrix. Although they were unaware of bone- resorbing cells, they did observe compound cells (perhaps referring to cells with multiple nuclei). They suggested these might be the precursors of bone-forming cells, which they called "osteal" cells to distinguish them from "lacunal" cells (what we now call osteocytes) (Fig. 1.4). They found osteal cells aligned parallel to bone lamellae in regions of bone formation, and suggested that osteal cells could in some cases become lacunal cells. Tomes and De Morgan described the arrangement of osteal bone-forming cells in a single layer on the Haversian canal wall:

> The manner in which the Haversian spaces become gradually occupied by Haversian systems is particularly interesting…osteal cells arrange themselves in single file within the Haversian space with intermediate lines of transparent tissue, and here and there a lacunal cell; the process commencing at the surface of the Haversian space, and extending gradually inwards till the system is completed. In fact the soft tissue takes the permanent form previous to the addition of the salts of bone Lamination is nothing more than a definite linear arrangement of the osteal cells with their outlines permanently retained in the perfect bone [23].

Tomes and De Morgan also realized that the processes of bone removal and replacement could occur at different rates, and that these rates were affected by age and also by pathologic or inflammatory processes. They pointed out that absorption and reproduction (as they termed these processes) occur in subjects of all ages, both young and old, but Haversian spaces are less numerous and smaller in older bone, signaling a diminished frequency of remodeling and reduced capacity for bone formation.

Tomes and De Morgan were thus aware of several salient features of bone remodeling, the importance of which was emphasized and solidified into concepts by Harold Frost almost 100 years later [23]. For instance, their descriptions of bone reaction in response to inflammation, associated with rapidly escalated resorption rates, probably based on some of Howship's earlier observations, calls to mind Frost's Regional Acceleratory Phenomenon (RAP) [24, 25]. Moreover, Tomes and De Morgan, realizing that the source for bone-resorbing cells must be somehow related to the vascularized soft tissues, suggested that there were most likely three different surfaces on which bone resorption can be initiated: periosteal, endosteal, and Haversian/intracortical. The term "endosteum," attributed to Walshi by Quain, had not come into general usage and was still called the medullary membrane [19].

By the 1850s, the scientific community was aware of the tissue-level structure of bone, and that primary bone was replaced by secondary bone through resorption and reformation, resulting in a Haversian system. In general, there was a feeling that bone resorption and formation were linked (coupled) somehow, and that these processes were altered by aging. The recognition of the significance of these concepts, both for understanding the shape and adaptation of bone, and for understanding bone loss, would come later, driven by the discovery of techniques to dynamically assess the remodeling process.

In 1736, John Belchier discovered that developing bone in livestock could be stained in vivo by feeding the animals madder, a dye derived from the Eurasian herb *Rubia tinctorum*. This fortuitous discovery happened when he observed red streaks in the bones of some pork that he was having for dinner. He investigated and found that the pig had been fed waste plant material from a dye shop. During the ensuing two centuries, Belchier's discovery provided impetus for several qualitative studies of bone growth and fracture healing. In the late 1700s, using madder-fed pigs, the famous English surgeon–anatomist John Hunter was the first to grasp the significance of what we now call modeling in the skeleton. Hunter's experiments demonstrated that bones grow radially by endosteal resorption and periosteal apposition. In the 1800s, synthetic alizarin replaced the natural dye and was found to be equally effective as a bone label. These dyes were found to bind to hydroxyapatite as it is being formed, and therefore preferentially stain the interface between osteoid and mineralized bone.

The modern use of such labels in human bone histomorphometry began in the 1950s when Milch reported that the common antibiotic tetracycline was incorporated into actively mineralizing bone surfaces and fluoresced yellow when histologic sections were observed under ultraviolet light [26] (Fig. 1.5). Along with Milch, Frost and others recognized the significance of this finding: for the first time, the dynamics of human bone modeling and remodeling could be safely studied [27]. Soon, tetracycline-based histologic analysis of bone remodeling was being widely used to study bone physiology.

It turns out that "natural" fluorochrome labeling of our bones has been going on for some time, even without our knowledge! Skeletons excavated from a cemetery in southern Sudan dating from the fourth to sixth centuries AD were found to fluoresce yellow–green when placed under UV light. These people were agriculturists who stored grains until they could be eaten. When the stored grains got damp, they provided a rich environment for the growth of Streptomycetes, a bacterium that produces tetracycline. Thus, by eating the grain, the people were not only ingesting a natural antibiotic (which may have kept them healthier), but also labeling their bones in a way that can be useful to archeologists and bone biologists today in trying to understand dynamic bone remodeling in prehistoric populations [28].

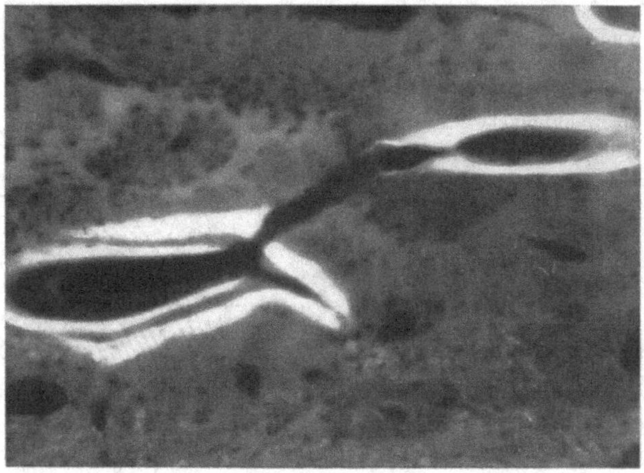

Fig. 1.5 Photomicrographs depicting a fluorescent haversian system [26]

Harold Frost (1921–2004), referenced above for being one of the first to use intravital labels to histologically explore aspects of bone remodeling, is considered one of the most influential theoreticians in skeletal biology of the 1900s [25, 27]. While best known for his work on remodeling, he has influenced broader fields of bone biology through his works and theories. Frost developed techniques to make measures on non-decalcified bone sections—and laid the foundation for modern-day bone histomorphometry. He established histology as a tool to diagnose metabolic bone disease (using rib biopsies), provided some of the first evidence that estrogens suppress bone remodeling, and was the first to histologically identify microcracks in bone. His theories included the basic multicellular unit, the "mechanostat," and the "intermediate organization" of bone [29, 30]. Frost was also one of the first to propose a direct connection between muscle and bone in their coadaptation. His long-time colleague Michael Parfit (who himself was highly influential in the field) said it best

> I often disagreed with Hal and think that he was wrong about many small things (which of us was not?), but he was right about the big things [31].

1.3 History of Concepts Related to Bone Adaption

The expansion of knowledge concerning bone mechanics and adaptation occurred during the same periods as those about structure and remodeling, although mostly by other individuals. As described elegantly by Roesler, there are three key concepts that arose in the nineteenth century related to bone adaption [32, 33]:

1. Optimization of strength with respect to weight
2. Alignment of trabeculae with principal stress directions
3. Self-regulation of bone structure by cells responding to a mechanical stimulus

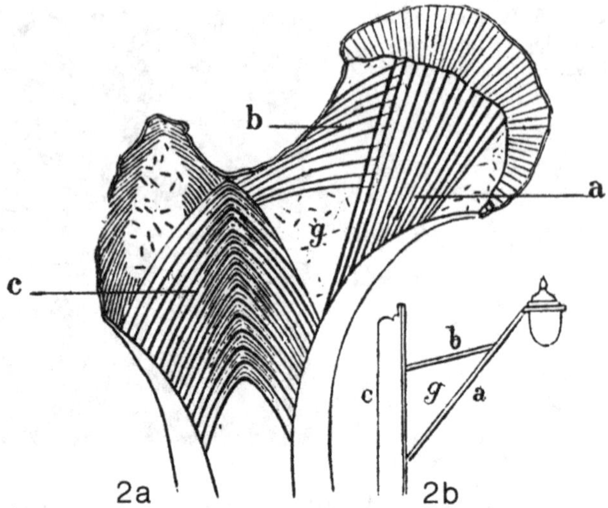

Fig. 1.6 Sketch showing Ward's analogy between the trabecular arrangement in the neck of the femur and a streetlamp bracket [37]

Julius Wolff wrote about the link between bone mechanics and architecture in the late 1800s and is routinely credited for this idea (Wolff's Law), but he was not the first to observe it. Galileo, in 1638, pointed out the mechanical implications of the shapes of bones [34], as did other early writers [7]. Bell, in the early 1800s, observed that cancellous bone structure had "reference to the forces acting on the bone" [35]. Both Bell and Bourgery apparently recognized that trabecular bone architecture is influenced by mechanical forces and thought that the bone maximized strength relative to the amount of material used [36]. This relationship is the first important concept to recognize: optimization of bone strength (or stiffness) with respect to its weight. This is important because the density of bone tissue is roughly twice that of other tissues in the body making the metabolic cost of carrying extra bone high. While Bell and Bourgery may not have ever considered this specifically, their description of the exquisite structure of cancellous bone was a foundation for the concept.

Around the same time as Bell and Bourgery, Ward compared the trabecular arrangement in the neck of the femur to a bracket designed to hold a street lamp to its post [37] (Fig. 1.6). We now apply the name Ward's triangle to the region of sparse trabecular struts corresponding to the central opening in this "bracket." Ward's comparison is significant because it is an early recognition that bone structure is analogous to structures engineered by people and that bone does not remain where it is not needed to support the structure.

The second key concept arose from a legendary conversation between Karl Culmann and Hermann von Meyer. The story goes that 1 day in 1866 Prof. Culmann, a well-known engineer, visited the laboratory of Prof. Meyer, an equally well-known anatomist, and happened to see a specimen of a human proximal femur that had been sectioned longitudinally to reveal its pattern of trabecular arcades. Culmann was immediately struck by the similarity of this pattern to that of the

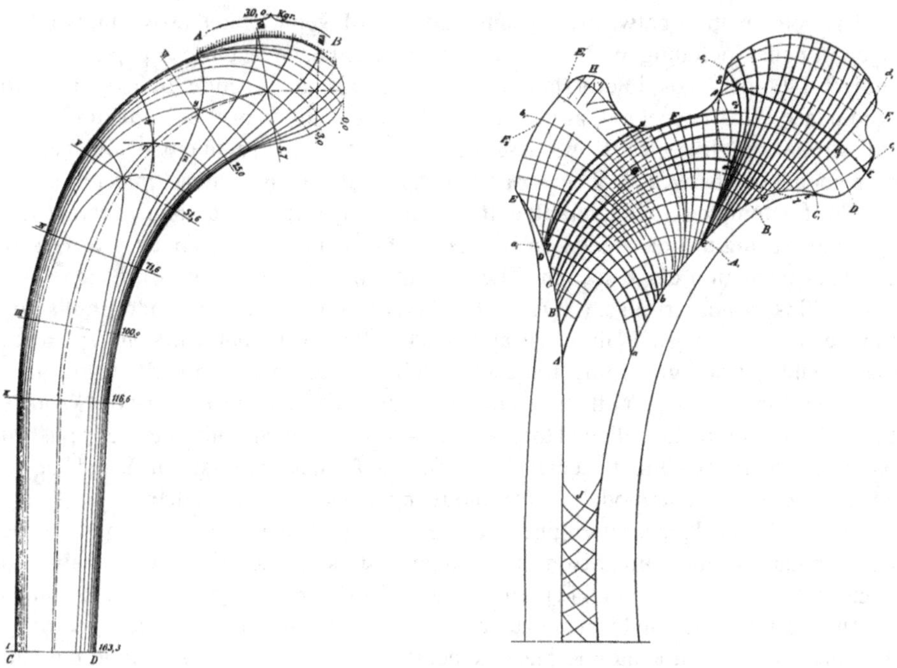

Fig. 1.7 Principal stress trajectories in a Fairbairn crane as calculated by Culmann (Left). The trabecular arrangement in a proximal human femur as sketched by von Meyer (Right) [38]

principal stresses within a crane structure which he had analyzed (Fig. 1.7). It is said that Culmann cried out "That's my crane!" [38]. The crane was actually named after Fairbairn, the Scots engineer who designed it. Its flowing, bone-like shape is typical of the trial-and-error designs common during the early Industrial Revolution in Great Britain. It is interesting to note that elasticity theory and the science of strength of materials were rapidly developing during this time period. Augustin Cauchy had, around 1822, discovered the existence of principal stresses, but decades passed before engineers began to make practical use of the concept. Culmann was one of the first to understand and use the concept of principal stress directions; he was one of several engineers from the continent who led the way in developing a mathematical theory of mechanics, which eventually refined the experimental approaches of British and American engineers. He had apparently been applying his graphical methods for calculating principal stress directions to Fairbairn's crane. In any event, the similarity between the trabecular arcades and the stress trajectories is indeed striking. Clearly, however, the shapes and loads are substantially different for the crane and the femur. Not only that, but the femur is a discontinuous collection of bony struts filled with marrow, while the crane is a homogeneous structure. Principal stresses do not exist for a discontinuous structure. The significance of this was overlooked for a long time [39].

The conversation between Culmann and von Meyer was noted by others interested in bone, including Wolff who soon thereafter began to publish papers on this subject [40, 41]. His interpretations of cancellous architecture quickly rose to dominance. Wolff accepted, at face value, the implications of the Culmann–Meyer stress-trajectory hypothesis and asserted that there was a mathematically perfect correspondence between trabecular architecture and principal stress trajectories.

Wolff attempted to explain the trabecular structure within many pathologic specimens using this basic idea. In 1892, he published his famous book, Das Gesetz der Transformation der Knochen (The Law of Bone Transformation), which summarized his earlier writings. (NOTE—this book has been translated into English by Maquet and Furlong). It must be appreciated, however, that while he spoke of mathematical rules governing trabecular architecture, he was not an engineer or mathematician, and never attempted to formulate a mathematical theory of "Wolff's Law." Furthermore, he believed that bone was formed interstitially (i.e., cells pushed out new bone from within the calcified matrix—a physical impossibility), and had no understanding of bone modeling or remodeling as we now know them.

In 1917, John Koch, an engineer as well as a physician, resolved some of the issues regarding the comparison of Culmann's stress trajectories for the Fairbairn crane with a human femur [42]. He calculated the stress trajectories in a human femur based on a reasonably accurate load applied to the femoral head, and geometry obtained from actual anatomic measurements. In doing so, he was able to demonstrate that trabecular orientations are, after all, similar to mathematical calculations of stress trajectories for a homogeneous structure of the same shape (Fig. 1.8).

We must return now to the nineteenth century to note the emergence of the third key concept regarding the adaptation of bone to mechanical loading. In the time of Wolff, Claude Bernard (1813–1878) introduced the idea of homeostasis to physiology: the concept that physiologic systems self-regulate so as to keep a critical internal variable constant. However, it was many years before the full significance of this concept came to be appreciated and accepted. Thus, it was still a fledgling idea when, in 1881, W. Roux, influenced by Darwin and others, hypothesized that

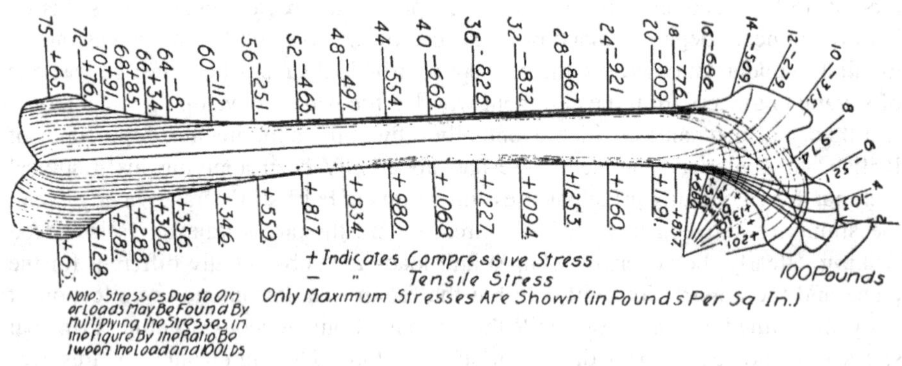

Fig. 1.8 Koch's computation of the principal stress trajectories and values in a human femur. Note the applied load vector on the femoral head [42]

organisms possess the ability to adapt to changes in their living conditions. This differs both from evolution, which involves long-term, genetic changes between one generation and the next, and from adjustments to internal, physiologic changes. What Roux was introducing was the ability of an organism to change in response to interactions with the external environment. Roux specifically proposed that the ability of bone to align trabeculae with stress trajectories was accomplished by cells forming and resorbing bone according to variations in a functional stimulus, in this case, a mechanical stimulus. While his imperfect understanding of biology and mechanics led him to make many mistakes in developing his theory, his concept of self-regulated functional adaptation governed by cell sensitivity to a functional stimulus was extremely important.

1.4 History of Concepts Related to Osteon Function

The idea that bone (and specifically trabeculae) was aligned in the direction of stress trajectories was important to the formative stages of thinking about osteon arrangement and function. Because it was felt that compact bone was simply a condensation of trabeculae, the trajectorial theory was applied to the microstructural components of compact bone; this concept underlies early ideas of osteonal function [43].

Todd and Bowman first raised the issue of the function of Haversian systems, pointing out that Haversian systems are "most artfully arranged for the mechanical ends in view" [10]. What those "ends" were was not explicitly stated. The issue appears not to have been raised in the literature again until 1905, when Gebhardt (1870–1918) discussed the structure and function of osteonal bone in detail [44]. Based on his own observations, he suggested that osteons were suited to resist mechanical stresses and were oriented according to stress trajectories, just like trabeculae. He further noted that the fibrils within concentric osteonal lamellae are arranged in a spiral and that these spirals alternated in adjacent lamellae. He also noticed differences in the angle at which the lamellar fibers were deposited. This was analogous to the different orientations and arrangements of trabeculae, which had been so widely publicized, and so it was quite natural that he should attribute the cause of the differing orientations to stress trajectories. Gebhardt reinforced the concept that the arrangement and structure of osteons continued on a finer scale the mechanical architecture of trabecular bone.

This idea ran into more or less the same criticisms that the trajectorial theory had, i.e., violations of the assumption of homogeneity and isotropy in bone material and the significant variation in the architecture of osteonal lamellae [45]. Benninghoff (1890–1953) attempted to overcome this argument by suggesting that osteon fragments were important to maintain an average homogeneity in bone, providing equal resistance in all planes [46]. As we now know that bone is anisotropic, this is clearly wrong and was shown so shortly after he made the suggestion. Nevertheless, Benninghoff and others, based on "split-line studies of doubtful accuracy," were willing to confirm that osteon orientation coincided with lines of mechanical stress [46, 47]. This body of data led Murray, by 1936, to suggest that

> ... there is every reason for supposing that they [osteons] do continue on a finer scale the mechanical architecture of the coarser structures [48].

The real question was whether the coincidence of osteonal direction with "lines of stress" was incidental or not. That is, if osteons align with stress trajectories, does the mechanical environment itself provoke osteonal remodeling? One gets the feeling in reading the work of these early bone morphologists that the question they were asking and answering was a teleological one: Osteons served a mechanical function once they were present; therefore, they were formed solely to serve a mechanical function. The predisposing factors to osteonal remodeling were not dis-cussed, and the failure to deal with this question ultimately led to Amprino's suggestion that although osteons might serve some mechanical function once formed, remodeling to secondary osteons was caused by the body's need for additional calcium. The same year that Murray published his classic work on bones, Amprino and Bairati published an article that suggested that osteonal remodeling was so variable between individuals and with aging that it could not be solely a function of mechanical need, which should be less variable from individual to individual [49]. By about 1950, Amprino concluded that bone remodeling is at least partly the result of the body's need for calcium, the ingestion and absorption of which was quite variable. Amprino suggested that the proof of his ideas was that secondary osteonal bone never became as highly mineralized as primary bone (a questionable conjecture) and there-fore could not function to increase the mechanical competence of the skel-eton. This "proof," of course, assumes that mechanical competence is dictated solely by the strength and stiffness of bone, ignoring other important mechanical parameters such as toughness. This error in reasoning is still commonly made. The evolution of Amprino's ideas is elegantly summarized in a tribute to him by his student Marotti [50].

As a part of the argument over the mechanical role of osteons, there was also the feeling, expressed by Weidenreich (1873–1948) and others, that bone tissue and even bone cells were not the elements responsive to mechanical stress [51]. Based on the observation that bone was always formed around spaces that contained blood vessels, they suggested that the vascular system itself responded to mechanical alterations, and the bone was simply deposited wherever vessels were present. The observations were correct, but the conclusions were not. Murray's summary of this line of thought illustrates the errors:

> Thus everything which changes the calibre or arrangement of the vessels be-comes expressed also in the architecture of the bone: the calibre and arrangement of emitte depends on that of the vessels. But mechanical forces influ-ence the calibre and arrangement of vessels very readily, tension causing them to become narrower and pulling them in the direction of the force. Thus those emitte which are laid down in this direction develop thicker walls and others remodel themselves so as to alter their direction. Pressure acts similarly; constant pressure, by obliterating the vessels, leads to atrophy of the bone, while intermittent pressure, causing alterations in the blood content of vessels, is favourable [48].

Jores and Petersen separately argued just the opposite point of view [52, 53]. They believed that bone and bone cells were the sensors for mechanical alterations. It is apparent that Murray sided with this latter view, even suggesting that the matrix itself somehow might sense changes in the mechanical milieu and cause remodeling to occur through changes in the molecular organization of the tissue [48].

Literature suggests that at this time (1920–1930s) there was much argument about the "functional unit" of bone, a discussion that was thereafter apparently forgotten for about 30 years, until Frost revived the idea in his quantum concept and skeletal intermediary organization [29]. Murray discussed the idea of a functional unit in bone at some length, finally concluding that the composite material itself, rather than the osteon (which he called only the morphological unit of description) comprised the functional unit of bone [48]. His ideas about the functional unit are not so far from correct, although he apparently failed to understand that the osteon might represent the packet of bone material in secondary bone.

It is interesting that the idea that bone may be sensitive to microscopic damage had already been broached by this time. In 1931, Greig suggested that local damage to bone might cause a release of histamine, stimulating blood flow locally and leading to a remodeling event [54]. As bone is removed and reformed, new and vital bone tissue would be deposited in the region where it is most needed, i.e., the damaged area. As with many other aspects highlighted in this chapter, evidence of microdamage was eventually confirmed by Frost in the 1960s [55].

By the mid-1930s many of the ideas prevalent today were present in some form: the response of bone to mechanical stress, and specifically the mechanical function of osteons; the functional unit of bone tissue; the importance of blood supply to secondary bone remodeling; the possible role of microdamage in site-specific initiation of bone remodeling; and the importance of cyclic loading in initiating these events [52, 56]. Our understanding of cortical bone mechanics was pushed forward in the late 1930s with the work of Maj, who used both animal and cadaveric bone to show that although there was an association between the breaking load of the bone and its porosity, it was disproportional [57]. This led Maj to conclude that although porosity was important, there were other factors in determining cortical bone mechanical properties—he proposed collagen fiber orientation/density as likely a key one. Subsequent studies on the mechanical connection occurred over the coming decades, led by work from Gaynor Evans and others in the 1950s–1970s. This work, along with that of investigators such as Ascenzi, made a definitive connection between osteons and mechanical properties but also began to highlight the role of microcracks, mineralization, collagen fiber orientation, and cement lines in defining aspects of strong and weak bones [58].

1.5 Summary

The foundation of musculoskeletal discovery began with its structure. In the chapters that follow, how the structure is formed, altered, and physiologically adapted is discussed with a particular focus on the mechanical consequences.

References

1. Martin RB, Burr DB. Structure, function, and adaptation of compact bone. Raven Press; 1989.
2. Bostock J. An elementary system of physiology. Boston: Wells and Lilly; 1825.
3. Havers C. Osteologia Nova. London: Samuel Smith; 1691.
4. Leeuwenhoek AV. Microscopical observations of the structure of teeth and other bones: made and communicated, in a letter by Mr. Anthony Leeuwenhoeck. Philos Trans R Soc Lond. 1677;12:1002–3. https://doi.org/10.1098/rstl.1677.0052.
5. Albinus BS. De constructione ossium. Academicarum Annotationum. 1754;3:23–4.
6. Leeuwenhoeck AV. Several observations on the texture of bone of animals compared with that of wood: on the bark of trees: on the little scales found on the cuticula, etc. Philos Trans R Soc Lond. 1693;17:838.
7. Monro A. The anatomy of the human bones, nerves and lacteal sac and duct. Dublin: Printed for W. Whitestone; 1776.
8. Winslow JB. Exposition Anatomique de la Structure du Corps Humain. Paris: G Desprez and J Desessartz; 1732.
9. Hamel MD. Memoire on the bones. Memoire de l'Academie des Science; 1739. p. 1–7.
10. Todd RB, Bowman W. The physiological anatomy and physiology of man. London: Parker; 1845.
11. Meckel J. Manual of general anatomy. London: G. Henderson; 1837.
12. Bichat X. Anatomie Generale. Paris: Brosson; 1812.
13. Lieutaud M. Essais Anatomiques. Paris: Pierre-Michel Huart; 1742.
14. Gagliardi D. Anatome Ossium Novis Inventis Illustratae. Rome: I. I. Komarek; 1723.
15. Lasane MD. Memoire sur 'organisation des os. Memoire de Paris; 1751. p. 98–120.
16. Bruns VV. Lehrbuch der Allgemeinen. Anatomie des Menschen. F. Vieweg und sohn, Braunschweig; 1841
17. Cruveilhier J. The anatomy of the human body. New York: Harper and Brothers; 1844.
18. Quekett J. On the intimate structure of bone, as composing the skeleton in the four great classes of animals, viz., mammals, birds, reptiles, and fishes, with remarks on the great value of knowledge of such structure in determining the affinities of minute fragments of organic remains. Trans Microsc Soc London. 1849;2:46–64.
19. Quain J. Human Anatomy. Philadelphia: Lea and Blanchard; 1849.
20. Peaslee ER. Human histology. Philadelphia: Blanchard and Lee; 1857.
21. du Monceau HLD. IV. Observations and experiments with madder-root, which has the faculty of tinging the bones of living animals of a red colour, by M. Du Hamel du Monceau, F. R. S. & c. communicated in a letter to sir Hans Sloane, Bart. Pr. R. S. Translated from the french by T. S. M. D. F. R. S. Philos Trans R Soc Lond. 1740;41:390–406. https://doi.org/10.1098/rstl.1739.0070.
22. Howship J. Microscopical observations on the structure of bone. Med Chir Trans. 1816;7:382–98.
23. Tomes MD. Observations on the structure and development of bone. Philos Trans R Soc Lond. 1853;143:109.
24. Frost HM. The regional acceleratory phenomenon: a review. Henry Ford Hosp Med J. 1983;31:3–9.
25. Frost HM. Bone remodeling and its relation to metabolic bone disease. Phys Ther. 1973; https://doi.org/10.1093/ptj/53.10.1146.
26. Milch RA, Rall DP, Tobie JE. Fluorescence of tetracycline antibiotics in bone. J Bone Jt Surg Am. 1958;40-A(4):897–910.
27. Frost HM, Villaneuva AR, Roth H. Measurement of bone formation in a 57 year old man by means of tetracyclines. Henry Ford Hosp Méd Bull. 1960;8:239–54.
28. Bassett EJ, Keith MS, Armelagos GJ, et al. Tetracycline-labeled human bone from ancient Sudanese Nubia (A.D. 350). Science. 1980;209:1532–4.
29. Frost HM. Intermediary organization of the skeleton. Boca Raton: CRC Press; 1986.

30. Frost HM. The mechanostat: a proposed pathogenic mechanism of osteoporoses and the bone mass effects of mechanical and nonmechanical agents. Bone Miner. 1987;2:73–85.
31. Turner C, Burr D, Jee WSS, et al. Tribute to Harold M. Frost M.D. J Musculoskelet Neuronal Interact. 2004;4:347–56.
32. Roesler H. The history of some fundamental concepts in bone biomechanics. J Biomech. 1987;20:1025–34. https://doi.org/10.1016/0021-9290(87)90020-0.
33. Roesler H. Some historical remarks on the theory of cancellous bone structue (Wolff's law). In: Cowin SC, editor. Mechanical properties of bone. New York: American Society of Mechanical Engineers; 1981. p. 27–42.
34. Galileo G. Discorsi e dimonstrazioni matematiche, intorno a due nuove scienze attentanti alla meccanica ed a muovementi locali. Madison: University of Wisconsin Press; 1638.
35. Bell C. Animal mechanics, or proofs of design in the animal frame. Cambridge, MA: Morrill Wyman; 1827.
36. Bourgery JM. Traite Complet de l'Anatomie de l'Homme. I Osteologie; 1832.
37. Ward FO. Outlines of human osteology; 1838.
38. Thompson DW. On growth and form. London: Cambridge University Press; 1942.
39. Cowin SC. The false premise of Wolff's law. Forma. 1997;12:247–62.
40. Wolff J. Uber die bedeutung der architektur der spongiosa. Zentralblatt fur die Medizinische Wissenschaft VI Jahrgang S 223–234; 1869.
41. Wolff J. Uber die innere architektur der knochen und ihre bedeutung fur die frage vom knochenwachstum. Virchows Archiv fuer Pathologische Anatomie und Physiologie. 1870;50: 389–453.
42. Koch JC. The laws of bone architecture. Am J Anat. 1917;21:177–298. https://doi.org/10.1002/aja.1000210202.
43. Freiberg AH. Wolffs law and the functional pathogenesis of deformity. Am J Méd Sci. 1902;124:956–71. https://doi.org/10.1097/00000441-190212000-00003.
44. Gebhardt W. Über funktionell wichtige Anordnungsweisen der feineren und gröberen Bauelemente des Wirbeltierknochens II. Spezieller Teil. 1. Der Bau der Haversschen Lamellensysteme und seine funktionelle Bedeutung. Roux Arch. 1905;20:187–322.
45. Burkhardt L. Über den Aufbau der menschlichen Osteone. Verh Anat Ges (Jena), Erg-H zu Anat. 1929;38:97–102.
46. Benninghoff A. Über die Anpassung der Knochenkompakta an geänderte Beanspruchung. Anat Anz. 1927;63:289–99.
47. Dowgjallo ND. Die Struktur der Compacta des Unterkiefers bei normalem und reduziertem Alveolarfortsatz. Anat Entwickl Gesch. 1932;97:55–67. https://doi.org/10.1007/bf02134377.
48. Murray PDF. Bones; a study of the development and structure of the vertebrate skeleton. Cambridge: The University Press; 1936.
49. Amprino R, Bairati A. Contributo allo studio del valore funzionale della struttura della sostanza compatta della osse. Chir Organi Mov. 1936;21:527–41.
50. Amprino R. A contribution to the functional meaning of the substitution of primary by secondary bone tissue. Acta Anat (Basel). 2008;5:291–300. https://doi.org/10.1159/000140331.
51. Weidenreich F. Knochenstudien: II. Teil: Über Sehnenverknöcherungen und Faktoren der Knochenbildung. Z Anat Entwicklungsgesch. 1923;69:558–97.
52. Jores L. Experimentelle Untersuchungen über die Einwirkung mechanischen Druckes auf den Knochen. Beitr Path Anat. 1920;66:433.
53. Petersen H. Die Organe des Skeletsystems. Handbuch der Mikroskopischen Anatomie des Menschen, vol. 2, pt. II, W. von Möllendorf, ed. 1930.
54. Greig DM. Clinical observations on the surgical pathology of bone. London: Oliver and Boyd, Edinburgh; 1931.
55. Frost HL. Presence of microscopic cracks in vivo in bone. Henry Ford Hosp Med J. 1960;8:25–35.

56. Loeschcke H, Weinnoldt H. Über den Einfluß von Druck und Entspannung auf das Knochenwachstum des Hirnschädels. Beitr Path Anat. 1922;70:406–39.
57. Maj G. Studio sulle variazioni individuali e topografiche della resistenza meccanica del tessuto osseo diafisario umano in diverse eta. Arch Ital Anat Embriol. 1942;47:612–33.
58. Evans FG. Mechanical properties of bone. Thomas Springfield; 1973.

Chapter 2
Functional Musculoskeletal Anatomy

Abstract This chapter describes how engineering principles can be used to estimate joint forces. Principles of static analysis are reviewed, with examples of static analysis applied to the hip and elbow joints and to the analysis of joint forces. Applications to indeterminant problems of joint mechanics are presented and utilized to analyze complex joints.

What You Will Learn
1. How to estimate and analyze forces between bones and across joints.
2. The principles of static equilibrium and how they help determine forces in the musculoskeletal system.

2.1 Introduction

The skeleton is, first and foremost, a mechanical organ. Its primary functions are to transmit forces from one part of the body to another and to protect other organs (e.g., the brain) from mechanical forces that could damage them. Therefore, the principal biologic role of skeletal tissues is to bear loads with limited amounts of deformation. To appreciate the mechanical attributes that these tissues must have to perform this role, it is necessary to learn something about the forces that whole bones normally carry. In most cases, these forces result from loads being passed from the part of the body in contact with an environmental surface (e.g., the heel on the ground when walking) through one or more bones to the applied or supported load (e.g., the torso). In addition to the forces transmitted in bone-to-bone contact, muscle and ligament forces act directly on the bones, and these forces (especially the muscle forces) are large and important.

Most muscle, ligament, and bone-to-bone forces act in or near the body's major diarthrodial joints (closed, low-friction systems such as the knee). The purpose of this chapter is to explain how conventional engineering analysis may be used to estimate joint forces and to provide some practical experience in solving such problems.

M. R. Allen, J. Wallace, *Skeletal Tissue Mechanics*,
https://doi.org/10.1007/978-3-031-90609-1_2

2.2 An Illustrative Experimental Model

A major advantage of conventional static and dynamic analyses is that the forces (the physical action of one body on another) and moments (the tendency of a force to cause rigid body rotation) imposed on bones can be estimated with reasonable precision from noninvasive external measurements. Actual physical measurement of internal loading is more challenging, for both experimental and ethical reasons. Although physical measurements of skeletal loading have been made in live subjects (we will explore an example later in this chapter), these types of studies are relatively rare and always limited in scope. One way around the ethical dilemma posed by invasive measurement is to apply life-like loading environments to cadaveric specimens donated for scientific research, thereby enabling direct, detailed measurement of internal kinetics and their physical consequences within the tissues themselves. Orthopedic research has a long and rich history using this approach and much has been learned about normal skeletal loading and the biomechanical consequences of pathology, surgical reconstruction, and joint replacement. Before tackling classic static analyses, we will examine some datasets taken from a series of cadaver experiments focused on the lower extremity, an exercise that will lend further appreciation of the large forces routinely imposed on our skeletons.

The data presented in Fig. 2.1 were gathered using an apparatus that reproduces the kinematics and kinetics of normal gait in cadaver lower extremities [1]. One set of computer-controlled actuations reproduces the motions of the proximal tibia as recorded in a standard gait lab while a second set simultaneously reproduces the actions of six major, extrinsic muscle groups of the foot and ankle. These muscles reside in the shank and include the triceps surae (gastrocnemius and soleus), tibialis posterior, tibialis anterior, peronei (peroneus longus and brevis) flexor hallucis longus, and flexor digitorum longus. The vertical axis of the plot is force, given in units of body weight (BW) for easy comparison, and the horizontal axis represents increments of time when the foot is in contact with the ground during normal walking. The vertical ground reaction force is given by the heavy dark double-peaked plot and reflects the classic force–time curve for gait, whereby the first peak is a component of the braking force occurring shortly after heel strike and the second peak a component of the push-off force to propel the body forward. The shape and magnitude of this plot confirm life-like loading equivalent to that routinely measured in standard gait labs.

The remaining plots in the graph represent the internal tissue forces present during normal walking. Peak tension in the Achilles tendon (a function of triceps surae contraction needed to propel the body forward) is approximately 1.8× BW while the peak tensile force carried in the planar fascia, a tough band of ligamentous tissue that spans the sole of the foot, is roughly equivalent to body weight [2]. Perhaps most striking are the plots of joint reaction force measured in the tibiotalar (ankle) and first metatarsophalangeal (MTP; at the base of the great toe) joints. The routine load transmitted through the ankle is over 4.0× BW [1] while that carried through a very small digital joint in the first ray of the foot is almost equal to BW [3]. This is a remarkable feat (no pun intended). With each step we take every day,

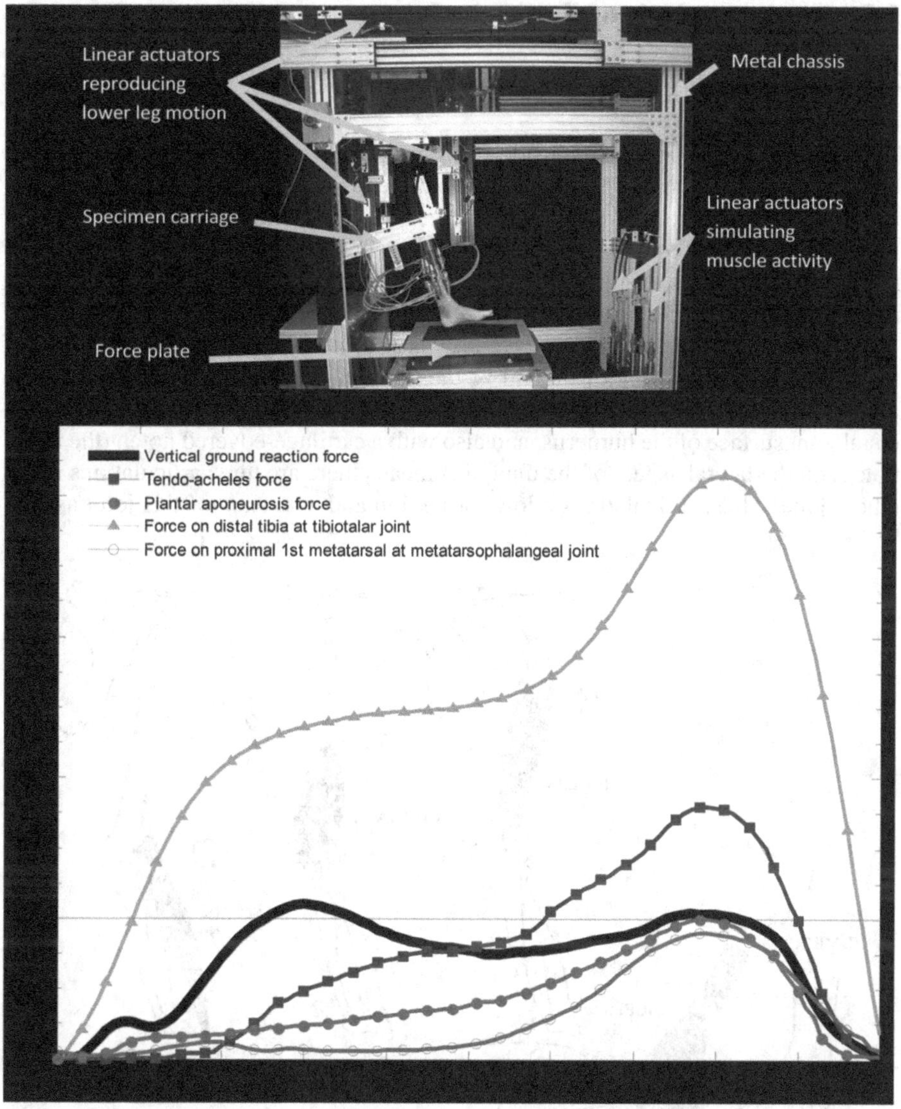

Fig. 2.1 Top: Image of an experimental apparatus enabling detailed measurements of biomechanical phenomena in skeletal tissues under normal or pathologic conditions. Bottom: Plots (force versus time) of the forces carried in select skeletal tissues during normal walking

both our hard and soft skeletal tissues are exposed to tensile and compressive forces considerably greater than body weight and this is only for walking; the forces are much larger during more strenuous activities like running and jumping. The mechanical explanation for these high forces is examined in detail within the next sections of this chapter, while future chapters examine the mechanisms enabling such an incredibly strong and durable system.

2.3 Static Analysis of Forces in Joints

2.3.1 *Forces in the Elbow Joint*

To extend this subject, let us consider the human elbow joint. We choose this joint because it can be accurately modeled as a simple hinge, and because it is familiar to us, conforming to the image that we acquired in grade school of the skeleton as a system of levers. Although we use this simplicity in solving for the forces, we should spend a few moments examining the anatomy to appreciate the simplifying assumptions being made in this analysis.

A lateral view of the bones of the elbow is shown in Fig. 2.2a. The olecranon of the ulna wraps around the distal end of the humerus to form the major part of the elbow's hinge joint. The proximal joint surface of the radius articulates with the distal joint surface of the humerus, and also with a cartilage-covered notch (the radial notch) on the lateral aspect of the ulna; this means there are three articulations in the elbow joint. These articulations allow for flexion and extension of this joint as well

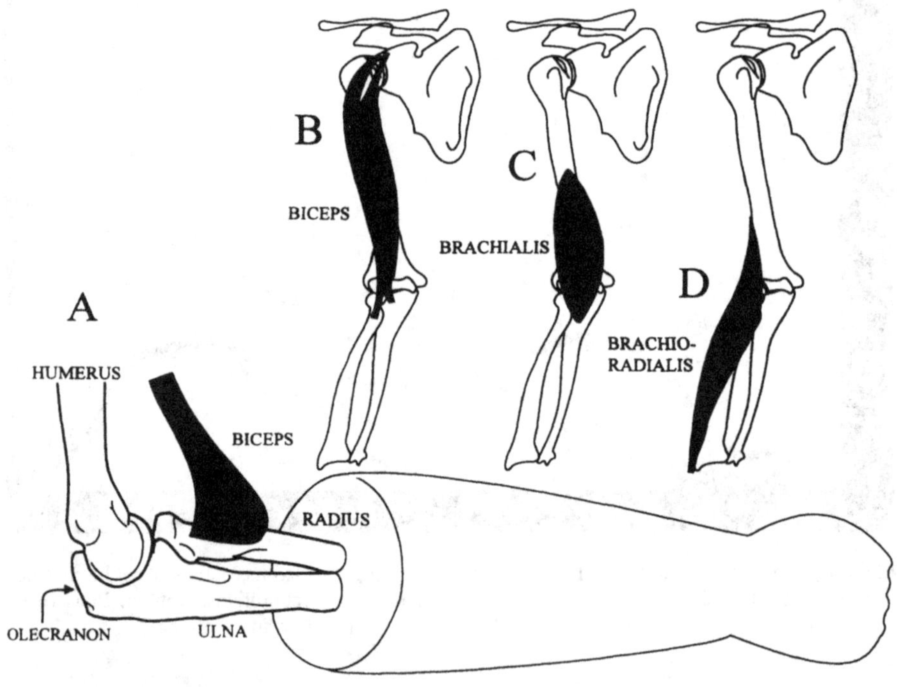

Fig. 2.2 Anatomical drawings of the elbow joint. (**a**) Overall view of the flexed joint and the biceps inserting primarily on the radius. (**b–d**) Approximate locations of the biceps, brachialis, and brachioradialis muscles

as pronation and supination of the forearm. The radius and ulna are bound together by the interosseous membrane along the central part of their length and by several ligaments at their proximal and distal ends (not shown). We assume that these structures cause the two forearm bones to act as one structural unit. We also assume that the wrist joint is stabilized by its musculature so that the hand and forearm flex about the elbow as a simple hinge joint.

When the elbow is flexed with the palm upward, there are primarily two muscles that are active: the biceps and the brachialis (Fig. 2.2b, c). The biceps has two heads (i.e., points of origin). The short head originates on the coracoid process of the scapula, while the long head runs through the shoulder joint to attach to the superior lip of the glenoid fossa (the scapular part of the shoulder joint). The biceps divides distally into a tendon that inserts on the proximal radius and an aponeurosis (tough band of connective tissue), which blends with other muscles in the proximal forearm (Fig. 2.2b). The brachialis originates along the anterior surface of the central humerus and inserts on the proximal ulna (Fig. 2.2c). When the forearm is flexed rapidly, or a large force must be applied during flexion, the brachioradialis also acts between the distal humerus and the distal radius (Fig. 2.2d). Other muscles also cross the elbow and participate in flexion. However, to simplify this analysis, and because the present problem is for a static situation, we assume that the biceps and brachialis muscles act as one in holding the elbow in a flexed position against the action of a weight held in the hand. We want to find the force required in the "biceps–brachialis muscle" to support the flexed forearm and the total joint reaction force exerted on the radius and ulna by the distal end of the humerus. These forces are expressed as multiples of the weight in the hand.

To solve this problem and others like it, we will assume a state of static equilibrium: all forces and moments acting on the system are balanced and through Newton's First Law of Motion, the object remains at rest. To proceed, three steps are necessary:

1. Draw a free-body diagram—the body of interest is isolated and all forces and moments acting on the body are drawn.
2. Write the equations of static equilibrium—the sum of all forces and moments acting on the body of interest are equal to zero.
3. Solve these equations simultaneously to obtain the unknown forces.

We simplify the analysis by assuming that it is two-dimensional problem and that the forces all act in a sagittal plane containing the humerus and the forearm. The free-body diagram is constructed by isolating the structure being analyzed such that the internal forces to be determined (e.g., the force exerted on the radius and ulna by the distal end of the humerus) are exposed and all forces acting on the structure are represented by vectors (either an unknown force with an unknown direction, or force components typically acting in the x- and y-directions). The free-body diagram for the elbow problem is shown in Fig. 2.3. The forearm is made a free body by "amputating" through the elbow joint, whose internal forces we wish to reveal, and cutting through the biceps muscle. The forearm lies in the x–y plane. The weight in the hand, the muscle force, and the joint reaction force are represented by the

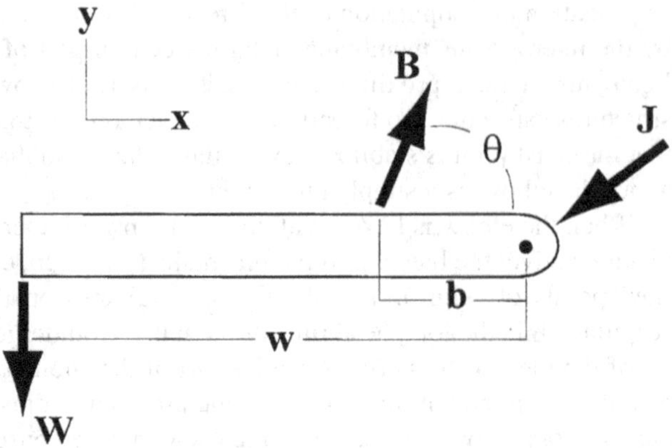

Fig. 2.3 Free-body diagram for the elbow force calculation. (J) is the joint reaction force, (W) is the weight in the hand, and (B) is the biceps force

vectors W, B, and J, respectively.[1] The joint force is assumed to pass through a fixed center of rotation for the joint, shown by a dot, and is shown acting toward the forearm because it represents the actions of the humerus on the forearm and we assume that the two bones are pushing against one another. If this assumption is incorrect, our math will tell us this by resulting in a negative force. Likewise, the muscle must be acting in tension (the only way muscles can act), and is shown pulling at the forearm. The insertion point of the muscle is b meters (m) along the forearm from the joint center, and the center of gravity of the weight in the hand is w meters from this point. The equations for static equilibrium in two dimensions are

$$\begin{aligned} \sum M &= 0 \\ \sum F_X &= 0 \\ \sum F_Y &= 0 \end{aligned} \tag{2.1}$$

where F_X and F_Y are force components in the x- and y-directions and M stands for moments about an arbitrary point (calculated as the magnitude of the force times the perpendicular distance between the point about which the moment is calculated and the line of action of the force). Based on the right-hand rule, if the force tends to cause clockwise (CW) rotation about the point of interest, the moment it generates is negative (likewise, counterclockwise (CCW) rotation generates a positive moment). In two dimensions, there are three such equations and we can therefore solve for three unknowns (force components or directions).

For the elbow problem, the biceps force can be represented as x- and y-components using simple trigonometry (the y-component of the force is Bsinθ

[1] In this chapter, vector quantities are represented in bold type and scalars (such as vector magnitudes) in ordinary type.

acting up and the x-component is Bcosθ acting right). Similarly, the unknown joint reaction force can be represented as x- and y-components (acting left and down, respectively, based on the direction of the original vector). By summing moments about the joint center, the equilibrium equations are

$$\sum M = wW - bB\sin\theta = 0 \tag{2.2}$$

$$\sum F_X = B\cos\theta - J_X = 0 \tag{2.3}$$

$$\sum F_Y = B\sin\theta - W - J_Y = 0 \tag{2.4}$$

Note that moments are considered positive CCW and forces are considered positive if they act in the direction of the positive x- and y-axes (right and up, respectively). Also note that the x-component of B and both components of J pass though the joint center and therefore do not create a moment about that point (the perpendicular distance between the point and the line of action of the force = 0). Solving these equations yields

$$B = wW/(b\sin\theta) \tag{2.5}$$

$$J_X = B\cos\theta \tag{2.6}$$

$$J_Y = B\sin\theta - W \tag{2.7}$$

If $\theta = 75°$, $w = 0.35$ m, and $b = 0.04$ m, then $B = 9.1$ W, $J_X = 2.3$ W, and $J_Y = 7.8$ W. From trigonometry, the total magnitude of the joint reaction force is $J = \left(J_x^2 + J_y^2\right)^{1/2} = 8.1\,W$, and its orientation is arctan $(J_y/J_x) = 74°$ with respect to the x-axis. Thus, the muscle and joint reaction forces are eight to nine times greater than the weight held in the hand. This result is typical of virtually all the joints in the body in that the skeleton works at a mechanical disadvantage in terms of force. As a consequence, the forces acting on bones are high relative to the forces applied by (or to) the environment.

There is, of course, a good reason for this. Muscles can only contract about 30% of their length. In the case of the biceps, the overall length is about 25 cm, so the maximum contraction is 7–8 cm. The lever action of the forearm magnifies this distance by the ratio w/b shown in Fig. 2.3, enabling the hand to move much further (and also much faster). Of equal importance, the muscles insert proximally on the radius and ulna and do not create a "web" of flesh across the front of the elbow. Therefore, magnification of motion enables larger movements to be accomplished with a compact structure. The price that is paid for this compactness and magnification of motion is high forces in the muscles, across the joint surfaces, and within the bones.

Box 2.1: Giovanni Borelli on the Movement of Animals

Giovanni Alfonso Borelli (1608–1679) was the greatest of the early biomechanicians. He held the chair in mathematics at Pisa in Italy, where he was a close friend of Malpighi, who was the professor of theoretical medicine. Together, they did much to persuade seventeenth-century physicians of the importance of physics in understanding medicine and physiology. Borelli's treatise, De Motu Animalium (On the Movement of Animals) [4], has been translated into English and is a marvelous testament to his genius and ability to explain musculoskeletal mechanics clearly and simply. In the preface to his translation, Maquet points out that later biomechanicians unwittingly duplicated several of Borelli's important discoveries. While some of his results contain errors, it must be remembered that Newton's laws were not published until 1687. Thus, although Borelli understood very well the principle of balancing moments about a fulcrum, he could not have been expected to fully understand static equilibrium based on Newton's laws. One of Borelli's greatest achievements was the discovery that animals' joints work at a mechanical disadvantage in terms of force. He discusses this in the following excerpt (Fig. 2.4).

> Proposition VIII
>
> It is commonly thought that Nature raises considerable weights by using the machines of the muscles with a weak moving force.
>
> The magnitude of the vital force of the muscles … sustains, raises, and moves not only an arm or a leg, but the whole animal machine, enabling it even to dance. Besides the mass of the animal, heavy enough by itself, this force carries, pulls and pushes considerable weights.

Fig. 2.4 Diagram of elbow force problem from Borelli's On the Movement of Animals [4]

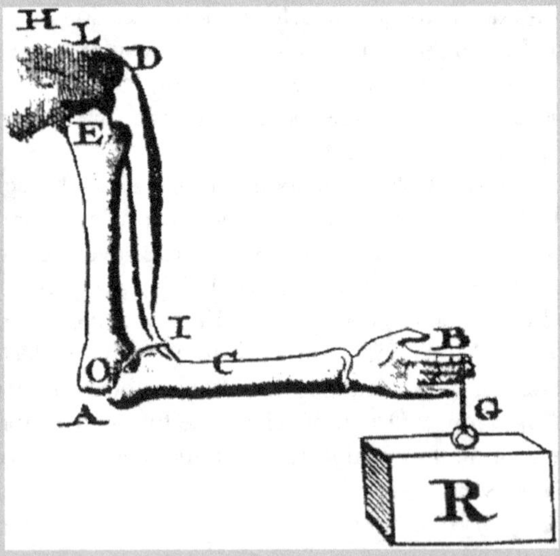

(continued)

Box 2.1 (continued)

Aristotle…did not recognize the muscles but imagined spirits which pull and push the limbs. [He] remarked how difficult it would be for the huge mass of an elephant to be moved…by tenuous spirit or wind. He met the difficulty by saying that Nature moves the joints and limbs of the animal by using very small force…and said that the operation is carried out by way of a lever. Therefore, it is not surprising that huge weights can be moved…by a small force. Lucretius used the same example… .Galen also says that a tendon is like a lever. He thinks that, consequently, a small force of the animal faculty can pull and move heavy weights.

This general opinion seems to be so likely that, to my knowledge, not surprisingly, it has been questioned by nobody. Who indeed would be stupid enough to look for a machine [in the body] to move a very light weight with a great force, i.e., not to save forces but rather to spend forces? And if this is rightly considered as stupid, how is it possible that wise Nature, everywhere looking for economy, simplicity and facility, builds with great efficiency in animals machines to move, not heavy weights with a small force, but on the contrary, light weights with almost boundless force? …I shall demonstrate that multiple and different machines actually are used in the motions of animals but that light weights are carried by large and strong force rather than heavy weights being supported by small force.

2.3.2 Forces in the Hip Joint

The hip joint is the articulation between the femur and the acetabulum of the pelvis. The hip is one of the most important joints in the body from a medical perspective. Especially in aging individuals, overall health is promoted by the exercise that goes with walking. The ability to walk depends on having a healthy, pain-free hip. Two kinds of diseases, both very common among the elderly, preferentially affect the hip. Osteoarthritis is the primary reason that about 500,000 total hip replacement procedures are performed each year in the United States. Osteoporosis in the femoral neck results in several hundred thousand hip fractures each year, virtually all of which require surgical treatment with a metal fixation device or hip replacement. To better understand the mechanics of the hip and the demands on the implants used to correct its problems, it is important to know the forces across the hip during walking and their determining factors.

When one is walking, the lower extremities and other parts of the body are moving and changing direction (i.e., accelerating), so the conditions of static equilibrium are not satisfied ($F = \mathrm{ma} = 0$). However, most people do not usually walk very vigorously, so the accelerations are small relative to the forces produced by muscle pulls and gravity. Therefore, the problem is usually solved as though the person were simply standing on one leg, assuming that this approximates the conditions during the "single leg stance phase" of gait, that is, when all the weight is being carried by one leg. In addition to this assumption, we assume that the problem is two dimensional, in the frontal plane, and that only one muscle is acting

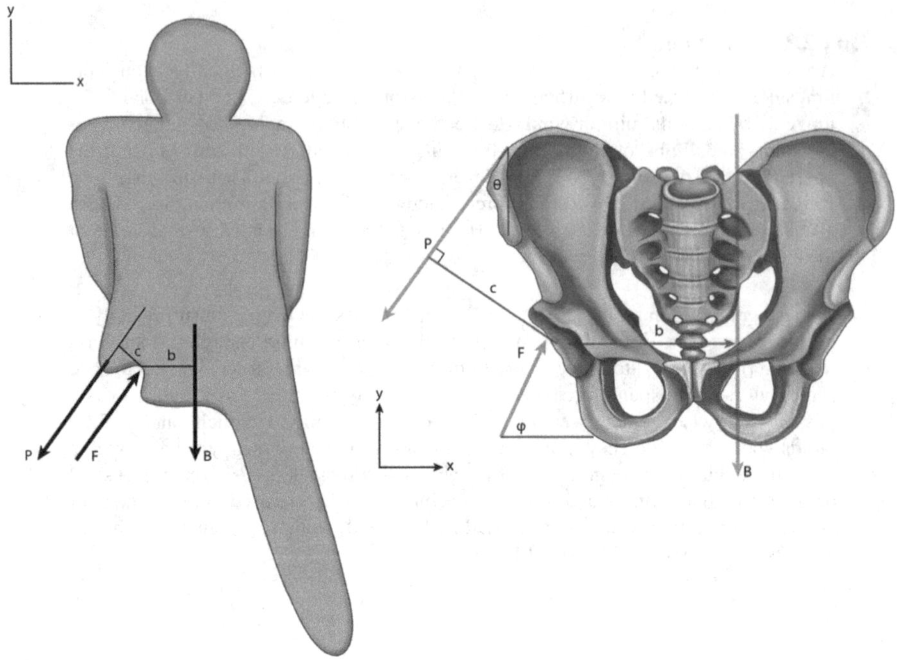

Fig. 2.5 Free-body diagram for the calculation of the hip joint force while walking. The "amputated" leg is the one supporting the body. The leg in the free-body diagram is not in contact with the ground. (*B*) is the weight of the body (minus the amputated leg), (*P*) is the abductor muscle force, and (*F*) is the joint reaction force

across the joint and pulling on the leg. If you try to stand on one leg, the force pulling your center of gravity downward tends to rotate your torso toward the medial side of the leg you are standing on. The muscles that resist this movement are the same ones, which you use to abduct your thigh when lying down (i.e., move your lower limb away from the body's center line). Again, several muscles act to do this, but we can lump them all together and simply call them the "abductor muscles."

Again, the first step in calculating the forces in the joint is to draw a free-body diagram, "amputating" through the joint and muscle in question to "reveal" the force vectors acting there. In this case, we delete the lower extremity and study the remaining portion of the body (alternatively, we could analyze just the leg and consider the ground reaction force). Figure 2.5 shows this situation. It is important to understand that the subject was standing on the leg that has been "amputated," and that the other leg, of which only a portion is shown, was not touching the ground. The abductor muscle force is represented by the force vector P (again acting in tension away from the body). The joint reaction force acting on the middle of the acetabulum in the pelvis is F (again assuming the femoral head and acetabulum are pushing against one another). The weight of the body, represented by the vector B, is the entire body weight, W, minus the weight of the leg supporting the body (since it has been removed from the problem). Because each lower extremity weighs about 1/6 W, we let B = 5 W/6. The full body weight would act through the centerline of the body but with the leg removed, this force B acts downward slightly to the right of the centerline. Using the same conventions for forces and moments and summing moments about the center of the acetabula, we have

$$\sum M = cP - bB = 0 \tag{2.8}$$

$$P = (b/c)B = (b/c)(5/6)W \tag{2.9}$$

The lengths of the moment arms *b* and *c* have been estimated from antero-posterior pelvic radiographs. It was found that the b/c ratio ranges between 2 and 3.5. Following the lead of Frankel and Burstein [5], we choose a conservative value of 2.4 and obtain the convenient result that $P = 2$ W. That is, the force required in the abductor muscles to balance the body on the head of the weight-bearing femur during the stance phase of gait is twofold body weight.

To solve the rest of the problem, we write the equations for force equilibrium:

$$\sum F_X = F_X - P_X\ F_X - 2W sin\theta = 0 \tag{2.10}$$

$$\sum F_Y = F_Y - P_Y - B\ F_y - 2W cos\theta - 5W/6 = 0 \tag{2.11}$$

Assume $\theta \approx 30°$ is the angle that the abductor muscle line of action makes with the y-axis. Because sin 30° = 0.5, the components of the joint reaction force *F* are

$$F_X = W \text{ and } F_Y = (2\cos 30° + 5/6)W = 2.57W \tag{2.12}$$

The horizontal force of the femur on the pelvis is equal to body weight, and the vertical force is 2.5 times as much. The total joint reaction force is $F = \left(F_X^2 + F_Y^2\right)^{1/2} = 2.75W$. This force acts at an angle to the horizontal $\phi = \arctan(F_y/F_x) = 68.7°$.

We have seen that the magnitude of the forces in the hip joint depends critically on the ratio of the body weight moment arm to the abductor muscle moment arm. Thus, anything that increases the former or decreases the latter increases the abductor muscle force required for gait and consequently the force on the head of the femur as well. People with short femoral necks have higher hip forces, other things being equal. More significantly, people with a wide pelvis also have larger hip forces. This means that women typically have larger hip forces than men because a women's pelvis accommodates a birth canal [6]. This fact may be one reason that women have more hip fractures and hip replacements because of arthritis than men do. It is also conceivable that this places women at a biomechanical disadvantage with respect to some athletic activities, although studies do not always show sex-specific differences in the biomechanics of running, particularly endurance running [7].

Box 2.2: Measurement of Hip Joint Forces In Vivo

To verify the estimates of hip joint forces made using free-body calculations, several groups of investigators have implanted devices that allowed measurement of hip joint forces into hip replacement patients [8–16]. When humans are used as experimental subjects, ethical concerns demand that the subject's health not be endangered in any way. This ethical mandate has extensive technical consequences for measurement of hip joint forces. To begin with, a healthy human subject cannot be used; experimentation must be restricted to a subject with a diseased or injured hip, who will be undergoing replacement of the proximal femur with a prosthesis by way of treatment for this condition. This condition provides an opportunity to use an instrumented prosthesis (Fig. 2.6) that can measure joint forces instead of a standard prosthesis. However, the instrumented prosthesis must be as strong and as durable as a standard one, which imposes important constraints on the instrumentation design. Moreover, wires cannot be run from the prosthesis to the surface of the skin, where they could pose a risk of infection or other problems. Consequently, the instrument must be fully contained within the prosthesis, with a working space of a few cubic centimeters, and it must be capable of transmitting data to a receiver outside the body while the patient walks.

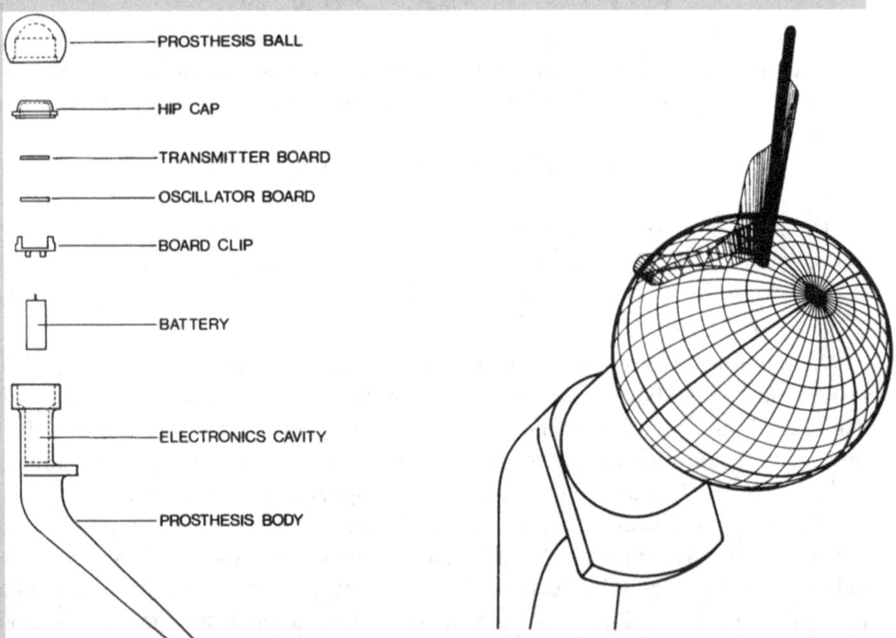

Fig. 2.6 Left: Exploded view of telemetry electronics inside an instrumented hip prosthesis [11]. Right: Three-dimensional plot of hip joint force vector during a gait cycle (with crutches). Orientations and lengths of lines emanating from the surface of the prosthesis head indicate force direction and magnitude, respectively [11]

(continued)

Box 2.2 (continued)

The solution to this technical problem has been to use strain gauges (3–12 in number) to measure strains on the interior surfaces of the prosthesis, and microelectronic circuitry to process the strain signal, which is broadcast as an FM signal to a receiver held against the skin. The device must be fully calibrated in the laboratory before insertion into the subject, so that the measured strains can be converted to force components on the ball or head of the prosthesis. The power source in the prosthesis used by [11] Davy et al. was a battery located in the stem of the prosthesis and activated by a magnetic switch turned on by a magnet held outside the patient's leg (Fig. 2.6). Because batteries carry a slight risk of releasing toxic components, and because they have a finite life, other investigators [14] used an inductive power source. A magnetic coil wrapped around the subject's leg induced current in a receiver coil within the prosthesis.

One might question whether a patient who has a degenerated joint, has recently had surgery on it, and is in a laboratory environment, is capable of walking "normally"—but these are the best data available. Findings from all studies are consistent with one another and with the theoretical estimates. During the single-leg-stance phase of gait, hip joint forces between 2.5 and 3.3 times body weight were measured. The highest force recorded was in an individual, who was attempting static single-leg balancing; during recovery from a momentary loss of balance, a force of 5.5 body weight was measured in the hip [11].

2.3.3 Clinical Significance of High Joint Forces

Because diarthrodial joints work at a mechanical force disadvantage so that limbs can move far and rapidly with short muscle contractions, high forces are produced in the tissues of the bones and joints. Normally, these tissues carry their loads without causing pain, but various diseases and injuries can damage the tissues so that the deformations associated with loading are painful. To some degree, the pain is proportional to the amount of force carried by the tissues; in other words,

$$\text{Pain} = \text{Force} \times \text{Disease}$$

There are no nerves in cartilage, and the source of joint pain is poorly understood. Often, the physician is not able to do much about the disease, so the first consideration in controlling pain may be to reduce the forces in the joint. For example, the patient can lose weight, or walk with a cane, and thereby reduce the forces transmitted across the joint.

Box 2.3: Reproducing Joint Forces in the Laboratory

As introduced previously, cadaveric specimens are frequently used to model normal, pathologic, or reconstructed joints. For these laboratory simulations, researchers often rely on static calculations to approximate the loading conditions experienced by the joint in life. For example, a typical experiment examining the stability of a new hip prosthesis might use cadaveric hip joints implanted with the new component and then mounted into a testing machine for loading. But in what direction should the load be applied? What should the load magnitude be? Static analysis can be used to answer these questions.

Once the investigator has determined the direction and magnitude of the joint reaction force for a given condition, it is simply a matter of aligning the specimen such that the vector of the joint force coincides with the axis of the load delivery system (Fig. 2.7, left). Many important biomechanical studies have been executed in just this fashion. However, like static analysis itself, this type of simulation incorporates several assumptions and the results should be interpreted with caution.

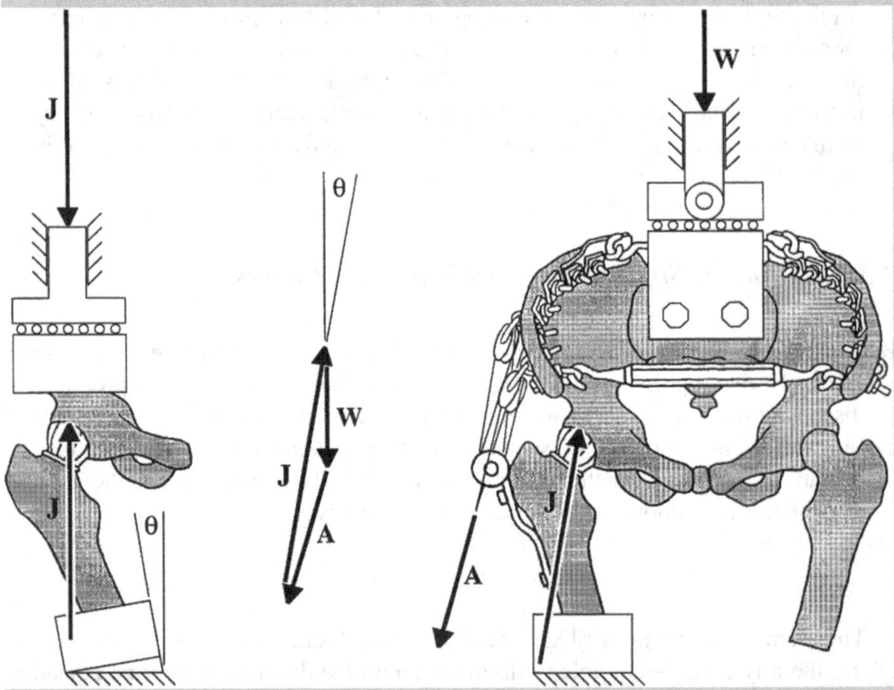

Fig. 2.7 The left and right diagrams show two different experimental setups for simulating hip joint loading using cadaver bones. The vector diagram in the middle is for diagram on the right [17]

(continued)

Box 2.3 (continued)

Let us revisit the hip example. A more physiologically correct model might use the whole pelvis and incorporate abductor muscle forces to reproduce the in vivo loading environment (Fig. 2.7, middle and right). In this case, the proper joint reaction vector is produced by simulating abductor contraction using a cable system attached at the muscle origin and insertion sites. Shortening the cable produces a moment about the hip that is opposed at the sacrum. In theory, when the vertically directed force at the sacrum equals the upper-body weight, the joint reaction force at the hip will be the same as that in the extracted hip joint model and as that produced during single-leg stance in life.

In a comparison of these two laboratory methods [17], whole intact pelvises were first loaded under simulated muscle action and the contact pressure distribution on the articular surfaces was recorded. The hips were then extracted from the pelvis and reloaded. The direction and magnitude of the joint reaction force was identical for both cases, but the distribution of contact pressure within the joint was quite different. Why? Remember, static analysis assumes rigid bodies, but musculoskeletal tissues do not always behave as rigid bodies. In the intact model, the bones of the pelvis were free to flex and deform under load, while in the simpler joint model, the pelvis became more rigid because of its altered geometry and the constraints placed upon it.

2.4 Indeterminate Joint Problems

In addition to primarily examining the static case, the problems we have considered so far were limited to two-dimensional analyses in which it was possible to solve for three unknowns. To reduce the number of unknowns to the number of equations, we had to combine the function of some muscles while completely ignoring others. We could have obtained more equations by extending our analysis to three dimensions, but then we would have had to consider additional muscles as well. All real joints present static equilibrium problems that are mathematically indeterminate because there are always more unknowns than equilibrium equations. In principal, this physically means that there is more than one solution for the set of unknown forces that maintain equilibrium. To solve for the most realistic set of joint and muscle forces, it is necessary to find more equations to use in the solution.

To see an example of how this might be done, we return again to the static elbow problem. Previously, we noted that the primary flexors of the forearm are the brachialis and the biceps, and we combined their actions into a single force vector. Now, we separate these muscle forces. In addition, we include the third muscle, the brachioradialis (see Fig. 2.2). Table 2.1 shows the moment arms of the three muscles with respect to the center of rotation of the elbow when it is flexed at 90° [18]. It also shows the physiologic cross-sectional area (PCA) of each muscle. It is approximately true that the maximal force that a muscle can exert is proportional to its

Table 2.1 Elbow muscle data

Muscle	Moment arm, cm	PCA, cm^2
Biceps	$a = 4.6$	$\alpha = 4.6$
Brachialis	$b = 3.4$	$\beta = 7.0$
Brachioradialis	$c = 7.5$	$\gamma = 1.5$

Data from Ref. [19]

physiologic cross-sectional area. Another way to say this is that muscles have an upper limit to the stress (generated force/area over which it acts) that they can generate within themselves. Measured maximum stresses vary from 20 to 100 N/cm^2, depending on conditions (isometric or dynamic, parallel or pennate fibers, etc.). We will use 40 N/cm^2 as a rule of thumb for the maximal stress [19, 20].

With all three muscles in the problem, there are five unknowns (three muscle force magnitudes and two components of joint force) but still only three equilibrium equations. We need two other relationships between the variables to obtain a solution. One way to get such additional equations would be to assume that all three muscles work together to equalize their individual stresses. That is, each muscle force is proportional to the muscle's physiologic cross-sectional area. Then, the three muscle forces are

$$\begin{aligned} A &= k\alpha \text{ (biceps)} \\ B &= k\beta \text{ (brachialis)} \\ C &= k\gamma \text{ (brachioradialis)} \end{aligned} \quad (2.13)$$

where α, β, and γ are the respective cross-sectional areas and $k = 40$ N/cm^2 is the constant muscle stress. This may be rewritten as

$$A/\alpha = B/\beta = C/\gamma \quad (2.14)$$

which constitutes the two additional equations that were needed. Using these to substitute for B and C in the moment equilibrium equation,

$$\sum M = aA + bB + cC - wW = 0 \quad (2.15)$$

one has

$$\begin{aligned} A &= [\alpha w/(a\alpha + b\beta + c\gamma)]\mathrm{W} = 2.86\mathrm{W} \\ B &= (\beta/\alpha)A = 4.36\mathrm{W} \\ C &= (\gamma/\alpha)A = 0.93\mathrm{W} \end{aligned} \quad (2.16)$$

This solution indicates that the biceps, the muscle with the intermediate cross-sectional area and moment arm, also exerts the intermediate force. The brachioradialis, which has the largest moment arm and the smallest cross-sectional area, exerts the least force, and vice versa for the brachialis. Whether this has anything to do with the actual distribution of forces in the muscles is debatable! Many other equally plausible, but not necessarily so mathematically trivial, "additional conditions" may be imagined. For example, the force distribution may shift around to give different muscle fibers a rest, or to keep the weight from tipping sideways in the hand, or both. A number of investigators have speculated that the body resolves the indeterminacy of such problems by acting to optimize or minimize some biologically important variable, such as the total energy required by all the muscles acting. Collins (1995) [21] provides additional discussion of this subject.

Box 2.4: Surgical Reconstruction of the Hip Joint

Total hip arthroplasty, which involves replacement of the ball and socket of the hip joint with manufactured components, is one of the most successful surgical operations developed in the 1900s. The procedure has relieved joint pain and increased the quality of life for millions of people. Before the development of successful hip replacement surgery by Sir John Charnley [22], patients with debilitating degenerative arthritis of the hip were forced to either suffer with the condition or undergo hip joint fusion (called arthrodesis). As described in Sect. 2.2, the loads imposed on bones and joints are dramatic as a result of the mechanical disadvantage under which they function. Forces at the hip can easily exceed fivefold body weight. In addition, the hip joints of an average individual experience about a million load cycles (steps) each year. Thus, total hip replacement components should be both strong and durable.

A total hip replacement consists of two parts: a femoral component, the ball, and an acetabular component, the socket (see Fig. 2.8). Femoral components are made from high-strength alloys such as cobalt–chrome or titanium, and consist of a highly polished head atop an intramedullary stem. The stem is inserted into the canal of the femur and is usually fixed with an acrylic plastic (polymethylmethacrylate, PMMA) called bone cement. This material does not actually cement the implant to the bone, however; it simply fills space so that the fit does not have to be exact, "grouting" the implant in place. More recent designs utilize a porous stem surface designed for fixation by bone ingrowth. The acetabular component is fabricated from ultrahigh molecular weight polyethylene, with or without a metal backing. A design with a metal backing is illustrated. Other combinations of materials for the bearing surface are sometimes used, including metal-on-metal, ceramic-on-ceramic, and ceramic-on-polyethylene. The PMMA and metal-on-polyethylene bearing surfaces innovations were important factors enabling Charnley to develop a system that featured good material and structural strength, good resistance to fatigue damage, and low friction at the bearing surface. Careful surgical technique remains important, but with current methods most patients, being elderly and relatively sedentary, can expect satisfactory performance for 15 years or more.

(continued)

Box 2.4 (continued)

Despite the operation's overall success, failures do occur, more frequently in young, active patients. The high forces imposed on the joint make maintenance of implant fixation difficult; component loosening and migration are not uncommon. High joint reaction forces also produce relatively high frictional forces at the metal–plastic interface. Over time, frictional wear debris, both metallic and polyethylene, elicits biologic responses that resorb bone and may further compromise implant stability. Improving the fixation and wear characteristics of total joint components is a major focus of orthopedic research.

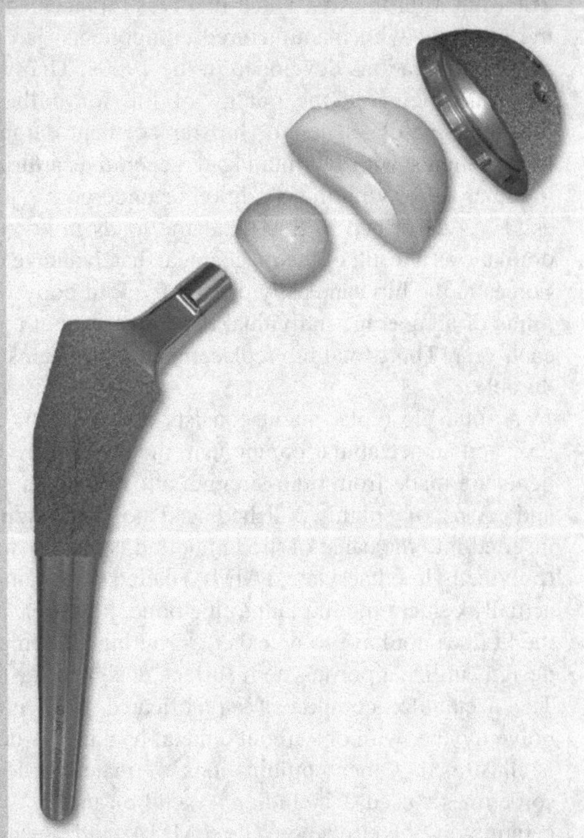

Fig. 2.8 A modern hip joint replacement prosthesis. The spherical ceramic ball articulates with an ultrahigh molecular weight polyethylene socket fitted into a metal cup. The long metal stem is placed inside the medullary canal of the femur. The proximal stem and socket backing have porous surfaces designed for bone ingrowth in lieu of using polymethylmethacrylate (PMMA) for fixation to the bone

2.5 Equine Fetlock Forces

Horses, and in particular racehorses, place extraordinarily high loads on their limbs. This is especially true of the forelimbs, which are thought to carry about 60% of the animal's weight. Many racehorses are seriously injured each year by mechanical failures of the structures in the distal forelimb. Because it is so difficult to repair these failures in such a way that the horse can stand and walk during healing, and because horses must be erect and mobile to survive, many animals die as a result of injuries that would not be life-threatening in humans or other smaller animals.

Figure 2.9 shows a sketch of the anatomy of the distal forelimb of the horse. Remember, the horse does not have feet the way humans do; it walks on the tips of its "fingers." The other bones of the foot (or hand) serve to lengthen the leg. P1 and P2 mark the first and second phalangeal bones in the figure; the third phalangeal bone is inside the hoof. The equine third metacarpal bone (MC3), which is also called the cannon bone, has become extremely stout, and carries loads from the phalangeal complex up to the carpal bones and the radius, seen at the top of the diagram. (The second and fourth metacarpal bones are vestigial struts stuck to the sides of MC3; the other metacarpal bones no longer exist.) The joint between MC3 and P1 is called the fetlock; its biomechanical importance derives largely from the fact that the bones above it rise straight up to the torso, but the phalangeal bones below it angles sharply forward in a springy cantilever arrangement. The joint between P1 and P2 is called the pastern joint.[2]

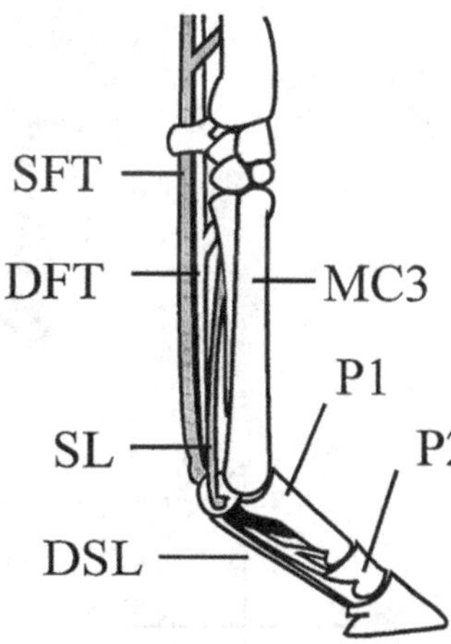

Fig. 2.9 Equine distal forelimb anatomy, sagittal view. *SFT* superficial digital flexor tendon, *DFT* deep digital flexor tendon, *SL* suspensory ligament, *DSL* distal sesmoidean ligaments; MC3, third metacarpal bone; P1, proximal phalanx; P2, middle phalanx

[2]P1 is also called the pastern bone. The name apparently comes from the fact that to pasture a horse without benefit of fences, ranchers would hobble it with a short rope connecting the P1 bones.

Behind the third metacarpus (MC3), two tendons come down from the superficial and deep digital flexor muscles. The superficial flexor tendon (SFT) inserts on P2, and the deep flexor tendon (DFT) inserts on P3 inside the hoof. On their way to these attachment sites, these tendons pass over two small bones located just behind the distal end of the cannon bone. These bones are called the proximal sesamoid bones; their cartilaginous dorsal (anterior) surfaces articulate with the palmar (posterior) distal condyles of the cannon bone. The fetlock's sesamoid bones function very much like the patella in your knee, moving the line of action of the tendons away from the joint center so that they exert a larger moment about it. In addition to these tendons, there is a set of four ligaments that help stabilize the fetlock. The suspensory ligament (SL) runs from the proximal posterior surface of MC3 down to the sesamoid bones. From there, three distal sesamoidian ligaments connect to P1 and P2. These four suspensory ligaments, along with the sesamoid bones, are often called the suspensory apparatus. The cannon bone brings the weight of the horse down to the fetlock, which sits in the suspensory apparatus in much the same way as a person lies in a hammock. The flexor muscles and tendons help carry this load, with the sesamoid bones providing a bearing surface to protect the distal cannon bone, as well as increasing the moment arms of the tendons and ligaments, as previously mentioned.

To analyze the forces in the fetlock joint, we construct the free-body diagram shown in Fig. 2.10. The analysis is two dimensional in the sagittal plane. The "amputation" has been done in such a way that the free body includes the sesamoid bones and everything distal to the cannon bone. These structures are all assumed to

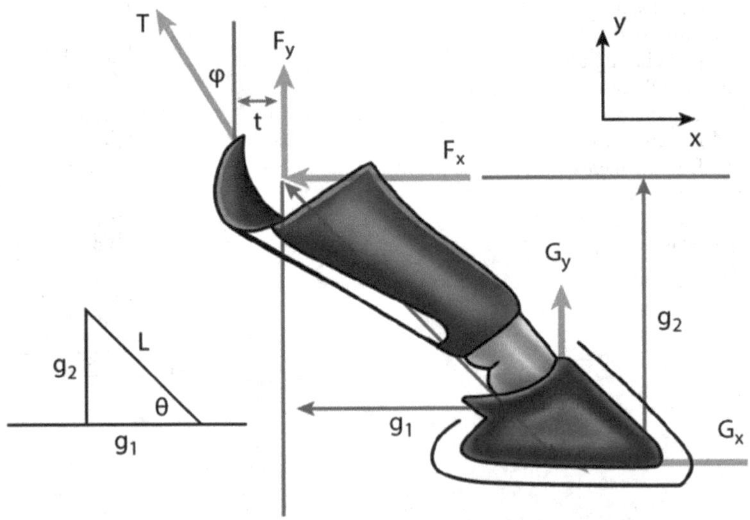

Fig. 2.10 Upper: Free-body diagram for forces in the equine fetlock joint. *Gx* and *Gy*, components of the ground reaction force; *T* tendon force; F_x and F_y, components of the joint force. Lower: Diagram for the moment arms, g1 and g2, of the ground reaction force. L, distance from the ground contact to the joint center; Θ, angle of the phalanges with respect to the ground

be rigidly connected; the pastern joint is assumed to be fixed. The forces resulting from the flexor tendons and the suspensory ligament are all combined into a vector, T, which has components in the x- and y-directions (left and up, respectively). Considering the large weight of the horse, the weight of the free body is ignored. The remaining forces acting on the free body are the two components of the ground reaction force, *G*, and the two components of the joint reaction force, *F*. *F* is assumed to act at the joint's center of rotation. Taking moments about this center, we have

$$\sum M = g_1 G_y - g_2 G_x - tT_y = 0 \tag{2.17}$$

where g_1 and g_2 are as defined in the lower portion of Fig. 2.10 and *t* is the moment arm of the vertical component of *T*. It is assumed that the x-component of *T* has a negligible moment arm with respect to the joint center (Fig. 2.10).

The force equilibrium equations are

$$\sum F_X = -T_x - G_x + F_x = 0 \tag{2.18}$$

$$\sum F_Y = T_y + G_y - F_y = 0 \tag{2.19}$$

These equations may be solved for various ground reaction forces and anatomical angles (ϕ and θ). First, let us assume that *G* and *T* are entirely vertical (e.g., when the horse is standing). Then, $\phi = 0$, $T = (g_1/t)G = \mathrm{GL}\cos\theta/t$ and $F = T + G$. If the horse weighs 500 kg (roughly 5000 N), and 60% of that is divided equally between the two forelimbs, then $G = G_y = 1500$ N. For $L = 0.2$ m, $\theta = 40°$, and $t = 0.05$ m:

$$T = (1500)\ (0.2)\ (0.766)/0.05 = 4596\ \mathrm{N}$$

The force carried by the tendons and ligament behind the cannon bone is nearly equal to the entire weight of the animal! In fact, if the angle θ sags to 33°, *T* becomes 5000 N. The joint reaction force is, of course, also vertical, and is $F = T + G = 6096$ N, or 1.2 times body weight.

Now suppose that the ground reaction force is still entirely vertical, but the angle ϕ is 20°. Then, from the moment equation,

$$T_y = (g_1/t)G = GL\cos\theta/t \tag{2.20}$$

so that

$$T = T_y/\cos\varphi = (GL/t)(\cos\theta/\cos\varphi) = 4891\mathrm{N} \tag{2.21}$$

and from the force equations

$$F_y = T_y + G = [1 + (L/t)\cos\theta]G = 6096\text{N} \tag{2.22}$$

$$F_x = T_x \sin\varphi = 1673\text{N} \tag{2.23}$$

The resultant joint force is 6321 N, and it acts at an angle of 15° to the vertical.

If this topic is of particular interest to you, you may want to read papers by Bartel et al. [23] and Riemersma et al. [24]. A related work on the evolution of equine locomotion by Thomason [25] is also of interest.

2.6 A Note on Dynamic Joint Loading

The overriding take-home message in this chapter is that that forces carried in bones and joints are magnified by the small lever arms of our muscles, producing internal joint forces many times the external forces imposed by the environment. It may seem odd to engineer a system to increase loads on joints, but from an evolutionary standpoint, our limbs and joints evolved for range of motion and speed of movement. The natural trade-off for this is short lever arms and increased forces on the joints (which was not such a problem, until we began to live so long, causing small amounts of damage that accumulate over time to progress to significant age-related joint deterioration, or osteoarthritis, in some people). Because it is difficult to do otherwise, all of the examples used in this chapter have been for static loading, e.g., holding a ball in your hand or standing on one leg. Under normal conditions, however, we do not hold still for very long (think of a 3-year-old child!). Even simplified dynamic models are significantly more complex than a static model (and therefore are beyond the scope of this chapter). However, it should be noted that the dynamic forces in muscles (and joints) can be many times greater than the static estimates—which are themselves many times larger than the forces applied to the hand or foot. Mechanical support for these very high loads (and the resulting high stresses) is provided by bone, cartilage, tendon, and ligament, skeletal tissues that are examined throughout the remainder of this book.

2.7 Summary and Further Reading

This chapter has considered the analysis of forces between one bone and another across joints. We have seen that such forces are always several times the external force being supported. This occurs because skeletal levers are arranged to magnify movement rather than force. Thus, muscle forces must be substantially greater than the forces being applied to the external environment. A person turning over in bed produces substantial hip joint forces, without supporting body weight, through the required muscle actions. Similarly, marine mammals such as dolphins and whales, whose buoyancy spares them the need to resist gravity, nevertheless have massive skeletons to transmit the muscle forces produced in swimming. The forces acting on bones can be estimated using the principles of static equilibrium. This method can provide very useful information, but significant approximations must be made to solve such problems. In general, the joints of vertebrate animals are statically indeterminate; that is, the number of unknowns (muscles acting to move or stabilize the joint) is almost always greater than the number of equations to be solved.

Suggestions for further reading in this field begin with the classic textbook Biomechanics of Human Motion [26]. Another well-known textbook, aimed at the medical community, is Basic Biomechanics of the Musculoskeletal System by Nordin and Frankel [27]. Steindler's Kinesiology of the Human Body Under Normal and Pathological Conditions [28] is excellent for its concentration on anatomy and pathomechanics. Early studies of the biomechanics of the human hip joint by Inman; McLeish and Charnley; and Rydell's paper on measurement of hip joint forces using an instrumented prosthesis, will give you a greater appreciation for the complexities of experimental work in this area. Animal Mechanics provides an approach to this topic well suited to zoologists. If you are historically inclined, by all means seek out Maquet's translation of Giovanni Borelli's wonderful seventeenth-century work, De Motu Animalium (On the Movement of Animals) [4, 9, 29–31].

References

1. Sharkey NA, Hamel AJ. A dynamic cadaver model of the stance phase of gait: performance characteristics and kinetic validation. Clin Biomech. 1998;13:420–33.
2. Erdemir A, Piazza SJ, Hamel AJ, Fauth AR, Sharkey NA. Dynamic loading of the plantar aponeurosis in walking. J Bone Joint Surg. 2004;86A(3):546–52.
3. Kirane YM, Michelson JD, Sharkey NA. Contribution of the flexor hallucis longus to loading of the first metatarsal and first metarsophalangeal joint. Foot Ankle Int. 2008;29(4):367–77.
4. Borelli GA. On the movement of animals (1989 translation by P. Maquet). Berlin: Springer-Verlag; 1989.
5. Frankel VH, Burstein AH. Orthopaedic biomechanics. Philadelphia: Lea & Febiger; 1973.
6. Burr DB, Gerven DPV, Gustav BL. Sexual dimorphism and mechanics of the human hip: a multivariate assessment. Am J Phys Anthropol. 1977;47:273–8.
7. Atwater AE. Gender differences in distance running. In: Cavanaugh PR, editor. Biomechanics of distance running. Champaign: Human Kinetics Books; 1990.
8. Rydell N. Forces in the hip joint. Part II. Intravital measurements. In: Kenedi RM, editor. Biomechanics and related bioengineering topics. Oxford: Pergamon Press; 1965. p. 351–7.
9. Rydell N. Forces acting in the femoral head prosthesis. Acta Orthop Scand Suppl. 1966;88:1–46.
10. English TA, Kilvington TM. In vivo records of hip loads using a femoral implant with telemetric output. J Biomed Eng. 1979;1:111–5.
11. Davy DT, Kotzar GM, Brown RH, Heiple KG, Goldberg VM, Heiple JG, Berrile J, Burstein AH. Telemetric force measurements across the hip after total arthroplasty. J Bone Joint Surg. 1988;70A:45–50.
12. Kotzar GM, Davy DT, Goldberg VM, Heiple KG, Berilla J, Heiple KG, Brown RH, Burstein AH. Telemeterized in vivo hip joint force data: a report on two patients after total hip surgery. J Orthop Res. 1991;9(5):621–33.

13. Bergmann G, Graichen F, Siraky J, Jendrzynski H, Rohlmann A. Multichannel strain gauge telemetry for orthopaedic implants. J Biomech. 1988;21:169–76.
14. Bergmann G, Graichen F, Rohlmann A. Hip joint loading during walking and running measured in two patients. J Biomech. 1993;26:969–90.
15. Bergmann G, Deuretzbacher G, Heller M, Graichen F, Rohlmann A, Strauss J, Duda GN. Hip contact forces and gait patterns from routine activities. J Biomech. 2001;34(7):859–71.
16. Taylor SJ, Perry JS, Meswania JM, Donaldson N, Walker PS, Cannon SR. Telemetry of forces from proximal femoral replacements and relevance to fixation. J Biomech. 1997;30(3):225–34.
17. Bay BK, Hamel AJ, Olsen SA, Sharkey NA. Statically equivalent load and support conditions produce different hip joint contact pressures and periacetabular strains. J Biomech. 1997a;30: 193–6.
18. Winter DA. Biomechanics and motor control of human movement. New York: Wiley; 1990.
19. Morris CB. The measurement of the strength of muscle relative to the cross section. Res Q. 1948;19:295–303.
20. Maughan RJ, Watson JS, Weir J. Strength and cross-sectional area of human skeletal muscle. J Physiol Lond. 1983;338:37–49.
21. Collins JJ. The redundant nature of locomotor optimization laws. J Biomech. 1995;28:251–67.
22. Charnley J. Arthroplasty of the hip: a new operation (reprinted from Lancet, pp. 1129–1132, 1961). Clin Orthop Relat Res. 1973;95:4–8.
23. Bartel DL, Schryver HF, Lowe JE, Parker RA. Locomotion in the horse: a procedure for computing the internal forces of the digit. Am J Vet Res. 1978;39:1721–33.
24. Riemersma DJ, Schamhardt HC, Hartman W, Lammertink JLMA. Kinetics and kinematics of the equine hind limb: in vivo tendon loads and force plate measurements in ponies. Am J Vet Res. 1988;49:1344–52.
25. Thomason JJ. Functional interpretation of locomotory adaptations during equid evolution. In: Rayner JMV, Wootton RJ, editors. Biomechanics in evolution. Cambridge: Cambridge University Press; 1991. p. 213–27.
26. Williams M, Lissner HR. Biomechanics of human motion. Philadelphia: Saunders; 1977.
27. Nordin M, Frankel VH. Basic biomechanics of the musculoskeletal system. Philadelphia: Lea & Febiger; 1989.
28. Steindler A. Kinesiology of the human body under normal and pathological conditions. Springfield: Charles C. Thomas; 1955.
29. Inman VT. Functional aspects of the abductor muscles of the hip. J Bone Joint Surg. 1947;29A:607–19.
30. McLeish RD, Charnley J. Abduction forces in the one-legged stance. J Biomech. 1970;3:191–209.
31. Alexander RM. Animal mechanics. Seattle: University of Washington Press; 1968.

Chapter 3
Musculoskeletal Tissue Structure and Biology

Abstract The musculoskeletal system consists of bone, cartilage, ligaments, and tendons. Each of these tissues has unique components to facilitate their function within the body. A common feature among them is their hierarchical structure and how their structure supports their function. In this chapter, we will provide an overview of the cellular and matrix constituents of bone, cartilage, ligaments, and tendons.

What You Will Learn

1. The macro- and microstructure of each component in the musculoskeletal system.
2. The cell and extracellular components of the musculoskeletal system
3. How structure is determined by biology and determines mechanical function

3.1 Bone Structure

As more is learned about the mechanics of the skeleton, it becomes increasingly clear that mechanical demands heavily influence what happens biologically. Most of what we know about the basic cell- and tissue-level mechanisms that enable bone and its adjoining connective tissues to adapt to mechanical forces has been learned since the 1960s. Before then, skeletal research was focused on bone structure and chemistry. Calcium is a metal and phosphorus a nonmetal that, along with oxygen and hydrogen, are primary constituents of bone mineral. The serum concentrations of calcium and phosphorus are of great significance in physiology and medicine, and there was intense interest in the mid-twentieth century in how bone contributed to the homeostatic regulation of these elements. Physiologists and endocrinologists realized that bone mineral served as a calcium and phosphorus reservoir for the rest of the body. Bone biology was seen largely in terms of its role in transferring mineral to and from the skeleton. Moreover, it was recognized that bone's mechanical properties—its strength and stiffness—were, to a significant degree, determined by the bone's mineral content. Early work by John D. Currey of the University of York and others demonstrated the strong correlations between bone density and strength and, since

M. R. Allen, J. Wallace, *Skeletal Tissue Mechanics*,
https://doi.org/10.1007/978-3-031-90609-1_3

bone mineral content could be measured in living animals using x-ray methods, it appeared that the main factors that cause bone tissue to be strong or weak were well understood [1, 2]. The roles of collagen alone and of the mineral-collagen composite were investigated to a much lesser degree, partially because collagen content and conformation could not be noninvasively measured in living animals.

Exploring the microanatomy of bone and the cellular constituents responsible for maintaining its strength was also challenging. Preparation of undecalcified bone sections entailed the tedious process of cutting thick sections of cortical bone and grinding them by hand using grinding powder or sandpaper to a thickness of 100–150 μm, where accurate analysis could take place. With the advent of plastic embedding in the 1970s, it became possible to cut thin sections using a microtome or diamond saw, greatly improving and accelerating the technical preparation of microscope slides. This fueled research by Ascenzi, Bonucci, and others to study the microscopic structure of bone and relate this to its mechanical properties [3–6].

Bone is much more, however, than its mineralized components. It is a hierarchically organized material, often described as a fiber-reinforced composite material (Fig. 3.1). Fiber-reinforced composites are typically highly anisotropic, meaning they exhibit different mechanical properties in different directions. This allows increased strength in the predominant load direction and lower, yet most often

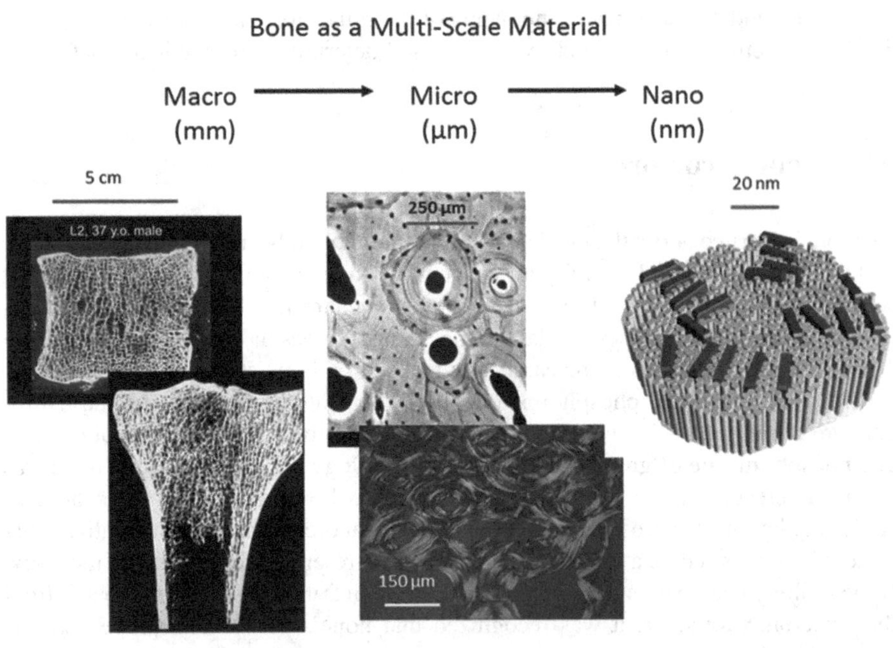

Fig. 3.1 Bone is arranged hierarchically according to structure at size scales ranging from nanometers (collagen and mineral) to microns (trabecular plates) to milli- and centimeters (the bone as an organ). Structure at each level imparts a unique mechanical adaptation, and the combination across length scales forms a structure capable of lasting decades without failing

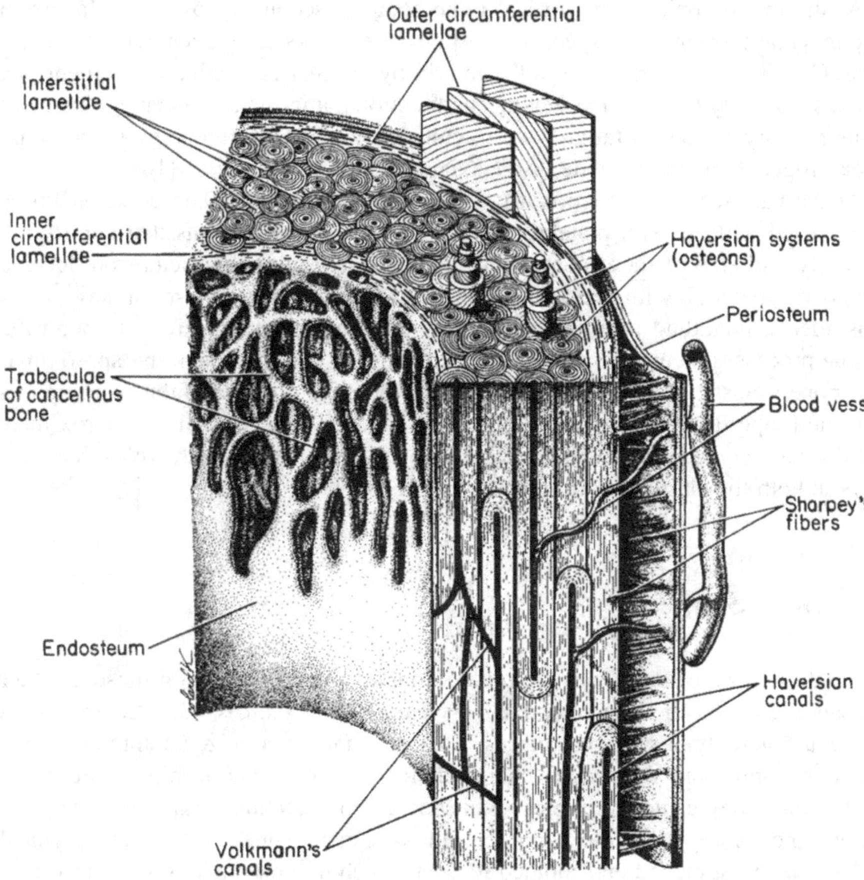

Fig. 3.2 Hemisection of bone showing important microstructural features from a typical long bone

sufficient, properties in alternative directions. For example, in the femur, the strongest direction is along the femoral shaft and the weaker, but functionally adequate, directions are circumferential and radial. This is a good design for a femur, where most of the loading is along the long axis of the bone.

At the structural level, bone is either dense compact (cortical) bone, or more porous trabecular (cancellous) bone (Fig. 3.2). The tissue that forms these microscopic architectures can be highly structured lamellar (sheetlike) or completely disorganized and nearly devoid of identifiable structure (woven). Mature bone microstructure is often layered, alternating between regions with highly oriented mineral (lamellae) and adjacent interlamellar regions. The regions with oriented crystals are 20–50% stiffer than the interlamellar regions, at least in longitudinal compression [7–9].

At the microscopic level, cortical bone contains secondary osteons (Haversian systems) that form in many species in response to a remodeling event within cortical bone. Osteons are separated from the matrix by an interface called the cement line that has relatively low shear strength. The functional importance of cement lines lies in their ability to control fatigue and fracture processes, absorb energy by stopping crack propagation, and provide viscous damping in compact bone [10].

At the nanoscale, the bone matrix is composed of cross-linked collagen fibrils interspersed with mineral platelets, with non-collagenous proteins that control the assembly and size of these components. This organizational structure produces a composite material with mechanical properties superior to those of any of its constituents, including a remarkable ability to resist structural failure through the fatigue processes incurred by a lifetime of repetitive loading. As a first approximation, bone's elastic properties are attributed to the mineral phase of the tissue and its inelastic properties to the organic (primarily collagen) matrix. This is approximate and does not consider either the interaction of the two phases or the roles that each plays in both the elastic and plastic deformation of bone tissue.

3.2 Bone Shape

While the shapes of our bones are quite variable, comparisons with those of other animals show that the architecture of a vertebrate skeleton is fairly conserved in nature and stereotypical across species. We can usually recognize a femur or a tibia, and we recognize that they have a characteristic positional relationship, regardless of what animal they came from. Yet within these broad architectural constraints, the skeletal morphology is quite variable, and we can imagine a rubber archetypical skeleton being stretched and molded to create such diverse skeletons as those of a flamingo, a bat, an elephant, a whale, or a human.

Not only is skeletal architecture conserved across species, but so too is the topology of individual bones within a species. Thus, a physical anthropologist can recognize an isolated bone, or even a fragment of a bone, as belonging to a human rather than one of the other primates. These constraints on architecture and topology give the impression that bones are determinate, static structures. On a scale of centimeters, this is approximately true. Yet the specific mechanical behavior of bones also relies on slight variations in the geometric distribution of bone, either within a cross section of compact bone, or by variations in trabecular architecture that correspond to the pattern of locomotion and the directions of the stresses and strains imposed by the animal and its environment. Differences in geometry (gross or macroscopic anatomy) can compensate for differences in the microanatomy (or nanoanatomy) that change the mechanical properties of the tissue. It can be readily appreciated that the strength of a whole bone depends on the material composing it, and its volume, cross-sectional area, three-dimensional shape, ratio of the smallest diameter to total length, and other shape characteristics. These gross geometric properties of the bone are complemented by the material properties of the tissues themselves. Tissue material properties also are variable in space and time, responding to mechanical stimuli. The complex interrelationship between the physical properties of bone—those determined by the mineral, collagen, non-collagenous proteins, and water—and the geometric properties together determine the overall mechanical behavior of a whole bone as a structure.

3.3 Types of Bone and Bone Tissue

3.3.1 Cancellous and Cortical Bone

If you were to slice open a dozen bones of various sorts and wash away the marrow with a high-pressure stream of water, it would be clear that the remaining bone is of two distinct kinds, differentiated by porosity (the volume fraction of bone relative to non-bone tissues). While in principle the porosity of bone can vary continuously from zero to 100%, most bone tissue is of either very low or very high porosity, with little bone of intermediate porosity (outside of pathology). These two types of bone are referred to as cortical and cancellous bone, respectively (Fig. 3.3).

Cancellous bone (also called trabecular bone) is porous bone found inside cuboidal bones (e.g., vertebrae, carpal bones), the flat bones (e.g., ilium, sternum), and the ends of long bones; its porosity is 75–95%, meaning there is typically more marrow than bone in a volume. The pores are interconnected and filled with marrow. The bone matrix is in the form of plates or rods called trabeculae, each about 100–150 μm thick. The arrangement of the trabeculae is variable, determined by factors such as mechanical load or pathology. Sometimes they appear to be organized into orthogonal arrays; often, they are more randomly arranged. Cancellous bone tissue has a much lower density (on a volume basis) than cortical bone and allows the skeleton to build large but lightweight conical volumes of bone at the ends of long bones that carry the large loads of the joints with low stress at the wide end (the metaphysis) and higher stresses in the narrower shaft (the diaphysis). This serves as a means for the bone to funnel the high loads imposed at the joint to the stronger and more dense cortical bone comprising the middle portion of the bone (i.e., the diaphysis). Cancellous bone also has a much larger surface area relative to cortical bone and has much higher remodeling rates (in most cases).

Compact (cortical) bone is the dense bone found in shafts of long bones. It also forms a cortex or shell around vertebral bodies and other cuboidal and flat bones. The outer surface of cortical bone is called the periosteal surface; the inner surface, next to the marrow, is called the endocortical (sometimes endosteal) surface. Cortical bone porosity is typically around 5–10%, although this increases with age and with pathological changes to the skeleton. The porosities in cortical bone can be categorized as follows:

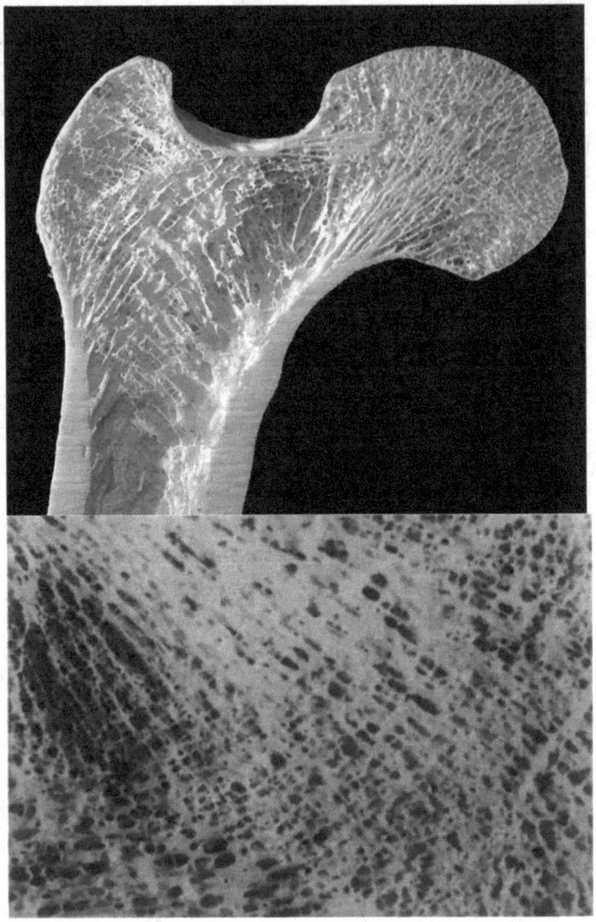

Fig. 3.3 (**a**) A sagittal section of the proximal end of a human femur provides an appreciation for the organization of cancellous bone and its relationship to the dense cortical bone surrounding it. (**b**) Cancellous bone structure from the distal end of the femur at higher magnification shows more clearly the integration of trabecular rods and plates. Field width, ~1 cm

- Haversian canals are approximately aligned to the long axis of the bone, contain capillaries and nerve fibers, and are about 50 μm in diameter (about the diameter of a human hair). Haversian canals were named after an English physician, Clopton Havers, although modern literature refers to these as central canals (referencing the fact that they are central to osteons).
- Volkmann's canals are short, transverse canals connecting Haversian canals to each other and to the inner and outer surfaces of the bone. These canals also contain blood vessels and nerve fibers. They were named after Richard von Volkmann (1830–1889), although modern literature refers to these as transverse canals.
- Resorption cavities are the temporary spaces created by osteoclasts in the initial stage of remodeling. Resorption cavities are about 200 μm in diameter. They contain a vascular channel and then various bone and marrow cells. With aging,

and other pathological conditions, some resorption spaces are permanent (they never refill with bone or refill incompletely, leaving a larger pore than a typical central canal).

- Depending on the scale of assessment, lacunae and canaliculi can be included in the assessment of porosity. Lacunae are the spaces where osteocytes reside. They are small, typically 300–400 μm^3 in volume, but there are about 26,000 lacunae per cubic millimeter (mm^3) of bone [11]. Canaliculi, the channels that house osteocyte cell processes, average ~250 nm in diameter. There are roughly 1×10^6 canaliculi per cubic millimeter of bone.

Cortical and cancellous bone are dynamic and thus their porosity may change as the result of a pathologic condition, with age, or in a normal adaptive response to a mechanical or physiologic stimulus. Cancellous bone may become more dense (with interventions that increase bone mass) and cortical bone may become more porous (with aging or pathologies such as chronic kidney disease). Such changes strongly affect bone's mechanical properties.

The nonmineralized spaces within bone contain marrow, a tissue composed of blood vessels, nerves, and various types of cells. The chief function of the marrow is to generate the principal cells present in the blood. The internal presence of marrow is a nearly universal feature of bones (the ossicles of the inner ear are an exception). The relationship between marrow and bone is biological and physical: they share common stem cells, and marrow never exists outside of bone. Bone can be made to form outside the normal skeleton (e.g., in a muscle) by implanting osteogenic materials, but when this happens, a space containing marrow automatically forms in the nodule of extraskeletal bone. Moreover, marrow itself is a highly osteogenic material and can stimulate bone formation if placed in an extraskeletal location [12].

There are two types of marrow in bone, each with a different function. Bone is a blood-forming organ, and red, or hematopoietic marrow within the marrow cavity is abundant during rapid growth when there is a need for an ever-increasing volume of red and white blood cells. Red marrow begins to be replaced by yellow (fatty) marrow around the age of 13 as the growth rate slows down, with the red marrow in the diaphysis of long bones being completely replaced by yellow marrow by the middle of the fourth decade of life. The cells in yellow marrow (adipocytes) and bone-forming cells (osteoblasts) arise from the same stem cell, creating what can appear to be an inverse relationship between the amount of fat and the amount of bone. In some pathologic conditions, the inverse relationship is very strong. Although some red marrow is retained in the ends of the long bones, and in areas characterized by large amounts of cancellous bone (e.g., vertebral bodies, iliac crest and ribs), aging gradually converts more and more of this marrow to yellow marrow.

The Haversian canal contains three components: blood vessels, nerves, and cells. There is little known about the types of vessels within the canals; most evidence suggests they are fenestrated capillaries that lack smooth muscle. The lack of smooth muscle indicates that blood flow through the cortex is determined by alteration to higher-order vessels in the marrow and periosteum. Adjacent to the vasculature are unmyelinated nerve fibers, thought to be part of the sympathetic system, although their role is unknown. Various cells have been identified in the canals, most often osteoblasts, their precursors, and their ancestors (lining cells).

The outer surface of cortical bone is covered by the periosteum, a dual layer of connective tissue (Fig. 3.2). The inner layer of the periosteum layer is called the cambium layer. It contains osteoblast precursor cells that can be engaged to form bone. The outer fibrous layer of the periosteum is mostly connective tissue. The absence of periosteum can significantly inhibit bone formation during conditions of healing, illustrating its importance in providing bone-forming cells.

The existence of nerves in bone tissue has been appreciated since the mid-1800s, when John Hilston observed that bone and joints were fed by the same nerves as nearby muscle and skin. It is now accepted that these are both myelinated and nonmyelinated fibers and that they are both sensory and autonomic (both sympathetic and parasympathetic). The periosteum has higher innervation than the marrow. Nearly all the known neurotransmitters (including but not limited to norepinephrine, leptin, serotonin, opioids) have effects on bone. Additionally, emerging data shows that the skeletal nerves can change in response to loading/unloading and disease.

Long bones are supplied with blood through one or more principal nutrient arteries. These pierce through the bone and divide into multiple branches within the marrow and eventually form capillary networks both in the marrow and central canals. Blood flow through bone is described as centrifugal (from inside to outside). Some data suggest this changes with aging and/or disease (to more centripetal from the periosteum inward). The vase capillary network has both H- and L-type capillaries, which differ in the types of cells adjacent to them as well as the factors that govern their permeability.

The existence of lymphatic vessels in bone has been debated due to the challenge of visualizing these delicate structures. Recent work, using light-sheet microscopy, which allows tissues to be examined with minimal disturbance, revealed human lymphatic vessels in bone. Preclinical studies show lymphatics in both periosteum and marrow and that manipulation of them (removal through radiation/intervention) can affect normal bone physiology.

3.3.2 Lamellar and Woven Bone Tissue

Examining bone at a finer scale reveals two types of bone tissue, lamellar and woven bone. Lamellar bone is a slowly formed, highly organized bone consisting of parallel layers or lamellae comprising an anisotropic matrix of collagen fibrils interdigitated with mineral plates. There may be a variety of patterns of lamellar arrangement at the level of the collagen fibril, and these likely each have important mechanical implications [13]. Microscopically, however, the lamellae create optically bright circumferential bands of bone, each 3–7 μm thick, that give an appearance like tree rings, each separated by an optically darker layer about 1 μm thick that is mechanically less stiff than the thicker lamellae [14]. The lamellae may be arranged around the

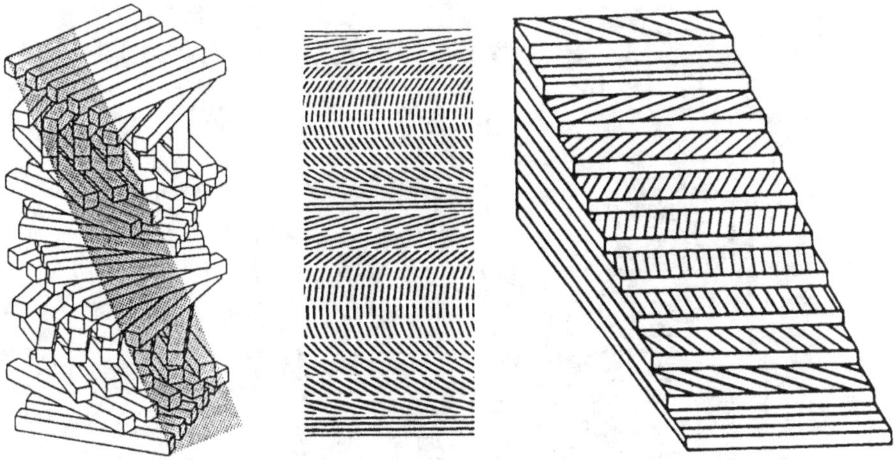

(A) (B) (C)

Fig. 3.4 Diagram of helicoidal plywood structure. (**a**) Three-dimensional structure. (**b**) Visual effect of arches seen when an oblique section (as indicated by the shaded plane in **a**) is cut. (**c**) The arches seen in an oblique section are formed by the helicoidal lamellae [16]

endocortical or the periosteal circumference of the bone (primary or circumferential lamellae), or may be arranged concentrically around individual vascular channels of osteons (osteonal or concentric lamellae).

Two fundamentally different kinds of "plywood" architecture coexist within the general complexity of bone's lamellar structure [15]. The first of these schemes corresponds to the classical view of lamellar structure: the collagen fibrils are parallel in each lamella and change direction by 90° at the lamellar interface. That is, as a lamella is built up, layers of collagen are put down in one orientation. Then, when the next lamella is started, the orientation of collagen suddenly changes so that they are laid down at right angles to the previous direction. In other regions of the bone, this orthogonal plywood-like structure is replaced by helicoidal plies (Fig. 3.4). In this scheme, the collagen fibrils continuously change their direction, so that in a sense there are no individual lamellae. However, the bone still shows a lamellar structure because as the orientation of the collagen rotates through 180° cycles, the fiber orientation repeats itself, and layers appear when a histologic section is examined microscopically. A helicoidal arrangement of collagen fibrils leads to a transversely isotropic structure at this level or organization as the elastic properties are similar in different directions [13]. In a landmark paper, Giraud-Guille showed that both architectures are present in human compact bone, but their relative distribution and interspecies variation have not been determined [15].

Both forms of lamellar structure give rise to birefringence, the capacity of some fibrous structures to interact with polarized light. When a histologic section of bone is transilluminated with polarized light and viewed through a polarizing filter oriented perpendicular to the vibration plane of this incident light, the section

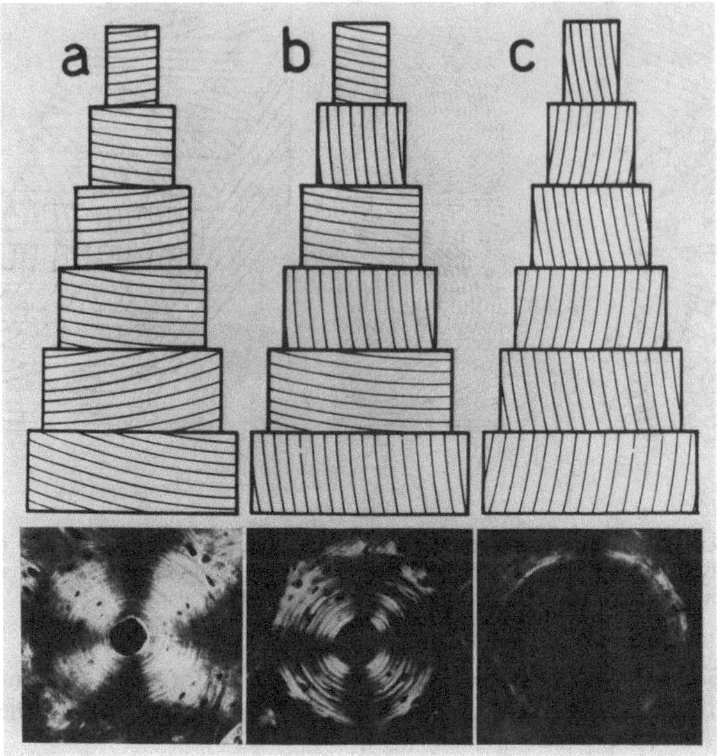

Fig. 3.5 Three osteon types as defined by Ascenzi and Bonucci. Photomicrographs at bottom show appearance in plane-polarized light; diagrams above show hypothesized fiber arrangements in successive lamellae. (**a**) Type T or transverse (i.e., circumferentially wrapped) fiber orientation, (**b**) type A or alternating fiber orientations, and (**c**) type L or longitudinal fiber orientation [5]

appears dark (i.e., the observer sees a "dark field"), except where collagen fibrils are parallel to the plane of the section. These collagen fibrils rotate the light's plane of polarization so it is no longer perpendicular to the viewing polarizing filter. Therefore, the light is not blocked and reaches the viewer's eyes. Thus, in a bone section observed in a polarizing microscope, transversely oriented fibrils are bright and longitudinally oriented fibrils are dark. In the 1960s, Ascenzi and coworkers used this phenomenon to categorize osteons as having bright, alternating, or dark lamellae (corresponding to their collagen fibers exhibiting mostly circumferential, alternating, or longitudinal orientation) [5] (Fig. 3.5). A dark "iron cross" pattern superimposed on the brighter osteons shows where the lamellae lie parallel to the polarization directions of the upper or lower polarizing filters.

Woven bone is a quickly formed, poorly organized tissue in which collagen fibrils and mineral crystals are randomly arranged. Woven bone may become more highly mineralized than lamellar bone, which, mechanically speaking, may help compensate for its lack of organization. Woven bone is common in a fracture callus and is

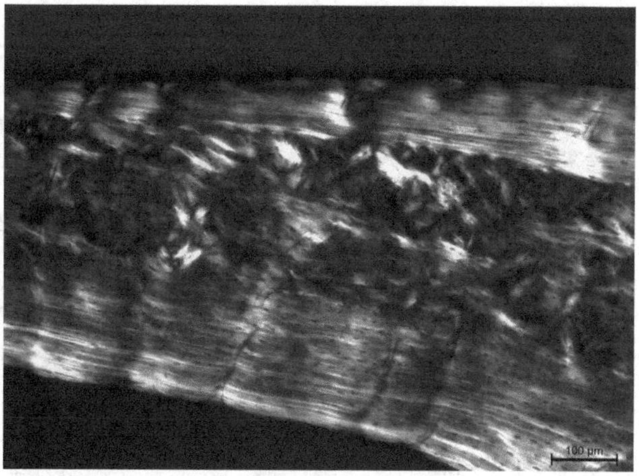

Fig. 3.6 Polarized light image from a rat tibia showing disorganized fine-fibered bone in mid-cortex. This bone appears almost like woven bone and is composed of very small collagen fibers, randomly arranged. Bone appearing more like primary lamellar bone is found along the periosteal and endocortical surfaces of the bone. Org mag 10×

formed early in the healing process to help stabilize the broken bone. In this case, woven bone can be considered calcified scar tissue that is rapidly formed for damage control and then remodeled and replaced over time with lamellar tissue reflective of the site's original organizational structure. The important generalization is that woven bone can be made more quickly than lamellar bone but has inferior mechanical properties. Rats and other small animals also have fine-fibered bone in the cortices of their bones. In this bone, the rather randomly arranged collagen fibrils are smaller and closer together than in woven bone, so the tissue appears to be more organized (Fig. 3.6). While the rate of formation clearly dictates lamellar fiber orientation/variation in woven versus lamellar bone, it also dictates orientation across species. Those that grow slower are more likely to have bone comprised of a larger proportion of oblique to transverse collagen fibers [17].

3.3.3 Primary and Secondary Bone

Bone tissue can be further characterized as primary or secondary bone. Primary bone is tissue laid down de novo in space or on an existing bone surface. The most common example is when new bone is laid down on the periosteal surface during growth, but any bone that has not been remodeled is considered primary bone. Since rodents do not normally undergo remodeling of cortical bone, most of their bone is considered primary. Primary bone may be of two general types:

Circumferential lamellar bone, in which the lamellae are parallel to the bone surface is shown in Fig. 3.7a–c. Blood vessels are incorporated into the lamellar structure such that each is surrounded by several circular lamellae, forming a primary osteon with a primary central canal. As opposed to secondary osteons, primary osteons do not have a cement line because they are not formed through the normal remodeling process (where the cement line is formed during the reversal phase). This makes them morphologically and mechanically quite different.

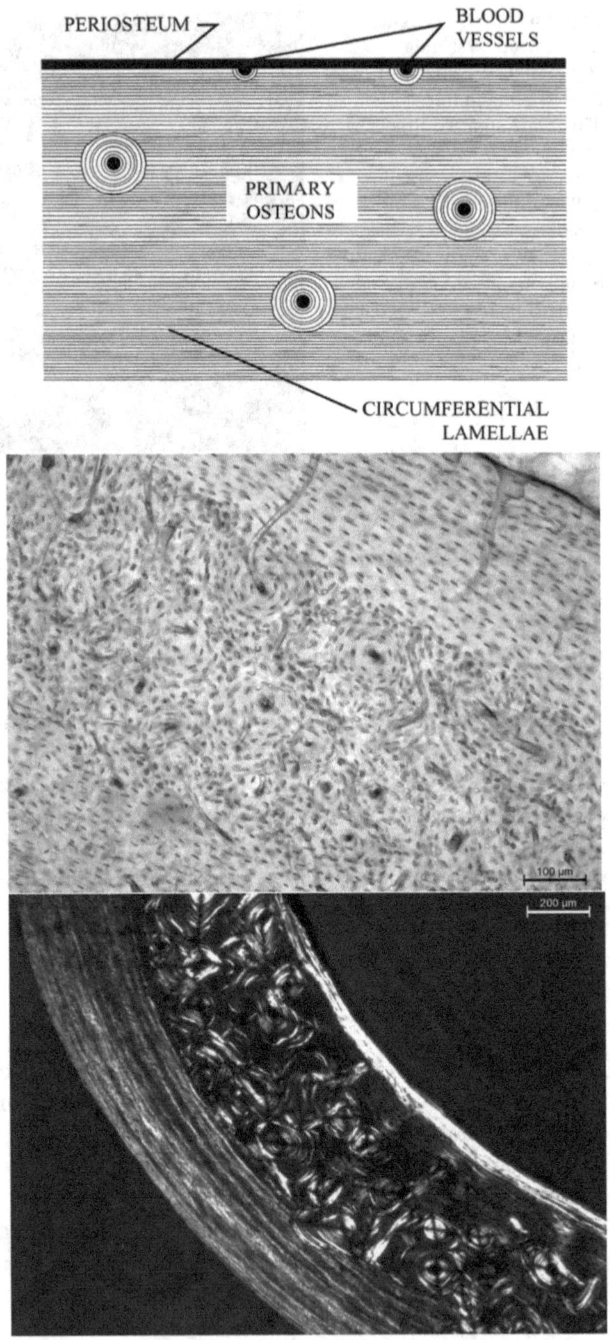

Fig. 3.7 (**a**) Sketch of primary circumferential lamellar bone structure. Primary osteons form when blood vessels on the bone surface become incorporated into the new periosteal bone. They usually have several concentric lamellae, but their cement line is not scalloped. (**b**) Photomicrograph of a cross section of a rabbit humerus showing primary lamellar bone running in parallel sheets around the periosteal circumference of the bone. Osteonal bone—demonstrating both primary and secondary osteons, can be found adjacent to the circumferential lamellar bone. (**c**) Bone from a squirrel monkey in polarized light clearly shows primary lamellar bone surrounding remodeled secondary osteonal bone

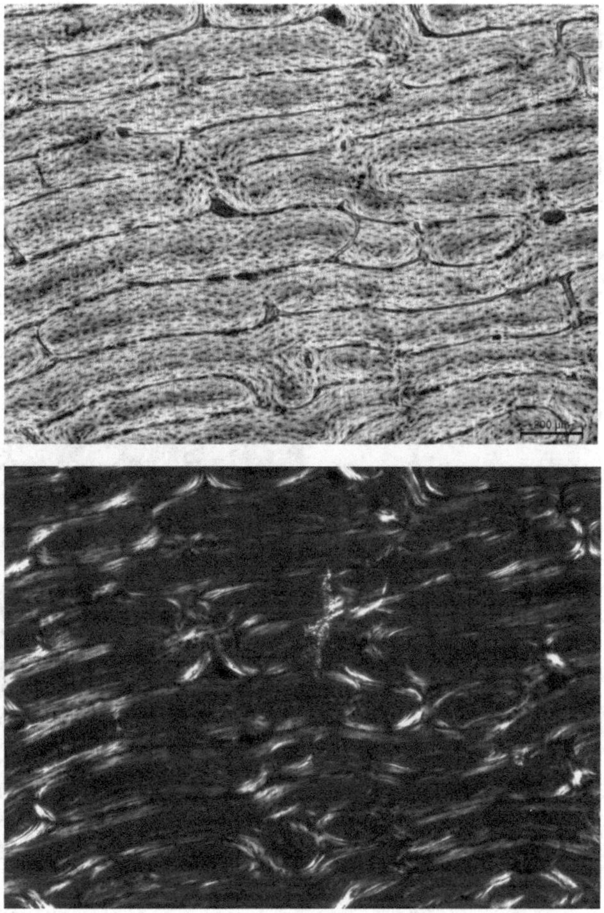

Fig. 3.8 (**a**) Photomicrograph of plexiform bone from a cow. Vascular channels run between the blocks of bone. (**b**) In polarized light, one can appreciate the structure of the plexiform bone, with non-lamellar bone forming the core (which is dark in this image), surrounded by lighter primary lamellar bone. Orig Mag 5×

Plexiform bone, in which the rate of formation is high like woven bone, has superior mechanical properties compared to woven bone. It typically occurs in large, fast-growing animals like cows but has been observed in humans during periods of rapid growth. As the name implies, this bone has intertwining vascular plexuses within a highly structured "brick wall" appearance (Fig. 3.8). Racehorses, which put enormous stress on some of their bones, have a similar kind of bone that may have exceptional fatigue resistance [18].

Secondary bone results from the resorption of existing bone and its replacement by new, lamellar bone. This process is known as remodeling (see Chap. 4). In cortical bone, secondary tissue consists of cylindrical structures known as secondary osteons or Haversian systems (Fig. 3.9a). These are about 200 μm in diameter, although this varies with age in predictable ways, and consist of about 16–20 cylindrical concentric lamellae surrounding a central canal [19–21]. Osteons are about 1–10 mm long and in the human femur they run at an angle of 11–17° with respect to the long axis of the bone [22].

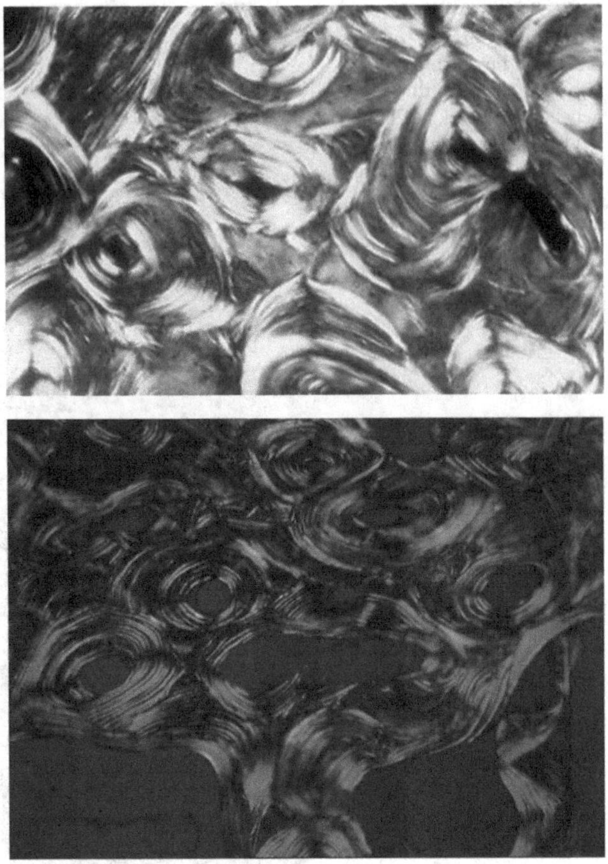

Fig. 3.9 (**a**) Higher magnification image in polarized light of remodeled secondary osteonal bone, taken from a chimpanzee femur. (**b**) In trabecular bone, new packets of bone are often formed from the surface and do not create an entire osteonal structure. The "half osteons" are termed hemi-osteons (arrow). One can also see complete secondary osteons within the individual trabeculae

The boundary between secondary osteons and the surrounding bone is known as the cement line. It is collagen-deficient; rather, it is composed largely of glycosaminoglycans, suggesting it is a residuum of the initial osteoid laid down during the reversal phase of the remodeling cycle. There are conflicting data on the mineralization level of the cement lines (data exist showing it is poorly mineralized [23] and highly mineralized [24]). Regardless of the composition, the cement line acts mechanically to trap cracks and slow crack growth in cortical tissue [10]. This mechanical phenotype suggests a low mineralization boundary as a viscous material would relieve locally high shear stresses and reduce the ability of a cement line to transmit energy to a growing crack. This is supported by the data showing that cement lines contain only 85–90% of the mineral as compared to nearby lamellae. Displacement (creep) of cement lines has been observed in human bone over long periods of torsional loading [25].

In adult humans, most compact bone is predominantly composed of secondary bone, which may include whole both osteons and the remnants of older osteons that have been partially remodeled (interstitial bone). Remodeling of individual trabeculae within cancellous bone, which occurs frequently in both children and adults, rarely produces a complete osteon because the 200 μm diameter of an osteon is too large to fit within most trabeculae of average diameter 100–150 μm, although they do exist. Instead, hemi-osteons are formed (Fig. 3.9b). One surface of the hemi-osteon borders the marrow cavity, rather than a central canal, but the remaining surface is separated from older bone within the trabecula by a cement line just as whole osteons are separated from their surroundings in compact bone. The role of the cement line in trabecular bone is less studied but likely serves the same mechanical purpose of dissipating energy as it does in cortical bone.

3.4 The Composition of Bone Tissue

At the ultrastructural level, bone is composed of organic and mineralized components, mainly consisting of a matrix of cross-linked type I collagen mineralized with nanocrystalline, carbonated apatite. About 65% of bone by weight is mineral, and 20–25% is collagen, mostly Type I. Although bone is a hard and stiff tissue, water composes about 10% of its structure and is essential to normal mechanical properties. Non-collagenous proteins make up the remainder, only about 1–2% of the bone, but have vital regulatory and mechanical functions that belie their relatively minor contribution to bone mass.

Collagen is a structural protein (molecule), widely distributed throughout the animal kingdom, that can spontaneously organize itself into strong fibers. Several dozen types of collagen have now been identified, although only a few of these are found in bone. The predominant fibrillar collagen in bone is Type I, which is also found in tendons, ligaments, and skin.

The primary collagen molecule is Type I, a glycine-rich triple helix, composed of two α-1 chains and one α-2 chain. As with any other protein, collagen synthesis begins with gene transcription in the nucleus followed by translation of the mRNA in the cytoplasm. Translated products are routed to the endoplasmic reticulum where the procollagen molecule is assembled and then extensively modified before being secreted into the extracellular matrix. Even after secretion, several enzymes continue to modify the procollagen molecule in preparation for assembly. Finally, the processed molecules, now simple triple-helical subunits, spontaneously self-assemble into collagen microfibrils (4 nm in diameter), which in turn assemble into sequentially larger strands of protein called subfibrils (20 nm in diameter) and fibrils (50–500 nm in diameter). Collagen fibrils can be visualized with electron microscopy and have a characteristic striated pattern resulting from the ordered assembly of the individual collagen molecules within the microfibrils. These aggregates have a periodic arrangement of parallel molecules spaced in a quarter-staggered array (having a ~67 nm repeating pattern) (Fig. 3.10). The mean diameter of collagen fibrils in osteoporotic bone is less than in healthy bone, and this may increase bone's fragility. Variations in spacing may also accompany certain diseases affecting collagen formation, such as osteogenesis imperfecta [26].

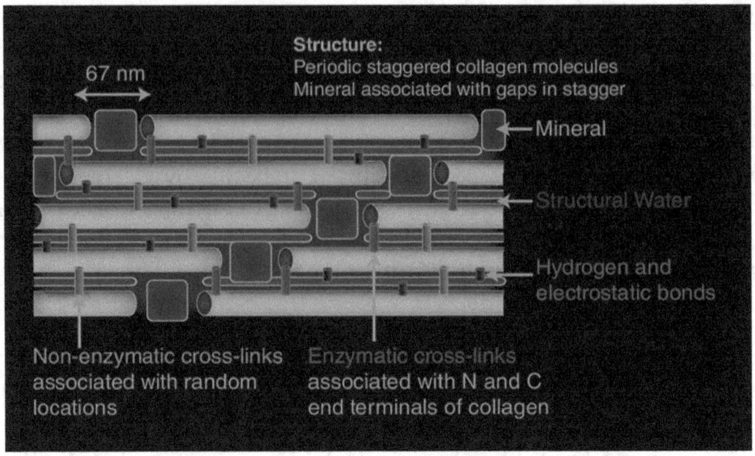

Fig. 3.10 A schematic illustration of the collagen/mineral arrangement within bone. Note the presence of collagen cross-links and water, both of which have a significant influence on the mechanical properties of the material

Chemical bonds, called cross-links, form bridges between the collagen molecules, stiffening the fibrils. Immature divalent cross-links are converted fairly rapidly (within weeks of bone deposition) to mature, or irreducible, trivalent cross-links by enzymatic processes. Cross-links formed by nonenzymatic processes can also be formed. Nonenzymatic reactions between sugars and proteins are common. After a series of Maillard reactions, the products are called advanced glycation end products (AGEs). Some of the AGEs create collagen cross-links that stiffen collagen fibrils in a manner similar to normal enzymatic cross-links.

Small amounts of Types III and V collagen can be found in the peri-lacunar region of the osteocytes and possibly contribute to the organization of the extracellular matrix and the construction of the Type I collagen fiber itself [27].

The amount and maturity of the Type I collagen in bone, and the manner in which it is cross-linked, can have a profound effect on the mechanical properties of bone tissue. Lower collagen content is a good predictor of reduced energy to fracture [28]. Increased density of mature, trivalent pyridinoline and deoxypyridinoline cross-links increases the compressive strength and stiffness of bone [29–32], but increases in these natural cross-links have not been demonstrated to change bone's fracture energy or ductility [33–36]. There is a smaller proportion of immature divalent cross-links in newly formed bone from individuals with osteoporosis [37], which could increase bone fragility and decrease bending strength.

Nonenzymatic cross-links may reduce bone toughness by reducing ultimate strain and post-yield deformation. They reduce collagen fibril diameter by dehydration, which could also contribute to increased fragility. AGEs naturally accumulate in bone with age. Bone from individuals over 70 years old has three times more nonenzymatic cross-links than bone from those less than 50 years. Diabetes causes

AGEs to accumulate in collagenous tissues, such as skin. Bone that is experimentally glycated in vitro exhibits brittle mechanical behavior [38]. Preclinical models of diabetes show inverse associations between levels of bone AGEs and mechanical properties; human data are less clear.

Mineral in bone consists almost entirely of highly substituted, poorly crystalline apatite mineral (Ca10(PO4)6(OH)2 that nucleates within the spaces between the ends of the collagen fibrils, as well as the spaces that run longitudinally between the fibrils. The individual crystals are rods or plates with hexagonal symmetry. As bone tissue matures, the mineral crystals grow, become more platelike, and orient themselves parallel to one another and to the collagen fibrils. Therefore, the measured average size of mineral crystals highly depends on tissue age. In both cancellous and cortical bone, the c-axis, or long axis, of the mineral and collagen aligns with the longitudinal axis of the structure, i.e., the trabecula or the osteon, respectively. When bone is fully mineralized, the crystals are sufficiently close to one another such that removal of the organic matrix by ashing or by chemical means leaves behind a solid piece of material rather than a pile of nanocrystalline "dust." It is unknown whether the crystals actually merge together into a single continuous phase such as in a stony material like granite, or are cemented together like sandstone, but their mechanical behavior in fully mineralized tissue seems to be similar to a continuous polycrystalline solid.

The hydroxyapatite found in bone contains many ionic substitutions (e.g., strontium, lead, carbonate, fluoride). These substitutions are governed by the composition of the body fluids and in turn affect the solubility and material behavior of the bone mineral. This means they are likely very heterogeneous across individuals and how this interacts with other aspects of bone to alter mechanical properties is unknown. As bone matures, mineral crystal size and perfection increases due to changes in ion substitutions and mineral stoichiometry [39]. These can substantially alter the hardness and brittleness of bone in either a positive or a negative manner. Substituting carbonate for the phosphate in hydroxyapatite makes the crystal more soluble, reduces the stability of the mineral lattice, and increases fragility. Different cation and anion substitutions in the apatite lattice can, in some cases, alter the mechanical properties of bone and in others affect the activity of osteoblasts and osteoclasts. Although not a strict substitution due to size constraints, citrate binding on the surface of the apatite crystal lattice prevents further crystal growth, leading to thinner nanocrystals that may render the overall structure more resistant to crack propagation [40, 41].

There is a known positive correlation between tissue mineralization and stiffness, but also an inverse relationship between these and energy absorption/toughness [41]. As the mineral in bone increases, the tissue becomes stiffer, stronger, and more brittle. The stiffer tissue deforms less under load and when the bone begins to fail, cracks will propagate with less energy than in less mineralized bone. Thus, a higher level of mineralization makes it more difficult to start a crack in the bone (it is stronger), but once a crack starts, it is more likely to propagate and break the bone (it is less tough). Lower mineralization may therefore be beneficial, but only to a point; under-mineralized bones, such as occur in rickets, are weak and readily break.

There are many non-collagenous proteins (NCPs) and glycans (sugar polymers) that have different functions in regulating the assembly and mineralization of the bone matrix. Non-collagenous molecules can be divided into several large groups that include the proteoglycans (versican, heparin sulfate, biglycan, and decorin); the glycoproteins (alkaline phosphatase, thrombospondin 1 and 2, fibronectin, vitronectin); the SIBLING proteins (osteopontin, bone sialoprotein, Dmp-1, MEPE); and several other ubiquitous and very important proteins, including osteocalcin and osteonectin. The proteoglycans have variable spatial locations with versican expressed highly in osteoid and prevalent during skeletal development. Decorin and biglycan regulate the construction of collagen fibrils, and their mineralization, and act in the differentiation of bone-forming cells (osteoblasts). The glycoproteins are active in bone formation and mineralization, but are also important during development in the early stages of bone formation and cell proliferation. The SIBLING family of proteins regulate bone mineralization, and several of them (Dmp-1, MEPE) are expressed highly by osteocytes. Osteocalcin and osteonectin (also called SPARC) modify mineral deposition. Osteocalcin is a marker of bone formation (or turnover), whereas osteonectin plays a role in osteoblast proliferation and helps to regulate growth factors in bone.

Water in bone exists in four different compartments, free/pore water, water loosely bound at the collagen/mineral interfaces, water tightly bound within collagen triple helices, and structural water within the mineral. Hydration has clear effects on bone mechanics, illustrated by the extreme condition of complete removal of water, which leaves bone stiff and brittle. More physiological changes to hydration, which occur with aging, disease, and even pharmacological treatment, can also modulate mechanical properties.

3.5 Bone Cells

Formation, removal, and maintenance of bone tissue is carried out by the cells that exist within and upon the bone matrix. There are three main bone cells: those that resorb bone (osteoclasts), those that form bone (osteoblasts), and those that orchestrate/maintain bone (osteocytes). Osteoclasts are closely related to macrophages, cells that migrate throughout all tissues of the body to remove debris and pathologic material. Osteoblasts are closely related to cells like fibroblasts, which produce structural molecules in other tissues. Osteocytes are terminally differentiated osteoblasts that live embedded in the mineralized matrix and have a distinct genetic signature that differs from an osteoblast. Despite their different lineages, osteoclasts and osteoblasts/osteocytes work together at discreet sites during bone remodeling by forming coordinated assemblies of cells called Bone Multicellular Units (BMUs).

Osteoclasts are the cells that resorb bone (Fig. 3.11). They are multinucleated cells (with 4–20 nuclei) that are derived from the monocyte/macrophage lineage under the stimulation of receptor activator of NFκB ligand (RANKL) and macrophage-colony stimulating factor (M-CSF). The molecule RANKL, which

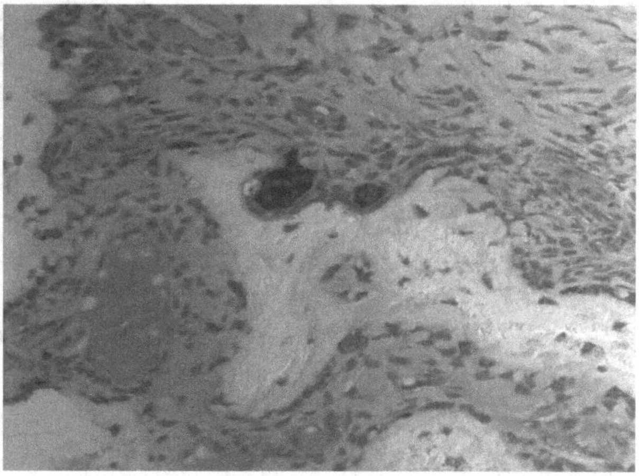

Fig. 3.11 Bone section stained with tartrate-resistant acid phosphatase (TRAP) showing multinucleated osteoclasts (red) on trabecular bone surfaces

binds to the RANK receptor, is produced by osteoblast, osteocytes, and numerous non-bone cells. Resorption occurs along a highly invaginated brush or ruffled border of the cell, which is sealed to the bone surface by a peripheral clear zone. Osteoclasts erode their way along or through bone at a rate of tens of micrometers per day by first demineralizing the adjacent bone through the activity of a proton pump (H+) and chloride ions, which leads to low pH and acidification of the bone mineral. This exposes the collagen, which is then decomposed by the enzymes cathepsin K and matrix metalloproteinases. These destructive chemicals are manufactured by the cell and transported to the ruffled border in portable intracellular chambers called secretory vesicles. Failure of the ruffled border to form properly, or failure of the polarization of the intracellular actin to form an actin ring around the clear zone, will prevent the osteoclast from resorbing bone. Once resorption is complete, most osteoclasts die by nuclear disintegration (apoptosis). Other osteoclasts may undergo fission (sometimes called osteocyte recycling), a process where they break apart and take on a unique genetic signature from osteoclasts and their mononuclear precursors. Osteoclasts that have undergone fission are sometimes referred to as osteomorphs. The role of osteomorphs in the overall bone remodeling process remains unclear.

Osteoblasts are mononuclear, cuboidal cells that produce osteoid, the organic portion (Type I collagen and non-collagenous proteins) of the bone matrix (Fig. 3.12). Osteoblasts develop from mesenchymal precursor cells from the bone marrow that differentiate under the early influence of bone morphogenetic proteins, Runx2, and osterix, which regulate osteoblast function by coordinating several other signaling pathways that include Wnt and BMPs. Full differentiation of osteoblasts also requires some level of mechanical stress, one reason that bone formation is impaired in cases of disuse or during spaceflight. Osteoblasts contain large nuclei and Golgi apparatus, and have a large endoplasmic reticulum, where procollagen and the other organic matrix molecules are produced. The boundary between osteoid and calcified bone is called the mineralization or calcification front. Osteoblasts

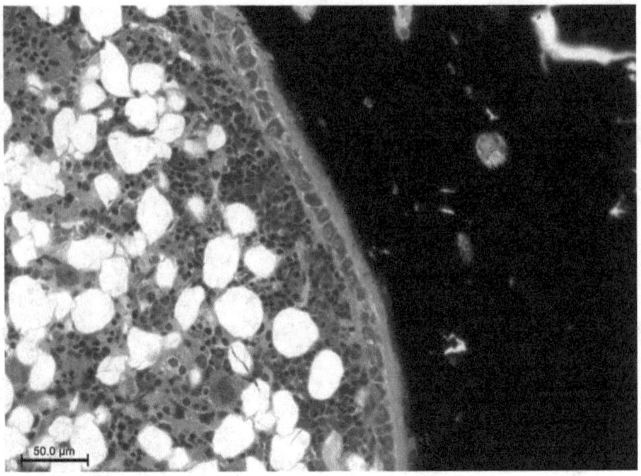

Fig. 3.12 Bone section showing cuboidal osteoblasts depositing osteoid (blue) on the surface of mineralized bone (black). Marrow cells (blue) and adipocytes (white) can be seen within the marrow cavity. Stained with McNeal's tetrachrome

express osteocalcin and alkaline phosphatase, which can be measured in blood serum. The concentrations of these molecules are related to the systemic rate of bone formation and mineralization, although there is heterogeneity in such rates throughout the skeleton. Osteoblasts have extensive communication with other cells on bone surfaces (lining cells) and within the bone matrix (osteocytes) through cellular processes that connect at gap junctions with processes from neighboring cells (Fig. 3.13). Gap junctions are intercellular channels composed of adjoining hemi-channels, each spanning the lipid bilayer of their respective cell and each composed of six identical protein molecules. The most abundant gap junction molecule in osteoblasts is connexin 43. The absence of this communication protein causes dysregulation of osteoblast differentiation and function.

Osteocytes, which make up 90–95% of all bone cells, are former osteoblasts that have become entombed in the bone that they and their neighbors have made. Osteocytes sit in cavities called lacunae and communicate with each other and with osteoblasts/bone lining cells via cell processes passing through narrow tunnels called canaliculi ("little canals") (Fig. 3.14). Processes from adjoining cells are connected by gap junctions, implying that networks of cells buried in bone tissue, resident on bone surfaces, or close by in the marrow communicate and exchange substances to a significant degree. This is one reason that osteocytes are thought to be important in the transduction of mechanical signals (mechanotransduction). Although the volume of lacunae and canaliculi is only about 1% of the overall bone volume, their surface area is enormous. The total skeletal surface area of canaliculi in an adult male skeleton is estimated to be 1200 μm^2, compared to 3.2 μm^2 for Haversian/Volkmann's canals and 9 μm^2 for cancellous bone surfaces [43]. This surface area is likely important in the process of maintaining calcium homeostasis.

It is generally thought that because osteocytes form a functional syncytium (a group of cells in which the cytoplasm of one cell is continuous with that of adjoining cells) that connects cells on the bone surface with those in bone marrow

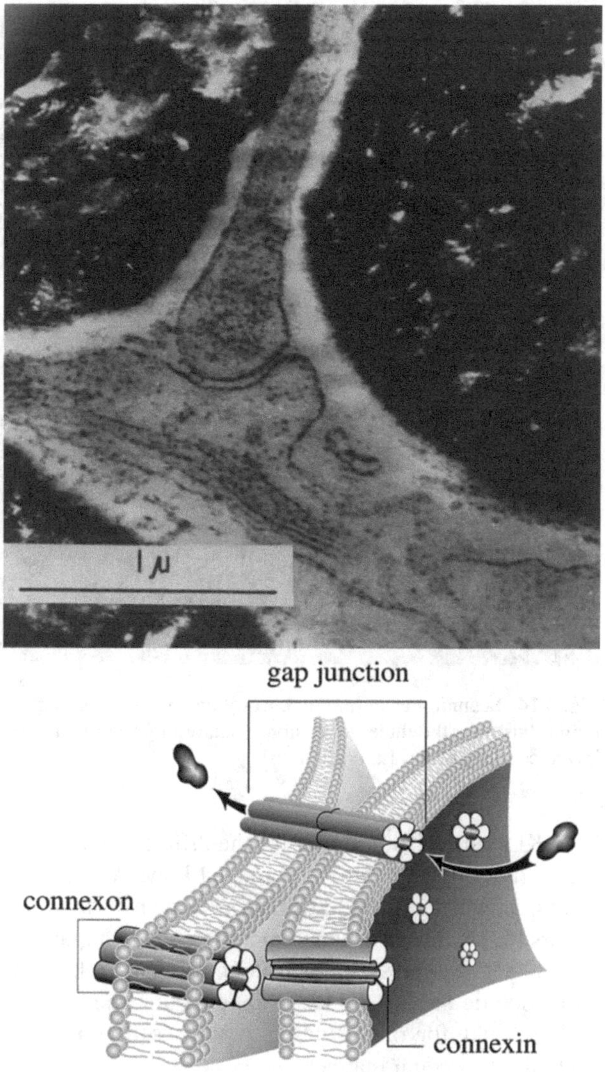

Fig. 3.13 (**a**) Electron photomicrograph shows a junction between two osteocyte processes [42]. (**b**) Osteocytes and osteoblasts communicate through gap junctions that are spanned by bundles of proteins called connexons, which are made up of individual connexins. Connexin-43 is one of the most important and regulates bone adaptation to mechanical signaling

and with those in blood vessels, they form a mechanosensory network that is able to detect and respond to changes in mechanical stress. It is likely that they detect load through strain-induced flow of fluid within the canaliculi and lacunae that deforms the cell processes. Circulating factors, such as calcium ions, may serve to amplify the signal. During normal physical activity, osteocytes produce a protein called sclerostin that inhibits Wnt signaling and suppresses bone formation. When high levels of mechanical loading are detected, sclerostin expression is suppressed, stimulating osteoblast differentiation and increasing bone formation [44]. In other situations, for example, when the bone is damaged or when the bone is unloaded, osteocytes begin to die by apoptosis. Either the apoptotic osteocytes [45], or those healthy osteocytes adjacent to the dying ones [46, 47], will begin to produce

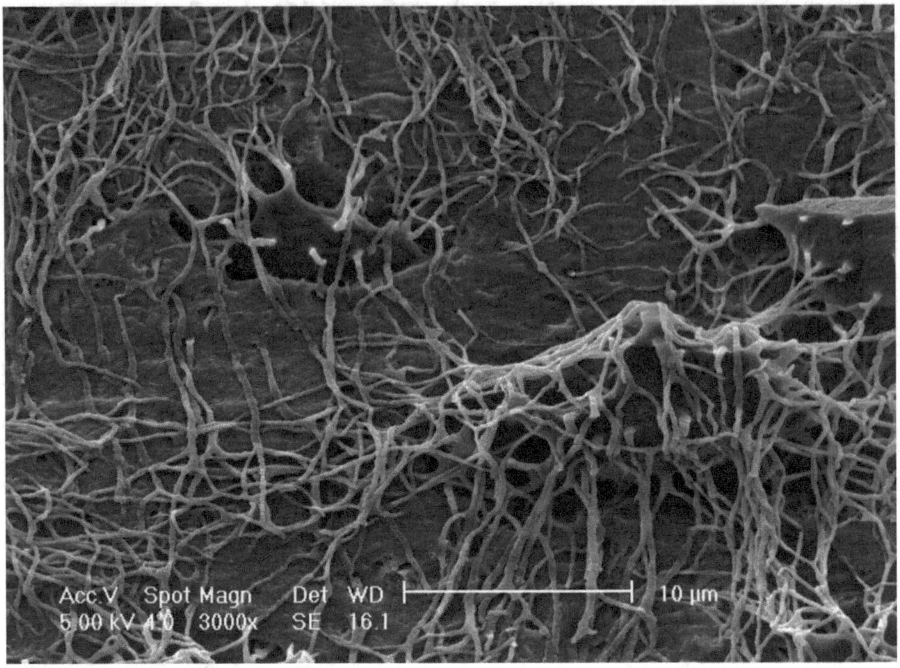

Fig. 3.14 Scanning electron microscopy imaging of a plastic-embedded, acid-etched bone preparation illustrates the three-dimensional nature of an individual osteocyte dendritic network. (SEM image courtesy of Dr. Lilian Plotkin)

RANKL, which is important for the differentiation and activation of osteoclasts that are recruited to remove the damaged bone. When osteocytes are removed entirely from bone through targeted ablation, bone resorption leading to significant bone loss occurs [48]. Similarly, when microdamage is imparted into bone, osteocytes nearest the damage express proapoptotic signals while those more distant produce antiapoptotic signals. Thus, it appears that the osteocyte syncytium is responsible in large part for orchestrating processes of both resorption and formation, in a spatially dependent manner.

Bone lining cells are, like osteocytes, "retired" (quiescent) osteoblasts. These are the osteoblasts that escaped being buried in newly formed bone and remained on the surface when bone formation ceased. As production of bone matrix stops, bone lining cells become quiescent and flattened against the bone surface, and for many years were not thought to form a continuous, gap-free barrier over the bone. This has been called into question recently with the description of a 'canopy' at trabecular remodeling sites, which functions to isolate the remodeling site and keep secreted factors in high local concentrations. Bone lining cells are thought to maintain communication with osteocytes and each other via gap-junctional processes, although how this would work in the context of the bone canopy is unclear. Like osteocytes, they are thought to be responsible for transfer of mineral into and out of bone.

3.6 Synovial Joints

Synovial and diarthroidal joints are those that move freely and contain synovial fluid, like the knee and elbow joints, as opposed to joints that do not, such as the "tether" joints between the vertebral bodies in the spine (Fig. 3.15). Synovial joints provide nearly frictionless motion between limb segments while transmitting relatively high loads between them. This is accomplished by the structure and composition of the tissues that comprise the joint.

The joint is surrounded by an articular capsule, lined by a synovial membrane. This membrane has cells that support the production of synovial fluid, a thick, viscous liquid that is essential for lubricating surfaces and providing nourishment for the cellular components. The bone surfaces within the joint capsule are covered by articular cartilage, a specialized form of hyaline cartilage. The purpose of this tissue is not to serve as a shock absorber but rather to provide a suitable surface for lubrication and wear prevention. The joint capsule is supported by ligaments and the surrounding musculotendinous tissue in holding the two bones of the joint together (i.e., "stabilizing" the joint).

Some joints contain fibrocartilage disks that help maintain the other joint components in appropriate positions and reduce stress by extending the load-bearing surfaces. Such disks are generally found in joints where there is substantial rotation or torsion but incongruous joint surfaces (e.g., the acromioclavicular joint). In the knee, these disks are called menisci. An important compositional difference between articular cartilage and menisci is that the latter contains type I instead of type II collagen.

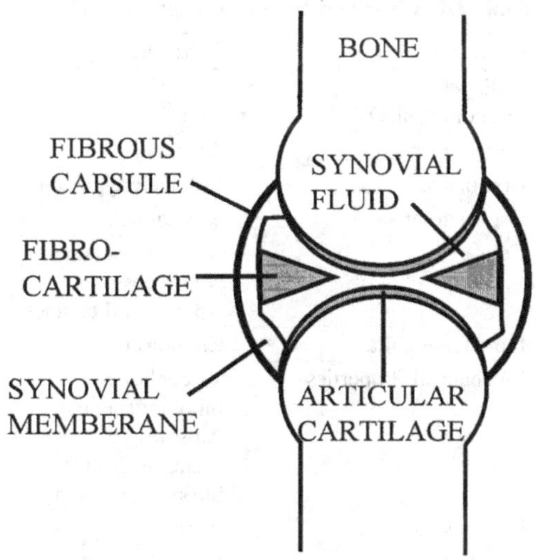

Fig. 3.15 Schematic representation of a synovial joint

3.7 Cartilage

Cartilage is not hierarchically organized but is heterogenous in its orientation (or lack of orientation) of the collagen fibers, and by the relative concentration of proteoglycans in different layers of the cartilage matrix. Collagen and proteoglycan content are to some extent inversely related, with the more superficial layers of the cartilage matrix closer to the articular surface, more highly enriched in collagen, whereas the deeper layers have less collagen and more proteoglycan. Because it is composed primarily of water, cartilage is a highly viscoelastic structure, but the movement of water into and out of the cartilage matrix also provides the primary means for cellular nutrition in this avascular tissue. Also, unlike bone, cartilage has a low metabolic rate and is somewhat hypoxic. Injury to cartilage is therefore more serious as the cartilage does not repair itself readily. Whereas in bone, turnover and renewal of the tissue is performed by a coordinated set of activities performed by different cells, in cartilage a single cell type performs both catabolic and anabolic functions. This may be one reason for its low capacity for repair. Table 3.1 provides a comparison of some similarities and differences between collagen and bone.

Like bone, cartilage comes in different forms, each suited to a particular application. There are three types of cartilage.

- Hyaline cartilage is the most prevalent type of cartilage in adults. It is found in the ventral ends of ribs, in the tracheal rings, and covering the joint surfaces of bones, where it is known as articular cartilage. In addition, the growth plates are composed of this cartilage.

Table 3.1 Characteristics of cartilage vs. bone

Characteristic	Cartilage	Bone
Collagen (predominantly)	Type II	Type I
Proteoglycan	Aggrecan	Non-aggregated proteoglycans
Composite	Collagen-aggrecan	Collagen-mineral
Organization	Non-lamellar	Lamellar
Cells	Not encapsulated No processes No cell–cell contact	Encapsulated Osteocytic processes Cell–cell contact (gap junctions)
Functional Unit	Chondrons	Bone multicellular unit (BMU)
Mechanical Properties	Viscoelastic Inhomogeneous Anisotropic Weak in shear Strong in tension	Viscoelastic Inhomogeneous Anisotropic Weak in shear Strong in compression
Response to Injury	Poor	Excellent
Metabolic Rate	Low	High
Remodeling/Repair	Single cell (chondrocyte)	Multicellular (osteoclast, osteoblast)

- Elastic cartilage is not found in bone. It is typically found in the external ear, eustachian tubes, and epiglottis. It has greater opacity, flexibility, and elasticity than hyaline cartilage and is yellowish rather than white. Its extracellular matrix is permeated with dense, branching elastic fibers unique to this type of cartilage.
- Fibrocartilage is the type of cartilage occurring in intervertebral disks, the pubic symphysis, menisci, and in the bony attachments of certain tendons. It also may form when hyaline cartilage is damaged.

The articular cartilage of synovial joints is hyaline, although some joints also contain fibrocartilaginous disks (e.g., the menisci of the human knee). The extracellular matrix macromolecular framework of articular cartilage is composed mostly of type II collagen and proteoglycans manufactured by cartilage cells (called chondrocytes), and then assembled outside the cell into a mesh of collagen interwoven with aggregated proteoglycan molecules called aggrecan. Like bone cells, chondrocytes live in chambers called lacunae. Each chondrocyte lacuna is surrounded by a distinct pericellular matrix (PCM), or glycocalyx, that together with its encapsulated cell is called a chondron. Chondrons often contain more than one chondrocyte.

Chondrocytes are encapsulated in a viscoelastic extracellular matrix of collagen fibers and proteoglycan molecules. The pericellular matrix surrounding the chondrocyte within the lacuna is enriched in binding proteins, such as anchorin, that help to anchor the chondrocytes to the surrounding substrate. Immediately adjacent to and surrounding the pericellular matrix is a territorial matrix containing large amounts of nonfibrillar Type VI collagen (which may also be found in the pericellular region as well) that may serve as a mechanotransducer of deformations during loading. The cells, their lacunae, and the territorial matrix are separated from each other by an interterritorial matrix composed of thicker fibers of Type II collagen, proteoglycans that are relatively enriched in keratin sulfate glycosaminoglycans, molecules that regulate collagen fiber assembly, and various growth factors (e.g., TGF-β and FGF) produced by the chondrocytes (Fig. 3.16). The chondrocytes also produce proteolytic enzymes, which contribute to turnover of the cartilage matrix. There is no direct cell–cell contact among chondrocytes, nor do these cells have processes to help them communicate with each other. The mechanisms for intercellular signal transduction, and how chondrocytes coordinate their activities, if they do, are unknown.

Cartilage matrix is ordinarily not mineralized and is about 70% (by mass) water. Approximately 60% of the dry weight of the matrix is collagen and 20% each is proteoglycans or other non-collagenous proteins. Chondrocytes make up less than 1% of the volume of articular cartilage in humans, although perhaps slightly more in other animals with smaller volumes of cartilage [49]. Type II collagen is the predominant species (about 90%) of collagen found in articular cartilage. The assembly of the Type II collagen molecule is very similar to that in bone in that three separate α-chains (but different ones from those in bone) form a triple helix, the procollagen molecule. When the ends of the molecule are cleaved, they organize themselves into a quarter-staggered array. Although Type II collagen is the most abundant in cartilage, there are several other collagen types also present (Table 3.2).

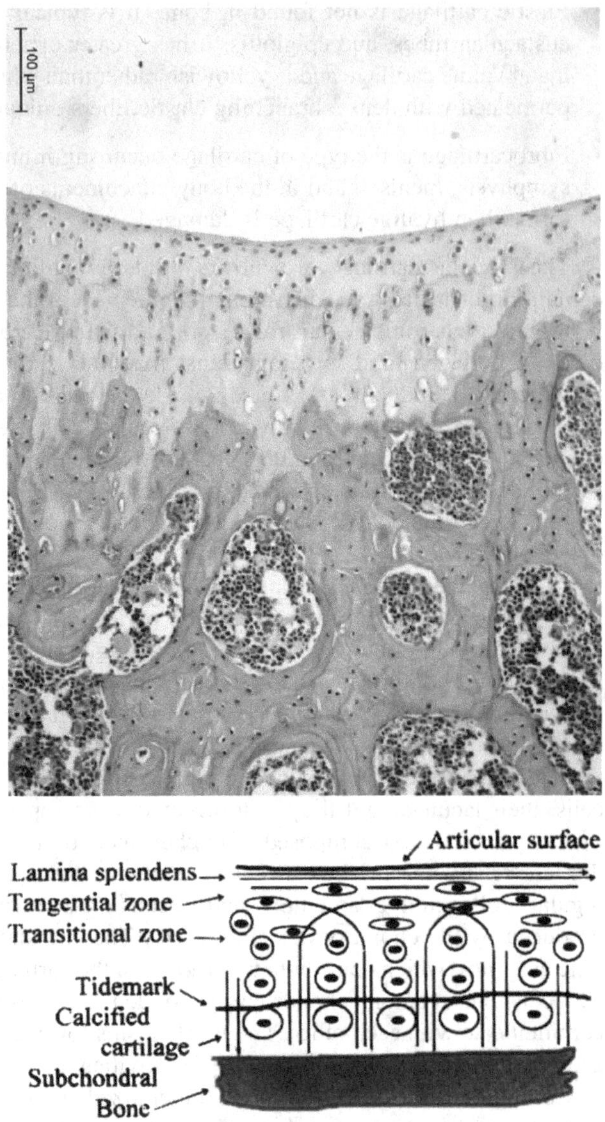

Fig. 3.16 (**a**) Histological section, stained with hematoxylin and eosin (H&E), showing the structure of mature articular cartilage. The tidemark can be seen as a darkly stained line within the cartilage, separating the transitional zone from the calcified cartilage. (**b**) Schematic representation of articular cartilage structure

Type XI collagen microfibrils form a core for the type II collagen fibril and control fibril growth. Larger Type XI triple helixes help to form the collagenous network of articular cartilage. Type IX collagen, which is a form of proteoglycan molecule, is found periodically along the Type II collagen fibril and is covalently cross-linked to it. It is probably important to the structural integrity of the collagen fibril assembly, and it has been suggested that disruption of the Type IX–Type II bond may be one factor leading to the degeneration of articular cartilage and a condition called osteoarthrosis. Osteoarthrosis is a term applied to joint degeneration that has a

Table 3.2 Types of collagen in articular cartilage and their functions

Type	Function
II	Structural and mechanical; primary constituent of articular cartilage
VI	Pericellular adhesion molecule; regulates protein-protein interactions
IX	Fibril association; stabilizes type II
X	Hypertrophic zone of growth plate; role in calcification postulated
XI	Core of type II; controls fibril assembly and growth
XII	Cohesion and stability of collagen and the extrafibrillar matrix
XIV	Cohesion and stability of collagen and the extrafibrillar matrix

mechanical rather than a primary inflammatory cause. This condition is more commonly referred to as osteoarthritis, even though the "itis" in that term implies that inflammation is the initiating cause.

Type VI collagen is found in the territorial region around the chondrocyte lacunae. It is connected by molecular links to Type II collagen and to the larger proteoglycans, providing stability to the cell–extracellular matrix system. This collagen fiber network is stabilized by the formation of cross-links, just as in bone. Cartilage cross-links are excreted in the urine and can be used as markers (though perhaps not very good ones) of cartilage degradation in osteoarthrosis. Type X collagen is found in small amounts within Type II collagen fibers. However, it is primarily found forming a network around hypertrophic chondrocytes in growth plates, where it is probably involved in controlling mineralization of the cartilage matrix. Interestingly, it is also produced in greater amounts during the process of cartilage degeneration. Types XII and XIV have also been recently discovered in cartilage and may provide cohesion between the extrafibrillar matrix and collagen fibers.

Proteoglycans in aggregate form are the other primary polymer in the matrix of cartilage. A core protein backbone provides the site for the attachment of glycosaminoglycans side chains (GAGs) (Fig. 3.17). The GAG chains consist of repeating disaccharide units in which a sugar is linked to a hexosamine. Chondroitin 4-sulfate, chondroitin 6-sulfate, and keratin sulfate, all of which carry a negative charge, are the three predominant GAGs found in cartilage. These molecules are covalently bound to the core protein to form the aggrecan molecule, which, when linked to a hyaluronic acid (hyaluronan) molecule, becomes the proteoglycan aggregate. Complete absence of the link protein weakens the bond between proteoglycans and hyaluronan and is perinatally lethal in mice. Cartilage matrix proteins (CMP), in particular matrilin-1 and -3, bind to the surface of the aggrecan molecule and may control the ability of the protein to bind to collagen.

The proteoglycan aggrecan serves an essential function in allowing the articular cartilage to have such unique mechanical properties. The proteoglycan monomers attached to the hyaluronic acid backbone are all highly negatively charged because of the chemical structure of the sulfated GAGs that compose them. This charge causes the aggrecan molecules to repel one another and branch out into something with the appearance of a bottlebrush. It also causes the aggrecan to be very

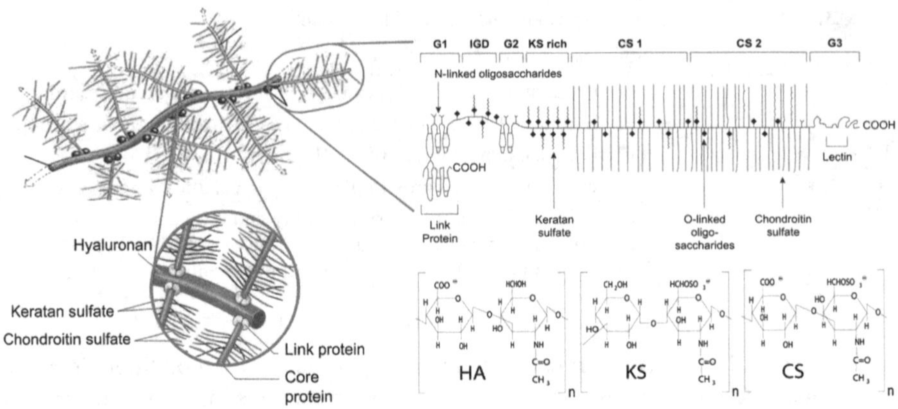

Fig. 3.17 Two-dimensional diagram of proteoglycan aggregate structure. It should be noted that branching of the core proteins from the hyaluronan backbone and the proteoglycan subunits from the core protein occurs in three dimensions

hydrophilic. This attraction for water molecules resists the tendency for water to flow out of the cartilage when it is compressed. As the "bristles" of the brush restrain the water, they are placed in tension, and when the compressive load on the cartilage is removed, the aggrecan molecules re-attract water to replace whatever has managed to escape. The large size and negative charge of the aggrecan molecules give cartilage its hydrophilic properties. Thus, compressive loads are borne by a volume of water that cannot escape, rather like a rubber water bottle filled with water. In this case, however, the water is constrained not by an outside membrane but by an internal structure that electrically confines the water molecules. Loss of water contributes to pathological changes to the structure and function of cartilage.

Although Type II collagen fibril size is in part regulated by the minor collagens, other molecules, some of them smaller proteoglycans, contribute to determining fibril size. One of the most prevalent and important is cartilage oligomeric matrix protein, or COMP, which is found in relative abundance in the interterritorial matrix (0.1% of wet tissue weight). COMP, or thrombospondin 5, is one of several thrombospondins (including TS-1 and TS-4) found in articular cartilage that all serve to regulate collagen fiber assembly. COMP binds to the surface of Type II and Type XI collagen and can accelerate the rate of fibril formation. Decorin is a smaller proteoglycan that "decorates" the collagen fibrils and changes fibril assembly. It probably functions to reduce fibril diameter, particularly close to the articular surface, where it is found in greater abundance. Fibromodulin and lumican are found in increasing amounts in the cartilage matrix away from the chondrocyte, binding to the collagen, and inhibiting fibril formation. The disruption of fibromodulin may be an early event in the osteoarthrotic process that disrupts the fibrillar "weave" of the cartilage resulting in overhydration and cartilage swelling. Other proteoglycans (epiphycan, versican, neurocan, mimecan, and perlecan) are found in varying amounts in different kinds of cartilage.

3.7.1 Organization of Articular Cartilage

Collagen and proteoglycans are dispersed differently through the thickness of the articular cartilage, with more collagen toward the joint surface and more proteoglycan in the deeper layers of cartilage matrix. This arrangement is mechanically advantageous. Large tensile stresses within the articular surfaces and at the edges of joint contact areas are resisted primarily by tangentially oriented collagen fibers. The compressive and hydrostatic stresses found in the deeper layers of cartilage are resisted by the incompressibility of water, which is held in place by the hydrophilic aggrecan molecules.

The matrix of collagen fibers in cartilage is not organized randomly. Instead, the structure varies from place to place in response to functional demands, some of which are mechanical. The surface layer, the lamina splendens, consists of closely packed, small diameter (~30 nm) fibrils, tangentially arranged. The lamina splendens acts as a barrier membrane to enzymes and large molecules such as hyaluronic acid. Thus, small molecules like glucose in the synovial fluid can penetrate the lamina to nourish the chondrocytes deeper in the cartilage, but large, lubricating molecules such as hyaluronic acid remain on the surface of the cartilage. This also protects the cartilage to some degree from immune responses. Mechanically, the horizontal fiber arrangement allows the surface to resist the high tensile and shear forces in and around the joint contact area. However, with age, gaps appear in the lamina, allowing synovial fluid to transport enzymes (e.g., collagenase or hyaluronidase) that can break down the cartilage matrix.

The lamina splendens is only a few micrometers thick and merges gradually into the tangential zone in which the collagen fibrils are slightly larger, although still arranged parallel to the articular surface. This layer is more cellular than the lamina, containing small, flattened chondrocytes that are not very metabolically active in synthesizing collagen, but catabolize proteoglycans more quickly than chondrocytes in other layers. They demonstrate a polarity with a smooth but invaginated surface facing the articular surface and cytoplasmic processes running into the deeper layers of cartilage. Typically, smaller proteoglycan aggregates are found in this layer. The tangential zone also contributes to the tensile stiffness of the cartilage matrix. When this layer becomes disrupted, as it does in the early stages of osteoarthrosis, it is less able to dissipate surface tensile stresses, exposing the underlying cartilage layers to increased compressive stresses.

The transitional zone (or radial zone) contains collagen fibers that are larger than those in the tangential zone and also are more widely spaced and obliquely or pseudorandomly arranged. Together with the larger proteoglycan aggregates found in this layer, and which repel water to make the cartilage matrix turgid, the more vertically oriented collagen fibers in this layer help to stiffen the matrix and reduce large strains [50]. The chondrocytes tend to line up in radial columns (i.e., perpendicular to the surface), particularly in the deeper layers of this zone. These chondrocyte columns are parallel to the surrounding collagen fibers, and the chondrocytes

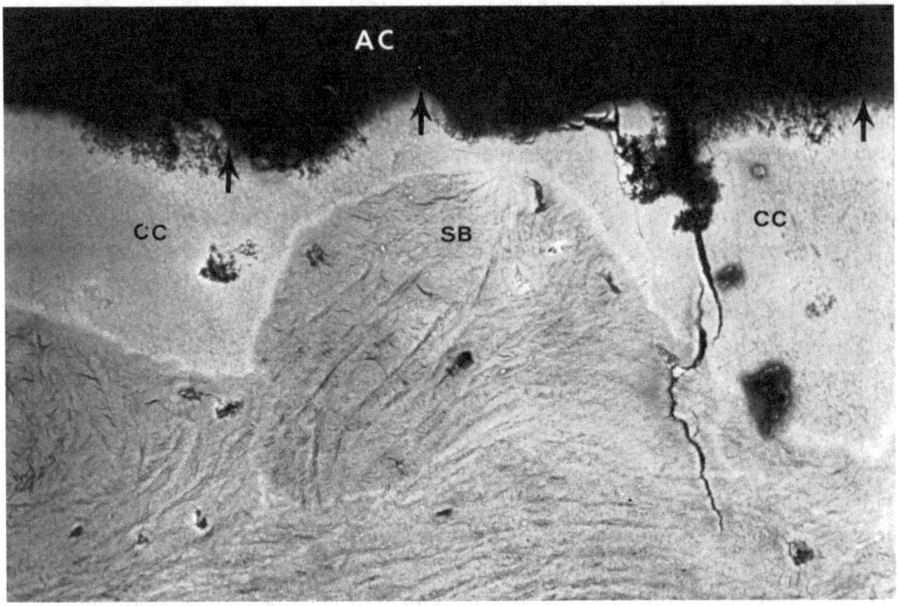

Fig. 3.18 Electron micrograph showing the tidemark (arrows) and osteochondral junction of the human femoral head. Note that the tidemark, which often appears as a well-defined line at lower magnification under the light microscope, is not definite. Darker areas within the calcified cartilage (CC) are nonmineralized regions or regions that are undergoing active mineralization. SB subchondral bone, CC calcified cartilage, AC articular cartilage

within them actively synthesize structural molecules that replenish the cartilage matrix. This is reflected by the greater amounts of endoplasmic reticulum and Golgi found within the chondrocytes in this layer.

The transitional zone ends at a structure called the tidemark, which is the remnant of the original calcification front in the growing bone and which continues to retain high levels of lipid and alkaline phosphatase, as one might expect in a mineralizing front. The calcified layer deep to this is mineralized with hydroxyapatite. Although it was thought at one time that the calcified cartilage might serve to make the change in elastic modulus between articular cartilage and bone more gradual, it now appears that the calcified cartilage is more highly mineralized than bone (Fig. 3.18). This likely increases shear stresses at the tidemark and at the osteochondral junction when the joint is loaded. Perhaps as an adaptation to this, collagen fibers from the transitional zone of the articular cartilage cross the tidemark and reinforce against shear between the stiff (calcified) and compliant (uncalcified, articular) cartilage layers.

Noncalcified articular cartilage lacks nerves and blood vessels (calcified cartilage has both). The chondrocytes are nourished by diffusion through the extracellular matrix, which is largely water. This passive transport of nutrients may be aided by flows induced by mechanical loading, but the dependence on passive transport rather than a blood supply limits the capacity of cartilage to repair large defects. Thus, although chondrocytes are constantly resupplying the extracellular matrix with collagen and proteoglycans as it degrades, the capacity to repair large amounts of damage is apparently absent.

3.8 Ligament and Tendon

Ligaments and tendons are the flexible structures that bind together the musculoskeletal system. They are often described together (as is done here) despite key anatomical and functional differences. Ligaments bind one bone to another to restrict their relative motions. Tendons provide the connecting link from a muscle to a bone. Although textbooks sometimes refer to ligaments as if they were distinct and easily identified anatomical structures, this is often untrue. Many ligaments represent thickenings or specializations within a joint capsule, and their margins may be blurred and indistinct. Moreover, there is considerable individual variation in ligaments, and a ligament that is well defined in one person may be obscure or absent in another. Similarly, textbooks often treat tendon as if it were a uniform generic material, when in fact there is considerable variation in the properties of different tendons that represent adaptations to specialized function. These variations are unlikely to be entirely genetic in origin; many may also derive from the different ways in which each individual habitually uses a particular part of their musculoskeletal system. As with bone, ligaments and tendons remodel to adapt them to the imposed loads.

Tendons and ligaments, along with other noncalcified, non-cartilaginous load-bearing structures of the musculoskeletal system (e.g., synovial joint capsules, aponeuroses, and retinacula), are all composed of dense connective tissue. These structures are characterized by an abundance of tightly packed collagen fibers, which give the tissue superior tensile strength. In tendons and ligaments, the collagen is packaged into bundles of parallel fibers aligned with the predominant force direction. Most collagen is type I with trace amounts of types II, III, V, VI, IX, XII, and XIV present at some locations. Ligaments and, to a lesser extent, tendons also contain highly extensible elastic fibers constructed from another fibrillar protein called elastin. Small amounts of the proteoglycan molecules biglycan, decorin, fibromodulin, lumican, and versican are also present in the extracellular matrix of tendon and ligament, many of which modulate collagen fibril formation and size. Aggrecan, the large hydrophilic molecule found in articular cartilage, is present in regions of the tissue that wrap around bony structures (e.g., tendon of posterior tibialis at the medial malleolus of the ankle; annular ligament that retains the head of the radius) and are routinely subjected to compressive force.

As is the case for bone and cartilage, dense connective tissues originate in the embryonic mesoderm, and the cells that populate and manufacture structural components needed to build and maintain these tissues are descended from pluripotent mesenchymal stem cells. As previously noted, mesenchymal stem cells differentiate

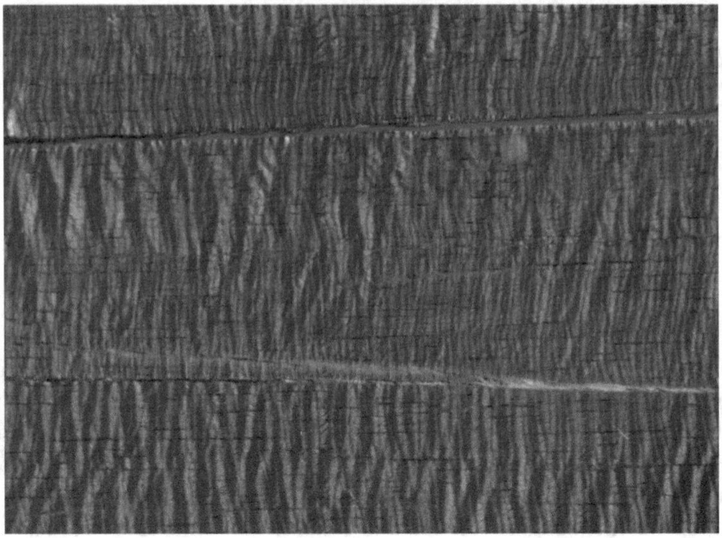

Fig. 3.19 Photomicrograph of equine tendon as viewed under polarized light illustrating the parallel arrays of collagen fibers interspersed with darker staining fibroblasts. The wavy or "crimp" pattern is characteristic of unloaded tendon. (Image provided courtesy of Dr. Brian Murphy, Univ of California, Davis. Orig mag 100×)

into osteocytes, chondrocytes, and adipocytes (fat cells). They are also the precursors to fibroblasts, the predominant cells found in tendons and ligaments. Fibrocartilage cells, intermediate cells possessing characteristics of both fibroblast and chondrocytes, can be found in regions where these structures attach to bone and at anatomic sites subjected to compressive forces. Fibroblasts in tendon and ligament tend to lie in longitudinal rows aligned with and interspersed among the collagen fibrils of the extracellular matrix (Fig. 3.19). Interestingly, fibroblasts, like osteocytes within bone, display long cellular processes that connect with processes from cells in nearby rows. Also, like osteocytes, these cells have been shown to communicate through gap junctions composed of connexin 43. It seems highly likely that local networks of interconnected cells sense the magnitudes and modes of loading in the immediate vicinity and respond accordingly to produce and maintain an appropriate extracellular matrix, but the transduction pathway(s) responsible are unknown.

Blood supply to the dense soft tissues is highly variable but in general less than that found in bone and far less than that in the more metabolically active tissues (e.g., muscle). Vessels and/or capillaries are predominately located on the surface of these structures and, to a lesser extent, within the midsubstance, or may be absent altogether depending upon structure and location. Vessels within tendon run longitudinally and may originate in the adjoining muscle or bone or permeate the tendon at intermittent locations along its length. Tendons and ligaments have a rich nerve supply that provides proprioceptive feedback for motor control and to prevent damage through overload. Pain receptors and nerve fibers involved in vasoregulation are also present in these tissues.

3.8.1 Ligament Structure and Composition

In its natural state, ligament is 55–65% water. Collagen comprises approximately 70–80% of the dry weight and elastin usually comprises another 10–15%, more in some specialized ligaments (e.g., nuchal (neck) ligaments). In most cases, proteoglycans comprise a very small percentage of the extracellular matrix in ligaments and tendons. It should be kept in mind that these proportions are general approximations; exact percentages vary according to ligament location and function.

Collagen, the major organic constituent of bone and cartilage, is by far the most abundant extracellular protein contained in a ligament. Fibrillar collagen (mainly type I and small amounts of types III and V) gives the ligament its high tensile strength and is synthesized by fibroblasts (Fig. 3.20).

Fibroblasts also possess the machinery required to break down and remove old collagen, an important function in response to injury. To do this, fibroblasts synthesize and secrete neutral collagenases, enzymes designed to cleave the collagen triple helix. Collagen turnover is a normal and continual process throughout life with the rate of synthesis and degradation dependent upon the levels of physical activity and imposed load [52].

In the absence of load, the collagen fibers demonstrate a wavy crimp pattern. As the structure is stretched, the crimp pattern disappears and the fibers become straight. The collagen fibers may be grossly aggregated into bundles that are oriented along the principal axis of stress experienced by the ligament. As a joint progresses through a range of motion, different portions of the ligament are stressed more than other portions at each joint position as a consequence of simple geometry. In many ligaments, the collagen bundles are not strictly parallel; this represents an adaptation by the ligament to provide optimal strength at all joint positions.

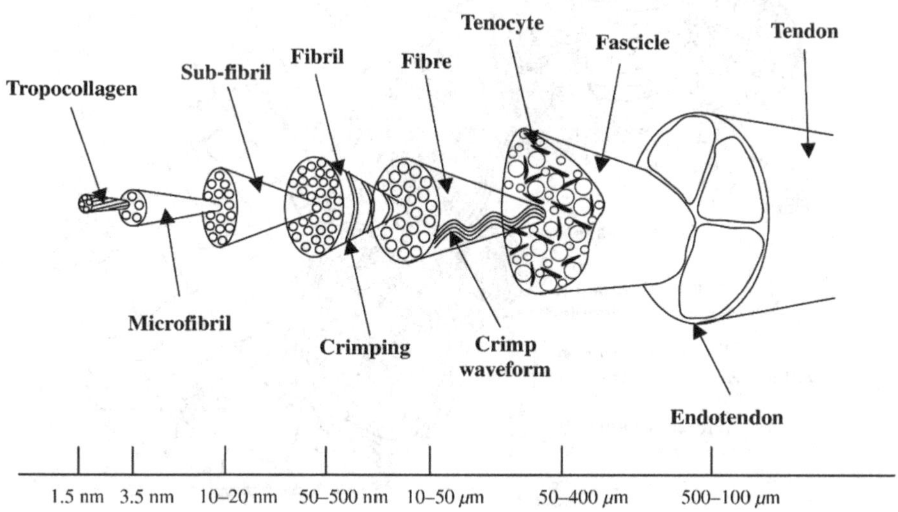

Fig. 3.20 The structure of a ligament at multiple length scales [51]

3.8.2 Tendon Structure and Composition

The composition of tendon is similar to that of ligament. Collagen comprises as much as 85% of tendon dry weight; of this, 95% is type I and 5% type III and/or type V [53]. Unlike ligament, elastin constitutes less than 3% of tendon dry weight. In most locations, the proteoglycan concentration is usually less than 2%; however, tendons that curve around bony surfaces experience compressive forces, and the cells within them respond by producing more proteoglycans. Increasing proteoglycan concentrations give the tendon a more cartilaginous quality that decreases friction and enhances motion.

The structural hierarchy of tendon resembles that of ligament, except that collagen fibers are arranged into discrete packets, which in turn are bundled together within fascicles. Primary fiber bundles and fascicles are bound together and supported by the endotenon, a loose connective tissue containing blood vessels, nerves, and lymphatics (Fig. 3.21). The endotenon is continuous with the epitenon, which encases the whole tendon. Long tendons such as the digital flexors are enclosed in a synovial sheath that nourishes the structure and enhances gliding. Fibroblasts are interspersed between collagen bundles and synthesize collagen for structural repair and maintenance. The fibroblasts in tendons are sometimes called tenocytes. On average, tendons contain fewer cells and are less metabolically active than ligaments.

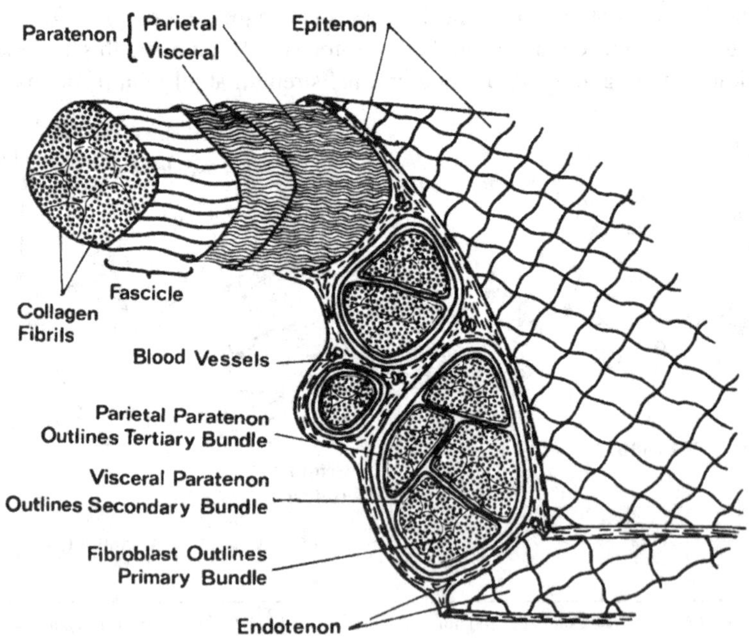

Fig. 3.21 The structure of a tendon at multiple length scales

3.9 Further Readings

Several excellent works can be suggested for further reading. These include Principles of Bone Biology [54], Orthopaedic Basic Science [55], and Osteoporosis [56]. Rheumatology provides excellent chapters reviewing bone, cartilage and tendon/ligament biology [57]. For more in-depth reading, one may turn to B. K. Hall's series of books titled Bone. An entire issue of the Journal of Biomechanics was devoted to bone as a hierarchical material [58]. For more specifics on cartilage, we recommend Osteoarthritis, second edition, by Brandt [59]. An extensive review of all aspects of ligament and tendon biology and mechanics may be found in the book Injury and Repair of the Musculoskeletal Soft Tissues, edited by Woo and Buckwalter [60]. Another excellent review of tendon biology is Structure-function relationships in tendons: a review [61].

References

1. Currey JD. The mechanical consequences of variation in the mineral content of bone. J Biomech. 1969;2:1–11.
2. Currey JD. The relationship between the stiffness and the mineral content of bone. J Biomech. 1969;2:477–80.
3. Ascenzi A, Bonucci E. The compressive properties of single osteons. Anat Rec. 1968;161:377–91.
4. Ascenzi A, Bonucci E. The shearing properties of single osteons. Anat Rec. 1972;172:499–510.
5. Ascenzi A, Bonucci E. The tensile properties of single osteons. Anat Rec. 1967;158:375–86.
6. Ascenzi A, Bonucci E. The ultimate tensile strength of single osteons. Acta Anat. 1964;58:160–83.
7. Donnelly E, Baker SP, Boskey A. Effects of surface roughness and maximum load on the mechanical properties of cancellous bone measured by nanoindentation. J Biomed Mater Res A. 2006;77:426–35.
8. Hengsberger S, Kulik A, Zysset P. Nanoindentation discriminates the elastic properties of individual human bone lamellae under dry and physiological conditions. Bone. 2002;30:178–84.
9. Xu J, Rho JY, Mishra S. Atomic force microscopy and nanoindentation characterization of human lamellar bone prepared by microtome sectioning and mechanical polishing technique. J Biomed Mater Res A. 2003;67:719–26.
10. Burr DB, Schaffler MB, Fredrickson RG. Composition of the cement line and its possible mechanical role as a local interface in human compact bone. J Biomech. 1988;21:939–45.
11. Mullender MG, van der Meer DD, Huiskes R, Lips P. Osteocyte density changes in aging and osteoporosis. Bone. 1996;18:109–13. https://doi.org/10.1016/8756-3282(95)00444-0.
12. Tavassoli M, Yoffey JM. Bone marrow structure and function. Inc, Liss; 1983.
13. Reisinger AG, Pahr DH, Zysset PK. Elastic anisotropy of bone lamellae as a function of fibril orientation pattern. Biomech Model Mechanobiol. 2011;10:67–77.
14. Tai K, Dao M, Suresh S, et al. Nanoscale heterogeneity promotes energy dissipation in bone. Nat Mater. 2007;6:454–62.
15. Giraud-Guille MM. Twisted plywood architecture of collagen fibrils on human compact bone osteons. Calcif Tissue Int. 1988;42:167–80.

16. Neville AC. Cuticle: organization. In: Invertebrates I, editor. Biology of the integument. Berlin: Springer-Verlag; 1984. p. 611–25.
17. Warshaw J, Bromage TG, Terranova CJ, Enlow DH. Collagen fiber orientation in primate long bones. Anat Rec. 2017;300:1189–207. https://doi.org/10.1002/ar.23571.
18. Stover SM, Pool RR, Martin RB, Morgan JP. Histologic features of the dorsal cortex of the third metacarpal bone mid-diaphysis during postnatal growth in thoroughbred horses. J Anat. 1992;181:455–69.
19. Currey JD. Three analogies to explain the mechanical properties of bone. Biorheology. 1964;2: 1–10.
20. Jowsey J. Studies of Haversian systems in man and some animals. J Anat. 1966;100:857–64.
21. Martin RB, Pickett JC, Zinaich S. Studies of skeletal remodeling in aging men. Clin Orthop Relat Res. 1980;149:268–82.
22. Cohen J, Harris WH. The three-dimensional anatomy of haversian systems. J Bone Joint Surg Am. 1958;40-A(2):419–34.
23. Schaffler MB, Burr DB, Frederickson RG. Morphology of the osteonal cement line in human bone. Anat Rec. 1987;217:223–8.
24. Skedros JG, Holmes JL, Vada EG. Cement lines of secondeary osteons in human bone are not mineral-deficient: new data in a historical perspective. Anat Rec A Discov Mol Cell Evol Biol. 2005;286(1):781–803.
25. Lakes R, Saha S. Cement line motion in bone. Science. 1979;204:501–3.
26. Wallace JM, Orr BG, Marini J, Holl MMB. Nanoscale morphology of type I collagen is altered in the Brtl mouse model of osteogenesis imperfect. J Struct Biol. 2011;173:146–52.
27. Olsen BR. New insights into the function of collagens from genetic analysis. Curr Biol. 1995;7: 720–7.
28. Ding M, Dalstra M, Danielsen C. Age variations in the properties of human tibial trabecular bone. J Bone Jt Surg [Br]. 1997;79-B:995.
29. Bailey AJ, Wotton SF, Sims T. Post-translational modifications in the collagen of human osteoporotic femoral head. Biochem Biophys Res Commun. 1992;185:801–5.
30. Banse X, Sims TJ, Bailey AJ. Mechanical properties of adult vertebral cancellous bone: correlation with collagen intermolecular cross-links. J Bone Miner Res. 2002;17:1621–8.
31. Lees S, Eyre DR, Barnard SM. BAPN dose dependence of mature crosslinking in bone matrix collagen of rabbit compact bone: corresponding variation of sonic velocity and equatorial diffraction spacing. Connect Tissue Res. 1990;24:95–105.
32. Oxlund H, Mosekilde L, Ortoft G. Reduced concentration of collagen reducible cross links in human trabecular bone with respect to age and osteoporosis. Bone. 1996;19:479–84.
33. Zioupos P, Currey JD, Hamer AJ. The role of collagen in the declining mechanical properties of aging human cortical bone. J Biomed Mater Res. 1999;45:108–16.
34. Wang X, Shen X, Li X. Age-related changes in the collagen network and toughness of bone. Bone. 2002;31:1–7.
35. Keaveny TM, Guo XE, Wachtel EF, et al. Trabecular bone exhibits fully linear elastic behavior and yields at low strain. J Biomech. 1994;27:1127–36.
36. Hernandez CJ, Tang SY, Baumbach B. Trabecular microfracture and the influence of pyridinium and non-enzymatic glycation-mediated collagen cross-links. Bone. 2005;37:825–32.
37. Bailey AJ, Knott L. Molecular changes in bone collagen in osteoporosis and osteoarthritis in the elderly. Exp Gerontol. 1999;34:337–51.
38. Vashishth D, Gibson GJ, Khoury JI, et al. Influence of nonenzymatic glycation on biomechanical properties of cortical bone. Bone. 2001;28:195–201. https://doi.org/10.1016/s8756-3282(00)00434-8.
39. Miller LM, Vairavamurthy V, Chance M. IN situ analysis of mineral content and crystallinity in bone using infrared micro-spectroscopy of the nu(4)PO(4)(3-) vibration. Biochim Biophys Acta. 2001;1527:11–9.

40. Hu Y-Y, Liu XP, Ma X, et al. Biomimetic self-assembling copolymer–hydroxyapatite nanocomposites with the nanocrystal size controlled by citrate. Chem Mater. 2011;23(9): 2481–90.
41. Currey JD. Incompatible mechanical properties in compact bone. J Theoret Biol. 2004;231: 569–80.
42. Cooper RR, Milgram JW, Robinson RA. Morphology of the osteon. J Bone Jt Surg. 1966.
43. Johnson LC. The kinetics of skeletal remodeling. Birth Defects Orig Artic Ser. 1966;2:66–142.
44. Robling AG, Niziolek PJ, Baldridge L. Mechanical stimulation of bone in vivo reduces osteocyte expression of Sost/Sclerostin. J Biol Chem. 2008;283:5866–75.
45. Kennedy OD, Herman BC, Laudier DM, et al. Activation of resorption in fatigue-loaded bone involves both apoptosis and active pro-osteoclastogenic signaling by distinct osteocyte populations. Bone. 2012;50:1115–22.
46. Nakashima T, Hayashi M, Fukunaga T. Evidence for osteocyte regulation of bone homeostasis through RANKL expression. Nat Med. 2011;17:1231–4.
47. Xiong J, Onal M, Jilka R. Matrix-embedded cells control osteoclast formation. Nat Med. 2011;17:1235–41.
48. Tatsumi S, Ishii K, Amizuka N. Targeted ablation of osteocytes induces osteoporosis with defective mechanotransduction. Cell Metab. 2007;5:464–75.
49. Buckwalter JA, Mankin HJ. Articular cartilage. Part I: tissue design and chondrocyte-matrix interactions. J Bone Jt Surg. 1997;79-A:600–11.
50. Shirazi R, Shirazi-Adl A, Hurtig M. Role of cartilage collagen fibrils networks in knee joint biomechanics under compression. J Biomech. 2008;41:3340–8.
51. Screen HRC, Bader DL, Lee DA, Shelton JC. Local strain measurement within tendon. Strain. 2004;40:157–63. https://doi.org/10.1111/j.1475-1305.2004.00164.x.
52. Kjaer M, Magnusson P, Krogsgaard M, et al. Extracellular matrix adaptation of tendon and skeletal muscle. J Anat. 2006;208:445–50.
53. Cetta G, Tenni R, Zanaboni G. Biochemical and morphological modifications in rabbit Achilles tendon during maturation and aging. J Biochem (Tokyo). 1982;204:61–7.
54. Bilezikian JP, Raisz LG, Rodan GA. Principles of bone biology. New York: Academic Press; 2019.
55. Simon SR. Orthopaedic basic science. Rosemont: American Academy of Orthopaedic Surgeons; 1994.
56. Marcus R, Feldman D, Kelsey J. Osteoporosis. New York: Academic Press; 2013.
57. Hochberg MC, Silman AJ, Smolen JS, et al. Rheumatology, 2-volume set. Amsterdam, Netherlands: Elsevier. 2023.
58. Ascenzi M-G, Reilly G. Bone tissue: hierarchical simulations for clinical applications. J Biomech. 2011;44:211–364.
59. Brandt KD, Doherty M, Lohmander LS. Osteoarthritis. Oxford: University Press; 2003.
60. Frank C, Woo S, Andriacchi T, et al. Normal ligament: structure, function, and composition. In: Woo SL-Y, Buckwalter JA, editors. Injury and repair of the musculoskeletal soft tissues. Park Ridge: American Academy of Orthopaedic Surgeons; 1988. p. 45–128.
61. Benjamin M, Kaiser E, Milz S. Structure-function relationships in tendons: a review. J Anat. 2008;212:211–28.

Chapter 4
Growth, Modeling, and Remodeling of Bone

Abstract Bone growth requires a coordination of processes to lengthen the mineralized matrix. This activity is centralized at the growth plate and involves both cartilage and bone cells. Bone modeling and remodeling, while sharing some similar features, are uniquely different processes. Bone modeling plays an important role during growth and serves mainly to reshape and grow bones. Bone remodeling occurs throughout life and serves a main function of renewing/replacing bone tissue. The process of bone remodeling is complex, can be estimated through bone histomorphometry, and can be modeled through mathematical calculations. Such calculations provide a means of hypothesis generation to assess/understand the physiologically complex nature of remodeling. Bone modeling and remodeling play a fundamental role in fracture healing, which integrates both physiological processes on an accelerated timeframe.

What you Will Learn

1. How bones grow longitudinally
2. The processes of bone modeling and remodeling
3. How remodeling alters structure and the mathematical models that can help understand bone remodeling
4. Histomorphometry analyses of bone remodeling and assumptions in the measures
5. The process of bone fracture repair

4.1 Introduction

Three processes determine the amount of bone we have, and for where it is found in our skeletons. Growth is defined as the accrual of bone mass through processes of bone formation during development. It does not account for the shapes of the bones, except for their length, and does not involve resorption of tissue. Modeling is a term used to define either bone formation or bone resorption, but not both, at a given site. Modeling modifies the structural geometry of the whole bone; this is really its primary function. Remodeling, on the other hand, is defined by the sequential

M. R. Allen, J. Wallace, *Skeletal Tissue Mechanics*,
https://doi.org/10.1007/978-3-031-90609-1_4

resorption and subsequent formation of bone at a single location. Although remodeling may modify the geometry of the bone as well, its primary function is to renew the bone tissue.

Remodeling is the only one of these processes that occurs significantly throughout one's lifetime. Growth is essentially restricted to children, and although modeling occurs in adulthood, it slows substantially. Because bone remodeling occurs throughout life, it plays a dominant role in determining the properties of skeleton. Consequently, remodeling is a primary determinant of the mechanical properties of bone, its resistance to fatigue failure, and its ability to function in a changing mechanical environment.

4.2 Longitudinal Growth

Longitudinal growth of bones occurs in a nonmineralized region of growth near, but not at, each end of the bone. This region is called the growth plate or physis. The physis separates the bony epiphysis from the bony metaphysis (Fig. 4.1).

New cartilage is constantly formed by chondrocytes within the growth plate. Cartilage is viscous enough that cells within it can extrude more cartilage matrix, increasing its volume (termed interstitial growth—growth from within). This would increase its thickness except that the cartilage on the metaphyseal side of the growth plate mineralizes and becomes part of the metaphyseal bone, so that the length of the metaphyseal and diaphyseal parts of the bone is increased, while the thickness of the growth plate remains relatively constant. The width of the growth plate region is wider in bones that grow quickly and narrower in bones that grow more slowly.

The growth plate is divided into zones, each representing a stage in the life cycle of its chondrocytes (Fig. 4.2). The number of these zones can vary, depending on the reference consulted. In the following description, you can imagine that you are

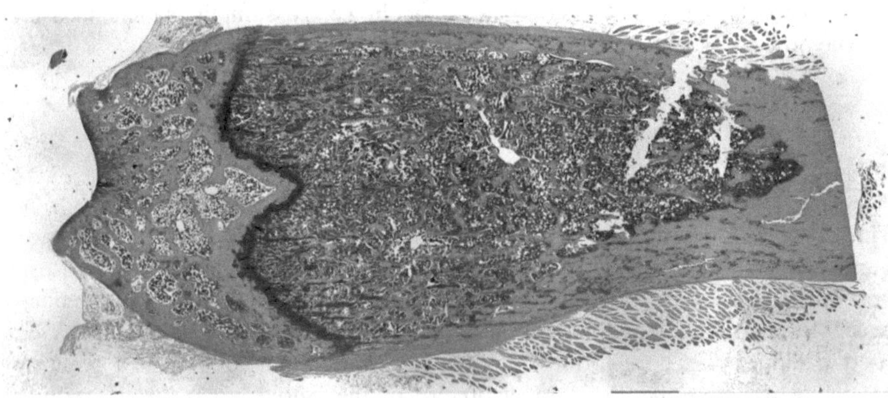

Fig. 4.1 Distal femur of a rat showing the separation of the epiphysis and the metaphysis by the growth plate (red). Stained with Safranin O Fast Green

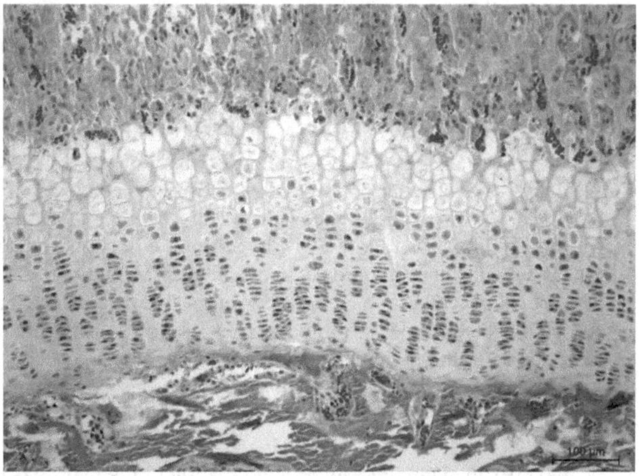

Fig. 4.2 Higher magnification showing the structure of the different zones of the growth plate. Rat proximal tibia

traveling down through these zones and into the metaphysis at a given point in time, leaving the upper surface of the growth plate behind. Alternatively, you can imagine that you are sitting at a point which is fixed in relation to the bone as a whole and watching your surroundings change as time passes, so that the top of the growth plate leaves you behind. Forming a clear mental image of this relativity between time and position will help you understand the process by which bones grow longitudinally. The processes that occur during this temporal sequence through the growth plate are largely directed by a protein called SOX 9, perhaps in collaboration with hedgehog proteins [1]. The hedgehog family of proteins is also important for the differentiation of the different tissues, and even the fingers, during limb development. The growth plate's zones are, from the top (i.e., the epiphyseal side) to the bottom, as follows:

Reserve or resting zone: Cells of moderate size are scattered irregularly throughout this zone, which is anchored to the bone of the epiphysis and receives nourishment from epiphyseal blood vessels. Some of these cells may be chondroprogenitor cells, but the expression of Sox9 suggests that many have differentiated to chondrocytes. These cells are not really "resting" but are dividing slowly to provide chondrocytes for the remainder of the growth plate. Type II collagen content is high in this region, and the fibrils are randomly arranged.

Proliferative zone: Here, the chondrocytes divide repeatedly, are close together, and arrange themselves in columns. They also become disklike, so that they look somewhat like a stack of coins. This is the region where most of the growth in length occurs in the bone, primarily due to cell proliferation rather than cellular activity itself [2]. Chondrocyte proliferation and maturation are permitted by suppression of type X collagen production by SOX 9, but at the same time SOX 9 stimulates the production of type II and type IX collagen, which allows for the assembly of the fibrillar network. Each cell produces about two times its own volume in new matrix during its lifetime [2]. The collagen fibers are aligned parallel to the cell columns. The chondrocytes produce large amounts of proteoglycan and other molecules needed for the mechanical integrity of the surrounding extracellular matrix.

Hypertrophic zone: As they stop proliferating, the chondrocytes are said to "mature." They accumulate glycogen in their cytoplasm and secrete copious amounts of matrix, increasing the volume of the surrounding substance. They also hypertrophy, increasing their volume and surface area by 4–10 times [3]. It is thought that the chondrocytes in this zone are in the early stages of apoptosis (programmed cell death). They stop producing cartilage matrix and begin producing molecules that prepare the adjacent cartilage for calcification.

Zone of provisional calcification: In this zone, the degenerating chondrocytes continue to hypertrophy. As they reach the limit of this zone, where the growth plate ends and the metaphysis begins, the chondrocytes die. Simultaneously, the surrounding cartilage matrix is calcified. The calcification mechanism is imperfectly understood, but a working hypothesis for this process is as follows. The hypertrophying chondrocytes synthesize type X collagen and a protein called chondrocalcin, which may help initiate calcification. Proteoglycans may also be disassembled by enzymes released by the degenerating cells. Mineral crystal formation commences within matrix vesicles, which are intact spherical membranes extruded from the hypertrophic chondrocytes. These vesicles contain the enzyme ATPase, which provides energy to transport calcium ions into the vesicle against a concentration gradient. In addition, other enzymes in the vesicles cleave phosphate and calcium from compounds to which they are bound, allowing them to form crystals. The initiation of apatite crystal growth is the most important function of the matrix vesicle. As the levels of calcium and phosphate increase, amorphous calcium–phosphate aggregates form, eventually spreading beyond the vesicle wall and accumulating more mineral by epitaxy (crystal growth that imitates the form of the substrate). At this point, apatite crystals begin to grow in such a fashion that they are integrated with the collagen molecules in the matrix, and water is displaced as the cartilage matrix mineralizes.

The anabolic and the catabolic processes that occur at the growth plate are coupled so that the width of the growth plate is normally controlled within limits. However, if mineralization of the distal regions is delayed, or if chondrocyte division and matrix synthesis in the proliferative region are stimulated, the width of the growth plate will increase. Disturbances to growth plate width—either narrowing or broadening, or to other aspects of its morphology—can be caused by hormonal deficiency, metabolic disorders, and genetic disease that can disturb growth plate development resulting in dysplasia (abnormality of development).

4.2.1 Development of Metaphyseal Trabeculae

The columns of enlarged lacunae left by the dead chondrocytes form the basis for tunnels, allowing blood vessels and cells from the metaphysis to gain access to the calcified cartilage at the bottom of the growth plate. As each chondrocyte hypertrophies, it dissolves some of the cartilage around it, so that only narrow walls, or septa, are left between the enlarged lacunae. Matrix vesicles which help to control the process of mineralization can be found within these septa. Cells called chondroclasts, another monocyte descendant, but perhaps not identical to osteoclasts, resorb the septa at the bottom of each chondrocyte column. Blood vessels from the metaphysis marrow then invade these tunnel-like spaces, which are separated by columns or "trabeculae" of calcified cartilage. Chondroclasts also resorb some of the calcified cartilage on the walls of these tunnels. Osteoblasts then lay down small amounts of bone on these calcified cartilage trabeculae so that early in growth, the trabecular struts may retain a core of calcified cartilage that can persist in some diseases in which resorption is impaired. This remodeling activity begins to convert calcified cartilage to bone and marks the transition from growth plate to metaphyseal "bone." At this early stage in the formation of trabeuclar bone, the tissue is called primary spongiosa; it still contains considerable calcified cartilage within the trabecular cores. As time goes by, and remodeling continues to replace the cartilaginous portions of the trabeculae with metaphyseal bone, the trabeculae become the secondary spongiosa. Eventually, the trabeculae become entirely bone. Those trabeculae in the center of the metaphysis are eventually entirely resorbed to form the medullary canal, with an arch of trabecular struts remaining to transmit loads from the central portion of the growth plate out to the cortices of the diaphysis.

4.2.2 Growth of the Physis

In the newborn, the physis is typically a flat, more or less circular plate. As the child grows and the bones become larger, the physis must increase in diameter. This increase is achieved by cell division at its circumference in a region called the zone of Ranvier. In addition, the physis loses its flat shape and becomes curved, often with a complex system of ridges and valleys. The physis, being composed of cartilage, is weak in shear and tension, and physeal injuries in children are relatively common; the epiphysis of the proximal femur sliding off the shaft at the physis (slipped capital femoral epiphysis) is a well-known example. One consequence of the development of an irregularly shaped contour to the physis in the growing child is that the complex interlocking geometry of the epiphysis and metaphysis confers some protection against failure of the physis in response to shearing forces.

Most long bones have two physes, one at each end (although some bones, such as the metacarpals and phalanges, have only one functional physis). Usually one of these contributes more than the other to longitudinal bone growth. For example, in the femur, the distal physis contributes more to growth, whereas in the tibia, the proximal physis is more active. Historically, the relative growth rates of the two physes of a long bone have been measured by comparing bones of different age with respect to some naturally occurring marker in the diaphysis, usually the entrance point of the nutrient artery through the cortex.

4.2.3 Closure of the Physes

As an individual matures, bones reach their adult length and the physis is no longer required. At this point, the physes "close" by ossification, connecting the epiphysis to the metaphysis with bone. The blood circulatory systems of the epiphysis and metaphysis, formerly independent, also unite. This region of bone remodels over many years, but the physeal "scar" or "ghost" (a plate of bone at the site of the former physis) may persist into old age.

Although it is often thought that closure of the epiphyses stops longitudinal growth, available evidence suggests that growth slows down and stops as a function of the increase in sex steroids during maturation and that closure of the physis follows cessation of growth [4]. Some animals (e.g., rodents such as rats and mice) slow growth with age but never completely stop growing and consequently their growth plates never fuse. In animals that lack an ossified epiphysis (many reptiles, e.g., crocodiles), there is no physeal closure and the bones continue to increase in length indefinitely, so that great longevity becomes associated with great size.

Closure of the physes of different bones, and even the proximal and distal ends of the same bone, occurs at different ages and in a specific sequence. This allows one to determine radiographically the skeletal age of an individual by observing which physes, usually those of the left hand and wrist, have closed. The hands and wrist are most useful because of the number of bones and the fact that the eight wrist bones, which are derived only from a primary center of ossification and therefore do not have epiphyses, appear radiographically at different ages and grow at different rates (the skeletal elements are present before they appear radiographically, but are completely cartilaginous and are not visible). Skeletal age is not the same as chronological age; the growth plates close earlier in some individuals than others. In girls, physes close several years earlier than in boys; this contributes to the shorter average stature of women compared to men and probably also to their lower overall bone density.

4.3 Modeling and Remodeling Processes

As the long bones develop, their shafts must grow in diameter as well as length. We have seen how growth in length throughout childhood is accomplished by continuation of endochondral ossification in the growth plate. Growth in diameter is accomplished by addition of new bone to the periosteal surface. However, as growth occurs, it is not enough just to add material to make the bone longer and larger. The bones must also be shaped in various ways. That means that bone must be removed in some places while it is added in others. This sculpting, involving osteoclastic activity (bone resorption) in some places and osteoblastic activity (bone formation) in others, has become known as modeling. In addition to spatially distinct formation or removal of matrix, the bone also has the capacity to "fix" regions of preexisting bone. A classic example, detailed later in this book, is the removal of fatigue damage, caused through the repeated fatigue loading of bone. Repair of the bones' internal architecture must occur both while the skeleton is developing and throughout the remainder of one's life. This repair involves removal and replacement of bone in particular places by the coupled actions of osteoclasts and osteoblasts working at the same site. This replacement process has become known as remodeling. Thus, modeling and remodeling refer to the actions of osteoblasts and osteoclasts in reshaping and replacing portions of the skeleton. They are distinguished from one another in several ways:

Modeling involves independent actions of osteoclasts and osteoblasts on related but typically different spatial locations. Remodeling involves sequential, "coupled" actions by the two types of cells within a given spatial location.

Modeling results in change of the bone's size, shape, or both; remodeling does not usually affect size or shape.

The rate of modeling is greatly reduced after skeletal maturity. Remodeling occurs throughout life, although it too is substantially reduced after growth stops; remodeling increases in the years before and during menopause.

Despite clear physiological differences in the processes of modeling and remodeling, some in the field lump them together and refer to them as "remodeling." For example, the literature has many statements about rodents undergoing cortical "remodeling." While it's true that rodents have robust modeling (on periosteal and endocortical surfaces), they do not typically undergo remodeling on these surfaces.

4.3.1 Modeling

Modeling is necessary because the longitudinal growth process does not produce a bone with the correct shape for an adult and because each growing child loads his or her skeleton in a somewhat different way, requiring each skeleton to be "customized." Modeling may be resorptive, where osteoclasts remove a region of bone, or formative, where osteoblasts form new bone on a given surface. To be considered modeling, the resorption or formation processes must occur along, without being preceded or followed by the other process.

As some long bones grow longer, the diameter of the cylinder of bone created beneath the growth plate must be reduced to create the diaphysis. This change is accomplished by osteoclasts working continuously on the periosteal surface of the metaphysis to cut the shaft down to size (Fig. 4.3, top). This action is most important in bones with a widely flaring metaphysis such as the proximal tibia; in other bones, diametric growth of the physis is coordinated with longitudinal growth in such a way that removal of bone in the lower metaphysis is unnecessary (e.g., metatarsals). Disruption of this process, which occurs in conditions of altered osteoclast activity, results in morphologically distinct metaphyseal regions [5].

As a child continues to grow, long bone shafts become larger in diameter, accomplished by slow, continuous addition of bone to the periosteal surface by osteoblasts. Simultaneously, bone is removed from the endosteal surface (Fig. 4.3, middle). This occurs throughout life, and differences in the rate of addition to periosteal surface bone and removal of endocortical surface bone account for sexual dimorphic bone size and cortical thickness [6]. In addition, the curvature of long bones is usually adjusted during growth. This is accomplished by increased formation and resorption on the sides of the bone such that the cross section "drifts" sideways relative to the ends of the bone (see Fig. 4.3, bottom). This is illustrated in the bones of the skull that must increase in size to accommodate the increasing size

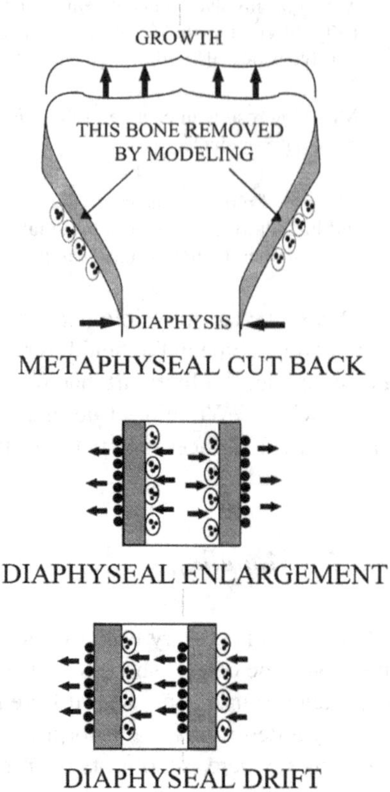

Fig. 4.3 Top: Resorptive modeling beneath the growth plate to form the diaphysis from the metaphysis. Middle: Formative (periosteal surface) and resorptive (endosteal surface) modeling to enlarge the diaphysis. Bottom: Modeling to "drift" the diaphysis to the left (thereby altering diaphyseal curvature)

of the brain during growth. This enlargement cannot be accomplished simply by addition of bone at the suture lines between the cranial bones because the plate's radius of curvature must also be adjusted. Modeling of these bones (resorption on the inner surface, formation on the outer surface) is required for proportionally correct increases in cranial size. Simultaneously, modeling transforms the facial features of a child into those of an adult. As one might imagine, similar processes must occur in the ribs to adjust the curvature of the rib cage as the individual's chest expands during growth. Modeling is also the basis for orthodonic tooth movement.

4.3.2 Remodeling

Remodeling removes a portion of older or damaged bone and replaces it with newly formed bone. This process repairs microscopic damage and prevents the accumulation of fatigue damage that could lead to fatigue fracture. It also likely replaces regions with high mineral content, high collagen cross-linking, and/or large collections of apoptotic osteocytes. It has been hypothesized that remodeling also mechanically "fine tunes" the skeleton to increase its mechanical efficiency.

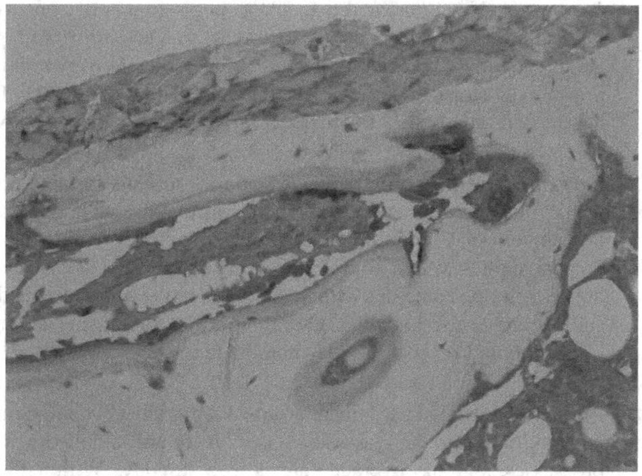

Fig. 4.4 Photomicrograph of a cutting cone in bone

Remodeling is accomplished by teams of osteoclasts and osteoblasts that work together in basic multicellular units or BMUs (Fig. 4.4). A BMU consists of a team of osteoclasts and osteoblasts. There are six total, but three principal, stages in a BMU's lifetime: activation (A), resorption (R), and formation (F). This is sometimes referred to as an A-R-F sequence. It can be contrasted with modeling which has activation (A) and resorption (R) or A and formation (F).

Activation occurs when a molecular or mechanical signal causes osteoclasts to form (by fusion of monocytes) and begin to remove bone. The osteoclasts resorb a volume of bone in the form of a ditch (on bone surfaces) or tunnel (in the midst of compact bone). After the region has been excavated, osteoblasts are differentiated from mesenchymal cells over a period of several days and begin to replace the resorbed tissue. Formation is much slower than resorption. Completed BMU's render new osteons (Haversian systems) roughly 3–9 mm in length. If one were embedded in a bone watching a BMU advance through a particular site, the time required to fully excavate the tunnel to its full diameter (i.e., the resorption period) would be about 3 weeks; while the time required to refill the resorption space with new bone (i.e., the formation or refilling period) would require about 3 months. The total remodeling period for any discreet location along the length of a progressing BMU is about 4–6 months. Because the bone is resorbed and formed in a packet, this has been termed the "quantum concept of bone remodeling" by Parfitt [7].

The remodeling process has six separate phases, which are always sequential in normal bone remodeling.

Activation. Differentiated cells must be recruited from precursor cell populations before any bone resorption or bone formation can occur. It is important to recognize that in the purest sense, this process occurs only at the point of origin of the BMU. The osteoclasts produced at this time appear to survive for weeks, and perhaps for the entire duration of the tunneling by the BMU. Thus, the "opening up" of a resorption space on a cross section is not actually "activation" because it does not involve the initial recruitment of osteoclasts, but rather the passage of osteoclasts through a particular plane of observation. The original activation process occurred earlier; in either case, however, the time required is about 3–5 days.

Resorption. Newly created osteoclasts begin to resorb bone, moving longitudinally within cortical bone at a rate of about 40 μm/day in a so-called cutting cone, an ellipsoidal surface about 200 μm in diameter and about 300 μm long. However, it is difficult to be certain where this region stops and the following region begins, so the length of each is debatable. Resorption on a bone surface occurs in the same way both downward and outward.

Reversal. The transition from osteoclastic to osteoblastic activity takes several days and results in a space lying between the resorptive region and the refilling region. The length of this region may vary considerably, depending on the lag between the resorptive and formative phases. In a completed cortical bone secondary osteon or trabecular hemiosteon, the cement line coincides with the location of the bone surface during the reversal period; the cement line is also known as the reversal line. In humans, about 30 days are required for the resorption and reversal periods combined.

Formation. Osteoblasts appear around the periphery of the tunnel formed by the osteoclasts, and refilling begins. The osteoblasts lay down lamellae at a decreasing rate as refilling progresses, but the average rate is 1–2 μm/day. Within the cortical bone osteons, the osteoblasts do not completely refill the tunnel but leave a central passageway for a vascular loop to support the metabolism of the BMU and the osteocytes in the bone matrix and to carry calcium and phosphate to and from the bone when necessary. This central canal is typically 40–50 μm in diameter in humans. The formation phase in adult humans averages about 3-4 months.

Mineralization. Following deposition of the unmineralized organic matrix of bone, or osteoid, mineral is deposited within and between the collagen fibers. This process is delayed by a period of time known as the mineralization lag time, which is normally about 10 days. This delay is manifested by a layer of osteoid between the osteoblasts and mineralized bone. Once begun, approximately 60-70% of the mineralization of osteoid occurs during the first three weeks [8, 9]; this is called primary mineralization. The remainder of the mineral is added at a decreasing rate for about a year during secondary mineralization [9]. The bone in new, incompletely mineralized osteons can exhibit very different mechanical properties from that in older osteons.

Quiescence. After the tunneling and refilling processes are completed, the osteoclasts disappear from the scene and the remaining osteoblasts become lining cells or disappear. A period of relative inactivity ensues during which the BMU and its associated cells carry on their mechanical, metabolic, and homeostatic functions.

Each of these six phases is location-, magnitude-, and rate-specific, so that alterations in the intensity or timing of any one produce an altered histologic morphology characteristic of specific skeletal disorders. For example, an increased lag between the bone formation and bone mineralization stages produces the excessive amounts of osteoid found in osteomalacia (a disorder characterized by decreased mineralization of bone).

Because the processes of resorption and formation are tightly linked within the BMU, they are said to be coupled. This concept refers only to the relationship of resorption and formation, and the cross-talk between osteoclasts and osteoblasts that must occur to coordinate the remodeling unit. Coupling between bone formation and bone resorption has made it extremely challenging to direct therapies that alter either side of the resorption-formation equation without affecting both sides simultaneously. Balance, on the other hand, refers to the relative amounts of bone that are resorbed or formed, typically at an individual BMU unit. If more bone is resorbed than formed, the bone balance is negative (i.e., bone is lost) and vice versa if the balance is in favor of formation. These terms are often confused and incorrectly used interchangeably, but concern very different aspects of the remodeling scheme. For instance, resorption and formation can be both coupled and unbalanced, as in the example above. However, there are some situations in which it is not clear if the processes are uncoupled or imbalanced. One of these would be the resorption that occurs along the endocortical surface of bone. This bone may eventually be replaced (suggesting coupling), but it may require a longer period than normal to complete.

4.4 Remodeling Cycle Duration

Figure 4.5 depicts select phases of a BMU remodeling unit in a schematic diagram. This figure also illustrates the relationship between the appearance of an osteonal BMU in longitudinal and cross sections and, in addition, the fact that the duration of events perceived by an observer in a particular bone cross section does not correspond to the duration of events that would be observed by someone moving along the bone with the BMU. For example, the moving observer would say that the resorption phase of the BMU lasts from the time that osteoclasts begin to excavate the tunnel until the tunneling ends, perhaps a centimeter down the bone shaft; this might take 200 days. A stationary observer at some arbitrary cross section along the way would say, however, that the time required for resorption is only the few days required for the osteoclasts to pass through the plane of the section and open a hole corresponding to the osteonal cement line. In virtually all discussion of osteonal BMU dynamics, time periods refer to those observed in cross sections, because it is such sections that are routinely observed and analyzed.

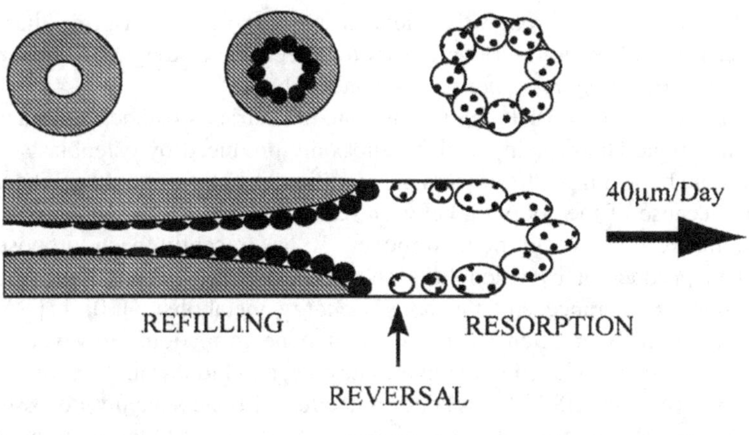

Fig. 4.5 Schematic diagram of an osteonal BMU. Larger, multinucleated cells to right are osteoclasts; smaller cells shown in black to left are osteoblasts. Cross-sectional views are shown above. Not to scale

As one can see from this, there is a period of time during which bone has been resorbed, but new bone has not formed at that location. This is called the remodeling transient, because in healthy bone, it is a transitory phase that will eventually be resolved when new bone is formed at that location. This transient also creates a remodeling space, which is all of the space in bone that is undergoing processes of bone remodeling that have not yet been completed. This space has been estimated at 5–10% of the bone volume. The size of the remodeling space is defined by the frequency with which new remodeling sites are started (the activation frequency, or Ac.f), the rate at which bone is removed or formed at those sites, and the overall length of time it takes for remodeling to be completed at those sites (the remodeling period). The remodeling transient can be disrupted by aging or other conditions such as disuse.

4.5 Biomarkers of Bone Remodeling

It is not always possible to obtain a piece of bone to measure the components of the bone remodeling process histologically. Even for animal studies, it may be desirable to follow changes in remodeling rates longitudinally during the course of an experiment. In these cases, the rates of bone resorption and formation can be estimated from blood or urine by measuring biochemical substances that are released from bone when it is resorbed, or that are produced by osteoblasts or osteoclasts during the bone remodeling process. This is an imperfect measure of bone turnover processes (as is histology as well, for different reasons that are described later) because it provides only a systemic picture of bone resorption and formation and is not location-specific. Moreover, many of the compounds are not specific to bone but found in tissues and organs other than bone. Their appearance in blood or urine may reflect processes occurring at those sites. Moreover, the measurement does not consider the fact that some substances may be reabsorbed without entering the blood stream or going through the kidneys. Nevertheless, measuring these compounds can provide a noninvasive means to screen for or diagnose disease, or to monitor changes during an experimental challenge.

The most commonly used measure of bone formation is osteocalcin (also sometimes called bone Gla-protein), which is not only produced by osteoblasts but also released from bone during the process of resorption by osteoclasts. In part because of this, and because of the coupling between resorption and formation, osteocalcin is really a marker of overall bone turnover. It has recently been proposed that osteocalcin production by osteoblasts enhances insulin secretion and so plays a role in glucose tolerance and the body's energy metabolism [10, 11]. Alkaline phosphatase is another often used marker of bone formation. However, alkaline phosphatase is also produced in the liver, intestine, and kidney, and so bone-specific alkaline phosphatase (BSAP) is a better reflection of bone-specific processes. Two pro-peptides of the type I collagen molecule, one from the C-terminal end (PICP) and one from the amino-terminal end, (PINP) are also now used. These are produced by both osteoblasts and fibroblasts, and so in part the picture they give may be altered by processes occurring in soft tissues. All these bone formation markers can be measured from blood serum.

The most useful resorption markers are those that are derived from parts of the collagen fiber or its cross-links when it is degraded by osteoclastic resorption. The difficulty with some of these, for example, hydroxyproline and the cross-link pyridinoline (PYD), is that they are found not just in bone (type I collagen) but also in cartilage (type II collagen) and/or skin and other tissues that contain different types of collagen. Deoxypyridinoline (DPD) is another cross-link but is more specific to bone (although it is also found in dentin). The most widely used resorption markers, therefore, are those more specific to type I collagen and those that come from the C- and N-terminal telopeptides of the type I collagen molecule (CTx and NTx). These can be measured from urine or serum, and although they are found in other tissues and organs that have type I collagen (e.g., skin), the highest contribution is from bone. Tartrate-resistant acid phosphatase (TRAP) is an enzyme produced by osteoclasts during the process of resorption. It is measured from blood plasma or serum and is also very commonly used, especially the isoform TRAP5b.

Although attempts have been made to use markers to evaluate the balance between resorption and formation, these attempts have not been very successful, in part because of the inherent variability and non-specificity of the markers, but probably also because of temporal differences in the processes of resorption and formation.

4.6 Bone Structural Units Are Produced by Remodeling

The packet of new bone produced by a BMU is called a bone structural unit, whether the remodeling occurs in cortical or trabecular bone or on an endosteal or intracortical surface/region. Thus, a secondary osteon in cortical bone is a bone structural unit, as is the hemi-osteon in cancellous bone.

4.6.1 Osteonal Remodeling

When a BMU tunnels through compact bone, it creates a secondary osteon. Figure 4.4 shows the leading portion of an osteonal BMU in longitudinal section, tunneling to the right. A few osteoclasts may be seen on the resorbing surface. At left, osteoblasts and osteoid line the refilling surfaces. The "head" or front end of a BMU contains a capillary "bud" to supply nutrients and remove by-products created by working osteoclasts and osteoblasts, and probably to supply additional progenitor cells as required. Because osteonal BMUs become isolated deep within the cortex, this vascular supply must be maintained. Therefore, the tunnel cannot be entirely refilled, and each BMU leaves a new central canal in the bone. This means that every

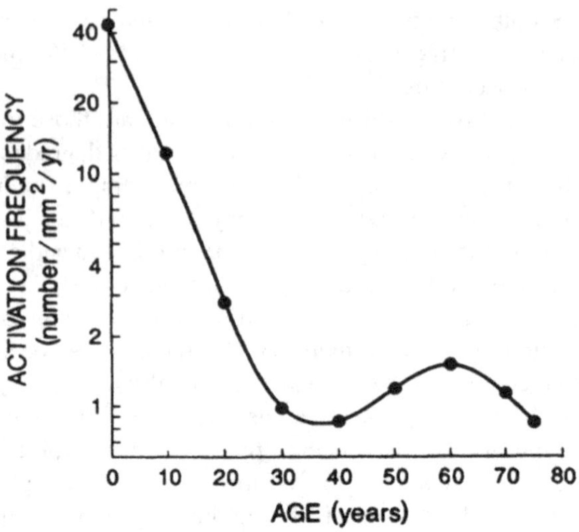

Fig. 4.6 BMU activation rate vs. age for human ribs [12, 13]

intracortical remodeling unit results in a small amount of bone loss (with the exception of an osteon that resorbs an existing central canal). Within each Haversian canal are vessels that connect with the vasculature in the medullary canal or on the periosteal surface. Haversian canals also contain nerve fibers that produce peptides that likely play a regulatory role in bone homeostasis.

In human adults, osteonal BMUs replace about 3% of compact bone each year. Figure 4.6 shows how the rate of remodeling changes with age in the human rib [12, 13]. It is very high in children, is reduced in young adults, then rises and falls again in older individuals.

4.6.2 *Trabecular Remodeling*

Trabeculae are remodeled similarly, except that the BMUs work on their surfaces, digging and refilling trenches. Because a BMU is about the same diameter as a trabecula, it cannot usually tunnel longitudinally within a trabecular strut without constantly breaking out along its sides. Therefore, BMUs trench out "half-moon"--shaped erosion pits, and refill these with new bone, creating a hemi-osteon. BMUs replace human adult trabecular bone at a relatively high rate, about 25–30% each year. In fact, the rate of bone turnover varies widely throughout the skeleton. Table 4.1 demonstrates this using data from various skeletal sites in young adult dogs and humans [14, 15]. It can also be seen that the remodeling rates in dogs are much higher than in humans.

Treatment with anabolic agents, such as parathyroid hormone, causes trabeculae to become quite thick. This could eventually cause a problem with fluids and nutrition reaching the cells in the deepest layers of the trabecular strut, but BMUs

Table 4.1 Bone turnover rates in young adult dogs and humans

Site	Dog cancellous BFR (%/yr)	Human cancellous BFR (%/year)	Human cortical BFR (%/year)
Lumbar vertebrae	205	0–14	11–20
Thoracic vertebrae	167	2–21	0–24
Cervical vertebrae	121	–	–
Mandible	105	–	–
Skull	60	8–14	4
Calcaneus	120	18–23	0–1
Proximal humerus	174	10–26	3–9
Proximal femur	138	16	1–11
Proximal radius	127	9–29	7–12
Proximal tibia	112	–	–
Carpals	31		
Iliac crest	–	10–37	6

These values are approximate. More detailed information can be found in the original references [14, 15]

tend to "tunnel" through overly thick trabeculae, creating two trabeculae where one existed before. This has the additional advantage of creating four surfaces for bone formation instead of two, allowing bone to be formed much more quickly. Besides these advantages, tunneling also alters the architecture of the cancellous structure in a way that increases trabecular connectivity and improves bone strength.

4.6.3 Endosteal and Periosteal Remodeling

Remodeling also can occur on endosteal and periosteal surfaces. Endosteal remodeling is similar to remodeling along trabecular surfaces, although the erosion sites are often not filled completely, or at all, causing a life-long gradual expansion of the marrow cavity once growth is complete. Long bones expand radially with age in adults as well as children. This expansion is the result of direct bone apposition to this surface (i.e., modeling in formation mode). Although periosteal remodeling was dismissed at one time as non-existent or not important, it is now clear that this does occur in the long bones with increasing frequency with age [16, 17]. In adults, this remodeling always results in scalloping of the periosteal surface and some bone loss.

4.6.4 Skeletal Envelopes and Bone Loss

Trabecular and endosteal surfaces are bathed in marrow, and remodeling that occurs on them is usually characterized by an excess of resorption relative to formation. This endo-trabecular deficit of bone replacement during remodeling is the reason for normal, age-related bone loss that occurs in men and women after age 30–35 and for the more rapid loss of bone in women after menopause. The periosteal envelope experiences formative modeling which expands the outer diameter of the bone and increases bone strength, such processes partially compensate for involutional (i.e., senile) bone loss. However, as mentioned above, remodeling can also occur on this surface in older people. Like the endosteal and trabecular surfaces, the balance is usually in favor of resorption and bone loss, and in the older individual, contributes to involutional loss. The osteonal, or Haversian, envelope also contributes to bone loss. A Haversian canal must be created for each new osteon, increasing the porosity of compact bone until a level of porosity is attained at which each new BMU removes an existing Haversian canal before creating a new one. The reason why a deficit in formation occurs during remodeling on surfaces adjacent to marrow is unknown. There is, however, an association between a remodeling deficit and the amount of fat in the marrow [18, 19]. With age, part of the deficit is probably indicative of the smaller number of stem cells available to differentiate into bone-forming osteoblasts.

A sudden transient increase in the rate of remodeling sometimes occurs in one or more bones. This is called a regional acceleratory phenomenon, or RAP. The cause is usually trauma to the bone. When this happens, a transient osteoporosis occurs because there is a lag between resorption and formation in each BMU. The osteoporosis will resolve itself over a remodeling period (about 4 months) as the additional BMUs complete their refilling (Fig. 4.5).

4.7 Analysis of Bone Remodeling

The dynamic biology of bones during remodeling can be measured directly using microscopy of bone samples collected from animals or people. Bone histomorphometry is used commonly in animal experiments to understand the effects of disease, drugs, mechanical loading, healing, age, and other factors on bone metabolism. Bone histomorphometry is also used as a diagnostic and research tool on samples collected from human patients or volunteers, also for the understanding of bone changes in disease, although (of course) there are fewer cases of human experiments because sample collection requires an iliac crest biopsy or collection at surgery. In the next part of this chapter, some of the methods used for standard histomorphometry will be introduced and then we will explore some mathematical models that help in the understanding of the dynamic remodeling process.

4.7.1 Mathematical Notation for Bone Remodeling

Frost [12, 20] was the first person to develop mathematical models for bone remodeling. Being the first, he was free to choose mathematical symbols for key variables. For example, he used the symbol "sigma" (σ) to represent the time required by each phase of the BMU to be completed at a particular location, Mf to denote the rate at which bone is added to a forming surface, and μ for the rate at which new BMUs are created in a bone.

Subsequently, Frost's mathematical notation was used by other investigators in reporting their experimental data. Eventually, however, concern arose because various investigators would use the same symbol to denote a variable measured or expressed in somewhat different ways. For example, the amount of bone surface on which formation was occurring could be expressed per unit volume of bone (i.e., in three dimensions) or per unit area of histologic cross section (i.e., two dimensions). To eliminate this problem, the research community developed a standard symbology for referring to bone remodeling variables [21]. For example, the area of bone surface covered with osteoid would be called OS in three dimensions or O.Pm in two dimensions. Here, the ".Pm" refers to the fact that surfaces show up as perimeters in a two-dimensional section. The resorption "sigma" became RS.P (resorption period), the reversal period became Rv.P, and the formation sigma became FP (formation period). The total of these three periods is the remodeling period (Rm.P). In the first edition of this book, a nonstandard nomenclature was used for remodeling parameters.

There are numerous standard measurements used to understand bone remodeling. Some of these are discussed in the following sections along with some of the assumptions that underlie their validity.

4.7.2 Histomorphometric Measurement of Remodeling

The concept of bone remodeling in quantum units is important because it recognizes a predictable biologic system that is particularly amenable to quantitative analysis. The utility of the quantitative histologic techniques introduced by Frost, and used today in clinical medicine as well as basic science research, is a direct result of these qualities.

Frost demonstrated that histologic sections could be analyzed to quantify bone remodeling. He originally did this for osteonal remodeling in the cortex of the human rib. Since then, his method has been extensively used in animal experiments and adapted for use on human bone biopsies from the iliac crest. The latter is used in preference to the rib because it is more accessible and contains more cancellous bone, where bone remodeling is more active and bone loss appears more quickly. We provide a brief overview of the measurement process used in evaluation of bone remodeling for three reasons. First, histomorphometric analysis requires the reduction of the mechanics of remodeling and osteon formation to a few well-defined and measurable steps, which leads to a clearer and more integrated understanding of the osteonal remodeling process. Also, this sort of quantitative analysis is an excellent way to identify points of inconsistency in our theories and frame pertinent questions for future experiments. Finally, an understanding of histomorphometric analysis contributes greatly to a clearer perception of the problems associated with diagnosing pathologic states and defining their etiology.

After reviewing the methods used to quantify bone remodeling in a histologic section, we shall continue to use mathematical modeling as a tool for learning more about how remodeling works and the effects it has on bone structure. Making mathematical models forces us to be quantitative as we test hypotheses about the unknown. Eventually, these models may lead to a much clearer understanding of skeletal biology and mechanics and the relationship between them. The mathematical analyses in this chapter should be considered "works in progress," rather than derivations of the kinds of mathematical results that have been tested over prolonged periods of time. When asked how they obtained their insights in the world of physics, Newton said "By thinking into them." And Einstein said "I grope." In bone biology, we use mathematical models to think our way into bone remodeling. However, we remain in the stage where critical evaluation of the ideas/concepts should be challenged. Join in the adventure, look skeptically at what follows, and try to find flaws and fix them!

4.7.3 Assumptions in Histomorphometry

A number of assumptions are made when attempting to quantify dynamic processes of bone remodeling, and it is important to understand what these are so that the information that derives from them is interpreted properly. The most important assumption is that remodeling is in a steady state, that is, that new BMUs are created at a constant rate and are functioning in ways that are not changing with time. Of course this implies, for the analysis of remodeling surfaces, that resorption and formation are coupled and in balance as well. Of course, this assumption of a steady state is never strictly true. Dynamic histomorphometry is usually employed to evaluate potential reasons that bone is not in a steady state, whether caused by a disease process or caused by experimental manipulation or genetic modifications of animals. Situations in which histomorphometry is used require the understanding that the investigator is measuring transient remodeling changes. A second assumption is the stereometric assumption that makes it possible to use two-dimensional (2D) microscope slides to calculate estimates for the three-dimensional (3D) biology [22].

The accuracy of the estimation of 3D measurements from 2D microscopic images depends on how well the practical measurements meet the mathematical assumptions that underlie unbiased stereology. In the analysis of bone, we regularly violate some of the assumptions of unbiased stereology; for example, unbiased stereology commonly assumes that the microscope slides are oriented randomly in the 3D object (the bone). In the practical stereometric analysis of cortical bone, it is not possible to randomize the orientation of the slides and they are always cut either directly parallel to or directly perpendicular to the cortical diaphysis. Always sampling the bone in the same direction violates the random orientation assumption of stereology and creates systematic errors in the conversion of 2D measurements into estimates of the 3D data. In practice, however, this doesn't create a problem as long as one compares "like-to-like," that is, data collected using parallel (or perpendicular) slides is compared to data from slides cut in the same way. A different issue arises in the collection of data from slides of cancellous bone, in which trabecular struts may run at any angle with respect to the plane of section. As a result, 2D data for cancellous bone cut systematically from different sites has different bias (systematic error) when used to estimate 3D information. The mathematical ways to correctly adjust 2D measurements to 3D [35] are important to understand, especially for cancellous bone; however the details of these methods are beyond the scope of the current discussion. In practice, many labs simply measure several sections and average the values across those sections to estimate volumetric measurements. Ideally, the sections are not consecutive, although this may not always be practical. A third assumption is that the bone structures (or cells) that are measured were created during the period of interest. In actuality, even with sequential fluorochrome labeling, measurements are capturing a single snapshot in time. So, for instance, the mean wall thickness (W.Th) is defined as the average distance from the cement or reversal line to the bone surface. Mean wall thickness can be measured directly from any section containing osteons; however, there isn't any guarantee that the osteonal walls being measured were formed during the same period. More subversively, perhaps they were formed well before the onset of the bone disease or before the experimental manipulation that you want to measure. Since W.Th is used to calculate how long it takes for an osteon to form, not knowing when the tissue was created can lead to errors in evaluating the rate at which new BMUs are formed.

An additional problem for making biological measurements from microscopy slides is that very recently, with the advent of drugs for osteoporosis that create severe suppression of bone turnover, the methods used for histomorphometric analysis sometimes break down. If bone remodeling is suppressed to the degree that double fluorochrome labels cannot be found, then rates of bone formation, and a host of other calculated variables that define the remodeling system, cannot be determined. As the analytical method is based on the existence of a remodeling steady state in which remodeling is occurring at a measurable rate, many experiments and some clinical trials cannot be accurately assessed using conventional histomorphometric techniques. Overcoming this problem presents an important area for research [23].

4.7.4 Histomorphometric Measurements

Histomorphometric variables can generally be divided into those that can be directly measured (primary variables) and those that must be calculated (derived variables). Primary measurements can be broken down into a few essential kinds of measurements that include area or volume, perimeter or length, widths, and number. Because some bones are larger than others, or have more or less surface area, these measurements must be normalized by using a referent. For example, the amount of mineralizing surface is usually expressed per total bone surface (MS/BS), as are other surface measurements, and the number of cells is also normalized in this way (e.g., N.Ob/BS for number of osteoblasts per unit of surface).

4.7.5 Cortical Bone

Histomorphometry for cortical bone uses the total area (Tt.Ar) of the cortical bone cross section as the normalizing reference. Each section is measured for resorption and formation perimeters and areas, as well as labeled osteonal surfaces and distances between double labels and the measurements are normalized by dividing by Tt.Ar (Fig. 4.7). It is also important to measure the number of labeled (forming) osteons per bone area (N.On.L) as this can be used in the calculation of bone formation rate (which will be a volume-based measure, BFR/BV, rather than a surface-based one).

The mineral apposition rate (MAR) is the rate that mineralized bone matrix is added to a bone and is calculated from the interlabel width (Ir.L.Wi, μm) and the interlabel period (Ir.L.P, days) as:

$$\mathrm{MAR} = \mathrm{Ir.L.Wi}/\mathrm{Ir.L.P}\ (\mu\mathrm{m}/\mathrm{day}) \tag{4.1}$$

The word "mineral" is attached to this variable because it is measured from fluorochrome labels, which lie at the junction between the osteoid "seam" (i.e., layer) and the layer of bone experiencing primary mineralization. In a steady-state situation, the mineral apposition rate is the same as the rate at which osteoblasts are laying down osteoid. Therefore, it is common to shorten the term to "apposition rate."

The area of mineralizing surface per bone surface (MS/BS) is estimated as the simple ratio of the total mineralizing surface (MS = sL.S/2 + dL.S) and the bone surface (BS):

$$\mathrm{MS}/\mathrm{BS} = (\mathrm{sL.S}/2 + \mathrm{dL.S})/\mathrm{BS} \tag{4.2}$$

The amount of single label surface is divided by two to account for labels being administered at two different times [21].

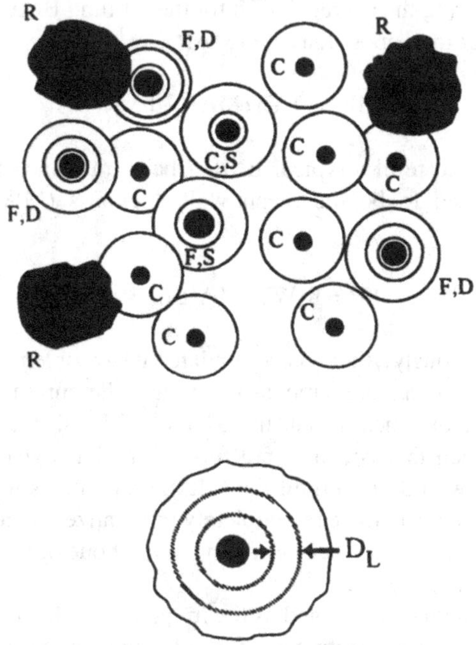

Fig. 4.7 (**a**) Upper: Schematic diagram of BMUs in a cross section of cortical bone. Black, irregularly shaped BMUs are resorption spaces [R]. BMUs with large canals are refilling; they have either a single label [F, S] or a double label [F, D]. Completed osteons have small canals [C]; some of these have a single label [C, S]. Lower: A single BMU in the last stages of refilling is depicted. It has a double label; the distance between the labels is measured as shown (Ir.L.Wi). (**b**) This photomicrograph shows double labeling with either calcein (green) or alizarin (red) on secondary osteons in rabbit bone. Note also the single alizarin labeled osteon on the left

The bone formation rate is often expressed per bone surface area. This is calculated from both MAR and from the total area (perimeter) of bone formation per bone surface (MS/BS) as:

$$\mathrm{BFR/BS} = (\mathrm{MAR})(\mathrm{MS/BS}) \tag{4.3}$$

Bone formation rate can also be expressed per bone area as

$$\mathrm{BFR/B.Ar} = (\mathrm{MAR})(\mathrm{MS/BS})(\mathrm{BS/B.Ar}) \tag{4.4}$$

These two examples of bone formation rate show how the histomorphometric parameters are combined to normalize a measurement in different ways.

The time that elapses between the production of osteoid by the osteoblasts and its initial mineralization is the mineralization lag time (Mlt). Histologically, this delay is represented by the thickness of the osteoid seam, O.Th (μm); this would be zero if the new bone mineralized immediately. The mineralization lag time may be determined by measuring the average O.Th for the refilling BMUs on the section and dividing the result by the apposition rate (MAR):

$$\text{Mlt} = \text{O.Th}/\text{MAR (days)} \quad (4.5)$$

The time required to refill a typical BMU (the formation period, FP in days) is then logically assumed to be the mean wall thickness (MWT) divided by the apposition rate:

$$\text{FP} = \text{MWT}/\text{MAR (days)} \quad (4.6)$$

As indicated previously, there is a fundamental problem in this part of the analysis. The measurements are supposed to reflect the current situation, but measuring completed osteons means that the results will be skewed toward the conditions that existed when the osteons were formed, perhaps years ago. One way to handle this problem would be to limit the Mlt measurements to completed osteons that microradiographs reveal to be incompletely mineralized. Another method would be to measure Mlt in those few osteons bearing at least one of the labels given for the analysis but no osteoid seam.

The activation frequency is defined as the frequency with which a new BMU will be initiated in a given cross section of bone. Its calculation relies heavily on the steady-state assumption. The times for each phase of BMU development are assumed to be constant, so that the number of BMUs activated each day is the same as the number entering the refilling phase. Applying this to refilling BMUs relates the number of forming osteons per bone area (defined by the number of osteons with osteoid, or N.O) to the activation frequency (Ac.f; number per bone area per time) and the formation period (FP) as:

$$\text{N.O}/\text{B.Ar} = (\text{Ac.f})(\text{FP}) \quad (4.7)$$

That is, the population of refilling BMUs per millimeter squared of bone section is equal to the birthrate per unit bone area (activation frequency, Ac.f) of refilling BMUs multiplied by their longevity, or the time required for refilling, FP. Rearranging this equation, one has:

$$\text{Ac.f} = [(\text{N.O}/\text{B.Ar})/\text{FP}](\text{refilling BMUs}/\text{mm2}/\text{yr}) \quad (4.8)$$

for the BMU activation frequency.

4.7.6 Estimating Bone Resorption Rate

The bone formation rate (BFR) is defined as the amount of new bone made per unit time, expressed as a fraction of the surface or volume of bone being considered. For example, if within each cubic millimeter of bone in a human femur the remodeling process formed a volume of bone equal to 0.05 mm^3 in 1 year, the bone formation rate would be 0.05 mm^3/mm^3 per year. Equivalently, if on a 1 mm^2 area of bone 0.05 mm^2/year was formed. Then the bone formation rate would be 0.05 mm^2/mm^2/year. Because fractions of area measured on a bone cross section are equivalent to volume fractions in the same region, this variable may be computed from the mineral apposition rate (MAR μm/day), osteoid perimeter (O.Pm μm of osteoid perimeter), and osteon number (O.N number of forming osteons per square millimeter) as:

$$\mathrm{BFR} = (\mathrm{MAR})(\mathrm{O.Pm})(\mathrm{O.N})\ \left(\mathrm{mm}^2/\mathrm{mm}^2/\mathrm{yr}\right) \tag{4.9}$$

In this equation, you need to make sure to use the correct units—don't forget that MAR and O.Pm are usually measured in micrometers/day and micrometers, respectively. As the amount of bone resorbed during osteonal remodeling is normally only slightly greater than that formed (the difference in osteonal bone resulting from the creation of new Haversian canals), the bone formation rate is often taken to be the same thing as the bone "turnover" or replacement rate.

There is no straightforward way to measure the osteoclastic equivalent of MAR, which one may call the erosion rate, BRs.R, but it may be approximated if the rate at which bone is being lost can be determined. Suppose, for example, that biopsies are obtained before and after an experimental treatment, and measurements of porosity indicate that 0.01 mm^2/mm^2 of bone was lost during the course of a year. This time, the rate of change of porosity, or bone balance, must equal the difference between the bone resorption rate and the bone formation rate:

$$\text{Bone balance} = \mathrm{BRs.R{-}BFR} \tag{4.10}$$

If we define the resorption perimeter (Rs.Pm) as the length of surface with active resorption, Rs.N as the number of resorbing osteons per square millimeter formed per year, and MMR as the linear rate of bone resorption at a surface (analogous to mineral apposition rate, MAR), then we can write an expression for osteonal resorption that is exactly analogous to that for formation:

$$\mathrm{BRs.R} = (\mathrm{MRR})(\mathrm{Rs.Pm})(\mathrm{Rs.N})\ \left(\mathrm{mm}^2/\mathrm{mm}^2/\mathrm{yr}\right). \tag{4.11}$$

The bone balance, therefore, is the difference between this new equation (Eq. 4.11) and the original equation for bone formation rate (Eq. 4.9). If the bone balance is known, then the mineral resorption rate (MRR) can be estimated from BRs.R-BFR, since all of their terms can be directly determined from the cross section except for MRR.

The change in bone volume also can be estimated as the difference between the depth of the resorption cavity (resorption depth, Rs.De), the amount of bone that has refilled it (W.Wi), the perimeter over which resorption occurred (Rs.Pm), and the number of resorption sites per area per time (Ac.F) as

$$\mathrm{BRs.R} = (\mathrm{W.Wi\text{-}Rs.De})(\mathrm{Rs.Pm})(\mathrm{Ac.F}) \tag{4.12}$$

Table 4.2 Bone remodeling data for rib osteonal BMUs

Variable	Dogs[a]	Rhesus monkeys[b]	Humans by age group, years[c]		
Age, years	2–3	"Adolescent"	1–9	30–39	70–89
On.N, #/mm^2	–	4.03 ± 1.64	2.7 ± 1.0	0.21 ± 0.54	0.66 ± 0.50
Rs.N, #/mm^2	0.75 ± 0.1	1.96 ± 1.15	2.0 ± 0.62	0.25 ± 0.33	0.84 ± 0.46
O.Pm, μm	–	200 ± 30	280 ± 40	250 ± 140	300 ± 72
MAR, μm/day	1.1 ± 0.05	1.33 ± 0.23	1.40 ± 0.46	1.1 ± 0.36	0.72 ± 0.20
FP, days	48 ± 2	54 ± 11	51 ± 47	73 ± 58	109 ± 55
Ac.f, #/mm^2/yr	29 ± 3	27.9 ± 13	19 ± 9	1.1 ± 0.5	2.2 ± 1.1
MWT, μm	50 ± 2	63 ± 2	68 ± 4	70 ± 5	66 ± 7
BFR/BV, %/yr	20 ± 4	35.9 ± 18.6	38 ± 24	1.8 ± 0.6	4.4 ± 1.5

[a]Mashiba et al. [24]
[b]Schock et al. [25]
[c]Frost [26]

Rs.De, W.Wi, and Rs.Pm are averaged over the completed BMUs in the section, first by measuring the distance between the reversal line and the original bone surface (which can be estimated in cancellous bone and the endosteal envelope, although not so easily in intracortical osteons), the second by measuring between the reversal line and the surface of new bone in a complete BMU, and third by measuring the total length of the BMU. The time rate of bone loss (or gain) for the entire cross section is the average of these values times the number that occur in the area (Ac.F) over time.

Table 4.2 shows typical values for a number of these histomorphometric variables as they have been measured in the ribs of humans, dogs, and monkeys. Clearly, there is variation between species, between bones (the rib remodels much more quickly than other cortical bone) and with age, so that numbers such as FP = 90 days, which we cited earlier as typical in humans, must be used with discretion.

4.7.7 *Label Escape Error*

When two fluorochrome labels are given and the bone is examined afterward, some of the labeled osteons only have one label. One reason for this is that some osteons began to refill after the first label was given, and others stopped refilling before the second label was given. Frost worked out the correction for this issue, which was further expanded by Martin [27]. In Fig. 4.8, time is represented on the horizontal axis, and the two vertical lines (L1 and L2) represent label times. The time between labels (labeled T_L in the graph) is Ir.L.P = L2 − L1. The diagonally distributed horizontal lines represent BMUs that began refilling at the time indicated by their left end and completed refilling at the time indicated by their right end. Thus, the lengths

Fig. 4.8 Label escape error. See text for details [27]

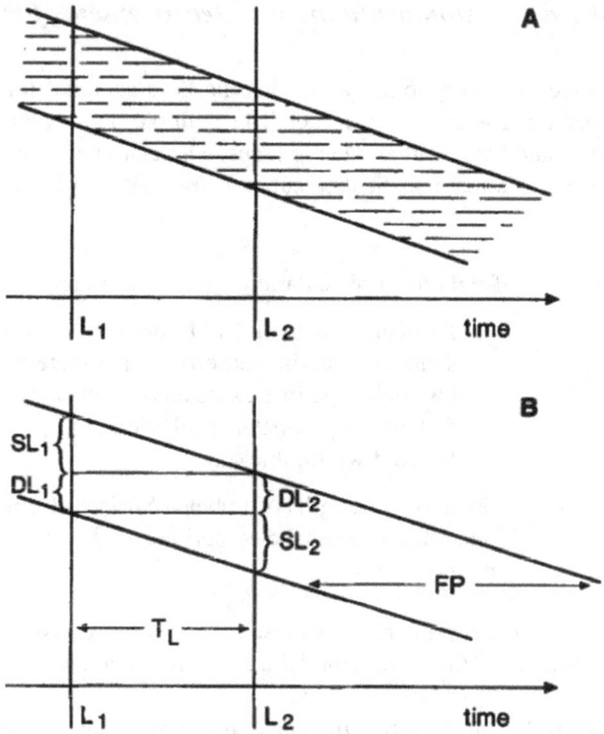

of the BMU lines are all FP. Doubly labeled BMUs have both vertical lines through them, and singly labeled BMUs have only one vertical line. Frost [28] showed that the ratio of the number of singly labeled osteons (sL.N) to those doubly labeled (dL. N) should be:

$$\mathrm{sL.N/dL.N} = 2(\mathrm{Ir.L.P})/(\mathrm{FP\text{-}Ir.L.P}) \tag{4.13}$$

and the "true" number of refilling osteons (N.O; number of osteons with osteoid) should be taken as:

$$\mathrm{N.O} = (1/2)\,\mathrm{sL.N} + \mathrm{dL.N} \tag{4.14}$$

There is experimental evidence that the sL.N/dL.N ratio is larger than predicted by Eq. 4.13 because some (or all) osteons pause during refilling. Such pauses may be exacerbated by age or mechanical disuse. This is an important detail of remodeling, which has been further analyzed but remains to be fully understood [29].

4.7.8 Histomorphometric Measurements for Cancellous Bone

Bone histomorphometry remains the gold standard for assessing remodeling. The technique is explained in detail in multiple publications [21, 23, 30, 31] so is not repeated here. Rather what is focused here are key variables that have been and can be used in mathematical calculations/estimates of bone remodeling processes.

Areas	
Total area, trabecular area, and osteoid area:	T.Ar, Tb.Ar, O.A

Perimeters or Lengths of Surfaces	
Trabecular perimeter, erosion perimeter:	Tb.S, Rs.S
Osteoclast perimeter, osteoid perimeter:	Oc.S, OS
Osteoblast perimeter, wall width:	Ob.S, W.Wi
Osteoid width, thickness:	O.Th

Perimeters or Lengths of Labeled Surface (from unstained sections)	
Single- and double-labeled perimeter:	sL.S, dL.S
Interlabel width:	Ir.L.W

From the primary variables, many useful derived variables are calculated to characterize the bone remodeling system, including:

Trabecular number, thickness and separation:	Tb.N, Tb.Th, Tb.S
Mineralizing surface per bone surface:	MS/BS
Mineral apposition rate:	MAR
Bone formation rate per bone surface or bone volume:	BFR/BS or BFR/BV
Mineralization lag time:	Mlt
Formation, resorption, and total remodeling periods:	FP, RP, Rm.P
Activation frequency:	Ac.f

The formulae for derived parameters for cancellous bone are usually the same as those presented previously for cortical bone. Detailed formulae for the histomorphometric calculations for cancellous bone are found in Parfitt et al. [21] and Li et al. [29].

Although histomorphometric measurements can be quite tedious to make, a very few measurements can fully characterize the remodeling system. One drawback is that the derived parameters can be quite variable. Each time you use several primary variables to calculate a derived variable, the inherent errors in each measured variable can combine to create even more variability in the calculated number. It is always a good idea to look critically at your result to see whether it makes any sense!

4.7.9 True vs Histologic BMU Activation Frequency in Cortical Bone

Activation frequency has two definitions. The first is the number of BMUs that appear on a cross section in a given time (Ac.f = BMUs/mm^2/yr). The second is the number of BMUs created per unit volume of bone per unit time (e.g., BMUs/mm^3/yr). We refer to this definition as the origination frequency (Or.f). Typically, BMUs are "born" when bone lining cells are stimulated to recruit osteoclasts to the site. These osteoclasts dig into the adjacent bone matrix, and in a few days create a tunnel or erosion cavity. The BMU continues to tunnel through the bone until some signal causes it to stop. What that signal or those signals may be is not known.

The first definition of activation frequency is the one we previously discussed; it arises when remodeling is quantified histomorphometrically. In this context, "activation" is not defined as the original creation of a BMU at a specific site along the bone, but as the appearance in the cross section of a functioning BMU in a particular stage of activity (Fig. 4.9).

As we have seen, the standard method of calculating activation frequency is to determine the population of refilling BMUs per mm^2 of section and divide this number by the refilling time (i.e., the "lifetime" of the refilling stage). This yields the "birthrate" of refilling BMUs in the section, that is, the rate at which the "front ends" of refilling regions enter the section. If all BMUs are similar with respect to the lengths of their resorbing and refilling regions, this "refilling region birthrate" is identical to that for the resorbing region, which is in turn identical to the rate at which BMUs themselves arrive at the cross section and begin to create a resorption cavity [12]. In other words, this definition involves the "creation" of sections of BMUs within the two-dimensional world of a histologic section, as opposed to the creation of the entire BMU in the three-dimensional world of the bone. We refer to the

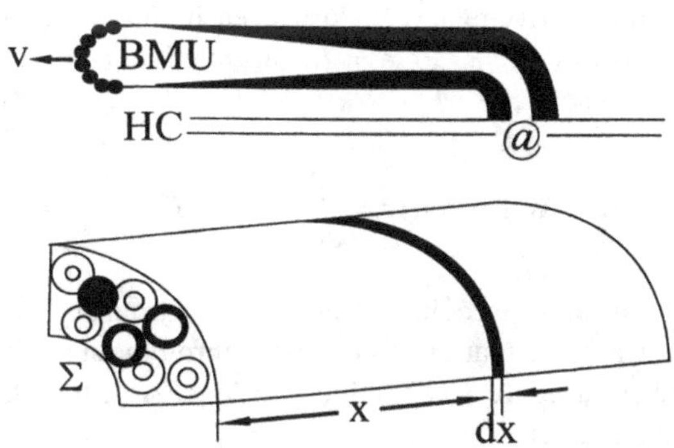

Fig. 4.9 Upper: An osteonal BMU has been activated on the wall of an existing Haversian canal (HC) at the site marked @ and is tunneling to the left with velocity v. Lower: A quadrant of cortex having cross-sectional area Ar is shown. A histologic section Σ at the left end is histomorphometrically analyzed to obtain Ac.f. BMU resorption cavities arrive at Σ after having originated at a distance x, within a zone dx

variable measured in this two-dimensional way as the activation frequency, Ac.f. As previously stated, units for Ac.f are typically given as BMUs/mm^2/yr. How is Ac.f, which is routinely measurable, related to the more biologically relevant three-dimensional Or.f, which cannot be measured?

Consider a region of cortical bone having cross-sectional area Ar and its long axis oriented in the x-direction (Fig. 4.9). If Ac.f is measured in a histologic section, Σ, made perpendicular to the bone at $x = 0$, we wish to know its relationship to Or.f in the bone as a whole. Let v be the mean velocity at which BMUs tunnel through the cortex, and let R be the mean BMU range, that is, the average distance that they dig before ceasing their activity. Now, those BMUs just arriving at Σ originated at various locations, xi, in the cortex, and at various times, T_i, in the past. The latter times may be called the ages of the BMUs arriving at $x = 0$. If all traveled at the same average velocity, v, then the originating distances and the ages are proportional to one another such that:

$$x_i = \nu T_i \tag{4.15}$$

Let Or.f(x, t) be the rate at which BMUs are being created at time t and position x along the bone. The number of BMUs originating within an infinitesimal section dx at time t in the past per day is:

$$dN = \text{Or.f}(x, t)\ \text{Ar}\ dx \tag{4.16}$$

where x is the location of dx and A is the area of the cross section. The number of BMUs that originated in dx arriving at Σ during each unit of time is:

$$dN = \text{Or.f}(x, T)\ \text{Ar}\ dx \tag{4.17}$$

where $T = x/v$ is the time required for a BMU to travel from dx to Σ. The total number of BMUs/mm^2 arriving at Σ during each unit of time is obtained by integrating Eq. 4.13 from x = 0 to $x = R$, the maximum distance from which a BMU could travel to reach Σ:

$$\text{Ac.f} = N/Ar = \int_0^R \text{Or.f}(x, T)dx \tag{4.18}$$

If Or.f varies in time and space, its histologic analog must be found by integrating Eq. 4.18. Let us now assume that Or.f is the same throughout the bone, varying only with time. Then substituting $dx = vdt$ and letting $L = R/v$ be the lifetime of a BMU's tunneling activity, one has:

$$\text{Ac.f} = \int_0^L \text{Or.f(t)}vdt \tag{4.19}$$

If Or.f is also constant in time, one has

$$Ac.f = (Or.f)\ v\ L \tag{4.20}$$

or,

$$Ac.f = (Or.f)R \tag{4.21}$$

That is, in a steady-state situation, the two-dimensional, histologically measured activation frequency is equal to the three-dimensional, true activation frequency multiplied by the mean BMU range.

This derivation has only considered BMUs arriving at Σ from one direction. When BMUs arriving from the other direction are considered, it might seem that Eq. 4.21 would yield values of Ac.f that are too small by half. That is not the case, however, because it can be assumed that half the BMUs represented by Or.f will tunnel away from Σ. Thus, the effects of ignoring both these details cancel one another. Another detail should be mentioned as well. As noted, sometimes activation results in a "double-headed" BMU, which arises at one site but forms two cutting cones that tunnel in opposite directions. For the purposes of this analysis, and in terms of the amount of remodeling that result from such incidents, these events should be counted as two activations.

If Ac.f and Or.f are related through the BMU range, *R*, the next question is, what is the value of R? This is not an easy question to answer. Because of the difficulty in sorting out anastomoses between Haversian canals and Volkmann's canals, it is difficult to trace the paths of BMUs from their point of origin to their terminus. However, Cohen and Harris [32] observed that "In approximately one centimeter of the length of the femur, an osteon and its descendants coursed from periosteal lamellae to endosteum." Tappen [33] was able to trace forming osteons back 5–6 mm from the cutting cone toward their origin. These observations suggest that a BMU may be expected to tunnel several millimeters before it ends. In that case, the true activation frequency, in BMUs/mm^3/yr, would be several times smaller than the histologically measured value, in BMUs/mm^2/yr.

In some early work, activation frequency was measured as just described, obtaining values of Ac.f, but the data were reported as Or.f, that is, as BMUs activated annually per cubic millimeter rather than per square millimeter. Other authors would simply report Ac.f. This may seem to be a small distinction, but it is important for two reasons. First, the origination frequency is probably only a fraction of that measured. This is not important when comparing groups in the same experiment, but it could be very important elsewhere, such as when three-dimensional mathematical models are constructed for remodeling phenomena. Second, ignoring the distinction deemphasizes important details of remodeling dynamics that are rarely discussed. Activation of BMUs at one site is only the beginning of an ongoing process that reaches out in both space and time to generate effects which may appear unrelated to the initial event. If BMUs travel several millimeters after they are activated, the lifetime of a BMU in a bone, as opposed to its lifetime in a section, is considerably longer than Frost's Rm.P. This means that transients caused by changes in activation frequency may take longer to play themselves out than one

Table 4.3 Remodeling periods for cortical bone: Cross-sectional and longitudinal values

	Cross-sectional (days)	Longitudinal (days)
Recruitment	5	5
Resorption	15	165
Reversal	0–20	0–20
Formation	90	260
Total[a]	95–125	265–285

[a]Note that the reason the numbers for each process do not add up to equal the total remodeling period is that these processes are occurring simultaneously

Rm.P. For example, activation of remodeling at a screwhole in the middle of a long bone sends waves of BMUs traveling up and down the diaphysis, so that its whole length is affected for months afterward [34]. Other events that alter activation frequency can be expected to produce effects of similar scope, and it is important to know their scale in time and space. It was estimated by Frost that such transients may require as much as four remodeling periods to resolve, but that is just an estimate, and it might be even longer.

These concepts, which rely on the differences between understanding the bone remodeling system in three dimensions rather than two dimensions, also underscore the reason that the stereological constructs used to measure bone histomorphometrically are only valid when used on the proper cross sections of bone. The measured periods for resorption, formation, and total remodeling are different when viewed in longitudinal section (which might show the entire lifespan of a BMU) than when viewed from a cross section, or even a series of cross sections. This is illustrated by a comparison of the times for various phases of bone remodeling, shown in Table 4.3.

The relationship between Or.f and Ac.f in cancellous bone is unclear. While investigators describe trabecular BMUs as migrating over trabecular surfaces how far an individual BMU may wander and in what direction before it merges with another, divides into two, or "dies" is poorly understood.

Note that the reason the numbers for each process do not add up to equal the total remodeling period is that these processes are occurring simultaneously.

4.8 Remodeling Concepts/Questions

4.8.1 Activation: What Initiates New BMUs?

Osteonal BMUs may originate on periosteal or endosteal bone surfaces, or along the walls of Haversian canals. Various activation signals are thought to be present in the skeleton, including chemical, mechanical, and electrical signals, but just how or where they affect cells and what sorts of intermediate messengers may exist are questions still being resolved.

The Rodan Theory. Rodan developed a theory of BMU activation based on the hypothesis that bone lining cells are responsible for initiating new BMUs [35]. That is not to say that lining cells differentiate to osteoclasts, but that they may be the target of various activation signals, responding in ways that lead to the appearance of osteoclasts. This idea springs from the observation that osteoblasts-turned-lining cells possess receptors for parathyroid hormone and 1-α,25-(OH)2 vitamin D3, but osteoclasts and their precursors apparently do not. Furthermore, lining cells respond to such signals in ways consistent with osteoclast activation. First, they change shape, contracting so as to expose the bone surface to the osteoclasts. Also, in cell culture mechanical and electrical stimuli increase cyclic adenosine monophosphate (cAMP) activity in osteoblast-like cells, which stimulates them to release prostaglandins, a potent intercellular messenger. Thus, the Rodan theory provides an interesting way for osteoclasts to respond to hormonal and physical signals to which they are not sensitive. It is also a theory that appeals to a sense of symmetry and simplicity; it provides a cytologic loop in which the final cell in the remodeling sequence becomes the initiator of the next remodeling cycle. Although insightful, it is not a theory that has fallen into favor.

Hormonal Activation Many hormones affect bone remodeling both directly and indirectly; these include parathyroid hormone (PTH), glucocorticoids, the active form of vitamin D (i.e., 1-α,25-(OH)2 vitamin D3), and calcitonin. The first two are usually thought of as leading to bone loss, and the last two as producing increased bone mass. In truth, however, we currently know very little about the effects of these hormones on the bone remodeling system, because so much of the available information has been obtained in tissue or cell culture experiments. Clearly, such experiments distort or destroy many characteristics of the skeletal intermediate organization in general and of BMUs in particular. Furthermore, many in vivo experiments have observed the effects of a hormone soon after it was administered, thereby missing effects related to the refilling stage of the BMUs activated by the treatment. Finally, much of the information obtained from experiments and clinical biopsies is of limited usefulness because it is difficult to sort out direct and indirect effects.

Activation Is Prime. Whatever causes activation of osteonal BMUs in a given situation, the body's ability to vary activation is much greater than the ability to control the numbers of osteoblasts or osteoclasts within the BMU or the rates at which they work. Figure 4.10 shows this by comparing the correlations between bone formation rate and apposition rate (MAR, a measure of osteoblast activity, top graph) and activation frequency (Ac.f, bottom graph). Clearly, the correlation between bone formation rate (BFR) and Ac.f is much higher than that between MAR and BFR. From this we deduce that changes in activation frequency have a greater effect on the amount of bone formed (or turned over) each year. Thus, effective (i.e., reasonably rapid and precise) responses to a situation calling for decreased bone mass, for example, cannot be achieved simply by decreasing osteoblast activity or stimulating osteoclasts; activation of many more BMUs is required. It turns out that activation is both the most important bone remodeling parameter and the one we know the least about.

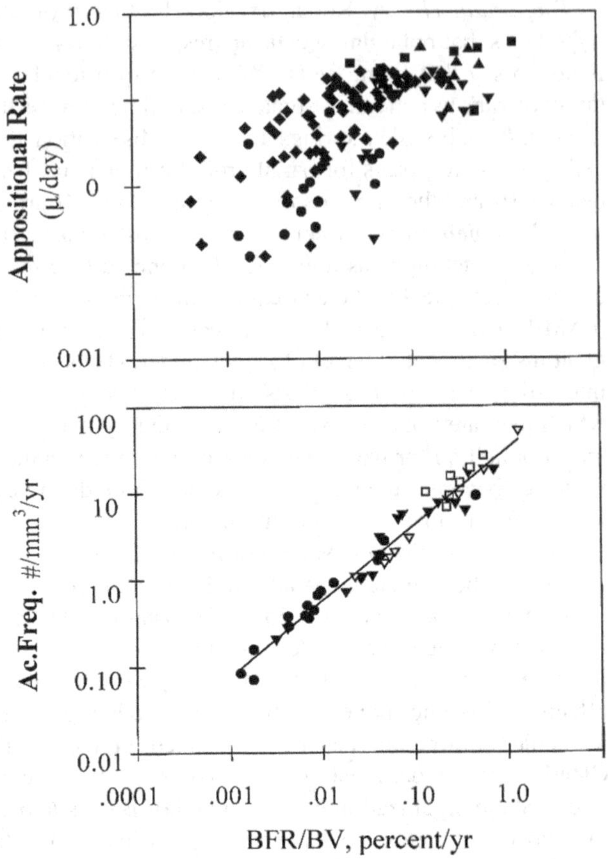

Fig. 4.10 Graphs of apposition rate [top] and activation frequency [bottom] vs. bone formation rate (BFR) in human rib specimens. Note the much tighter correlation in the lower plot. The different symbols represent specimens from various kinds of patients, including normal controls

4.8.2 Resorption: Out with the Old

Bone resorption is primarily the function of osteoclasts, which are mobile, multinucleated cells closely related to macrophages. Osteoclasts attach themselves to a bone surface, form a peripheral seal, and break down the bone within the sealed area by means of enzymes and other chemicals. The final stages of the dissolution of the bone tissue appear to be carried out by mononuclear cells following along behind the osteoclast, or by those formed through fission of multinucleated cells. Nevertheless, osteoclasts are the primary resorbing cells within the BMU. Although osteoclasts exhibit considerable mobility on trabecular bone surfaces, apparently ranging over resorption territories several times their contact area, in osteonal cutting cones, they seem to dig straight ahead in a tightly bunched configuration.

How fast is an osteoclast? In 1972, Jaworski and Lok measured the longitudinal advancement rate of osteonal cutting cones in dog ribs using double tetracycline labels [36]. By measuring the distance between the leading end of the two labels and

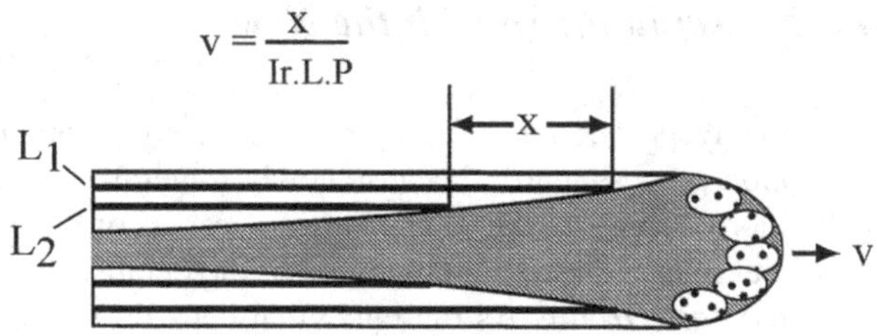

Fig. 4.11 Schematic diagram shows measurement of osteonal BMU advancement rate (v) in a longitudinal section. x is the distance between the leading edges of labels L1 and L2 given Ir.L.P days apart

dividing by the interlabel time, Ir.L.P, they determined that the region where the "refilling cone" began to mineralize advanced longitudinally some 39 ± 14 μm/day. Assuming the BMU's geometry does not change as it moves through the bone, the cutting cone should have a similar rate of advancement. The rate at which the osteoclasts dig radially can be determined from the longitudinal advancement rate and geometry of the cutting cone. As seen in Fig. 4.11, in longitudinal section, the cutting cone is ellipsoidal rather than hemispherical. In dog ribs, the major longitudinal radius is about 300 μm and the minor radius (i.e., the cement line radius) is about 100 μm. Thus, the time to open up the resorption space is 300/39 or about 8 days, and the rate of radial erosion is 100/8 or 12 μm/day. It is not clear just how the osteoclasts move in the process of opening the tunnel. One can imagine each cell starting at the apex of the cutting cone and digging radially, or all the cells digging straight ahead. So little is known of the dynamics of cells in the osteonal BMU that neither possibility has much evidence in its favor, but radial motions would seem to require that the cells either have short lifetimes or recirculate. Whether this is similar in humans in unknown.

How long does an osteoclast live? The average lifetime of osteoclasts is a rather arbitrary concept, for they appear to consist of an evolving syncytium of mononuclear cells that have fused. This was learned initially as a result of experiments in which dogs were injected with tritiated thymidine and the labeled cell nuclei traced [37, 38]. When this was done, the number of labeled osteoclast nuclei increased and then decreased, demonstrating that nuclei enter and leave the cell while it persists in its activity. The average life span of a labeled nucleus in canine osteoclasts is 11 days.

More recently, using live cell microscopy, this process has been observed directly [39]. Fusion of mononuclear cells to each other, with multinucleated cells, and fusion of individual multinucleated cells to each other has been observed. Fission has also been identified, with the osteoclast compartmentalizing one or more nuclei, which either split to become a "new" osteoclast, or die by apoptosis. This constant fission and fusion may serve, much like genetic reshuffling during reproduction, to maintain osteoclast vitality, almost like "hybrid vigor," by mixing different sets of nuclei.

4.8.3 Refilling: In with the New

In some respects, the refilling process is more amenable to analysis than the resorption process, partly because newly mineralizing bone surfaces can be labeled with fluorochrome markers. The basic processes of bone formation within BMUs are clear, but many questions remain about the details.

Osteoblast Origins As mesenchymal cells do not circulate, it seems that osteoblast precursors must come from a self-perpetuating pool of mesenchymal cells near the site of bone formation. Assuming that osteoblast precursor cells are present in the reversal region, there must be mitogenic and chemotactic factors to orchestrate their conversion to functioning osteoblasts on the refilling surface. Using a simple yet elegant experimental approach, Roberts et al. characterized the stages of osteoblast differentiation in terms of nuclear size and used mathematical models to reveal a differentiation pathway through four precursor cell types to the functional osteoblast [40]. They found that, in the rat periodontal ligament, the differentiation process is keyed to five light–dark cycles, so that the entire process requires 60 h or 2.5 days. Furthermore, the transition from stem cells to committed osteoprogenitors to pre-osteoblasts requires the presence of normal levels of mechanical stress (or strain) in the tissues; this has been confirmed in rats subjected to either simulated or actual weightlessness [41]. In an osteonal BMU advancing at 40 μm/day, the cutting cone would move forward 100 μm during the time required for osteoblasts to go through the differentiation process. Allowing some time for events preliminary to those observed by Roberts, these observations are entirely consistent with the fact that reversal regions are typically 100–200 μm long. Abnormally long reversal regions could mean that the osteoblast differentiation process has been compromised, but observations in dogs subjected to disuse suggest that global bone formation may increase with disuse—the number of forming osteons with osteoid actually increases—even though the balance at each individual BMU favors resorption [42]. The effects of disuse on osteoblast differentiation in osteonal BMUs have not yet been observed.

Labeling with tritiated thymidine has revealed that osteonal osteoblasts derive from spindle-shaped cells located near the vascular sinus in the resorption cavity, behind the advancing osteoclasts [38]. These spindle cells are believed to constantly divide to maintain their population while half the new cells differentiate into osteoblasts, a process consistent with Roberts' observations in periodontal ligament and trabecular bone. Furthermore, the differentiation process appears to carry each subsequent group of precursors further from the capillary bud. Once created, osteoblasts begin work as cube-shaped cells occupying an area of bone surface about 15 μm square, and become shorter, flatter, and extend over more bone surface as they work. Ultimately, some of them end up embedded in bone as osteocytes, and others as lining cells, each covering perhaps 300 μm^2 of the finished bone surface.

Jaworski and Hooper also showed that osteoblasts do not move along the filling cone, but remain stationary with respect to the surrounding bone, falling further and further behind the cutting cone [38]. Because the BMU is advancing about 40 μm/

day, and each osteoblast occupies about 15 μm of the closing cone's length, simple geometry suggests that about three new rings of cells must be created each day. If the radius of the resorption cavity is initially 100 μm, its circumference is about 600 μm, and each new ring could contain some 40 osteoblasts. Thus, about 120 osteoblasts must be produced daily to maintain the filling cone.

One or Many Generations? Is the initial ring of osteoblasts responsible for the complete refilling job at its section of the osteon, or do additional generations of osteoblasts follow to complete the work? If the former possibility obtains, then the lifetime of some osteoblasts must reach FP, or about 3 months. Alternatively, several generations of osteoblasts could do the job, each living a fraction of this time.

Frost described "resting" osteonal osteoid seams based on a qualitative change in their histologic appearance when stained with basic fuchsin [43]. He associated these characteristics with lack of tetracycline uptake but normal osteoid thickness, suggesting that both the elaboration of osteoid by osteoblasts and its mineralization ceased simultaneously. This could occur because one generation of osteoblasts has reached the end of its lifetime and another was being recruited to resume the work. If multiple generations of osteoblasts refill the osteon, both resting osteoid seams and lamellar structure could result from the transition from one generation of osteoblasts to another. The exact nature of lamellar structure is controversial, but if one merely accepts the view that completed Haversian systems contain a number of discrete layers of bone between the cement line and the canal, then the boundaries between these layers may somehow be associated with interruptions or pauses in the refilling process.

Johnson suggested that refilling discontinuities involves the replacement of generations of osteoblasts rather than pauses in the activity of the original cells; existing data may be used to test this hypothesis [44]. Using polarized light, Kelin and Frost counted the numbers of lamellae in 350 osteons in normal human ribs; they counted bright layers, which were separated by dark layers [45]. They found 7.9 ± 0.5 lamellar pairs per osteon. The mean wall thickness of an osteon is typically 80 μm. If this wall contains eight lamellar pairs, the average pair thickness is 10 μm. For a mean apposition rate of 1 μm/day, the time required to form the first lamellar pair is 10 days; this can be regarded as a maximum time, as we show that the initial accretion rate is normally much greater than 1 μm/day. Thus, the 11-day period of the Jaworski–Hooper experiment should have seen the labeled osteoblasts well into formation of the second lamellar pair, suggesting that lamellar structure is not correlated with separate generations of osteoblasts. On the other hand, this observation certainly does not prove that refilling is completed by only one generation of cells, and it leaves open the possibility that lamellar structure is associated with mere pauses in osteoblastic activity.

Tam and his co-workers showed that apposition rate measurements in rats yield results that are inversely proportional to the time between labels when this time is greater than 72 h [46]. This result could be explained by the occurrence of 3-day rest periods and longer work periods during refilling. As shown in Fig. 4.12, interlabel times longer than 3 days (e.g., A and B) could produce double-labeled BMUs that had rested between labels. These BMUs would underestimate apposition rate

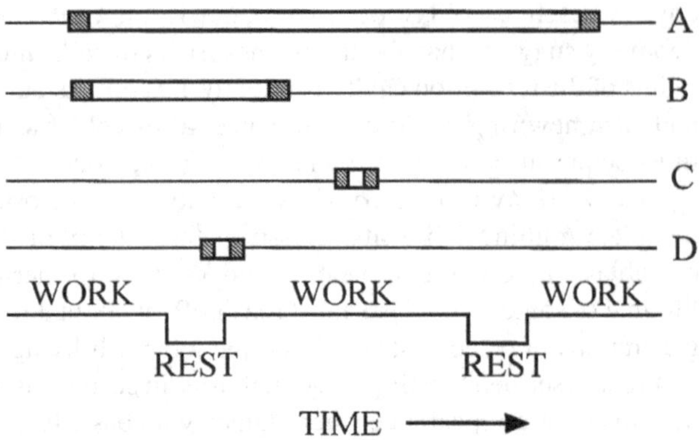

Fig. 4.12 Schematic diagram shows possible relationships between fluorochrome labels and putative osteoblast work-rest cycles in a refilling BMU. Labels of varying intervals are shown as shaded blocks at ends of label intervals on time lines A–D. Assumed work and rest periods for bone formation are shown at bottom

because they would overestimate the time actually devoted to making bone. Double-labeled BMUs obtained with interlabel times shorter than 3 days (e.g., C and D in Fig. 4.12) could not have rested and would thus give accurate results. A 3-day rest would be long enough to include Roberts' 60-h osteoblast differentiation time. Thus, the "rest period" could be either a "sabbatical" for a persistent population of osteoblasts, or a time in which a new generation of osteoblasts is produced.

Apposition Rate Slows During Refilling There has been some controversy regarding the uniformity of the rate at which osteonal BMUs refill. As refilling is initiated, it appears that there is rapid bone formation that "plasters over" the cusps of Howship's lacunae and fills in other eccentricities in the resorption cavity surface. Vincent originally observed that the rate of refilling slows down in proportion to the diameter of the remaining cavity [47]. Other investigators suggested that the apposition rate was constant over most of the refilling period or depended on the diameter of the original osteonal resorption cavity rather than on the age of the osteoblasts or the time since refilling began.

Manson and Waters performed an analysis that resolved the issue [48]. They postulated that the apposition rate, MAR, is not constant and directly proportional to the radius, Rd, of the refilling cavity. That is,

$$\mathrm{MAR} = d(\mathrm{Rd})/dt = -k\,\mathrm{Rd} \tag{4.22}$$

where k is a constant and the minus sign occurs because d(Rd)/dt is directed opposite to the radius vector. The solution to this equation is:

$$\mathrm{Rd} = (\mathrm{Cl.Rd})\ e\text{-kt} \tag{4.23}$$

where Cl.Rd is the cement line radius. Figure 4.13 shows this exponential model's prediction about how refilling would proceed. Formation period, FP, can be estimated by solving for *t* when Rd is the Haversian canal radius, H.Ca.Rd. One obtains:

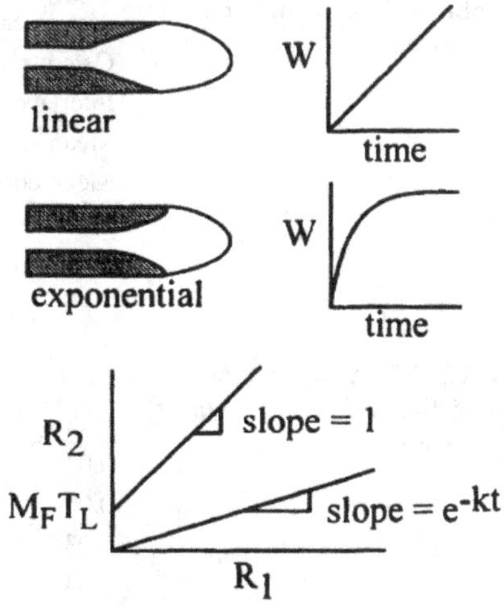

Fig. 4.13 Linear and exponential models for refilling compared. Upper left: The shapes of the refilling surfaces in longitudinal profile. Upper right: Osteon wall thickness, W, as a function of time at a single longitudinal location. Bottom: Plots of the radii of double fluorochrome labels have different slopes and intercepts in the two models

$$\text{FP} = [\ln(\text{Cl.Rd}) - \ln(\text{H.Ca.Rd})]/k \tag{4.24}$$

This expression is quite different from the conventional histomorphometric formula which, for a diaphyseal cross section, takes the form:

$$\text{FP} = (\text{Cl.Rd–H.Ca.Rd})/\text{MAR} \tag{4.25}$$

which assumes a constant apposition rate.

To test the model and obtain values for the constant k, Manson and Waters planned an experiment in which fluorochrome labels would be given at times T1 and T2, producing rings of radii R1 and R2, respectively, in those osteons that were doubly labeled. Consider now the result of plotting R2 vs. R1. If the apposition rate were constant, one would have:

$$\text{R2} = \text{R1–(MAR)(Ir.L.P)} \tag{4.26}$$

where Ir.L.P = T2 − T1 is the interlabel interval. The plot of R2 vs. R1 should produce a straight line with a slope of 1.0 and an intercept proportional to MAR and the interlabel time, Ir.L.P. From their exponential model, Manson and Waters obtained a distinctly different expression:

$$\text{R2} = \text{R1}\ \exp(-k\ [\text{Ir.L.P}]) \tag{4.27}$$

which indicates the plot should pass through the origin and have a slope determined by k and Ir.L.P. Upon executing their experiment in cats of two different ages, and analyzing several different bones, they found that their data clearly fit Eq. 4.27 much better than Eq. 4.26 (Table 4.4). The intercepts of the R2 vs. R1 plots were nearly zero, and the slopes were significantly different from unity. Thus, the data confirmed the hypothesis that the apposition rate declines exponentially during refilling.

Table 4.4 Apposition rate experiments

Age of cat, years	0.75	2.5
Osteons analyzed	101	169
Interlabel time, days	14	28
Mean slope	0.539	0.436
Mean correlation coefficient	0.933	0.943
k, days−1	0.0441	0.0296

Data from Manson and Waters [48]

Table 4.5 Apposition rate modeling: effect of k on other variables

	MAR, μm/day			
k, day^{-1}	Initial	Final	Mean	FPTF, days
0.01	1.0	0.2	0.5	160
0.02	2.0	0.4	1.0	80
0.03	3.0	0.6	1.5	53
0.04	4.0	0.8	2.0	40

4.8.4 Computation of FP

Although the formula for computing FP that arises from the Manson–Waters theory is different from the conventional histomorphometric formula, if the mean apposition rate is used in the conventional formula (Eqs. 4.6 and 4.25), the correct result is nevertheless obtained. The mean apposition rate is related to k by the expression:

$$\mathrm{MAR} = \mathrm{k(Cl.Rd\text{-}H.Ca.Rd)/[ln(Cl.Rd)\text{-}\,ln(H.Ca.Rd)]} \qquad (4.28)$$

Table 4.5 shows values of initial, final, and mean apposition rate, and FP for four possible values of *k*, assuming that the cement line and Haversian canal radii are 100 and 20 μm, respectively.

4.8.5 The Equal Lamellar Work Areas Principle

The fact that osteonal refilling proceeds at an exponentially decreasing rate leads to an important geometric characteristic of the filling cone. As we have seen, there are typically eight lamellar pairs in the osteonal wall, and their thickness is remarkably uniform. In an idealized filling cone, a nonlinear refilling rate leads to a horn-shaped rather than a truly conical surface (Fig. 4.14). Note that the appositional surface of each lamella increases as one proceeds from right to left in the figure (e.g., $x2 > x1$ and $s2 > s1$).

Because the apposition rate declines during refilling, the inner lamellae not only take longer to form than the outer lamellae but they also present more surface on which the cells can work. In fact, it can easily be shown that this longitudinal

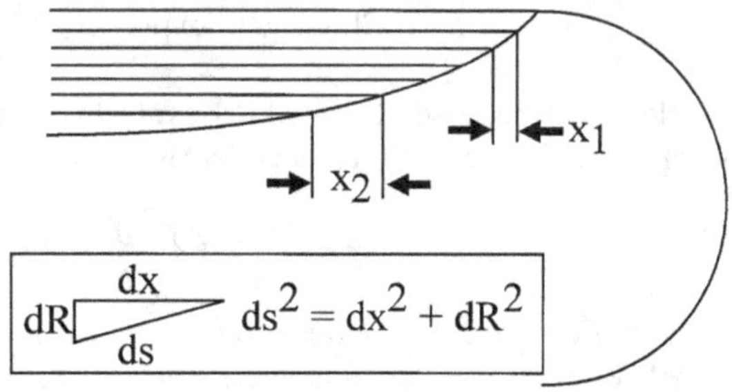

Fig. 4.14 Effect of nonlinear refilling on exposed surface of individual lamellae. Inner lamellae have a broader surface on which formation can occur, but osteoblasts at this point lay down bone more slowly. Inset: Relationship between ds, dx, and dR

extension of the lamellar work surface just compensates the loss of work area resulting from the contracting circumference. To see this, consider the calculation of the surface, dS, of a single lamellar work region bound by longitudinal positions x1 and x2. Because dS = 2 π R(x) ds and ds2 = dx2 + dR2, the surface integral is:

$$S = 2\pi \int_{x_1}^{x_2} R(x)\sqrt{dx^2 + dR^2} = 2\pi \int_{x_1}^{x_2} R(x)\sqrt{1 + (dR/dx)^2}dx \tag{4.29}$$

where x is the longitudinal coordinate and R is the radius of the filling cavity. For exponentially declining refilling where we use R as the current radius:

$$\mathrm{R}(t) = (\mathrm{Cl.Rd})\mathrm{e\text{-}kt} \tag{4.30}$$

This expression can be converted to one for R(x) by using the longitudinal advancement rate, v, of the osteon. That is, substituting

$$t = x/v \tag{4.31}$$

into Eq. 4.26 gives

$$R(x) = (\mathrm{Cl.Rd})e^{-kx/v}, \qquad d(R(x))/dx = -(k(\mathrm{Cl.Rd})/v)e^{-kx/v} \tag{4.32}$$

and substituting these expressions into the integral yields

$$S = 2\int_{x_1}^{x_2} (\mathrm{Cl.Rd})e^{-kx/v}\sqrt{1 - (k(\mathrm{Cl.Rd})/v)e^{-2kx/v}}dx \tag{4.33}$$

It can be shown that the square root term is almost equal to 1 for any value of x and typical values of k (0.02/day), Cl.Rd. (100 μm), and v (40 μm/day). Making such an approximation and performing the integration:

$$S = 2\mathrm{Cl.Rd}(v/k)[\exp(-kx_1/v) - \exp(-kx_2/v)] \tag{4.34}$$

Note that because R1 = Cl.Rd $\exp(-kx_1/v)$ and R2 = Cl.Rd $\exp(-kx_2/\mathrm{v})$, the term in brackets can be replaced by (R_1 − R_2)/Cl.Rd. Therefore:

$$S = 2v(R_1 - R_2)/k \tag{4.35}$$

Now (R1 − R2) is just the lamellar pair thickness, which is nearly constant for all lamellae. Thus, the lamellar work surface, *S*, turns out to be equal to a constant function of *k*, *v*, and lamellar thickness.

Equation 4.35 may be used to estimate the number of osteoblasts that work on each lamellar pair. Letting FP = 90 days, Cl.Rd = 100 μm, and H.Ca.Rd = 20 μm, Eq. 4.24 yields k = 0.0179/day. Then if v = 40 μm/day, and the lamellar pair thickness is 10 μm, $S = 2 \times 3.14 \times 40 \times 10/0.0179 = 140{,}000\ \mu m^2$. Osteoblasts that are rapidly forming bone have a "footprint" on the bone surface which is about 10 μm in diameter; those which are slowly forming bone are about 20 μm across [49]. Assuming, then, that the average osteoblast occupies approximately $15 \times 15 = 225\ \mu m^2$, the number of cells working on each lamella is 140,000/225 = 622.

4.8.6 Why Does the Apposition Rate Slow Down?

We might postulate four mechanisms for declining apposition rates during refilling: (1) A single generation of osteoblasts refills the cavity and its cells slow down, or spend more time resting as the refilling proceeds, from either internal or external factors. (2) Several generations of osteoblasts refill the cavity, and each generation becomes more indolent, owing to genetic or environmental factors. (3) Several generations of osteoblasts refill the cavity, and the time required for differentiation of each new generation is increased, causing increasingly longer "rest periods." (4) Both (2) and (3) occur.

Manson and Waters asserted that osteoblastic activity is enhanced when the cells are further from the blood vessels in the osteonal cavity. They did not offer a rationale for this assertion, which seems contrary to a straightforward notion of cell nutrition via diffusion from a capillary. It is, however, consistent with another observation by Roberts et al. concerning osteoblast differentiation [50]. When a periodontal blood vessel is too close to the bone-forming surface, osteoblast precursors are obliged to differentiate along a path that initially takes them away from the bone. They subsequently circle back to the bone surface. On the basis of this behavior, it could be postulated that as the osteonal resorption cavity is refilled, osteoblast precursors find it increasingly difficult to get enough "elbow room" between the blood vessel and the bone surface, and this affects their rate of differentiation or their subsequent ability to function.

This mechanism, based on spatial factors within the osteonal cavity, is external to the osteoblast; alternatively, it is possible that the decreased rate of osteoid synthesis is caused by factors internal to the cell. This distinction is important because internal actors would be likely to affect osteoblasts in osteonal and surface (i.e., periosteal, endosteal, or trabecular) BMUs alike, but external factors such as cell crowding may be peculiar to osteonal osteoblasts.

Martin and co-workers developed a model that assumes that variations in apposition rate are caused by factors within the osteoblast and compared the abilities of the three available models (linear refilling, slowing from external factors, slowing from internal factors) to describe apposition rates in osteonal and trabecular bone [51]. For osteons, the internal effects model yields essentially the same result as the Manson–Walters model. However, applying the internal effects model to trabecular bone by measuring R1 and R2 from the cement line:

$$\mathrm{R2} = \mathrm{R1}\ \exp(-k\ \mathrm{Ir.L.P}) + \mathrm{C}[1 - \exp(-k\ \mathrm{Ir.L.P})] \tag{4.36}$$

where k and C are constants. In this case, the slope term is similar to that in the Manson–Waters theory, but now there is also an intercept term. Thus, the three apposition rate models yield distinctly different predictions about R1 vs. R2 plots, as shown in Table 4.6.

These theoretical models were tested against R1–R2 data from both osteonal and trabecular BMUs in adult dogs. As shown in Table 4.7, it was found that osteonal apposition rates had slopes significantly less than 1 and intercepts close to 0, but trabecular BMUs had slopes very close to 1 and intercepts nearly equal to (MAR) (Ir. L.P). Thus, apposition rates were found to be constant in trabecular BMUs but not in

Table 4.6 Apposition rate models: predictions about R1 vs. R2 plots

	Slope	Intercept
Linear refilling	1	(MAR)(Ir.L.P)
External factors	<1	0
Internal factors		
Osteonal BMUs	<1	0
Trabecular BMUs	<1	C (1 – (MAR)(Ir.L.P))

Table 4.7 Apposition rates in osteonal and trabecular canine BMUs

	No. of dogs	No. of BMUs	Correlation coefficient	Slope	Intercept	(MAR) (Ir. L.P)
Osteonal BMUs						
Rib	7	88	0.97 ± 0.02	0.77 ± 0.06	−5.3 ± 1.4	15
Humerus	5	140	0.95 ± 0.02	0.70 ± 0.12	−1.5 ± 3.3	14
Trabecular BMUs						
Ilium	5	67	0.99 ± 0.03	1.02 ± 0.05	9.6 ± 1.7	8

Data from Martin et al. [51]

osteons. If internal factors (i.e., internal to the osteoblast) were responsible for the nonlinear refilling in osteons, trabecular osteoblasts should be similarly affected; as they were not, it appears that external factors are responsible for reducing the rate of bone apposition as osteons refill.

In thinking about the external factors that could cause this effect, we may reiterate our previous suggestion that decreasing space between the cavity wall and the blood vessel interferes with osteoblast differentiation. Another possibility has to do with Roberts' observation that the differentiation of osteoblasts is stress dependent. One would expect to find a stress concentration at the front end of a BMU, with diminishing stresses along the refilling region. Perhaps this diminishment of bone strain interferes with osteoblast production or with the regulation of bone formation by the osteoblasts. We now know that sclerostin, a negative regulator of bone formation produced by osteocytes, is inhibited in response to increased levels of strain [52]. The suppression of sclerostin at the head of the cutting cone would be permissive for new bone formation. But as those strains declined along the refilling region of the BMU, the production of sclerostin by osteocytes would return to normal and reduce the amount of bone formation. This is just one regulator of bone formation; it is likely that there are others that might have similar effects (e.g., TGFβ).

Osteoblast Accounting. Another aspect of refilling that appears to be related to both lamellar structure and cell lifetime is the conversion of osteoblasts to osteocytes. Osteocytes are formed when osteoblasts bury themselves in osteoid. Jaworski and Hooper found that the first osteocytes appeared 9 days into the refilling period; this places the first osteocytes near the junction between the first and second lamellar pairs.

It has been argued that because the working surface contracts as osteonal refilling proceeds, some osteoblasts should be forced out into the new bone matrix or back into the resorption cavity [8]. Pursuing this assumption quantitatively, it appears that the number of displaced osteoblasts would be much greater than the number of osteocytes and bone lining cells in a unit length of completed osteon. We first consider what fraction of the osteoblasts become osteocytes using the scheme of cell dynamics presented here. During the formation of each lamellar pair, 1 of every 28 cells gets trapped and becomes an osteocyte. This figure is deduced as follows. Consider the *i*th lamellar pair, which consists of a "pipe" of bone having an outer radius R_o and an inner radius R_i. Segments of the length of this pipe may be thought of as being formed by successive rings of osteoblasts arranged around the BMU cavity. If $b = 15$ μm is the breadth of the work area of an individual osteoblast, then the volume of such a segment of bone formed by a single ring of osteoblasts is:

$$V = \pi b \left(R_o^2 - R_i^2\right) \tag{4.37}$$

The average density of osteocytes in cortical bone is 12,000–20,000/mm^3 [53, 54]. Let Ot.Dn be expressed as cells/μm^3 and let its average value be 16 × 10–6. Then the expected number of osteocytes (N.Ot) in the pipe segment of volume V is:

$$\text{N.Ot} = (\text{Ot.Dn})(\text{V}) = \pi\, b\, (\text{Ot.Dn})\left(R_o^2 - R_i^2\right) \quad (4.38)$$

The average circumference of the work surface during formation of the pipe segment is:

$$C = 2\pi[(R_o + R_i)/2] \quad (4.39)$$

Assuming there is an osteoblast every b μm along this circumference, the average number of cells in the osteoblast ring would be:

$$\text{N.Ob} = \pi(\text{R}_o + \text{R}_i)/b \quad (4.40)$$

Combining and simplifying Eqs. 4.38 and 4.40, the ratio of the number of osteocytes buried in the pipe segment to the number of osteoblasts that formed the segment is:

$$\text{N.Ot/N.Ob} = (\text{Ot.Dn})\ (R_o – R_i)\ b2 \quad (4.41)$$

Note that $R_o - R_i$ is just the lamellar pair width, which can be assumed to be about 10 μm in all cases. Therefore, the fraction of osteoblasts converted to osteocytes is also constant for all lamellae and works out to be N.Ot/N.Ob = (16 × 10–6)(15)2 (10) = 0.036 = 1/28.

Assuming the cement line radius was 100 μm, and the osteoblasts averaged 10 μm in diameter at the start of refilling, the original ring of osteoblasts would have contained about 2π × 100/10 = 63 cells. Then only a couple of these cells could be expected to end up as osteocytes. If the radius of the completed Haversian canal is 20 μm, and it is covered by bone lining cells having a diameter of about 20 μm, there would be about 2π or 6 lining cells left from the original ring of 63 osteoblasts. That leaves about 55 osteoblasts unaccounted for, assuming that a single generation of cells accomplishes the refilling. One interesting possibility is that these cells are "surfing" on the advancing wave of new bone, which is, translating forward in the wake of the advancing resorption cavity, which would reduce the numbers of new osteoblasts that have to be produced. Alternatively, some of these cells may be forced off the shrinking refilling surface and die, or become lining cells. What actually happens is unknown.

4.9 Long-Term Effects of Remodeling Implications for the Aging Skeleton

Activation frequency in the human rib declines from very high levels in childhood to a minimum at about age 35 years and then rises and falls again at the end of an average lifespan (see Fig. 4.6). Unfortunately, such data are not available for long bones such as the femur or for trabecular bone. It is assumed, however, that this curve is generally descriptive of Ac.f vs. age, although the magnitudes of the values would be different in other bones.

Anyone who has studied dynamic systems theory will be struck by the similarity of this curve to that describing the behavior of a damped oscillation. One way that such oscillations can occur is when a feedback system is "overdriven," or operated with too much feedback. Whether the variation in activation rate has anything to do with an overdriven regulatory system remains to be learned, but it is certain that the shape and scale of this curve have a number of important consequences.

One of the most obvious of these consequences is that the rate of addition of new osteons, and the rate of turnover of cortical bone, is much higher in children than in adults. Most cortical bone is created by modeling processes associated with growth. The high values of Ac.f in children ensure that most of this bone is rapidly converted to secondary osteonal bone. One may postulate that later, during adulthood, Ac.f is maintained at some minimal level by systemic factors and is increased only under conditions of mechanical or metabolic need.

It appears that the difference between the values of Ac.f in children and adults is greater than the difference in levels of activity or mechanical usage in these two groups. The elevated activation rate is generally assumed to be a growth hormone-driven phenomenon, but part of the elevation may be caused by age-related variations in material properties of the bone matrix. The rapid growth and high rate of remodeling of children's bone cause it to be undermineralized. This deficit results in a relatively low elastic modulus, producing higher strains for a given load or level of stress. Nunamaker et al. have shown that peak strains in racehorses approach 5000 microstrain but are reduced to less than 3000 microstrain after maturity [55]. Strain levels this high may cause increased fatigue damage and thus increased activation of new BMUs.

After declining to a minimum at about age 35, the BMU activation rate increases again to about twice the minimal value in men and women in their sixties. This increase tends to amplify any deficit in remodeling at the BMU level that occurs in adults as well as that which may be associated with aging itself. The final decline in Ac.f in people in their eighties and nineties tends to reduce the effects of negative bone balance at each remodeling sites.

4.9.1 Osteonal Overlapping

Osteon accumulation is limited by the fact that osteons soon begin to overlap one another as more and more are added. When an osteon overlaps another, it is not prevented from "obliterating" (i.e., refilling) its predecessor's Haversian canal. While in general new osteons create new canals, they may also destroy an old canal. Age-related changes in porosity therefore are affected by the probability that old canals will be obliterated.

The analysis of osteon accumulation and overlapping is a geometric problem not unlike the one astrogeologists have encountered in trying to estimate planetary age from the numbers and overlapping of craters seen on the surface. Physical anthropologists have used osteon population density to estimate the age at death of skeletal remains. A mathematical theory can help clarify the nature of age-related osteoporosis in cortical bone. It can also allow one to use a simple census of osteons to estimate activation and bone turnover rates—information that normally can only be obtained from vital labeling of living animals.

4.9.2 The Random Remodeling Assumption

Because osteons are quite long in comparison to their diameter, their overlapping must be determined as much by the "steering" of the BMU as by its point of origin. It has been hypothesized that osteonal orientation is controlled by the principal strain direction, but it is also reasonable to think that it is affected by variations in the material properties or chemical composition of the bone which appears before the advancing osteoclasts. That is, regions of increased mineral content, patches of osteoid, volumes of damaged tissue and canals could conceivably alter the path. As all these factors may be expected to be at least a little different for the new BMU than for its predecessors, one would expect the steering of an osteon to be different from its nearby older neighbors, leading to quasirandom positioning of each new osteon. This "local randomness" does not imply that new osteons are randomly positioned with respect to the whole cross section but only with respect to local organization. Such local randomness is an underlying assumption in most analyses of the effects of osteonal remodeling on such tissue properties as porosity and mean tissue age.

4.9.3 Mathematical Theories

The problem of determining the average age of a region of bone (tissue age, or mean skeletal age) is very closely linked to that of osteon accumulation and has been discussed in some detail by Hattner and Frost [56]. Tissue age depends on the rate of remodeling and the degree to which new osteons obliterate those previously formed. We now examine the effects of the accumulation of osteons inside a cortical bone cross section using probability theory.

Suppose that within an area Tt.Ar of cross section a new resorption space is equally likely to appear at any location (Fig. 4.15). The area of existing Haversian canals that this resorption space can be expected to obliterate is, on the average:

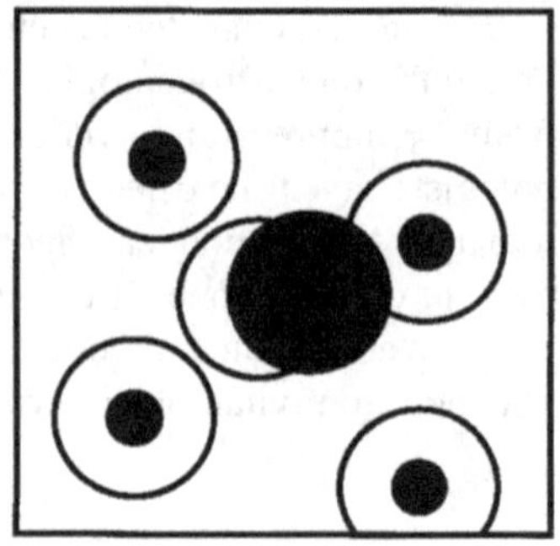

Fig. 4.15 A new resorption space (black circle) appears at a random location in a field of osteons in a cross section having total area Tt.Ar

$$\mathrm{Ar} = (\mathrm{Rs.Ar})(\mathrm{Po}) \tag{4.42}$$

where Rs.Ar is the area of the resorption space (which will be equivalent to the area of the final osteon, On.Ar) and Po is the porosity (total area of Haversian canals/total area, Po = H.Ca.Ar/Tt.Ar) of the region. The resorption space will refill to leave a Haversian canal of area H.Ca.Ar. The net change in void area to be expected is, when averaged over many such incidents:

$$\mathrm{d(H.Ca.Ar)} = \mathrm{H.Ca.Ar} – (\mathrm{On.Ar})(\mathrm{Po}) \tag{4.43}$$

The number of new osteons added to the section per square millimeter per year is just the activation frequency, Ac.f. If one defines an osteon area ratio, f, as H.Ca.Ar/On.Ar, then in an interval of time dt one has:

$$\mathrm{d[(N.On)(H.Ca.Ar)]} = (\mathrm{f\text{-}Po})\ (\mathrm{Ac.F})\ (\mathrm{On.Ar})(\mathrm{Tt.Ar})\ dt \tag{4.44}$$

where Tt.Ar is the total area of the region and N.On is the total number of osteons in the region. Dividing by (Tt.Ar)dt, one has:

$$\mathrm{d(Po)/dt} = (\mathrm{f–Po})(\mathrm{Ac.f})(\mathrm{On.Ar}) \tag{4.45}$$

where $d(\mathrm{Po})/dt$ = (d/dt)[(N.On)(H.Ca.Ar)/Tt.Ar] is the time rate of change of porosity resulting from Haversian canals. If f, Ac.f, and On.Ar can be assumed to be constant, Eq. 4.45 has the simple exponential solution:

$$\mathrm{Po(t)} = \mathrm{f}–(\mathrm{f–Po0})\exp(-(\mathrm{Ac.f})(\mathrm{On.Ar})\mathrm{t}) \tag{4.46}$$

where Po0 is the initial porosity. When this equation is plotted, it is apparent that as t becomes large, the porosity of the bone does not increase indefinitely, but asymptotically approaches f, which may be viewed as the porosity of each osteon. The equation may be used to study the ways in which osteon dimensions and activation frequency influence the rate of increase of porosity.

Equation 4.46 can be converted to a formula for the number of Haversian canals per unit area (H.Ca.N/Tt.Ar.Tt) by substituting f = H.Ca.Ar/On.Ar and Po = [(N.On)(H.Ca.Ar)/Tt.Ar.Tt]. The result is:

$$\text{H.Ca.N}/\text{Tt.Ar} = 1/(\text{On.Ar}) – (1/\text{On.Ar} – [\text{N.On}/\text{Tt.Ar.Tt}]0\,)\exp(\text{-}(\text{Ac.f})(\text{On.Ar})t) \tag{4.47}$$

or, if the initial osteon density, [N.On/Tt.Ar]0, is negligible:

$$\text{H.Ca.N}/\text{Tt.Ar} = [1 - \exp(-\,(\text{Ac.f})(\text{On.Ar})t)]/\text{On.Ar} \tag{4.48}$$

This variable also is limited in its growth, and one can see that the maximum number of Haversian canals (equal to the number of osteons) that can be fit into total area Tt.Ar is equal to Tt.Ar/ On.Ar—exactly the number of times the osteon area divides into the total area. The initial rate at which osteon density increases turns out to be a function only of activation frequency; subsequently, smaller osteons cause N.On/Ar to grow at a faster rate.

Also of interest is the fractional area of a section converted to secondary osteons or their fragments after a given period of time. This fraction, On.Ar/Tt.Ar, can be calculated as follows. Suppose an increment of resorption space of area d(Rs.Ar) appears in a field of area Tt.Ar of which On.Ar/Tt.Ar has already been remodeled. The expected portion of On.Ar/Tt.Ar that would fall on the already remodeled bone and be removed by the new resorption space is:

$$\text{dA} = \text{d}(\text{Rs.Ar})(\text{On.Ar}/\text{Tt.Ar}) \tag{4.49}$$

The expected increase in On.Ar is then the fraction of the new space that falls onto the unremodeled portion of the tissue:

$$\text{d}(\text{On.Ar}) = (1\text{-On.Ar}/\text{Tt.Ar})[\text{d}(\text{Rs.Ar})] \tag{4.50}$$

In an increment of time dt, the number of new resorption spaces appearing on the section would be (Ac.f)(Tt.Ar)dt and the area of each would be On.Ar, so that d(Rs.Ar) = (Ac.f)(Tt.Ar)(On.Ar)dt. Over this period:

$$d[\text{On.Ar}] = (1 - [\text{On.Ar}/\text{Tt.Ar}])(\text{Ac.f})(\text{Tt.Ar})(\text{On.Ar})dt \tag{4.51}$$

Reorganizing the equation, one may rewrite this as:

$$d[\text{On.Ar}/\text{Tt.Ar}]/dt = (1 - [\text{On.Ar}/\text{Tt.Ar}])(\text{Ac.f})(\text{On.Ar}) \tag{4.52}$$

Assuming that On.Ar/Tt.Ar.Tt was initially zero, this differential equation has the solution:

$$\text{On.Ar}/\text{Tt.Ar} = 1 - \exp(-(\text{Ac.f})(\text{On.Ar})t) \tag{4.53}$$

where t is the time during which the remodeling has occurred. On.Ar/Tt.Ar asymptotically approaches 100%, with the initial rate of increase being equal to the product of activation frequency and osteon size.

The fraction of the cross-sectional area occupied by primary (i.e., unremodeled) bone would be

$$1 - \text{On.Ar}/\text{Tt.Ar} = \exp(-(\text{Ac.f})(\text{On.Ar})t) \tag{4.54}$$

Naturally, this decreases at the same rate that On.Ar/Tt.Ar increases, eventually reaching zero.

This kind of analysis shows the essential ways in which mechanically important variables such as porosity and osteon density depend on remodeling parameters such as osteon dimensions and activation frequency. If the models can be substantiated by comparison with experimental data, they have the advantage that, unlike empirical correlations between the variables (e.g., polynomial curves), they lead to an understanding of the mechanisms underlying the phenomena.

The derivations of these equations assumed that On.Ar, H.Ca.Ar, and Ac.f are constant. Are these assumptions valid? H.Ca.Ar and On.Ar do not vary much over the course of a lifetime, but we have seen that the activation frequency can vary greatly, contrary to the assumption made in deriving the equations. Ac.f may be 20–30 times greater in childhood than in adults and differs across skeletal sites. This problem can be treated by using computer models that incorporate age- and site-corrected values of Ac.f into the aforementioned integrals. On the other hand, Eqs. 4.46, 4.47, 4.53, and 4.54 are useful in showing the basic exponential nature of osteonal accumulation, and they may be reasonably accurate in those regions of cortical bone that are created in late childhood, so that the variations in Ac.f are relatively small over an extended period of time. In general, the outer cortex of most long bones is formed in the latter stages of growth, and many bones "drift" in a particular direction so that much of the adult cross section contains little or no bone from early childhood. For example, most of the adult rib cross section is created after age 12 years [57]. By comparing the foregoing equations to experimental data, one can find out how this kind of theory holds up in the face of age variations in Ac.f and On.Ar.

There are two important applications for these relationships. First, the increase in osteon numbers with age means that age-at-death may be predictable on the basis of the osteon population density or the fraction of the cortex occupied by secondary osteons. Second, the activation frequency may be estimated from the rate at which these variables increase with age. One may also compare the values of these variables with those determined by histomorphometric methods (which have their own limitations, as we have discussed.)

4.9.4 *Comparison of the Theoretical Results with Experimental Data*

Kerley pioneered the use of osteon population density measurements to determine age from skeletal remains [58]. Figure 4.16 contains osteon density data that he published for the midshaft of the femur. These data are means for four sites evenly spaced around the outer half of the cortex. The best least-squares regression fit of these data to Eq. 4.48 yields the curve shown in Fig. 4.17. The form of the theoretical curve matches the data very well, and the parameters of the curve allow one to deduce the mean values of On.Ar and Ac.f. (The radius of the osteon, On.Rd, is calculated from On.Ar assuming the osteons are circular.) The results of this analysis are shown in the upper part of Table 4.8, along with results for the tibia and fibula obtained from other graphs in Kerley's paper.

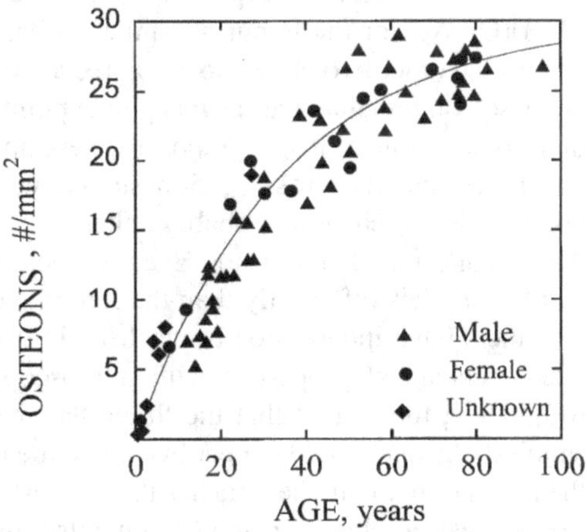

Fig. 4.16 Osteon density vs. age for a sample of human femurs [58]

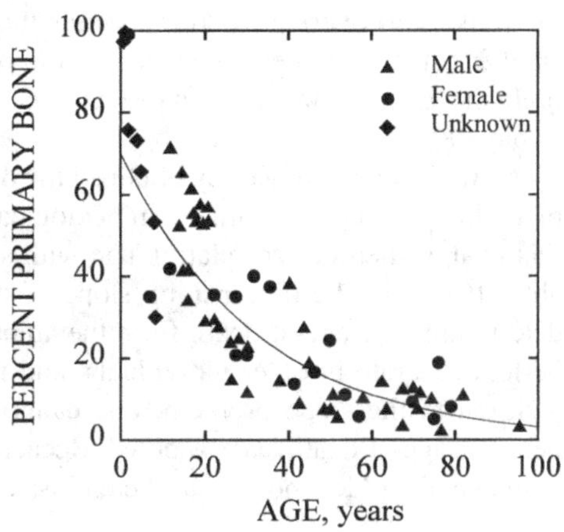

Fig. 4.17 Percent of cortex that is primary bone vs. age for a sample of human femurs [58]

Table 4.8 Bone remodeling parameters calculated from experimental data

Parameter	Femur	Fibula	Tibia
Osteon density data			
Ac.f(On.Ar), #/yr	0.029	0.026	0.018
Ac.f, #/mm^2/yr	0.53	0.46	0.38
On.Rd., μm	133	137	122
Fraction primary bone data			
Ac.f(On.Ar), #/yr	0.031	0.036	0.026
Fraction primary	0.70	0.55	0.60

From Kerley [58]. Corrected according to data by Kerley and Ubelaker [59]

Kerley also published graphs relating age to the percent of cross-sectional area occupied by primary bone. Fitting Eq. 4.54 to these graphs yields a second estimate of (Ac.f)(On.Ar) for the same set of samples. Figure 4.17 shows the femoral example, and the second set of (Ac.f)(On.Ar) values appears in the bottom of Table 4.8. The form of Eq. 4.54 again matches the data, and the two estimates of (Ac.f)(On.Ar) for the femur are quite similar. The separate estimates of (Ac.f)(On.Ar) for the other two bones do not agree as well, but this may be partly because they are based on less than half as many data points. Also, while the points in the percent of primary bone vs. age graphs suggest an initiation value of 1, the regression concludes that the cross section started with a significant amount of secondary bone. This conclusion probably reflects the inaccuracy resulting from assuming Ac.f is constant when it is not. Even so, the consistently lower k of the tibia suggests that it remodels differently than the femur and fibula.

Others have independently reached similar conclusions. Evans and Bang correlated mechanical properties with microscopic structure in the same three bones [60]. They, too, found that the fibula has the largest osteons and the tibia has the smallest. As one would expect from this, the fibula had the fewest osteons/mm^2, but the femur rather than the tibia has the greatest osteon density. The tibia, however, has the greatest number of osteon fragments/mm^2, suggesting a difference in the way osteons are positioned in these two bones. That is, new osteons in the tibia seem to be more likely to overlap existing osteons than those in the femur, creating more and smaller osteon fragments between whole osteons. With aging, the femur exhibited both the greatest increase in osteon density and the highest Ac.f estimate (see Table 4.8).

As we have noted, Kerley obtained the data in Figs. 4.16 and 4.17 for the purpose of finding the age-at-death of archeological bone specimens. Others have pursued this goal, sometimes correlating age with several different histomorphometric variables [61–65]. The diminishing slope of the graphs in these figures suggests that determining the age-at-death from the amount of remodeling that has occurred will be less accurate in older individuals, and this is what Walker et al. found [66]. In addition to the slope problem, one cannot be sure that the bone age reflects the person's age-at-death, as was noted. Because the diaphyses of adults expand radially with age, the oldest bone near the endosteal surface is preferentially removed; thus, the average age of the remaining bone may be considerably younger than the person. These limitations serve to illustrate the complex relationship between aging and remodeling in cortical bone, which depends on many other factors as well, including nutrition, activity, genetics, body size, and personal behavior (e.g., smoking and drinking). The prevalence of anti-remodeling agents also will have dramatic effects on these concepts in future studies.

The activation frequencies in Table 4.8 are all lower than those for adult rib shown in Table 4.2. There is a paucity of histomorphometrically measured values (i.e., using tetracycline labels) for the activation frequency of osteonal BMUs in bones other than the rib. One reason for this is that Ac.f is so low in cortical bone of the extremities of adults that a great many sections must be examined to find enough remodeling sites for a valid measurement. The theory presented here offers a way to estimate Ac.f for such low-turnover regions and to compare remodeling in one bone with that in another.

4.9.5 Calculating Activation Frequency Without in Vivo Labels

Suppose that you dug up a mastodon bone in your garden and wanted to know what its average bone formation rate had been. Could you estimate this without benefit of double tetracycline labeling? It is possible to do this by counting the number of osteons/mm^2 in a section and assuming that the total number of osteons created is the product of the activation frequency and the age of the bone [57]. However, you have to find a way to estimate the number of osteons that were created but have since been obliterated by more recent remodeling, so that you can add them to the currently visible osteons to get the grand total. This is shown in the accompanying graph of osteon density (On.Dn) vs. bone age (t). As in Fig. 4.18, the curved line represents the visible osteons at any t; it has an asymptote at On.DnA. The straight line represents the total number of osteons created, On.DnT(t). The difference between the two lines is the number of osteons obliterated. Frost developed a simple algorithm for finding N.OnT when you have measured On.Dn and know the asymptotic value, On.DnA [67]. First, calculate $\alpha = \text{On.Dn}/\text{On.DnA}$. Then compute $\beta = (1 - \alpha 3.5) - 1$, and $\text{On.DnT} = \beta\ \text{On.Dn}$. If you know the age of the bone, t, the activation frequency is estimated as $\text{Ac.f} = \text{On.DnT}/t$ and the bone formation rate as $\text{BFR} = (\text{Ac.f})(\text{On.Ar})$, where On.Ar is the mean osteon area. Stout and Paine compared this method's results to measurements made with double tetracycline labels in human ribs [68]. They found that Frost's method gave similar but somewhat lower results, especially in people more than 40 years old. Similar techniques have been used to estimate remodeling rates in human patients [57], nonhuman primates [69, 70], and modern archeological populations [71].

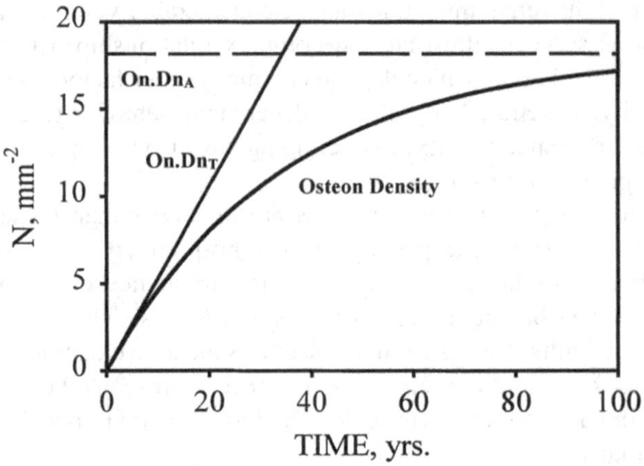

Fig. 4.18 Theoretical plot of On.Dn (osteons/mm^2) vs. t (time) for a bone cross section having a constant activation frequency. The curve On.Dn(t) is that obtained from Eq. 4.48. It has an asymptote at On.Dn = On.DnA = 1/On.Ar. The line On.DnT(t) represents the total number of osteons/mm2 that have been created and would be seen if new osteons never obliterated old ones. Its slope is the activation frequency, Ac.f. The number of observed osteons, On.Dn, is always less than On.DnT and On.DnA

Frost's algorithm is empirical, based on the asymptotic nature of On.Dn(t) as seen in graphs like those of Kerley [58]. One may derive an alternative procedure on the basis of our analysis of randomly overlapping osteonal remodeling. According to Eq. 4.48, the asymptote of the On.Dn(t) curve is On.DnA = 1/On.Ar. Thus, On.DnA can be actually be obtained by measuring On.Ar in the specimen of interest, assuming that this area does not change appreciably with age. Solving Eq. 4.48 for Ac.f, one obtains:

$$\mathrm{Ac.f} = -\ln[1 - (\mathrm{On.Ar})(\mathrm{On.Dn})]/[(\mathrm{On.Dn})\ (t)] \tag{4.55}$$

This equation can be solved for the activation frequency if On.Ar and On.Dn are measured directly from the section and the age of the bone, *t*, is known.

An important difficulty in calculating activation frequency using either of these methods is that the chronological age of an individual often exceeds the age of the bone (t) because of drifts that occur during growth. Based on fluorochrome labeling studies, Wu et al. calculated an adjustment of 12.5 years for the middle of the sixth rib to account for the effective "birth" date of the compact bone seen in an adult; i.e., t = person's age −12.5 years [57]. The correction for other human bones is unknown, and the data necessary to calculate it are not available. Therefore, studies using these techniques often assume the bone age is the subject's age or some other constant; this makes comparison between studies difficult, but maintains consistency within a study.

4.10 Fracture Healing

Fracture is the most obvious sign that the mechanical functional capacity of the skeleton has been exceeded. It may be argued that much of skeletal physiology is directed at preventing fracture. Even so, fractures are relatively common and usually occur because of trauma: an accidental overload substantially exceeds the normal range of loading to which the bone has adapted during its growth and development.

Pathologic fractures are caused by normal loading of a bone weakened by disease. The most common pathologic fractures are caused by osteoporosis in the elderly. Osteoporotic fractures are more common in women than men. Bone tumors are another important cause of pathologic fractures. The tissue in a bone tumor is variable in its mechanical properties, but it will almost always be different than the surrounding bone and therefore serve as a stress concentrator. Removal of a bone tumor obviously also places the patient at risk for fracture. Predicting and preventing pathologic fractures is an important issue in orthopedic medicine.

Once a fracture occurs, how does it heal? Today, people seek medical attention for the simplest of fractures, but the basic process of fracture healing obviously evolved as a natural phenomenon. Healed fractures are commonly seen in archaeological skeletons of prehistoric people and in wild animals. Therefore, fracture healing must be a self-controlling process that ordinarily proceeds on its own to reunite the broken bone. The remainder of this chapter is devoted to describing this process.

4.10.1 Basic Concepts

As we have seen, in the embryo and growing child, bone is formed directly by osteoblasts (intramembranous bone formation) and indirectly by the calcification of cartilage and replacement by bone (endochondral ossification). In the healthy adult, the latter process is absent, and bone is formed only by osteoblasts on existing bone surfaces. When a fracture occurs, however, adults as well as children are able to make bone by both intramembranous and endochondral osteogenesis, leading to much more rapid repair of the damaged bone. Unlike skin and other soft tissues, wound healing in bone does not produce a permanent scar. Eventually, the initial repair tissues are replaced by normal lamellar bone. Therefore, while the woven bone and calcified cartilage in the healing fracture can be considered "scar tissue," they are temporary, and the tissues in the healed bone are ultimately indistinguishable from those in the original bone.

A fracture that fails to heal is called a non-union. Frequently, non-unions produce a jointlike cartilaginous structure called a pseudoarthrosis (literally, "false joint"); these pathologic joints are lubricated with synovial fluid. A fracture that does not heal in the expected length of time, but which is still progressing, is called a delayed union. The length of time required to heal a particular fracture depends primarily on

Table 4.9 Typical healing times for common fractures

Bone	Typical healing time, weeks
Distal radius	6
Humeral shaft	12
Tibial shaft	18
Femoral neck	24

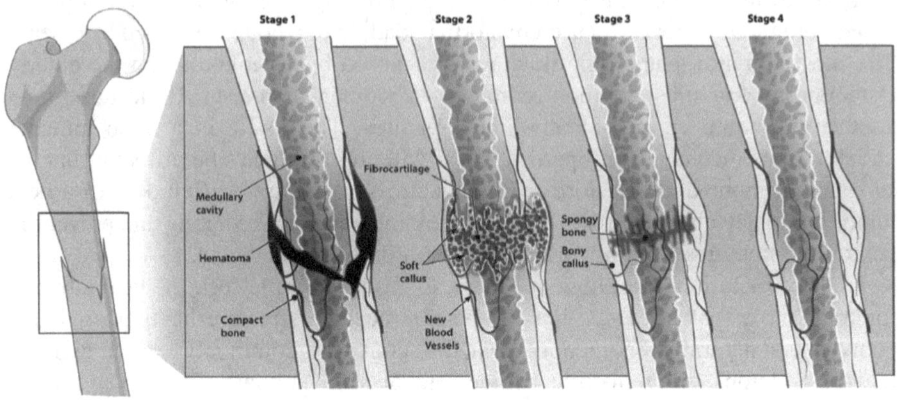

Fig. 4.19 Stages of a healing fracture. (**a**) Initially a hematoma forms at the fracture site, bringing cells to the site for repair. (**b**) These cells rapidly form a bridge across the fracture gap composed of both woven bone. (**c**) A cartilage callus forms to further stabilize the fracture. (**d**) The cartilage callus will gradually calcify through a process of endochondral ossification, further stabilizing the fracture. (**e**) The calcified cartilage gradually remodels to re-establish the integrity of the marrow cavity and both periosteal and endocortical surfaces will model to re-establish the original shape of the bone [72]

its anatomical site, among other things. Ribs, clavicles, and other bones of the torso heal in a few weeks; appendicular bones generally require several months (Table 4.9).

4.10.2 Important Tissues in Fracture Healing

Two tissues provide the cells responsible for the healing of a typical fracture: the periosteum and the marrow of the injured bone. Ordinarily, the osteoblasts of the adult periosteum produce lamellar bone at a relatively slow rate, so that long bones expand very slowly with age. However, even a slight amount of trauma stimulates the periosteum to produce more osteoblasts, which manufacture large amounts of woven bone in a short amount of time. In the case of a fracture, this periosteal woven bone arches over the fracture line and fuses to bridge the gap and makes what is known as the periosteal callus (Fig. 4.19).

The endosteal surfaces of bones are not actually covered by a membrane. Instead, they are lined with bone lining cells; this is true whether or not the endosteal space contains trabecular bone. In places, the endosteal surface may be experiencing ordinary remodeling, but most of the adult endosteal surface is quiescent. The bone lining cells may be stimulated to initiate bone remodeling by various factors, including mechanical ones, but such remodeling does not play a role in the initial fracture healing response. Instead, trauma to the marrow stimulates the formation of the medullary version of the periosteal callus. It has been shown that, even without fracture of the cortex, mechanical disruption of the marrow activates the differentiation of osteoblasts and the formation of woven bone within the medullary canal [73]. This bone formation is a prerequisite to the regeneration of the hematopoietic function of the marrow and so serves a dual role when fracture occurs. The periosteal callus tends to be more obvious, but the medullary callus is usually considered more important in achieving initial union.

Obviously, the jagged ends of the broken bone, as well as other trauma to the fracture site, can be expected to break many blood vessels. The resulting hematoma was once thought to be the primary source of the cells responsible for fracture healing. It is now thought that most of these cells come from the marrow and periosteum, but under the influence of inflammatory mediators produced in the hematoma.

4.10.3 Three Biological Phases

Healing of a fracture is often described sequentially as having inflammatory, reparative, and remodeling phases which, taken together, form a self-driven sequence that leads to a healed fracture.

Inflammatory Phase. This initial phase serves to immobilize the fractured bone and activates the cells responsible for repair. Immobilization is promoted by pain and by swelling, which in many cases serves to hydrostatically splint the fracture. In addition to hematoma formation, this phase is characterized by vasodilation, serum exudation, and infiltration by inflammatory cells. It lasts 3–7 days. It is important to realize that fracture repair is not accomplished by existing osteoblasts and chondroblasts but depends on the creation of an entirely new work force of cells from a relatively small population of stem cells. The mobilization of this work force requires a variety of chemical mediators and ancillary cell types that are poorly understood. However, it is clear that this process is similar to other biologic cascades. Once set in motion, each step is predicated on what happened earlier, and if a mistake is made, the process may make a wrong turn rather than stopping, with no way to recover the conditions at the point of the error.

It follows that many if not most non-unions or delayed unions are caused by a defect in the initial mobilization of the repair process, and once the problem becomes apparent, there is no way to correct it by manipulating the current group of cells. This is consistent with the fact that the most effective treatment for non-unions is to replace the tissues in the fracture site with a bone graft, which essentially initiates an entirely new repair process. Alternatively, electrical signals (see following) may be able to start a new cascade. Interruption of the normal vascular supply to the bone by the fracture also results in death of the osteocytes in the bone matrix. The significance of this is not clear, but it may initiate the remodeling activity that is important in the last stage of fracture healing.

Reparative Phase In this phase of healing, which lasts about 1 month, the periosteal and medullary calluses are formed by osteoblasts from the periosteum and marrow. The stem cells for medullary osteoblasts are thought to be pluripotential mesenchymal cells in the marrow stroma. These cells are also stem cells for fibroblasts, chondroblasts, and some marrow cells. Mesenchymal cell proliferation and differentiation are accompanied by intense vascular proliferation. The resulting osteoblasts produce woven bone at a high rate.

The fracture site exhibits some degree of endochondral ossification as well as direct bone formation by osteoblasts. This is thought to be related to inadequate blood supply in some portions of the fracture region. In the absence of adequate vascularization, stem cells are thought to differentiate into chondroblasts rather than osteoblasts. An important contributing factor to inadequate vascularization can be insufficient immobilization of the fracture fragments. Alternatively, the nature of the stresses at a site within the callus may affect the differentiation pathway [74]. In any event, the chondrogenic regions of the callus, which tend to be near the fracture line, usually calcify and are replaced by lamellar bone, just as the woven bone regions are. If this does not happen, a pseudoarthrosis may develop.

Regardless of how much of the callus is cartilaginous, it takes a week or two for the woven bone or cartilage to accumulate substantial mineral. This early, rather compliant callus is known as the provisional callus. As calcification proceeds, the callus material becomes increasingly rigid; it is then referred to as the bony callus. The rigidity of the callus is enhanced by the fact that it has a larger cross-sectional area and moment of inertia than the original bone cortex. This means that even though the material in the bony callus is less rigid (and strong) than cortical bone, the rigidity (and strength) of the callus structure can equal or exceed that of the intact bone. When this occurs, bony union has been achieved, completing the reparative phase.

Remodeling Phase Once bony union has been achieved, the broken bone is approximately as strong as the intact bone was, but often it has greater mass than the original bone, and is therefore less mechanically efficient. Modeling and remodeling of the fracture site gradually restore the original contour and internal structure of the bone, although radiographic evidence of the fracture may persist for many years. As the bone remodels, it is able to maintain its strength with less material than in the earlier phases of healing, thereby increasing its mechanical efficiency. The medullary and periosteal calluses are removed, and the remaining material (woven bone or calcified cartilage) is replaced by secondary lamellar bone (cortical or trabecular, as the site dictates). If the fracture has healed with incorrect angulation, this may be partially corrected as well by modeling drifts. In general, modeling and remodeling proceed much more rapidly in children than in adults, and this applies to the remodeling phase of fracture healing as well.

In considering the capacity of a malaligned healed fracture to correct itself, there are two important geometric principles to remember. First, an angulation in the plane of motion of the adjacent joint corrects itself better than one out of the joint plane. Second, rotational errors do not correct themselves very well.

4.10.4 Clinical Assessment of Healing Fractures

It is very important for the physician treating a fracture patient to be able to judge the strength and stiffness of the healing bone. If the cast is removed before the fracture is mechanically competent to bear the loads and moments applied to it, there is risk of refracture or malunion. On the other hand, if the cast is left on too long, there is risk of excessive disuse osteoporosis and contracture of soft tissues about the adjacent joints, not to mention the inconvenience and expense of needless immobilization. The usual clinical methods to assess fracture healing are inspection of radiographs and manipulation to see how stiff the healing fracture feels; both of these are subjective and imprecise. Several groups of investigators have experimented with methods to assess fracture healing in a more quantitative way.

If the fracture has been fixed with an external fixation frame, application of a strain gauge to the fixation frame can allow bone stiffness to be estimated [75, 76]. A second method for measuring the mechanical characteristics of a healing fracture involves the propagation of sound or stress waves through the tissues. The velocity of sound in an elastic bar is $c =$ (E/ρ)1/2 where E is the elastic modulus and ρ is the density of the tissue. The resonant frequency of a vibration in an elastic bar depends on the ratio of c to the length of the bar. When a vibration traveling along the bar reaches a segment having different material properties, a portion of the wave is reflected backward. The magnitudes of the transmitted and reflected waves depend on the ratio of the acoustic impedance, $z = c\ \rho$, of the two materials. These dependencies on the elastic modulus of the tissues are potential means for probing the stiffness of the fracture callus. For example, the velocity of an ultrasonic wave across the fracture site can be measured [77]. As the fracture heals, the wave velocity increases; healing can be assessed by comparison with the contralateral, uninjured bone (Fig. 4.20) Although this is capable of probing the mechanical attributes of a healing fracture, in practice it is cumbersome and imprecise to have come into general clinical use, and their applications have been limited to research.

4.10.5 Four Biomechanical Stages

A standard way to study fracture healing in experimental animals is to mechanically test the healing bone in torsion and compare the result with the stiffness of an intact bone. When this was done with rabbit tibial fractures [78], four stages of healing were identified (Fig. 4.21).

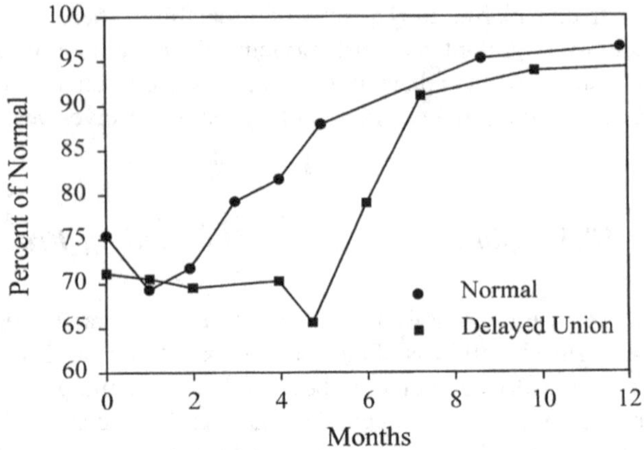

Fig. 4.20 Sound velocity, normalized by the values in the contralateral bone, measured through the tissues in a healing fracture is plotted against healing time. Examples shown represent human tibias with normal (circles) and delayed (squares) healing. (Adapted from [77])

Stage I
Virtually no stiffness is seen, and failure occurs through the original fracture line. This stage corresponds to the inflammatory and early reparative (i.e., provisional callus) phases of healing (days 1–26 in the rabbit tibia).

Stage II
Substantial stiffness is now encountered, but failure still occurs through the original fracture line. This stage corresponds to the middle of the reparative phase when the callus has become bony but still is not so large or calcified as it will become (days 27–49 in the rabbit tibia).

Stage III
There is no further increase in stiffness, but failure is now partially through the original fracture line and partially through intact bone (days 49–56 in the rabbit tibia).

Stage IV
Failure occurs through the intact bone rather than the callus at the fracture site (after day 56 in the rabbit tibia) (Fig. 4.21).

Stages III and IV correspond to the later portion of the reparative phase and the early remodeling phase, when the full bony callus exists but its mineral content and strength are still increasing. Another interesting observation from these mechanical studies of healing fractures is that there is typically a slight reduction in the strength of the fracture callus during stage IV of healing. It is reasonable to think that this corresponds to the beginning of the remodeling phase, when the resorptive portion of BMU activity has begun but substantial refilling of the osteonal cavities is yet to occur.

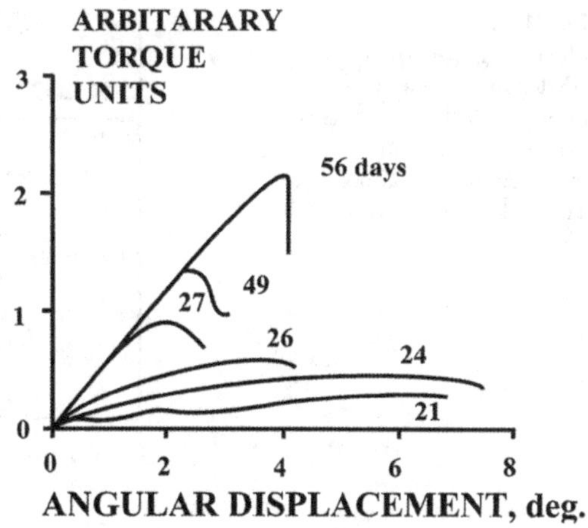

Fig. 4.21 Torsional stiffness of healing rabbit fractures as a function of days of healing [78]

4.10.6 Stability of Fixation

The importance of achieving good stability or immobilization of a fracture depends very much on the anatomical site. For example, ribs heal well even with a great deal of motion—rib non-unions are unknown. On the other hand, tibias and other large bones are very prone to non-union as the result of excessive motion of the fragments. This knowledge must be balanced against the fact that some motion is beneficial: there seems to be an inverse relationship between the amount of motion and callus size, up to a point that has not been defined. In the same vein, applying loads across the fracture (i.e., weight-bearing) serves to strengthen the callus, but obviously too much weight can lead to refracture or excessive motion and non-union. There are no clear rules about how much weight-bearing or motion is safe.

4.10.7 Primary Union

If a fracture is rigidly fixed so that the bone ends are closely approximated, healing may occur by osteonal remodeling across the fracture line (Fig. 4.22). This is known as primary fracture healing. (Therefore, secondary healing is ordinary fracture healing as previously described.) Usually the only way that primary healing can happen is when a plate or screw is used very effectively to internally “fix” (rigidly fasten together the fragments of) the fracture. In these cases, there is almost no periosteal callus.

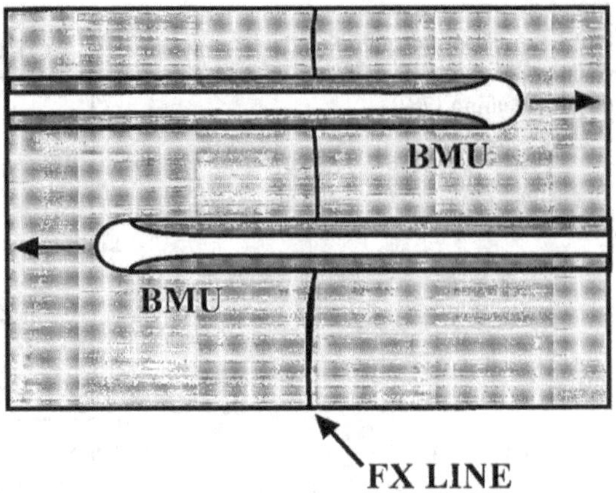

Fig. 4.22 Sketch illustrating how osteonal BMUs may tunnel across a tightly fixed fracture line or osteotomy (saw cut) interface in cortical bone. This is known as primary healing

4.10.8 Posttraumatic Osteoporosis

It is commonly observed clinically that the intact portions of the fractured bone become osteoporotic as healing occurs. This generalized osteopenia of the intact regions, called posttraumatic osteoporosis or posttraumatic bone atrophy, is caused by two factors. First, in addition to the healing response, the fracture causes a remodeling RAP throughout the bone, so that osteonal BMUs riddle the entire cortex with resorption cavities. The second factor is the removal of mechanical loading from the fractured bone. As is seen later, disuse also activates remodeling. If the fracture is well fixed and sufficient load-bearing is resumed, the resorption spaces will refill and the osteopenia will be transient. To the degree that the bone does not resume load-bearing, the atrophy persists.

4.10.9 Electrical Phenomena and Fracture Healing

When bones are mechanically deformed, they become electrically polarized via two mechanisms: (1) the ionic fluids in the calcified matrix are forced to move about, creating streaming potentials [79], and (2) the collagen molecules in the bone tissue are piezoelectric, so that they exhibit a dipole moment when strained [80]. The relative roles of these two phenomena in producing stress-generated potentials in bone are unclear, but it is well established, for example, that when a bone is bent, the concave surface becomes negatively charged and the convex surface becomes positively charged. Furthermore, when a bone is fractured, the fracture site becomes negatively charged with respect to the remainder of the bone. Experiments have shown that bone formation is enhanced in the vicinity of a cathode, and the optimal current for this effect is about 20 microamperes (μA).

Since the 1970s, electrical stimulation has been used as a treatment for non-unions. There are two basic methods of producing such currents [81]. Thin wires were placed into the fracture site to carry direct current, with an anode pad on the skin nearby. Bassett and co-worker passed currents through external coils to induce currents in the tissues noninvasively [82]. Both methods are reported to be about 70–85% successful at healing non-unions, but they require good immobilization and do not work in cases of true synovial pseudarthrosis. Encouraging studies by these investigators and others led to development of several commercially available devices, federally approved for use in cases of stubborn non-unions. The exact mechanisms responsible for enhanced healing have not yet been fully elucidated, and the degree to which electrical stimulation enhances clinical outcomes remains controversial and difficult to define even after several years of clinical use [83].

4.11 Summary and Further Reading

In this chapter, we have tried to present a deeper understanding of bone remodeling. In the process, we have introduced you to several ways to mathematically model components of this phenomenon and its effects. We have seen how Frost originally formulated the remodeling process so that it could be histomorphometrically quantified. In all this, we caution readers to remember that any mathematical approach is based on assumptions that the morphology and physiology can be expressed quantitatively, even though the histologic picture is one of considerable variations. The reason that this variability does not negate the usefulness of mathematical analysis is that, to be useful, the result must only correctly describe the average behavior of the system. The truth of this is aptly illustrated in the case of the histomorphometric measurement of remodeling dynamics and in the work of Manson and Waters [48]. In both cases, conceptual models of osteonal remodeling that assume simple geometric and physiologic features are found to be very successful in describing the average behavior of large numbers of BMUs. It does not matter, for example, that the osteonal apposition rate is not exactly an exponential function, but seems to have an especially rapid early phase. This phenomenon is not apparent when the experimental data are compared to an exponentially varying apposition rate, so it may be regarded as a second-order effect. Second-order effects and histologic variability mark the limitations of remodeling theories, but do not prevent them from being useful tools for understanding physiology and disease.

Further reading should begin with Frost's 1964 book, Mathematical Elements of Lamellar Bone Remodeling [12], which laid the groundwork for this subject. Bone Histomorphometry: Techniques and Interpretation, edited by Recker [31], contains several chapters on the details of quantitative histomorphometric analysis, as well as a chapter by Parfitt that remains an excellent resource for the details of bone remodeling. The recent update by Recker et al. that deals with problematic cases of very low bone turnover is also worth consulting [23]. The standard reference for histomorphometric technique and nomenclature remains another work by Parfitt, who chaired the ASBMR Task Force to standardize nomenclature [21]; updates to this classic work were published in 2013 [30].

References

1. Leung VYL, Gao B, Leung KKH, et al. SOX9 governs differentiation stage-specific gene expression in growth plate chondrocytes via direct concomitant transactivation and repression. PLoS Genet. 2011;7:e1002356. https://doi.org/10.1371/journal.pgen.1002356.
2. Hunziker EB, Schenk RK, Cruz-Orive LM. Quantitation of chondrocyte performance in growth-plate cartilage during longitudinal bone growth. J Bone Jt Surg. 1987;69:162–73. https://doi.org/10.2106/00004623-198769020-00002.
3. Horton W. Morphology of connective tissue: cartilage. In: McCusick V, editor. Connective tissue and its heritable disorders. Wiley-Liss Inc; 1993. p. 73–84.
4. Parfitt AM. Misconceptions (1): epiphyseal fusion causes cessation of growth. Bone. 2002;30: 337–9. https://doi.org/10.1016/s8756-3282(01)00668-8.
5. Whyte MP, McAlister WH, Novack DV, et al. Bisphosphonate-induced Osteopetrosis: novel bone modeling defects, Metaphyseal osteopenia, and Osteosclerosis fractures after drug exposure ceases. J Bone Miner Res. 2008;23:1698–707. https://doi.org/10.1359/jbmr.080511.
6. Seeman E. Periosteal bone formation – a neglected determinant of bone strength. N Engl J Med. 2003;349:320–3. https://doi.org/10.1056/nejmp038101.
7. Parfitt AM. Quantum concept of bone remodeling and turnover: implications for the pathogenesis of osteoporosis. Calcif Tissue Int. 1979;28:1–5.
8. Parfitt AM. The physiologic and clinical significance of bone histomorphometric data. Boca Raton: CRC Press; 1983.
9. Fuchs RK, Allen MR, Ruppel ME, et al. In situ examination of the time-course for secondary mineralization of Haversian bone using synchrotron Fourier transform infrared microspectroscopy. Matrix Biol. 2008;27:34–41.
10. Ferron M, Wei J, Yoshizawa T, et al. Insulin signaling in osteoblasts integrates bone remodeling and energy metabolism. Cell. 2010;142:296–308.
11. Fuzele K, Riddle RC, DiGirolamo DJ, et al. Insulin receptor signaling in osteoblasts regulates postnatal bone acquisition and body composition. Cell. 2010;142:309–19.
12. Frost HM. Mathematical elements of lamellar bone remodelling. Springfield: Thomas; 1964.
13. Martin RB. On the significance of remodeling space and activation rate changes in bone remodeling. Bone. 1991;12:391–400. https://doi.org/10.1016/8756-3282(91)90028-h.
14. Pødenphant J, Engel U. Regional variations in histomorphometric bone dynamics from the skeleton of an osteoporotic woman. Calcif Tissue Int. 1987;40:184–8. https://doi.org/10.1007/bf02556619.
15. Kimmel DB, Jee WSS. A quantitative histologic study of bone turnover in young adult beagles. Anat Rec. 1982;203:31–45.
16. Orwoll ES. Perspective: toward an expanded understanding of the role of the periosteum in skeletal health. J Bone Miner Res. 2003;18:949–54.
17. Bliziotes M, Sibonga JD, Turner RT, Orwoll E. Periosteal remodeling at the femoral neck in nonhuman primates. J Bone Miner Res. 2006;21:1060–7.
18. Minaire P, Edouard C, Arlot M, Meunier P. Marrow changes in paraplegic patients. Calcif Tissue Int. 1984;36:338–40.
19. Martin RB, Chow BD, Lucas PA. Bone marrow fat content in relation to bone remodeling and serum chemistry in intact and ovariectomized dogs. Calcif Tissue Int. 1990;46:189–94.

20. Frost HM. Preparation of thin undecalcified bone sections by a rapid manual method. Stain Technol. 1958;33:273.
21. Parfitt AM, Drezner MK, Glorieux FH, et al. Bone histomorphometry: standardization of nomenclature, symbols, and units. J Bone Miner Res. 1987;2:595–610.
22. Mouton PR. Unbiased stereology: a concise guide. Baltimore: Johns Hopkins University Press; 2011.
23. Recker RR, Kimmel DB, Dempster D, et al. Issues in modern bone histomorphometry. Bone. 2011;49:955–64.
24. Mashiba T, Hirano T, Turner CH, et al. Suppressed bone turnover by bisphosphonates increases microdamage accumulation and reduces some biomechanical properties in dog rib. J Bone Miner Res. 2010;15:613–20. https://doi.org/10.1359/jbmr.2000.15.4.613.
25. Schock CC, Noyes FR, Villanueva AR. Measurement of haversian bone remodeling by means of tetracycline labeling in rib of rhesus monkeys. Henry Ford Hosp Med Bull. 1972;20:131–44.
26. Frost HM. Tetracycline-based histological analysis of bone remodeling. Calcif Tissue Res. 1969;3:211–37.
27. Martin RB. Label escape theory revisited: the effects of resting periods and section thickness. Bone. 1989;10:255–64. https://doi.org/10.1016/8756-3282(89)90062-8.
28. Frost HM. Bone histomorphometry: correction of the labeling "escape error". Boca Raton: CRC Press; 1983.
29. Li XJ, Jee WSS, Chow SY, Woodbury DM. Adaptation of cancellous bone to aging and immobilization: a single photon absorptiometry and histomorphometric study. Anat Rec. 1990;227:12–24.
30. Dempster DW, Compston JE, Drezner MK, et al. Standardized nomenclature, symbols, and units for bone histomorphometry: a 2012 update of the report of the ASBMR Histomorphometry Nomenclature Committee. J Bone Miner Res. 2013;28:2–17. https://doi.org/10.1002/jbmr.1805.
31. Recker RR. Bone histomorphometry, techniques and interpretation. Boca Raton: CRC Press; 1983.
32. Cohen J, Harris WH. The three-dimensional anatomy of haversian systems. J Bone Joint Surg. 1958;40:419–34.
33. Tappen NC. Three dimensional studies of resorption spaces and developing osteons. Am J Anat. 1977;149:301–32.
34. Martin RB. Osteonal remodeling in response to screw implantation in canine femora. J Orthop Res. 1987;5:445–52.
35. Rodan GA. Introduction to bone biology. Bone. 1992;13:S3–6.
36. Jaworski ZF, Lok E. The rate of osteoclastic bone erosion in Haversian remodeling sites of adult dog's rib. Calcif Tissue Res. 1972;10:103–12.
37. Jaworski ZFG, Duck B, Sekaly G. Kinetics of osteoclasts and their nuclei in evolving secondary haversian systems. J Anat. 1981;133:397–405.
38. Jaworski JFG, Hooper C. Study of cell kinetics within evolving secondary haversian systems. J Anat. 1980;131:91–102.
39. Ineke DCJ, Vermeer JAF, Bloemen V, et al. Osteoclast fusion and fission. Calcif Tissue Int. 2012;90:515–22.
40. Roberts WE, Mozsary PG, Klingler E. Nuclear size as a cell-kinetic marker for osteoblast differentiation. Am J Anat. 1982;165:373–84.
41. Barou O, Lafage-Proust MH, Palle S, et al. Effects of immobilization on preosteoblast proliferation assessed histomorphometrically using BRDU uptake on epon-embedded rat bone (abstract). Bone. 1996;19:131S.
42. Li CY, Price C, Delisser K, et al. Long-term disuse osteoporosis seems less sensitive to bisphosphonate treatment than other osteoporosis. J Bone Miner Res. 2005;20:117–24.
43. Frost HM. Observations on osteoid seams: the existence of a resting state. Henry Ford Hosp Med Bull. 1960;8:220–4.

44. Johnson LC. Morphologic analysis in pathology: the kinetics of disease and general biology in bone. Boston: Little, Brown; 1964.
45. Kelin M, Frost HM. Evidence of periodic changes in the rate of formation of individual osteons in human bone. Henry Ford Hosp Med Bull. 1964;12:565–72.
46. Tam CS, Harrison JE, Heersche JNM, et al. Short term variation in the rate of apposition of mineralized bone matrix in small animals. In: Jee WSS, Parfitt AM, editors. Bone histomorphometry. Third international workshop. Paris: Societe Nouvelle Publications Medicales et Dentaires; 1980.
47. Vincent J. Recherches sur la Constitution de l'Os Adult. Brussels: Arscia; 1955.
48. Manson JD, Waters NE. Observations on the rate of maturation of the cat osteon. J Anat (London). 1965;99:539–49.
49. Holtrop ME. Light and electron microscopic structure of bone-forming cells. Caldwell: Telford Press; 1990.
50. Roberts WE, Wood HB, Burk DT. Vascularly oriented differentiation gradient of osteoblast precursor cells in rat periodontal ligament. J Periodontal Res. 1987;22:461–7.
51. Martin RB, Dannucci GA, Hood SJ. Bone apposition rate differences in osteonal and trabecular bone. Trans Orthop Res Soc. 1987;12
52. Robling AG, Niziolek PJ, Baldridge L. Mechanical stimulation of bone in vivo reduces osteocyte expression of Sost/Sclerostin. J Biol Chem. 2008;283:5866–75.
53. Boyde A. Scanning electron microscope studies of bone. In: Bourne GH, editor. The biochemistry and physiology of bone. New York: Academic; 1972.
54. Mullender MG, Meer DDVD, Huiskes R, Lips P. Osteocyte density changes in aging and osteoporosis. Bone. 1996;18:109–13.
55. Nunamaker DM, Butterweck DM, Provost MT. Fatigue fractures in thoroughbred racehorses: relationships with age, peak bone strain, and training. J Orthop Res. 1990;8:604–11.
56. Hattner R, Frost HM. Mean skeletal age: its calculation and theoretical effects on skeletal tracer physiology and on the physical characteristics of bone. Henry Ford Hosp Med Bull. 1963;11: 201–16.
57. Wu K, Schubeck KE, Frost HM, Villanueva AR. Haversian bone formation rates determined by a new method in a mastodon, and in human diabetes mellitus and osteoporosis. Calcif Tissue Int. 1970;22:204–19.
58. Kerley ER. The microscopic determination of age in human bone. Am J Phys Anthropol. 1965;23:149–64.
59. Kerley ER, Ubelaker DH. Revision in the microscopic method of estimating age at death in human cortical bone. Am J Phys Anthropol. 1978;49:545–6.
60. Evans FG, Bang S. Differences and relationships between the physical properties and microscopic structure of human femoral, tibial and fibular cortical bone. Am J Anat. 1967;120:79–88.
61. Ahlqvist J, Damsten O. A modification of Kerley's method for the microscopic determination of age in human bone. J Forensic Sci. 1969;14:205–13.
62. Bouvier M, Ubelaker DH. A comparison of two methods for the microscopic determination of age at death. Am J Phys Anthropol. 1977;46:391–4.
63. Thompson DD. The core technique in the determination of age at death in skeletons. J Forensic Sci. 1979;44:902–15.
64. Stout SD, Gehlert SJ. The relative accuracy and reliability of histologic aging methods. Forensic Sci Int. 1980;15:181–90.
65. Ubelaker DH. Human skeletal remains. Washington, DC: Smithsonian Institution Press; 1989.
66. Walker RA, Lovejoy CO, Meindl RS. Histomorphological and geometric properties of human femoral cortex in individuals over 50: implications for histomorphological determination of age-at-death. Am J Hum Biol. 1994;6:659–67.
67. Frost HM. Secondary osteon populations: an algorithm for estimating the missing osteons. Yearb Phys Anthropol. 1987;30:239–54.
68. Stout SD, Paine RR. Brief communication: bone remodeling rates: a test of an algorithm for estimating missing osteons. Am J Phys Anthropol. 1994;93:123–9.

69. Streeter M, Stout S, Trinkaus E, Burr D. Brief communication: Bone remodeling rates in Pleistocene Humans are not slower than the rates observed in modern populations: a reexamination of Abbott et al. Am. J. Phys. Anthropol. 1996;141:315–8.
70. Burr DB. Estimated intracortical bone turnover in the femur of growing macaques: implications for their use as models in skeletal pathology. Anat Rec. 1992;232:180–9.
71. Burr DB, Ruff CB, Thompson DD. Patterns of skeletal histologic change through time: comparison of an archaic native American population with modern populations. Anat Rec. 1990;226:307–13.
72. Li J, Kacena MA, Stocum DL. Chapter 12: Fracture healing. In: Allen MR, Burr DB, editors. Basic and applied bone biology. 2nd ed. Academic; 2019. p. 235–53.
73. Amsel S, Maniatis A, Tavassoli M, Crosby WH. The significance of intramedullary cancellous bone in the repair of bone marrow tissue. Anat Rec. 1969;164:101–11.
74. Carter DR, Blenman PR, Beaupre GS. Correlations between mechanical stress history and tissue differentiation in initial fracture healing. J Orthop Res. 1988;6:736–48.
75. Burny F, Donkerwolcke M, Bourgois R, et al. Twenty years experience in fracture healing measurements with strain gauges. Orthopedics. 1984;7:1823–9.
76. Cunningham JL, Evans M, Kenwright J. Measurement of fracture movement in patients treated with unilateral external skeletal fixation. J Biomed Eng. 1989;11:118–22.
77. Gerlanc M, Haddad D, Hyatt GW, et al. Ultrasonic study of normal and fractured bone. Clin Orthop Relat Res. 1975;111:175–80.
78. White AA, Panjabi MM, Southwick WO. The four biomechanical stages of fracture repair. J Bone Joint Surg. 1977;59:188–92.
79. Eriksson C. Streaming potentials and other water dependent effects in mineralized tissues. Ann N Y Acad Sci. 1974;238:321–38.
80. Fukada E, Yasuda I. On the piezoelectric effect of bone. J Phys Soc Jpn. 1957;10:1158–69.
81. Brighton CT, Friedenberg ZB, Black J (1979) Evaluation of the use of constant current in the treatment of nonunion. In: Black J, Pollack SR C.T. Brighton, Solomon R. (eds) Electrical properties of bone and cartilage. Experimental effects and clinical applications. Grune and Stratton, New York, pp. 519–546.
82. Bassett CAL, Mitchell SN, Gaston SR. Treatment of ununited tibial diaphyseal fractures with pulsing electromagnetic fields. J Bone Joint Surg. 1981;63:511–23.
83. Mollon B, Silva V, Busse JW, et al. Electrical stimulation for long-bone fracture-healing: a meta-analysis of randomized controlled trials. J Bone Joint Surg. 2008;90-A:2322–30.

Chapter 5
Mechanical Properties of Bone

Abstract In previous chapters, we have learned about the heterogeneous and dynamic structure of the material inside bones and that muscle forces acting on the skeleton are large relative to the forces exerted on bone by the outside world. By bringing these two aspects of skeletal tissue mechanics together, this chapter focuses on the factors that contribute to tissue-level and whole bone mechanical properties.

What You Will Learn

1. How to apply general engineering principles to normalize away the impacts of bone mass and measure estimated material-level mechanical properties.
2. How the mechanical properties of cortical and cancellous bone are influenced by structure and microstructure.
3. What it means to describe bone as an elastic (and mildly viscoelastic), anisotropic, composite material.
4. How mixture theories are introduced and can be used to formulate models for the mechanical behavior of bone.

5.1 Introduction

In previous chapters, we have learned about the heterogeneous and dynamic structure of the material inside bones. We also learned that muscle forces acting on the skeleton are quite large relative to the forces exerted on it by the outside world. It is now time to bring these two aspects of skeletal tissue mechanics together: to learn how strong bones are relative to the loads they must bear, and how the mechanical properties of bones depend on their structure.

The mechanical properties of bones are governed by the same engineering principles as those of man-made structures. However, the ability of vertebrates to adapt their bone structure to imposed loading, in terms of both the evolution of a genetic plan for the skeleton and the ability to acutely modify this plan by cellular activity, results in a structure that is highly complex and, when healthy, effective and efficient. The purpose of this chapter is to explain how the various attributes of whole

M. R. Allen, J. Wallace, *Skeletal Tissue Mechanics*,
https://doi.org/10.1007/978-3-031-90609-1_5

bones combine to determine their *monotonic* strength and stiffness. (The term monotonic refers to the behavior of a structure loaded steadily to failure, in distinction to repetitive or *fatigue* loading, which is considered in Chap. 7.)

5.2 Strength and Stiffness of a Structure

To determine the mechanical properties of a structure—or, more typically, a major part of a structure, such as a beam in a bridge or the shaft of a femur—engineers conduct a *load–deformation* test. The structure is loaded in a manner appropriate for its actual function—tension, compression, torsion, or bending—and the force developed in the structure to resist its deformation is recorded (where load is plotted as a function of deformation). Figure 5.1 shows a schematic representation of such a test, in this case a monotonic compressive test to failure. Typically, load and deformation are linearly proportional for some time (i.e., a straight line), until the *proportional limit* is reached; then, the slope of the load–deformation curve is reduced. The point where this begins to happen is known as the *yield point* because the structure begins to "yield" or *plastically* deform. Before the yield point, the deformation is *elastic*, meaning that the structure will return to its original shape when unloaded. If the structure is unloaded after exceeding the yield point, it will not return to its original

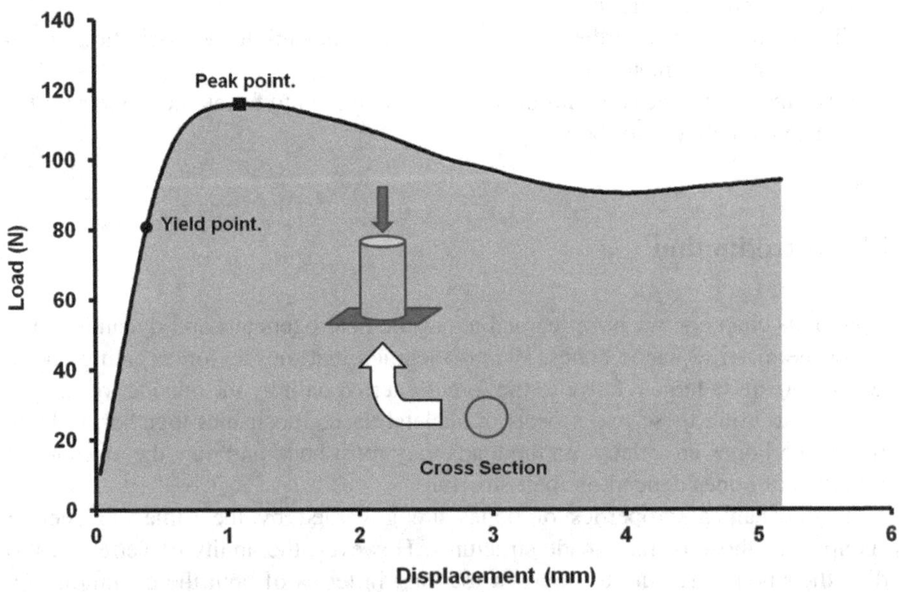

Fig. 5.1 A compressive load–deformation test. The peak load point is the largest load carried by the specimen. The slope of the linear portion of the curve (to the yield point) is the stiffness. The area under the curve (orange) is the energy absorbed during the test. The yield point is determined using the 0.2% strain offset method

shape, reflective of the plastic deformation that has occurred. The convention used to define where the structure begins to yield (i.e., the onset of nonlinear behavior) is defined below. If the load continues to increase past yield, a peak or *ultimate* load is reached; in some materials (e.g., metals), this may not be the point at which a catastrophic failure (or fracture) occurs. The point marking the latter event is known as the *failure point*. The *strength* of the structure may be defined as the load at the yield or failure points, or as the ultimate load, depending on the circumstances. There is no absolute definition of strength, but if it is not otherwise specified, it may be taken as the maximum or ultimate load.

The *stiffness* or *rigidity* of the structure is the load required to deform it a given amount. This is analogous to the slope of the linear region of the load–deformation curve. *Compliance* is the reciprocal of stiffness, and it is a measure of the ease of deforming the structure (i.e., the amount of deformation for a given applied load). The deformations at the yield and failure points are measures of the extent to which the structure may be deformed (and the difference between yield and failure is the post-yield deformation, a property often used to assess brittleness).

The area under the load–deformation curve is another important mechanical property known as *work*. *Work to yield* is energy that is elastically stored within the structure and returned upon unloading. *Work to failure* is the energy necessary to cause catastrophic failure or fracture. As with deformation, the difference between these two is the post-yield work.

5.3 Stress and Strain of a Material

The mechanical properties of a structure depend on its mass (how much material is present), its geometry (organization in 3D space), and the properties of the material itself. There are ways to directly measure properties of the material, but engineering principles can be used to estimate material properties by "normalizing out" the contributions of mass and geometry. In doing so, the load–deformation curve is transformed into a stress–strain curve. In its simplest definition, force is normalized by the area over which it acts within a structure and becomes stress (load per unit area, as shown at upper left in Fig. 5.2). The more general definition of stress is also shown in this figure. If one imagines a small cube of material at a point inside a loaded structure, one can see that each surface of the cube has three loads acting on it: a *normal* load perpendicular to the surface, and two *shear* loads parallel to the surface. If these three loads are divided by the area of the cube's face, they become the three stresses associated with the direction of that face. The cube has six faces, but because there must be equal and opposite forces acting on the opposing faces to maintain *force and moment equilibrium* (see Chap. 2), there are $3 \times 3 = 9$ *stress components* at each point in a loaded material. The first three are normal stress components, and the other six are shear stresses (note that some of the shear stresses are equal to one another so that the cube is in *moment equilibrium*). As a result, the total number of independent stress components is ultimately reduced to six: three

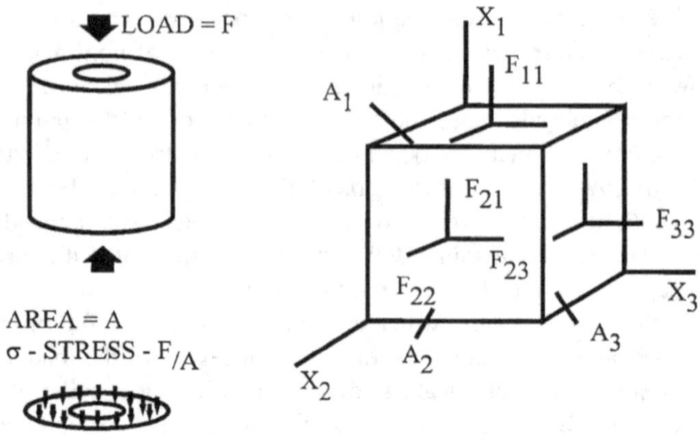

$$\sigma_1 = F_{11} / A_1 \qquad \sigma_4 = F_{23} / A_1 = F_{32} / A_3$$
$$\sigma_2 = F_{22} / A_2 \qquad \sigma_5 = F_{13} / A_1 = F_{31} / A_3$$
$$\sigma_3 = F_{33} / A_3 \qquad \sigma_6 = F_{12} / A_1 = F_{21} / A_2$$

Fig. 5.2 Simple and tensorial definitions of stress. The *cube at right* represents a free-body diagram of a small region within the loaded bone segment at *left*. *F* represents the normal and shear forces exerted on this region by the stresses in the surrounding bone. When divided by the areas of the cube's faces (A_1, A_2, A_3), these forces become the stress tensor components at the cube's location

normal and three shear. We represent these by the symbol s_i where $i = 1$ to 6. Stress has more components than a vector (which has only three) and is known as a second-order tensor (working with tensors is beyond the scope of this chapter). In the International System of Units (SI), stress is measured in Newtons/m^2 (force per unit area—N/m^2), also known as a *Pascal* (abbreviated Pa). A Pa is quite small and typical ultimate stress values for cortical bone would be around 200,000,000 Pa [200×10^6 Pa which is reported as 200 MPa (megapascals)].

Like force, deformation can be normalized as the fractional change in dimension of a body that is loaded to become *strain*. Strain is another second-order tensor. Figure 5.3 illustrates normal and shear strains (similar to normal and shear stress). One can show that there are, in general, six strain tensor components, three normal and three shear, which describe the deformation of a cube of material at any point in a loaded structure. We represent these by the symbol e_j, where $j = 1$ to 6. Strain is a dimensionless ratio (mm/mm) and can simply be reported as strain (e). Strains in bone are small and it is common to report in *microstrain* (0.000001 e or $1 \times 10^{-6}\, e$ or μe). For example, a strain of 0.02 mm/mm would be the same as 2% e or 20,000 μe, a dangerously large strain in a bone.

When a structure is loaded in tension, it becomes longer in the load direction but becomes narrower in the transverse directions (the opposite phenomenon occurs in a compressed specimen). A good example of this is seen in the stretching of a rubber band. The ratio of the transverse strain component to the longitudinal strain component is known as *Poisson's ratio*. For geometric reasons, this ratio is positive and

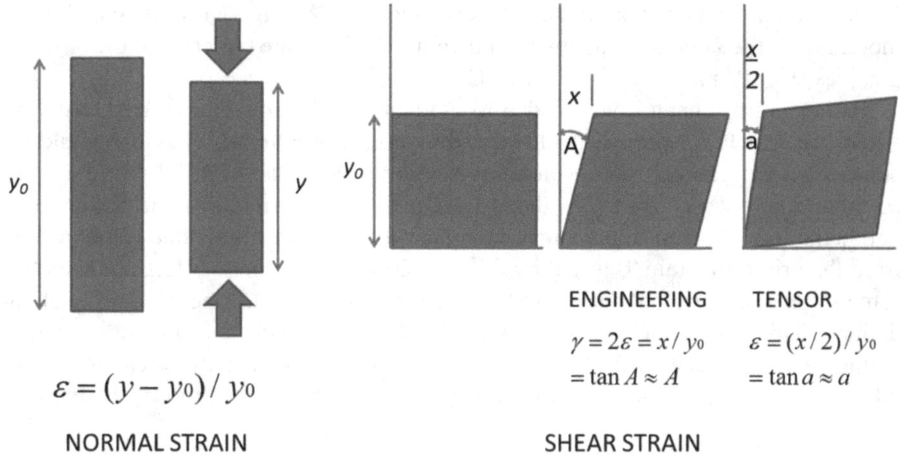

Fig. 5.3 Normal *(left)* and shear *(right)* strains are defined. A represents the angle through which the side of the shear specimen is rotated. Strains are dimensionless; shear strains may be expressed as radians

must be less than 0.5 in ordinary materials. For perfectly incompressible materials (e.g., water), Poisson's ratio is 0.5. Typical values for solids, including bone, are about 0.3. Cork, another biological material, has a Poisson's ratio that is nearly 0, making it perfectly compressible.

5.4 Strength and Modulus of a Material

A load–deformation curve may be converted to a stress–strain curve by using appropriate formulas to normalize load to stress and deformation to strain. Once complete, all the strength terms previously defined for the structure (yield, ultimate, and failure strengths, along with structural stiffness) can be redefined for the material in terms of stresses (yield, ultimate, and failure stress, along with elastic modulus). The area under the stress–strain curve to failure is analogous to the structural work to failure and is called the modulus of toughness. Toughness to yield is referred to as the modulus or resilience. Because of the assumptions made when using engineering principles to normalize to stress and strain (e.g., linear elastic behavior), post-yield toughness is typically not reported.

As was the case for force and deformation, stress is linearly proportional to strain prior to the onset of yielding. The proportional law for simple materials was discovered by Robert Hooke and is named in his honor. In its simplest form, Hooke's law relates one component of stress to its corresponding strain component (e.g., stress and strain in longitudinal tension). The slope of this line is the modulus of the material. For normal loading (tension or compression), this is the *elastic modulus* or

Young's modulus. For shear loading, this is the *shear modulus*. The normal and shear moduli have the same units as stress, and values for bone are typically of the order of gigapascals (10^9 pascals), abbreviated GPa.

Earlier in this chapter, we alluded to the fact that the convention used to define where the structure begins to yield (i.e., the onset of nonlinear behavior) would be defined below. There is no absolute convention used to define yield, but the most common approach is known as the 0.2% strain offset method. A line is drawn to define the linear region of the stress–strain curve. The start of this line is then offset from the origin (0 strain, 0 force) by 0.2% or 2000 μe. This new offset line with the same slope as the original will strike the stress/strain curve and where it does, yield is defined. This point can then be mapped back to the force–deformation curve to define yield there as well. Note that this is only one method to define yield and while others exist (e.g., 5% loss in secant stiffness), these other methods are far less common.

5.5 Determinants of the Mechanical Properties of a Whole Bone

Generally speaking, for a particular mode and rate of loading, the strength (or stiffness) of a structure is a product of (1) its shape and size, and (2) the strength (or stiffness) of the material within. For example, in compression, tension, or shear, the stress in a specimen of cross-sectional area A and bearing a load L is

$$s = L/A \tag{5.1}$$

When this stress reaches the failure stress for the material, the structure will fail. Therefore, by simply rearranging this equation and using the subscript f to represent the failure condition, one obtains an equation for the compressive load that would lead to failure in the diaphysis of a long bone:

$$L_f = As_f \tag{5.2}$$

where L_f is the failure load, and s_f is the material strength in the appropriate mode of loading. In this case, the shape/size factor is simply the cross-sectional area, A. Analogous equations describe the strength of the diaphysis in other modes of loading. When a circular cylindrical structure is twisted by a torque T, the shear stress varies linearly with distance, r, from the center of the cross section and is given by

$$s_s = Tr/J \tag{5.3}$$

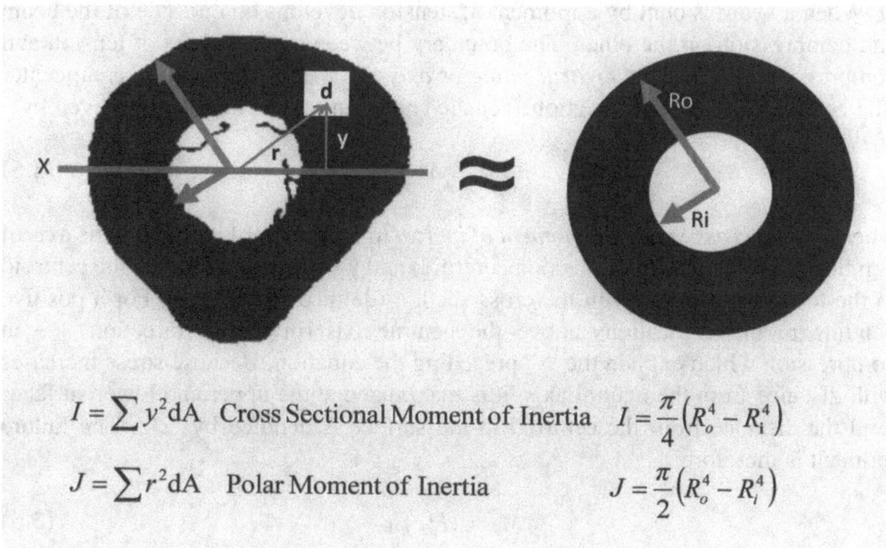

Fig. 5.4 Cross-sectional and polar moments of inertia are defined for a bone cross section. The cross section is divided into many small elements of area, dA. The *x*-axis is assumed to be the neutral axis for bending. Equations are for *I* and *J* of a cylindrical tube

where *J* is the *polar moment of inertia* (Fig. 5.4). This property is a measure of the distribution of area relative to the centroid of the section (for a circle, its center). Consider that the cross section of the bone cortex is divided into elements of area ?*A*, of which one is shown. *J* is the sum of the products of each ?*A* and the square of its radial distance from the centroid. Rearranging Eq. (5.3), the failure torque of a circular cylindrical long bone diaphysis loaded in torsion is

$$T_{\mathrm{f}} = (J/r)s_{\mathrm{sf}} \tag{5.4}$$

where r is now the periosteal radius, and s_{sf} is the torsional (or shear) strength of the bone material. The ratio *J*/*r* may be called the torsional "shape and size factor" for a circular cylinder.

For the hollow circular cylinder approximation of the femur, the torsional stiffness of the structure is exactly "*GJ*," where "*G*" is the shear modulus of the material and "*J*" is the polar moment of inertia. For a real bone with a cross-sectional shape other than a circle (or circular tube), however, the torsional rigidity is always *smaller* because the cross sections "warp" when a torque is applied. For *any* cross section, the equation relating torque (T_f), angle (?), and length (l) is: $T_f = (GK)\left(\frac{?}{l}\right)$, where "*K*" is a factor dependent on the shape and size of the cross section. For a circular (or circular tubular) section like in Fig. 5.4, *K* is equal to the polar moment of inertia *J*. For any other cross section, the factor *K* is smaller than *J* and can actually be a very small fraction of *J* for some cross sections.

When a beam is bent by a moment M, tension develops on one side of the beam and compression on the other. The boundary between these regions of tension and compression is called the *neutral plane* or *axis* and passes through the areal center (the centroid) of any cross section. Tensile and compressive stresses are given by

$$s = -My/I \tag{5.5}$$

where I is the *cross-sectional moment of inertia* (distribution of area about the axis of bending passing through the section centroid) and y is the distance from the centroid to the locations of interest in the cross section (defined in Fig. 5.4). For a positive bending moment, locations above the neutral axis (in the $+y$ direction) are in compression which explain the "-" preceding the equation. Because stress increases with distance from the neutral axis, it is maximized at the upper and lower surfaces (and the distance from the centroid to the surface is denoted by "c"). The failure moment is therefore:

$$M_{\mathrm{f}} = (I/c)s_{\mathrm{bf}} \tag{5.6}$$

where s_{bf} is the failure stress for bending. Here the "shape and size factor" is I/c.

Bones may frequently depart from conditions of elliptical or circular geometry—which are the basis for many of these equations. For example, torsion of a bone having a cross section that departs significantly from axial symmetry, or bending in which the load vector does not pass through the centroid of the cross section, would not accurately follow these descriptions. Even so, it generally remains possible to separate the effects of material properties (which depend on histologic structure) from those of the shape and size of the structure (which depend on anatomy). That is the important point insofar as this discussion is concerned.

Another limitation of this approach is that it assumes that skeletal "shape and size" characteristics include only macroscopic anatomical features and not microscopic features such as cancellous or osteonal architecture. It is assumed that the mechanical effects of microscopic features can be embodied in the material properties. This strategy, used in *continuum mechanics*, is consistent with the approach taken when *finite element modeling* is used to study skeletal stresses. The gross anatomy of the bone is modeled by the "shape" of the model, and variations in microscopic architecture are satisfied by assigning different material properties to individual regions (elements) within the model. The alternative is to attempt to formulate the shape and size of the microscopic structures, determine their behaviors, and then assign properties to the macroscopic structure accordingly. Although not yet in widespread use, newer high-resolution imaging techniques combined with multiscale FEA modeling can provide a platform for this computationally expensive approach. The more traditional approach favored here is to simply define material properties on a macroscopic scale, combining the mechanical effects of variations in microscopic architecture with those in microscopic material properties. This approach is also consistent with the common practice of measuring the material properties of bone using specimens small enough to define an anatomical region, but large compared to the size and distribution of microscopic features. For example, in the case of cancellous bone, test specimens are made large enough to contain perhaps 10–50 trabeculae [1].

5.6 Mechanical Failure of Whole Bones

In considering the mechanical requirements of whole bones, several conflicting demands must be balanced. In normal daily life, the skeleton provides struts (bones) that the muscles can work against to create motion. The stiffer these struts are, the more efficient is the action of the muscles. If bones are less stiff, then part of the mechanical output of the muscles is used in deforming the bones, rather than creating useful motion. This wastes metabolic energy. In this context, bones of greater stiffness are desirable. On the other hand, if the individual is in a situation where there is risk of bone fracture from a fall or a blow, less stiff (more compliant) bones would be preferred. This is because, for a given strength, the more compliant structure is able to absorb a greater amount of energy before reaching its failure load (Fig. 5.5). An important point is that this deformation may cause damage to the tissue (dissipating energy rather than leading to catastrophic failure), but the bone has a biological mechanism to repair that damage. Another consideration is weight: mineralized tissues are twice as dense as other tissues and require considerably more metabolic energy to carry around. The mechanical characteristics of bones, therefore, represent a compromise between the need for stiffness to make muscle actions efficient, the need for compliance to absorb energy and avoid fracture, and the need for minimal skeletal weight. This balance between stiffness and the need for energy absorption to prevent fracture may be one explanation for the inverse relationship often found between stiffness and modulus of toughness in bone [2].

Different bones have different functions, and the best solution to the conflicting mechanical demands need not be the same for every bone [3]. The bones of the inner ear need to be extremely stiff to efficiently transmit sound waves; the fact that they

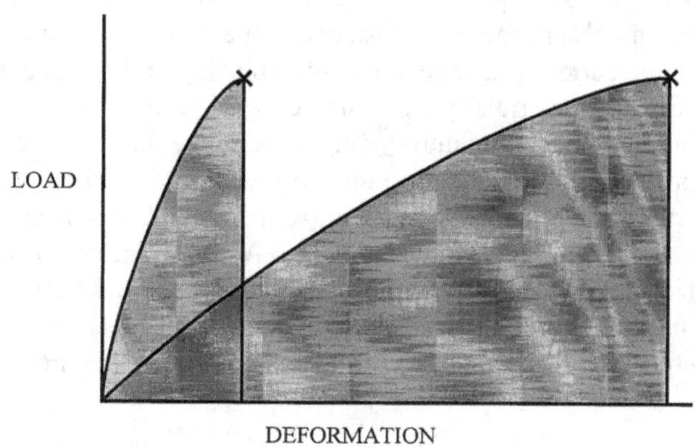

Fig. 5.5 If two bones have the same strength (failure load), the more compliant one (at *right*) requires more energy to break it

could absorb very little energy before fracture is unimportant because they are protected from extrinsic loads. At the other extreme, the material in the bones of children is much more compliant than those of adults. Currey and Butler [4] found the femurs of 2- to 8-year-old children to be 68% as stiff, requiring 45% more energy to break in bending, than those of 26- to 48-year-old adults. This increased compliance decreases the efficiency of children's muscle action, but it confers more protection against falling and crashing into things.

The energy required to break a bone is small compared with the energy encountered in daily activities. For example, the energy required to break the shaft of an adult tibia or femur is only about 15 J. The energy released when a 70-kg person falls to the ground from a standing position is about 500 J. The ability to absorb this energy using eccentric muscle contractions and deformations of soft tissues usually prevents fractures from occurring in such a fall. Soft tissue deformation absorbs energy during an accident in two ways. First, loads applied perpendicular to the tissues compress them and propagate stress waves up and down the body. The tissues effective in this way are skin, fat, and muscle (plus fur on animals and clothing on humans). Second, tissues such as fascia, tendons, ligaments, joint capsules, and contracted muscle brace bones against bending by supporting part of the tensile forces. These tissues absorb energy as they are stretched, but they must be nearly as stiff as the bone to be effective. If these energy-absorbing mechanisms are impaired by surprise, restrictions, or incapacity, fracture occurs more easily. The propensity of elderly persons to fracture their bones reflects not only the weakness of their bones but also their increased risk of falling and the inability of their muscles, ligaments, and soft tissues to adequately absorb the applied energy, which instead is transmitted to the bones.

5.7 Material Properties of Cortical Bone: Osteons

Osteonal remodeling of compact bone reduces its flexural fatigue strength and resistance to tension [5]. Compression, shear, and bending strength are also reduced. This section explores the reasons for these reductions and otherwise examines the mechanical properties of osteonal bone. Osteons can affect the mechanical properties of the cortex in several ways: by replacing highly mineralized bone matrix with less calcified material, by increasing cortical porosity, possibly by altering collagen fibril orientation, and by introducing cement line interfaces that have different mechanical properties. We first consider the mechanical properties of single osteons and then consider cortical bone as a composite of numerous such structural units. In a series of papers beginning in 1964, Antonio Ascenzi, Ermanno Bonucci, and their colleagues presented experimental information about the relationship between structure and mechanical function in individual secondary osteons. Taken together, their papers constitute the major part of our knowledge in this area.

5.7.1 Tensile Properties

In their initial work, Ascenzi and Bonucci [6] examined the tensile properties of the secondary osteon. Undecalcified longitudinal sections, 20–50 μm thick, were prepared from bovine and human femurs. A sharp needle was used to dissect free the lamina between the Haversian canal and the cement line on one side of an individual osteon. This procedure produced an approximately rectangular segment of the osteonal wall about 50 μm wide and 100–300 μm long. In their initial experiments, these osteonal segments were simply broken in quasistatic tension, with only the failure load recorded. Later, load–deformation curves were recorded when a microwave extensometer had been developed to measure the extremely small deformations encountered at this scale [7].

The specimens were categorized as either longitudinal (type L) or alternating (type A) in their lamellar structure (*c* or *b*, respectively, in Fig. 3.5). Microradiography was used to classify the degree of mineralization as either low or fully calcified. Both dry and wet bone were tested, but only the wet bone data are considered here. Figure 5.6 shows typical curves for secondary osteons with different lamellar orientations and degrees of mineralization. It was found that:

- The tensile strength of osteonal wall segments was about 114 MPa, a figure that lies in the midrange of data for larger test specimens composed of primary and secondary bone.
- Type L osteons had greater tensile strength and lower ultimate deformations than did type A osteons mineralized to the same degree.

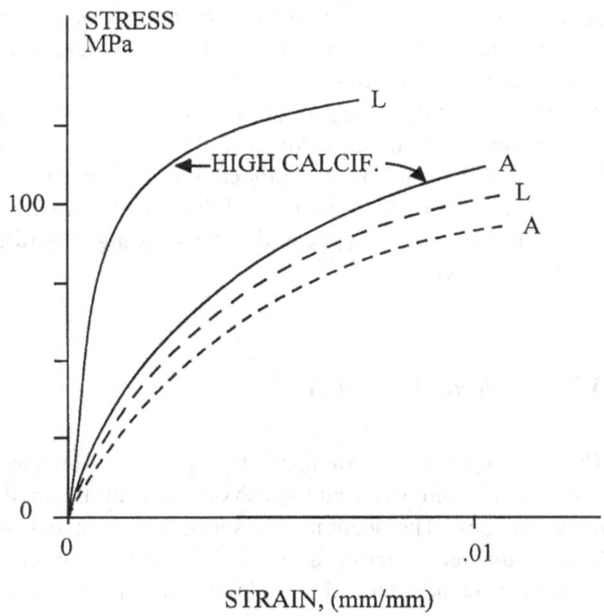

Fig. 5.6 Typical stress–strain curves for osteon segments in tension. *Solid curves*, fully calcified osteons; *dashed curves*, osteons in the initial stages of calcification. *L* type L osteon, *A* type A osteon (see Fig. 3.5) [7]

- The degree of calcification of type A osteons had little effect on the tensile strength of their wall segments. The less mineralized type L osteons were more plastic and deformable than highly mineralized type L osteons, which were extremely stiff. The properties of poorly mineralized type L osteons and all type A osteons (regardless of how well mineralized) were nearly the same (Fig. 5.6).

No differences were found between human and ox osteons, but the authors remarked on the considerable variation in the results for apparently similar osteons. No specific effect of osteocyte lacunae could be mathematically formulated, but examination of the published photographs of failed specimens indicates that lacunae and other irregularities in the lamellar structure may have played an important role in the results. The specimens did not have the reduced central "waisted" cross section that is preferred in tensile specimens. This factor surely led to stress concentrations at the points where the specimens were gripped, and the published photographs show breakage at these points.

5.7.2 Compressive Properties

Next, Ascenzi and Bonucci [8] studied the compressive properties of single osteons using specimens from a human femur. They had by now developed a technique for cutting cylindrical segments from osteons contained in cross sections 500 μm thick. These minute cylinders were 180–200 μm in diameter and contained the Haversian canal and all but the outermost lamellae of the osteon. The specimens were classified as to mineralization and lamellar structure as before, except that now osteons of circumferential structure (type T in Fig. 3.5) were also studied. In this case, they found that type T osteons were strongest and stiffest. Type A osteons were stronger than type L, but there was no difference in their elastic moduli. As before, the more fully mineralized osteons were more elastic, stiff, and brittle, regardless of their type. The specimens uniformly exhibited fracture planes at an angle of 30–35° to the longitudinal direction, regardless of the lamellar structure or degree of calcification. Thus, the failure mode appeared to be shearing, regardless of the orientation of the collagen fibrils.

5.7.3 Shear Properties

The next experiments examined shear properties of osteons [9]. A 160-μm-diameter cylindrical punch was used to push out the central lamellae from osteons in 300-μm-thick sections. The section was supported from below by a plate containing an 800-μm-diameter circular hole, so that the test region involved the cement line. It should also be noted that the test involved significant bending as well as longitudinal

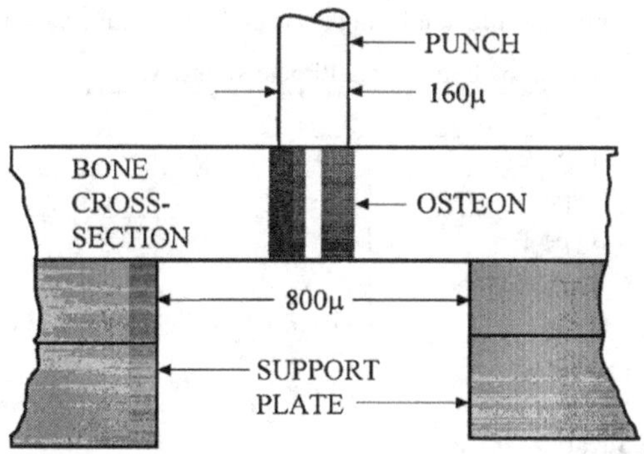

Fig. 5.7 Diagram of relative dimensions of the specimen and test apparatus used in Ascenzi and Bonucci's [9] shear experiments. Note that the 300-μm-thick bone section must have been subjected to significant bending loads

shear (Fig. 5.7). The osteons did not fail at the cement line, but along the margins of the punch. Because there would have been a stress concentration at this boundary, one ought not necessarily accept Ascenzi and Bonucci's assertion that this proves that the cement line is stronger than the interstitial lamellae, but their results are not consistent with a weak cement line. Type L osteons were the weakest, and types A and T were of equal strength. Again, strength and stiffness increased with calcification in each type. The usefulness of the data from this experiment is limited by the fact that the shear test arrangement departed significantly from "simple shearing"; that is, the lateral distance between the edge of the punch and that of the support plate hole was relatively large. Unless one is willing to make a custom apparatus for each osteon, this condition is a necessary practical compromise.

There is another report of a similar experiment. Frost et al. [10] arranged a simple shear test without regard to individual osteons or cement lines. Using 70-μm-thick sections, a 900-μm-diameter punch, and a "slightly larger" hole in the support plate, they pushed out areas containing several osteons and osteon fragments. The average stress for longitudinal shearing of bone from a human femur was about 74 MPa, considerably above the range reported by Ascenzi and Bonucci.

5.7.4 Combined Loading Effects

Ascenzi and Bonucci's analyses are theoretically instructive but oversimplistic in considering only "pure" loading in compression, tension, or shear. In reality, most bone is loaded in bending and incorporates combinations of these types of stresses. Computational analyses under combined loading conditions [11] suggest that collagen fibril orientation has little effect on bending stiffness of bone if collagen fibrils are oriented at $<60°$ with respect to the direction of load (i.e., the longitudinal axis). Thus, bending stiffness is actually compromised to some extent by the presence of only transverse lamellae, or a predominance of Type T osteons. For bending, then, overall mechanical properties are maximized when collagen fibrils are oriented at 30° or less to the axis of loading. Torsional stiffness, on the other hand, is maximized

Table 5.1 Mechanical properties of different types of osteons

Mode of loading	Ultimate stress, MPa	Elastic modulus, GPa	Ultimate strain, %
Tension			
Type L	114 ± 17	11.7 ± 5.8	6.8 ± 2.9
Type A	94 ± 15	5.5 ± 2.6	10.3 ± 4.0
Type T	–	–	–
Compression			
Type L	110 ± 10	6.3 ± 1.8	2.5 ± 0.4
Type A	134 ± 9	7.4 ± 1.6	2.1 ± 0.5
Type T	164 ± 12	9.3 ± 1.6	1.9 ± 0.3
Shear			
Type L	46 ± 7	3.3 ± 0.5[a]	4.9 ± 1.1[b]
Type A	55 ± 3	4.1 ± 0.4[a]	4.6 ± 0.6[b]
Type T	57 ± 6	4.1 ± 0.4[a]	4.6 ± 0.6[b]

[a]The shear modulus is defined as the slope of the linear portion of the stress–strain curve
[b]The strain in the shear test is defined as the ratio of punch advancement to section thickness, expressed as a percent

when collagen orientation is within 15–30° of either the longitudinal or transverse axis of the bone. In general, and somewhat contrary to the results of Ascenzi and Bonucci, Type A osteons are not well adapted to resist torsional moments.

These analyses show that when loading conditions are unpredictable, as they often are, it is beneficial to the bone for collagen to be oriented nearly longitudinally or at least within 15° of the loading axis. Axial stiffness is reduced by about 40% as collagen fibril bundles alter orientation from 0° to 90° with respect to the long axis under conditions of bending. This is consistent with the analysis performed by Katz [12] using the Hashin-Rosen model, and with the measurements by Rho et al. [13]. Although these analyses verified the measurements of Ascenzi under pure loading conditions, we can see that under more realistic conditions the adaptation of collagen fibril orientation is much more complex than the often quoted maxim that "longitudinal collagen fibrils bundles are best in tension, and transversely oriented bundles are best in compression."

The composite results of Ascenzi and Bonucci's experiments are summarized in Table 5.1. Variations in the mechanical properties of osteons with different collagen fibril orientations suggest that they are individually adapted to locally enhance the ability of bone to support a particular type of stress. The lamellae of type L osteons are stronger in tension, but it is relatively easy for their inner lamellae to slip longitudinally relative to the outer lamellae. This observation fits nicely with the frequently reported tendency of osteonal bone tested in longitudinal tension to show "osteon pullout" [14, 15]. Because type A and type T osteons are weaker in tension but shear less readily in the longitudinal direction, it is not clear whether macroscopic specimens in which these osteonal types predominate should be weaker or stronger in tension. However, it is clear that in bending, Type L and Type A osteons with fibrils oriented less than 60° to the loading axis will maximize the modulus.

As mentioned in Chap. 3, it is more likely that collagen fibril bundles are structured in a twisted plywood configuration, rather than completely in a parallel or alternating format between lamellae. From a mechanical standpoint, the twisted plywood configurations should enhance bone's ability to resist shear stresses developed in torsion and compression loading, because of their multidirectional organization. The twisted plywood configuration would be less than half as resistant as a parallel fiber arrangement to axial tensile forces [16], but bone is rarely, if ever, subjected to purely axial tensile forces in vivo. Moreover, even if it were, osteons with only longitudinally oriented fibrils would be highly susceptible to failure if the bone were loaded in any other orientation because the fibers would allow a preferential pathway for the growth of small fracture planes. Therefore, the twisted plywood organization may be the bone's adaptive response to withstanding unpredictable loads in many different directions. Bone's rather complex organization at the microscale provides an elegant example of evolutionary processes and the plasticity of bone at this scale of structure.

5.8 Effects of Osteons on Cortical Bone Mechanical Properties

We first consider the effects of osteons per se, that is, the simple addition of osteons to primary bone, on mechanical properties. Subsequently, the effects of intra-osteonal variations in collagen fibril orientation (i.e., adding type L, A, or T osteons) are examined.

Walmsley and Smith [17] suggested that Haversian systems weakened cortical bone. Soon after, Currey [18] found that completely remodeled cow bone was 35% weaker in tension than primary bone. This reduction in strength was attributed to the relative deficiency of mineral in more recently formed Haversian systems and the additional porosity caused by new Haversian canals. The former concept was reinforced when Hert and coworkers [19] found that fully mineralized osteonal cow bone was as strong in compression as primary cow bone. A decade later, Reilly and Burstein [20] tested Haversian and primary cow bone in both tension and compression, and found that osteonal remodeling decreased the strength by about 20% in both modes of loading. There was, however, no change in the elastic modulus. A year later, Reilly and Burstein [21] presented additional data showing that osteonal remodeling weakened cow bone by about 11% in tension and by about 9% in shear.

Margel-Robertson and Smith [22] studied the effects of Haversian remodeling on the three-point bending properties of bovine femur and the compressive properties of porcine mandible. Although they did not present their data in enough detail to confirm statistically significant effects, they reported that remodeling reduced strength in both tests. Their bending data are unique and support the hypothesis that remodeling reduces the strength of cortical bone in all modes of loading.

Carter and colleagues [23, 24] found that osteonal remodeling weakened bovine bone with respect to both monotonic and fatigue loading. However, their analysis indicated that this was not simply caused by reduced bone density (that is, increased porosity and reduced mineralization). The osteonal bone matrix seemed to contain another source of weakness. This additional factor may be the cement line, which has been reported as either more compliant [25] or less compliant [26] than surrounding bone tissue. Either of these cement line configurations could be a source of additional weakness, although either one is also likely to improve fatigue life under cyclic loading conditions (see Chap. 7).

Cortical bone exhibits substantial amounts of plastic deformation when loaded to failure in either tension or compression. The amount of plastic deformation is greatest when the bone is loaded parallel to the collagen fibrils, in which case it typically exceeds the elastic deformation [27]. Osteonal bone exhibits more plastic deformation than primary bone [28]. It is reasonable to speculate that cement line slippage or yielding accounts for much of this plastic ductility, but no one has compared its effects to those of porosity and mineralization.

These studies of the effects of osteonal remodeling on bone strength all used nonhuman bone because it is difficult to obtain human specimens of primary bone large enough to test (there are standards for specimen geometry which require that the minimum dimensions of the specimen be several times larger than those of its internal structural features). In the long bones of adults, and even adolescents, almost the entire cortex is occupied by several generations of secondary Haversian systems. Vincentelli and Grigorov [29] attempted to overcome this problem by selectively taking specimens from primary bone near the periosteal and endosteal surfaces of the human tibia. These specimens were tested in tension and compared with similar specimens taken from Haversian regions of the same bones. While there was some mixing of the two types of bone in various specimens, it was possible to analyze samples containing either more than 75% or less than 25% Haversian tissue. It was found that both the strength and the stiffness of the predominantly Haversian bone were less than that in primary bone (Table 5.2).

Ascenzi and Bonucci showed that the mechanical properties of individual osteons are affected by collagen orientation. Does osteon type—that is, different arrangements of collagen inside osteons—affect whole bone strength as well? In 1969, Evans and Vincentelli reported that the tensile strength of tibial bone was positively correlated with the number of type L osteons that it contained [30]. Because this work was done with embalmed material, they repeated the experiment with fresh human tibias [31]. They again found significantly greater tensile strengths associated with the presence of type L osteons, in concert with the greater tensile strength of

Table 5.2 Tensile strength and stiffness of human primary and secondary bone

Tissue type	Ultimate stress, MPa	Elastic modulus, GPa
Primary bone	161 ± 11	19.4 ± 2.4
Haversian bone	130 ± 14	17.6 ± 2.0

From Vincentelli and Grigorov (1985) [29]

single type L osteons. They also found, however, that bone with greater numbers of Type T osteons deformed more as it broke. This result may be related to the capacity of these osteons to exhibit longitudinal shear between their lamellae, and may be consistent with the observation that axial stiffness is reduced as collagen fibril bundles alter orientation from 0° to 90°. Increased stiffness values were similarly correlated with Type L osteons, but not significantly so.

Later, this team reported the results of similar experiments on the compressive properties of osteonal bone [32]. They did not find that compressive strength was inversely correlated to increased numbers of type L osteons, as they had expected from Ascenzi and Bonucci's work. Instead, they found that the greatest compressive strength was associated with osteons having a mixed dark–light appearance; these osteons were apparently intermediate between type L and type A. They also found that deformation at failure was greatest when more type A osteons were present, in contrast to the relative brittleness of single type A osteons when compared to type L (see Table 5.1). These results show that bone's macroscopic mechanical properties are not entirely predictable based simply on osteon fibril orientation. Nevertheless, this feature appears to play an important role in determining mechanical properties.

In addition to studying the mechanical effects of osteons having different collagen orientations, one may examine the effects of the average collagen fibril orientation throughout the bone matrix on the mechanical properties of primary or secondary cortical bone. Using birefringence in polarized light to estimate average collagen orientation in a whole specimen cross section, it has been shown that collagen orientation affects strength and stiffness at least as much as density, porosity, and mineralization [33–35].

If osteonal collagen fibril orientation affects bone strength, is this aspect of bone structure governed by mechanical loading? Gebhardt [36], followed by Olivo and coworkers [37, 38], was perhaps the first to postulate that osteonal collagen orientation is determined by mechanical stress. If this is the case, one would expect that osteons with different fiber orientations would be preferentially found in regions of bone carrying particular kinds of stress. Pursuing this hypothesis, Portigliatti-Barbos et al. [39, 40] demonstrated that the orientation of collagen in the shaft of the human femur is not random, but is correlated with tensile and compressive stresses determined by biomechanical analysis. These investigators quantified the brightness of 1-mm-square elemental areas within a complete femur cross section when viewed in polarized light. Thus, their analysis averaged the osteonal fibril orientation of intact osteons and interstitial lamellae, rather than characterizing osteons as belonging to one of three types. They found that regions carrying tensile stress tend to have more longitudinally oriented fibers and those carrying compressive stress contain more transversely oriented fibers. This is shown in Fig. 5.8, which contains computer-generated diagrams of human femur cross sections located 1 cm apart (upper and lower panels, respectively), showing collagen fiber orientation (left panels) and radiographic density (right panels). The diagonal lines approximate the position of the neutral axis for bending induced by loads at the hip [41]. The darker pixels of the left panels represent more longitudinally oriented collagen fibers which are concentrated in the latero-anterior regions carrying tensile stresses. The lighter pixels of the right panels represent greater volumetric mineralization, which are concentrated in

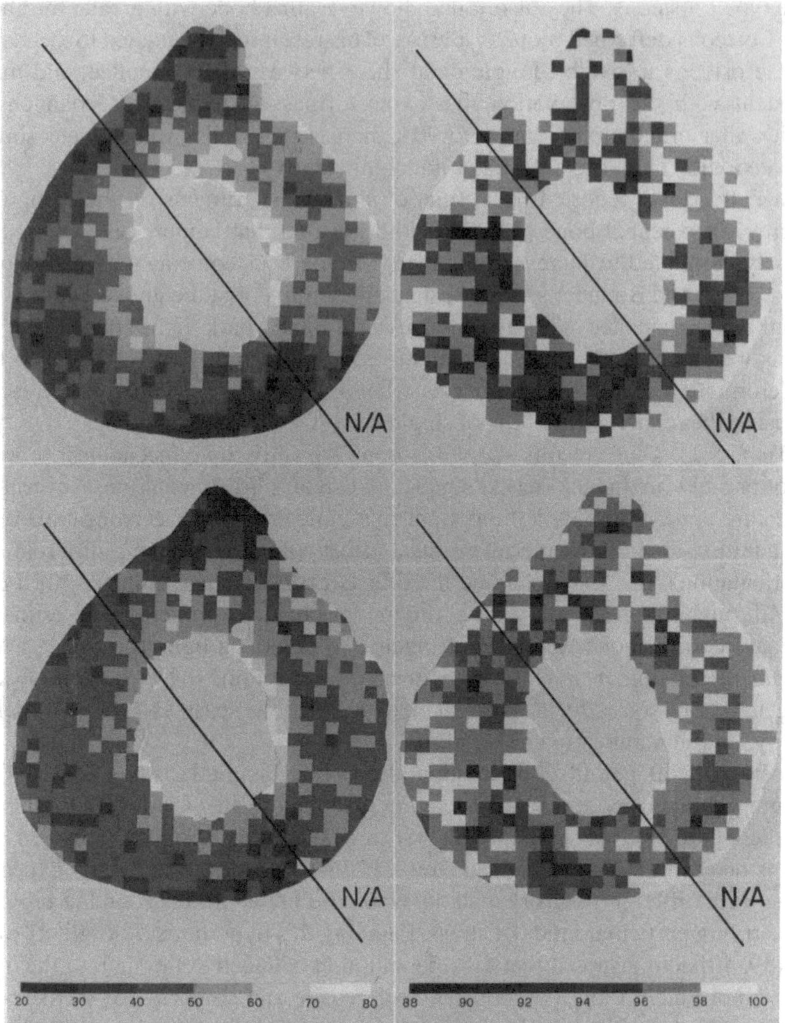

Fig. 5.8 Plots of distribution of collagen fibril orientation *(left panels)* and radiographic density *(right panels)* in two cross sections from a human femur. These sections were located 1 cm apart in the central diaphysis. Superimposed on the plots are the approximate locations of the neutral axis (*N/A*) for bending produced by loading at the hip [41]. The darker pixels of the left panels, which have more longitudinally oriented collagen fibers and greater tensile strength, are concentrated in the latero-anterior regions, carrying more tensile stresses. The lighter pixels of the *right panels*, representing greater volumetric mineralization, are more common in the medio-posterior regions on the compression side of the neutral axis [39]

the medio-posterior regions carrying more compressive stresses. Subsequently, others [42, 43] have correlated these variations in osteonal collagen fibril orientation to strain patterns determined in racehorses.

These observations are consistent with reports that porosity [44] and bone stiffness [45] also vary in a regular fashion with angular position in the femoral cortex. They support the compound hypothesis that (a) osteon collagen fibril orientation and mineralization help determine cortical bone's mechanical properties and (b) osteonal remodeling adapts internal structure to functional stresses. However, because osteonal mineralization in healthy bone is only supposed to be a function of time since formation, it is not clear why radiographic density should be greater in the compressive cortex, as is shown in the images from Portigliatti-Barbos. Is mineralization of osteoid enhanced by compressive stress? Is the rate of remodeling greater on the tensile side, so that its osteons are, on average, younger? Perhaps an attractive hypothesis for the Portigliatti-Barbos data is that radiographic density reflected decreased porosity rather than mineralization of the tissue. On the other hand, work by Goodwin and Sharkey [46] concluded that up to 20% of the variance in the tissue modulus of interstitial lamellae within adult tibiae and metatarsals could be explained by regional differences in the local strain environments routinely experienced by these bones. Moduli measurements from lamellae residing in regions experiencing the greatest compressive strains were 11% greater than those made on lamellae in more tensile regions. Such differences could be due to differences in mineralization, collagen orientation, or both.

5.8.1 Anisotropy of Cortical Bone Mechanical Properties

Cortical bone exhibits highly ordered internal arrangements of collagen fibrils and mineral crystals; this is true for both the plexiform and circumferential lamellar forms of primary bone, as well as osteonal bone. X-Ray diffraction techniques show that the average orientation of the mineralized collagen matrix is approximately parallel to the bone's long axis [47]. The fact that this causes marked differences in the mechanical properties associated with the longitudinal, radial, and circumferential directions was recognized long ago [48]. The radial and circumferential directions are visually distinguishable in both forms of primary bone, but heavily remodeled bone appears to the eye to be transversely isotropic. We now examine the effects of osteonal remodeling on the mechanical anisotropy of cortical bone.

5.8.2 Anisotropy Effects Determined by Load–Deformation Testing

Classic contributions in this area are the work of Reilly, Burstein, and Frankel [21, 28]. Using femurs, they measured the elastic moduli and strength of Haversian and primary bovine bone in different directions and compared them to those of

Table 5.3 Anisotropy of bovine and human bone

	Tension		Compression	
Elastic modulus, GPa	*Longitudinal*	*Transverse*	*Longitudinal*	*Transverse*
Human				
Haversian	17.9 ± 0.9	10.1 ± 2.4	18.2 ± 0.9	11.7 ± 1.0
Bovine				
Haversian	23.1 ± 3.2	10.4 ± 1.6	22.3 ± 4.6	10.1 ± 1.8
Primary	26.5 ± 5.4	11.0 ± 0.2	–	–
Ultimate stress, MPa	*Longitudinal*	*Transverse*	*Longitudinal*	*Transverse*
Human				
Haversian	135 ± 16	53 ± 11	105 ± 17	131 ± 21
Bovine				
Haversian	150 ± 11[a]	49 ± 7[a]	272 ± 3	146 ± 32
Primary	167 ± 9	55 ± 9[a]	–	–

[a]Standard deviations approximate because groups were averaged [21, 28]

human Haversian bone. Their data (summarized in Table 5.3) show that the strength and stiffness of bovine Haversian bone are similar to that of human bone when measured transverse to the long axis, but that cow bone gives consistently higher values in the longitudinal direction. In bovine bone, the ratios of elastic moduli in the longitudinal and transverse directions are similar in tension and compression and in primary and Haversian bone. These anisotropy ratios are less in human Haversian bone than in bovine Haversian bone (for both modes of loading) because the human bone is less stiff in the longitudinal direction. Both types of cow bone are stronger in compression than in tension, regardless of the loading direction. Human Haversian bone is stronger in compression when loaded transversely but not when loaded longitudinally. The strength anisotropy of Haversian bone from both species is more pronounced in tension than in compression, and it appears that bovine bone anisotropy is somewhat greater than that of human bone. Thus, Haversian cow bone is a reasonable model for human cortical bone, but it can be expected to be more anisotropic because of its enhanced strength and stiffness in the longitudinal direction.

Reilly and Burstein's data suggest that osteonal remodeling has two important effects on mechanical anisotropy. First, secondary osteons reduce tensile strength and stiffness more in the longitudinal direction than transversely, suggesting that the differences between bovine and human Haversian bone result from the more extensive intracortical remodeling of human bone. Second, they serve to transform the bone from an orthotropic material into a more transversely isotropic one. The "transverse" direction referred to in Table 5.3 is the circumferential direction. Reilly and Burstein also tested bovine specimens in the radial direction. Recalling the structure of plexiform bone, one would expect to obtain different results in these two directions when this tissue is tested, and that is what was observed. Tensile strength in the longitudinal, circumferential, and radial directions had proportionalities of about 3:1:0.4 in primary bone. In Haversian specimens, the radial tensile strength was increased so that the ratios were about 3:1:0.7, and the compression strength ratios were equal in the two transverse directions: 3:1:1. Although distinct differences were found to persist in the deformability and energy to failure in the two transverse directions of bovine Haversian bone, these data show that osteons make whole bone transversely isotropic with respect to strength.

5.9 Factors Contributing to Cortical Bone Mechanical Properties

In addition to the presence of osteons, a number of other factors affect the mechanical properties of cortical and osteonal bone. This section summarizes some significant contributors.

5.9.1 Porosity

Obviously, the introduction of holes in a structure should weaken it, and this is certainly true in bone. It is one thing to demonstrate this effect and quite another to explain in some detail the nature of the relationship. Empirical formulae have been developed relating porosity to the mechanical properties of various porous materials, but a comprehensive theory, useful over a broad range of porosities, has not been developed. These problems are explored later in this chapter.

We can define porosity as the ratio of void volume to total volume, and note that volumetric porosity is equivalent to that measured two dimensionally on a cross section [49]. Bone contains voids that range from a few nanometers to several hundred micrometers in diameter. Usually, the smallest of these (canaliculi and osteocyte lacunae) are not considered to affect bone's mechanical properties because, even though they probably do, they are rather uniformly distributed throughout all kinds of bone. They are therefore considered to be an intrinsic property of the bone material substance. In cortical bone, mechanical properties are considered to be affected by Haversian canals and related resorption cavities and vascular channels. In trabecular bone, they are affected by the marrow spaces between the trabeculae. Several investigators have proposed empirical relationships between mechanical properties and porosity. For example, Schaffler and Burr [50] found that the tensile elastic modulus (in GPa) of bovine cortical bone was related to porosity by the empirical equation

$$E = 33.9(1 - p)^{10.9} \tag{5.7}$$

(Note that $(1 - p)$ = bone volume fraction.) From a wide variety of cortical bone loaded in tension, Currey [51] obtained

$$E = 23.4(1 - p)^{5.7} \tag{5.8}$$

A comparison of these results suggests that the modulus of bone becomes more sensitive to changes in porosity as its value decreases. Schaffler and Burr's maximum porosity value was 7.8%; Currey's was 31%. Pursuing this, McElhaney et al. [52] found that the modulus of human skull bone (containing a very porous diploe layer) increased as the cube of (1 - p). Similarly, the modulus has been shown to increase as the apparent density raised to the 7.4th power for cortical bone, as the cube of density for cortical and cancellous bone samples pooled together, and as the square of density when a sample containing only cancellous bone was analyzed [50, 53, 54].

Porosity is implicitly regarded as a material feature of bone. Alternatively, one may think of porosity as a structural feature. If the porosity is uniformly distributed, the bone is loaded in simple tension or compression, and one is concerned only with macroscopic phenomena, the former point of view is convenient and acceptable. On the other hand, in a bent or twisted bone, voids located in the outer cortex, where stresses are high, weaken the bone more than those near the medullary canal. If there is a porosity gradient in the cortical wall, it makes a substantial difference whether the more porous region is in the outer, more highly stressed material or near the medullary canal. Bones such as the femur are frequently characterized by porosity gradients across the cortical wall. Smith and Walmsley [55] showed that gradients typical of those in elderly people can reduce the bending strength of a rectangular bone test specimen machined from a cortical wall by roughly 20%. However, the gradient is directed to enhance bone strength by placing most voids near the center of the bone. This is clear from Fig. 6.4, which compares the porosity distribution in the shaft of an elderly person's femur to that in a younger person. Not only are the void spaces concentrated near the medullary canal, but they are also preferentially located near the neutral axis.

5.9.2 *Mineralization*

As with porosity, it is intuitively clear that variations in the mineral content of bone should affect its mechanical properties, and experiments confirm this hypothesis. One may begin the analysis of this problem by breaking it into three parts according to the degree and cause of the mineralization difference. Because there is a delay in the mineralization of newly formed bone, bones that are being remodeled at a higher rate exhibit a reduced mineral content. This effect can be related to age, trauma, or metabolic disease. Reduced mineral content can also be caused by a lengthening of the osteonal refilling period because a delay in completion of new osteons means either that more osteoid or more void space is present for a given activation frequency. Much larger differences in mineralization may be produced by metabolic bone diseases, such as rickets or osteomalacia, in which the rate of turnover and refilling time are normal but the mineralization lag time is exaggerated.

It is important to be clear about how "mineralization" is defined. One may define the amount of mineral per unit volume of whole bone as *volumetric mineralization*, and the amount per volume of bone matrix, exclusive of the volume of the void spaces in the structure, as *specific mineralization*. Clearly, volumetric mineralization is a function of both porosity and mineralization of the organic matrix; specific mineralization discounts the effects of porosity and is therefore more indicative of the mineralization process itself. Increased rate of bone turnover and increased refilling period therefore create both a volumetric and a specific mineralization deficit, while the inhibition of mineralization of osteoid creates only a specific mineralization deficit. In reading the literature, one must be careful to understand which definition of "percent mineral" is being used. Similar comments could be applied to use of the term "density."

Most noninvasive determinations of mineral content (e.g., photon absorptiometry and computed tomography) measure areal or volumetric mineralization in some fashion, and that is why these procedures are not generally capable of separating diseases with reduced areal/volumetric mineralization (e.g., osteoporoses) from those that involve reduced specific mineralization (e.g., osteopenias such as osteomalacia). Specific mineralization is usually measured by ashing the bone and calculating its ash fraction.

When Vose and Kubala [56] correlated bending strength of human femurs with radiographic density, they were using a measure of specific mineralization because they pulverized and compressed the bone specimens before scanning by X-ray. They found that small differences in matrix mineralization made large differences in breaking stress: increasing the ash content from 63% to 71% caused the strength to rise 3.7-fold. Currey [57] became interested in this relationship and wondered why animals did not take better advantage of the apparently disproportionate enhancement of bone strength afforded by adding a few grams of mineral to each bone. To explain this conundrum, he postulated that more highly mineralized bone is weaker at higher strain rates and presented experimental data, suggesting that the optimal specific mineralization for cortical bone is 66–67%. The same year [58], he further hypothesized that the slope of the mineralization–stiffness (and strength) curve was especially steep because as mineralization became complete, the ends of the apatite crystals came into contact and fused. In a later publication, Currey and coworkers [59] concluded that 66–67% mineralization was the point of maximum toughness. A few years later, Bonfield and Clark [60] reached a similar conclusion.

Currey [61] related the elastic modulus of cortical bone from 17 vertebrate species, in both bending and tension, to porosity and specific mineralization (expressed in the form of calcium content). He found that these two variables together accounted for about 84% of the stiffness variation, but he did not discuss their relative effects. He observed that such high correlations with porosity and mineralization left little room for dependency on such histologic features as collagen orientation, but he also suggested that that was probably not true for strength.

Finally, Schaffler and Burr [50] found, for bovine cortical bone in tension,

$$E = 89.1A^{3.91} \tag{5.9}$$

where A is specific mineralization expressed as percent ash by mass. However, percent ash was not a significant predictor of E, perhaps because A varied so little, mineralization being tightly regulated.

For a mixed sample of human cancellous and cortical bone samples Hernandez and coworkers [62] found $E = 84.37(1 - p)^{2.58}a^{2.74}$ where a was the ash fraction of the tissue. These results are similar to previous results, but differ in detail as discussed by the authors.

5.9.3 *Density*

The material properties of bone are frequently discussed in terms of its "density." Used alone, this word would be expected to refer to the specific gravity of the solid matrix. The term *apparent density*, on the other hand, refers to the mass per unit volume of a region of bulk bone, which includes Haversian canals, marrow spaces, and other soft tissue spaces. Therefore, apparent density is a function of both the porosity and mineralization of the bone material. Carter and Hayes [53] showed that the compressive strength (s_{cf}) and elastic modulus of bone specimens having a wide range of apparent densities ($d = 0.5$–1.9 g/mL) were correlated with the square and cube, respectively, of density in g/mL:

$$s_{cf} = 68\overset{?}{e}\ 0.06 d^2 \text{ (MPa)} \tag{5.10}$$

$$E = 3790\overset{?}{e}\ 0.06 d^3 \text{ (GPa)} \tag{5.11}$$

In these equations, $e?$ is strain rate in s^{-1}. The latter relationship has been disputed by Rice et al. [54]. Through extensive multiple regression analysis of a large collection of data for cancellous bone, in which they attempted to control for differences in species, type of load, and load direction, Rice et al. found that both E and s_{cf} correlate better with d^2 than d^3. Other studies of cancellous bone have also obtained exponents closer to 2 than 3 [63, 64]. It appears that the reason for this discrepancy is that the data of Carter and Hayes included cortical as well as cancellous bone specimens, and these points increased the exponent from 2 to 3.

5.9.4 *Histological Architecture*

Two regions of bone having similar porosity and mineralization may still have quite different material properties because the organization of the solid matrix is highly variable. As noted in Chap. 3, compact bone may have a great variety of internal architectures. One may first distinguish between primary lamellar bone (mostly formed by modeling on periosteal or endosteal surfaces during growth) and lamellar bone composed of secondary osteons (formed by remodeling). We have seen that secondary bone is weaker than primary bone [21]. Primary cortical bone itself may also consist of two nonlamellar types of tissue. Plexiform bone is similar to lamellar bone but is laid down more rapidly in growing individuals [65–67]. Much of the data on primary bone's mechanical properties is based on this material because it occurs in cows, a ready source of material for testing. How these properties compare to the "pure" lamellar bone usually found in humans is unclear because remodeling occurs so early in humans that it is difficult to obtain sufficient material to test.

5.9.5 Collagen Fiber Orientation

Given a particular porosity and mineralization, and a particular trabecular or osteonal architecture, it remains possible for two regions of bone to have different material properties. There could be several reasons for this, but the next variable to be considered is the orientation of constituent collagen fibrils. This aspect of structure is apparently controlled both by cell functions and by extracellular physical processes. Frost [68] has called these processes "micromodeling."

Lamellar bone, with lamellae of parallel collagen fibrils, is stronger than woven bone in which the collagen fibrils are more or less randomly oriented. Furthermore, the predominant orientation of the collagen fibrils affects tissue strength in relation to the mode of loading: longitudinal fibrils promote strength in tension, while transverse fibrils are associated with strength in compression. This has been shown for individual osteons [7, 8] as well as bone in bulk, and it was seen in Fig. 5.8 that fibril orientations vary with position relative to the neutral axis of the femur, so that regions in habitual tension have more longitudinal fibers and those in compression have fewer [39]. In cortical bone, which generally exhibits little variability in porosity or mineralization, collagen orientation may be the predominant factor in determining tensile strength [33]. No studies have been found on the effects of collagen fibril orientation in cancellous bone, but it seems likely, for example, that trabeculae composed of woven bone must behave differently from those containing lamellar bone.

5.9.6 Microdamage

Another factor that may affect the material properties of bone is microdamage. It is known that fatigue microdamage reduces elastic modulus [24, 69, 70], and as techniques for measuring such damage improve, this may be added to the list of those things studied as determinants of bone's mechanical properties. When loaded repetitively in the laboratory, bone experiences fatigue damage and may fail from this like any other structural material [23, 71, 72]. The in vivo fatigue life of bone depends on the rate at which remodeling repairs fatigue damage. In Chap. 7, we consider the relationship between the creation of fatigue damage and its repair by remodeling and describe a model for stress fracture, which is what clinicians call the fatigue fracture that occurs when bone's repair process fails to keep up with damage formation.

5.9.7 Collagen Cross-Linking

As the collagen matrix in bone is deposited and subsequently mineralized, a tightly regulated process unfolds whereby enzymatic chemical cross-links form within the organic structure [73]. Initially, an immature linkage between two collagen molecules, these cross-links can mature into a more stable trivalent cross-link with time. Collagen cross-link quantity and maturity are critical to collagen strength and stability and clearly influence bone strength, stiffness, and deformability, but the specific and direct impacts of enzymatic cross-links are not fully understood. In an unregulated nonenzymatic process, reducing sugars, including glucose, can react with free amino groups in collagen to create a heterogeneous group of molecules known as advanced glycation end products (AGEs). Because all that is needed is sugar, protein, and a warm environment, these AGE-based cross-links can form more or less anywhere in the collagen network. Although not fully understood, these cross-links are known to accumulate with tissue age and are presumed to detrimentally impact bone toughness by stiffening the collagen structure and preventing the natural sliding that is meant to occur between molecules [74, 75].

5.9.8 Hydration

Water, the often overlooked third phase of the collagen/mineral composite in bone, is gaining interest for the roles it might play in determining the mechanical behavior of bone tissue [76]. Water occupies around 10–20% of bone's volume, and levels are highly influenced by factors including age, disease, and even pharmaceutical treatment. Bone water is generally considered to occupy four primary compartments. Free water, as its name implies, is water that is free to move withing the various porosities in the bone matrix. It is likely to play an important part in biological processes (e.g., cell signaling, mineral homeostasis [77, 78], but its contribution to mechanical properties is likely minimal. Bone's remaining water is bound to collagen, mineral, or both. Loosely bound water at the collagen mineral interface allows for the sliding of the interface and the efficient transfer of loads between collagen and mineral. This likely contributes to bone's mildly viscoelastic behavior ([79], see below). Water that is tightly bound within the triple-helical structure of collagen bridges between molecules and stabilizes the matrix [80]. Within hydroxyapatite itself, structural water helps to stabilize the mineral crystals [81].

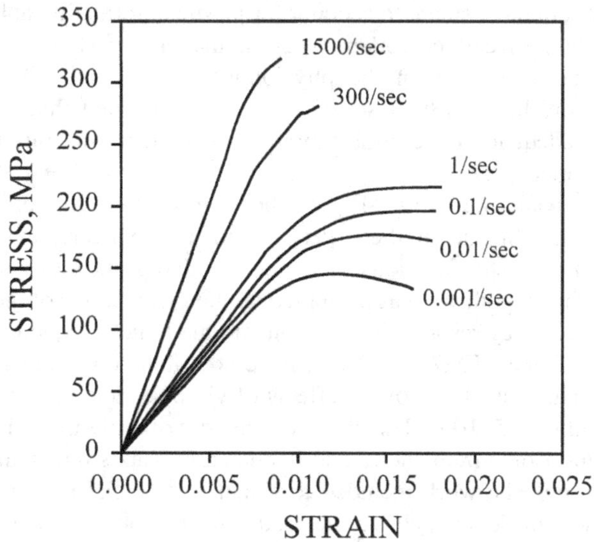

Fig. 5.9 Stress–strain curves for cow bone loaded in compression at various strain rates [82]

Likely due to the presence of water in the matrix, at least partially, cortical bone is a mildly viscoelastic material. One hallmark of viscoelasticity is that mechanical properties depend considerably on the rate of deformation. In addition to the flow of water, one may speculate that features of internal structure associated with remodeling and osteons could contribute to such behavior in several ways, including the presence of osteoid in varying amounts and the presence of cement lines. There is, however, little information available regarding the effects of osteons on bone's viscoelastic behavior.

The classic paper on bone's viscoelasticity is that of McElhaney [82]. He studied bovine femoral bone over a wide range of deformation speeds, building a pneumatic cannon to achieve an upper strain rate of 1500/s. In general, cortical bone becomes stronger, stiffer, and more brittle as strain rate increases (Fig. 5.9). Although he did not quantify the histologic structure of his specimens, McElhaney remarked that those loaded at high strain rates tended to fail along cement lines, while those loaded at low strain rates failed along shear planes passing through osteons. It is interesting, and perhaps functionally significant, that the energy absorbed to failure is greatest not at the lowest or highest strain rates, but in the range 0.01–1.0 /s. Assuming a bone the size of an adult femur is loaded in bending, this corresponds to a mean impact velocity much less than 1 m/s or 2 miles/h, well below that expected in trauma, even before the advent of modern transportation and weapons. Margel-Robertson and Smith [22] tested bovine femoral bone in compression at strain rates below this range and found a second local maximum in energy absorption at 0.00001–0.0001 s^{-1}. Others who have experimentally studied the viscoelastic properties of cortical bone include Smith and Keiper [83], Sedlin [84], Burstein and Frankel [85], Panjabi et al. [86], Crowninshield and Pope [87], and Bargren et al. [88]. Sedlin proposed a rheological model for cortical bone consisting of a three-parameter solid with a frictional element to account for plastic deformation at higher strains, but the data of Bargren and coworkers suggest that a two-element Kelvin model satisfactorily describes bone in the physiologic strain range. On the other hand, Laird and Kingsbury's [89] tests on cow bone indicated that a three-parameter model was inadequate to describe vibration phenomena in bone (frequencies of the order of kilohertz).

Ramaekers [90–94] published a series of papers on the rheological properties of bone. In one of these [91], he made the interesting suggestion that one of the viscoelastic functions of bone is to damp out the individual twitches of muscles. This hypothesis was prompted by the fact that one of the peaks in the shear damping frequency response corresponds to the twitch frequency of mammalian muscle.

Lakes, Katz, and Yoon have contributed a number of experiments and mathematical analyses of the effects of viscoelasticity on cortical bone mechanical properties [95–100]. This work considers primarily the deformation of bone at sonic or ultrasonic frequencies and is related to Katz's ultrasonic studies of stiffness anisotropy. The work included torsional and biaxial loading modes, and the suggestion was made that peaks in the frequency response of the material properties could have histologic and physiologic significance. Johnson and Katz [101] pointed out erroneous assumptions in these studies and attempted to correct them. They concluded that fluid flow through vascular channels is probably not important in cortical bone viscoelastic behavior; rather, it is the interactions of the fluids with the calcified matrix (which is not strictly solid, but contains minute water-filled spaces) that are important. Much remains to be done in extending these viscoelasticity studies to provide insights and predictive capabilities more relevant to bone fracture and remodeling.

In summary, some of the fundamental determinants of the mechanical properties of cortical bone's material include porosity and mineralization (which together determine apparent density), collagen orientation and cross-linking, and tissue hydration. These variables are strongly affected by histologic structure, starting with distinctions between primary and osteonal bone structure. Clearly, these factors are largely controlled by remodeling, particularly in adults. Fatigue damage may also intrinsically affect mechanical properties.

5.10 Material Properties of Cancellous Bone

We now turn our attention to cancellous bone. As one would expect from looking at its porous structure, cancellous bone is substantially weaker and more compliant than cortical bone. Table 5.4 gives some typical values of cancellous bone's compressive strength and elastic modulus. Note the considerable variability in these values. Although bone porosity is bimodally distributed (i.e., either cortical or cancellous, with few regions between), it varies continuously, and there are not sharp geometric, densitometric, or mechanical distinctions among various examples of cancellous bone or between dense cancellous bone and porous cortical bone.

Table 5.4 Compressive strength and stiffness of cancellous bone

	Ultimate stress, MPa	Modulus, MPa
Human		
[141] Yamada, 1973[a]	1.86–1.37	90–70
[142] Neil et al., 1983[b]	2.54 ± 0.62	272 ± 195
[143] Kuhn et al., 1989a[c]	5.6 ± 3.8	424 ± 208
[144] Rohl et al., 1991[d]	2.22 ± 1.42	489 ± 331
Canine		
[145] Vahey et al., 1987[e]	12.1 ± 5.7	434 ± 174
[143] Kuhn et al., 1989a[c]	7.12 ± 4.6	264 ± 132
[146] Norrdin et al., 1990[f]	9.60 ± 0.80	231 ± 22

[a]Means for people in their 40s and 60s, respectively; lumbar vertebrae
[b]Lumbar vertebrae from men aged 54–90 years, 12-mm-diameter cylinders, 25–30 mm long
[c]8-mm cubes from distal femur
[d]Proximal tibia, four men and three women, aged 42–76 years
[e]5-mm cubes from head and neck of the femur in two dogs
[f]21 vertebral specimens from seven dogs

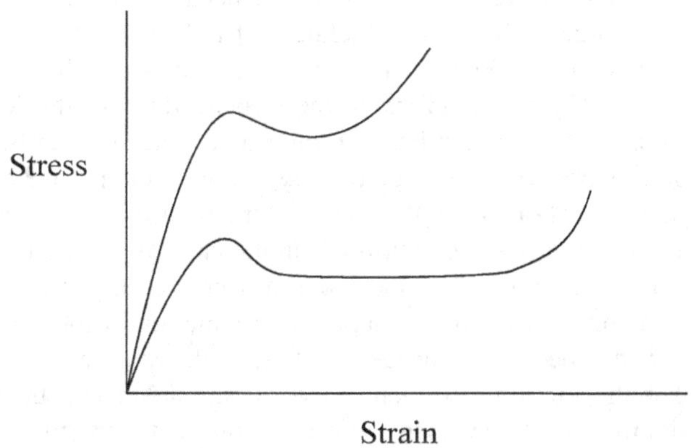

Fig. 5.10 Load–deformation curves for two specimens of cancellous bone having different porosities. Yield and ultimate loads are apparent in the early portion of each curve. Subsequently, reduction and plateauing of the load occur. As void spaces collapse, the specimen densifies and the load rises steeply

5.10.1 Stress–Strain Curves for Cancellous Bone

Figure 5.10 shows typical load–deformation curves for cubes or cylinders of cancellous bone loaded in compression. As the specimen compresses, it yields and reaches an ultimate stress, as does a cortical bone specimen. Thereafter, however, the load does not fall to zero because the specimen does not fracture into pieces. Instead, its void spaces collapse and its trabeculae compact against one another. As this process continues, the load is typically maintained, or may even rise far above the previous ultimate load. In addition to these behaviors, mechanical testing of cancellous bone is more problematic than testing cortical bone. When a compression test is done between platens and strains are based on platen displacement, errors can occur because of artifactual damage in the regions adjacent to the platens—damage caused by interruption of the trabecular network when the specimen is machined, friction against the platens, and stress concentrations. Other problems arise because supporting trabeculae have been removed along the sides of the specimen and the marrow in the void spaces is not confined as it is in the intact bone.

5.10.2 Strength in Tension vs Compression

Carter et al. [102], Bensusan et al. [63], and Rice et al. [54] all found that the failure stress of cancellous bone was the same in tension as compression. Kaplan et al. [103], on the other hand, found that cancellous bone was stronger in compression for any given apparent density. Cowin [104] suggests that this may be because the specimens in the Kaplan experiment came from a region particularly well adapted to compressive loading in the tested direction.

As with cortical bone, we treat cancellous bone here as a porous material used to construct whole bones and try to identify and quantify the features (e.g., apparent density) that govern its mechanical behavior. However, in cancellous bone, the situation is different and a bit tricky because of the relative sizes of its porosities.

In larger animals, the thickness of trabecular struts and plates is typically in the range of 100–300 μm, with the spacing between adjacent trabeculae on the order of 300–1500 μm. Therefore, a line passing through the femoral head of an elephant would encounter hundreds of trabeculae, and a similar line through the femoral head of a mouse would encounter only a few. Although we can imagine taking 1 cm^3 of cancellous bone from the elephant and treating it as a material of which the strength and modulus can be measured, in the case of the mouse the cancellous bone of the femoral head must be modeled as a structure rather than as a material, because slight geometric variations in shape and orientation of those few trabeculae could greatly alter the mechanical properties. One of the requirements of continuum mechanics is that the minimum dimension of a sample that is treated as a material must be significantly larger than the dimension of its structural subunits [105]. In the case of trabecular bone, the minimum continuum length has been estimated to be about five trabecular spacings (i.e., 5–10 mm, depending on the sample) [1, 106]. This is an order of magnitude greater than the continuum length for compact bone, where the porosity is much lower and the void spaces are much smaller.

A practical implication of this is that in speaking of cancellous bone as a porous material, we need to be thinking in terms of samples of about 1 cm^3 [107]. This leads to a conundrum. Cancellous bone can be quite heterogeneous, with its properties varying millimeter by millimeter and in penetration test data [108, 109]. A sample that is big enough to satisfy the requirements of the continuum length assumption, therefore, is quite likely to contain local heterogeneities in mechanical properties. When strength is measured on such a sample, the measurement often reflects the weakest portion of the sample. Consequently, local variations in mechanical properties tend to be overlooked in the results. Therefore, attempts to treat cancellous bone as a porous, continuum-level material must be evaluated cautiously. With this reservation in mind, we can proceed to examine cancellous bone as a porous material.

5.11 Determinants of Cancellous Bone Mechanical Properties

Insofar as we can treat cancellous bone porosity as a material property, we shall find that its mechanical properties are governed by the magnitude of the porosity (or apparent density), the anisotropy of the trabecular architecture, and the material properties of the tissue in the individual trabeculae.

The typical range of apparent density in cancellous bone is 1.0–1.4 g/cm^3, compared to about 1.8–2.0 g/cm^3 for compact bone. This decrease in apparent density primarily results from the presence of marrow-filled spaces; the density of trabecular bone's calcified matrix (or *tissue*) itself is similar to that of compact bone. Wolff [110], in formulating his ideas about bone's mechanically optimized structure, assumed that cortical and trabecular bone were both made of the same kind of tissue, so that the differences in their mechanical properties only resulted from the varying sizes and shapes of the soft tissue spaces. This concept has received more critical attention in recent years, however. There are apparently some important differences in the microstructure of the calcified bone matrix between cancellous and cortical bone, and these differences evidently are responsible for measured differences in material properties at the tissue level.

The tissue in cancellous bone is usually made in the same way, and of the same lamellar material, as cortical bone. In both cases, osteoblasts sit on a bone surface and emit an organic matrix that calcifies to produce an anisotropic, lamellar composite. There is little evidence that, at the lamellar level, this basic structure is fundamentally different in cortical and in cancellous bone. Studies of osteocyte lacunar sizes and numbers in cancellous and cortical bone have not been able to establish any clear differences. However, the mineral content and tissue density of cancellous bone have been found to be less than in cortical bone [111–113]. Consistent with this, the water content of cancellous tissue has been reported to be greater than that of cortical bone. These results are consistent with the higher rate of turnover in trabecular bone; the average age of the bone matrix is simply younger, so the tissue is less mineralized and contains more water.

Once one gets beyond the lamellar level and examines the geometry of the bone structural units created by the remodeling BMUs, another difference between cortical and cancellous bone tissues is apparent. Cortical bone osteons are parallel cylindrical structures that extend for several millimeters through the bone tissue. Their cement lines are closed surfaces. The trabecular equivalents, on the other hand, are dish-shaped "patches," or hemi-osteons, having cement lines that intersect the surface of the trabecula (Fig. 5.11). Therefore, cortical osteons act like reinforcing fibers running through the tissue, but trabeculae are composed of multiple dish-shaped patches. If a cement line is disrupted by stress in cortical bone, the disruption is contained within the bone, but in cancellous bone it may propagate to the surface, disconnecting the structural unit from the parent trabecula. Thus, remodeled cortical bone structure may be inherently more damage resistant than remodeled cancellous bone structure.

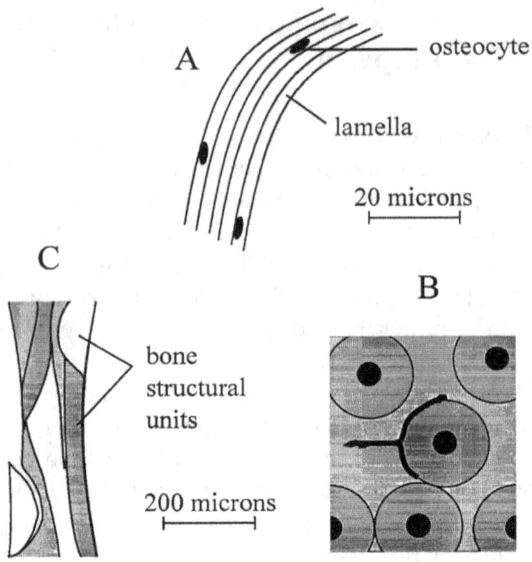

Fig. 5.11 (**a**) At the level of lamellae, there is little to distinguish between osteonal and trabecular bone tissue. (**b**) In osteonal bone, the structural units produced by remodeling are largely intact, relatively long, fiber-like osteons. Their cement lines are internal to the bone; cement line disruptions are self-contained and do not compromise longitudinal load-bearing. (**c**) In trabeculae, the structural units produced by remodeling are much smaller and dish-shaped with cement lines that, if disrupted, exfoliate the structural unit

These mineralization and structural issues lead to the question of whether cancellous tissue has the same mechanical properties as cortical tissue. There have been two ways of approaching this question. The first is to measure the property (e.g., elastic modulus) in about 1-cm^3 machined specimens of cortical and cancellous specimens and plot the results against the specimens' apparent density. If both kinds of specimen have the same tissue-level value of the property, they should fall on a single smooth curve. Rho and coworkers [114] did this with human bone, measuring the modulus ultrasonically (Fig. 5.12); they found that the modulus of the cortical bone specimens fell outside, and above, the confidence limits of the extrapolated regression line for the cancellous specimens. The triangular data points on the curve represent values ultrasonically measured in just the cancellous bone material. This was done by increasing the frequency of the sound so it only propagated through the trabecular tissue. These points fall within the extrapolated cancellous values. This graph provides strong evidence that cancellous bone has a reduced tissue-level modulus relative to cortical bone.

The other method for approaching the question is to test either individual trabeculae, or very small specimens machined from individual trabeculae, and compare the results with those from similarly sized specimens machined from cortical bone. Rho et al. [114] also did this, measuring the modulus both ultrasonically and in tension tests in whole trabeculae and microspecimens of cortical bone. The tension tests showed the modulus of cancellous material to be 56% of that of cortical bone (10.4 ± 3.5 vs 18.6 ± 3.5 GPa). The ultrasonically measured values also showed the cancellous material was less stiff (14.8 ± 1.4 vs 20.7 ± 1.9 GPa). Others have obtained different results when making the comparison between machined microspecimens of cancellous as well as cortical bone. In three-point bending, Kuhn et al. [115] found a larger modulus for cortical than cancellous bone (4.89 vs 3.81 GPa), but it was substantially less than typical values for large

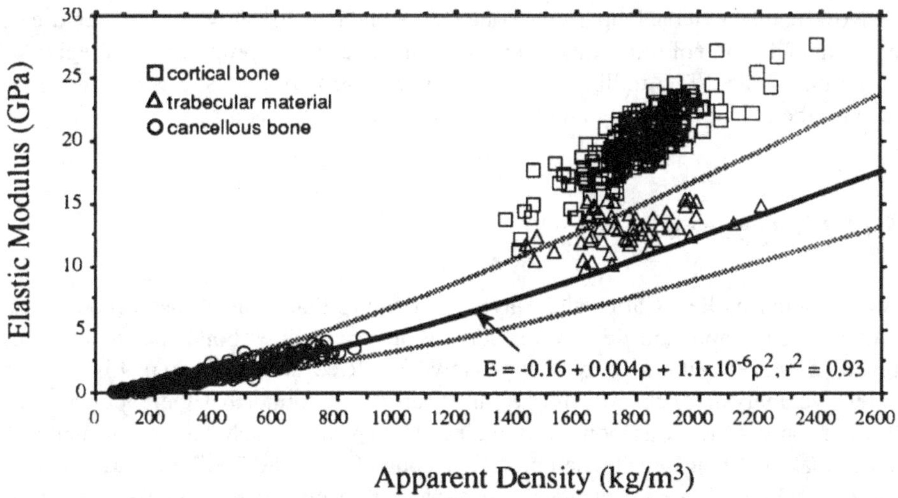

Fig. 5.12 Ultrasonically determined elastic modulus vs. apparent density for continuum-level cancellous (circles) and cortical (squares) bone specimens. A linear regression line and its 95% confidence limits are shown for the cancellous specimens. The fact that the cortical specimens have modulus values above the upper confidence limit for the cancellous data indicates that the cortical bone tissue has a greater modulus than that of cancellous bone. When the tissue modulus of cancellous bone was measured and plotted against the apparent density of the tissue, the data (triangles) fell inside the confidence limits [114]

cortical specimens (18–20 GPa). Choi et al. [116] obtained similar bending results and found that the modulus of cortical specimens decreased steadily with the dimensions of the specimens, from 14.9 GPa for a beam with a square cross section 1000 μm on each edge to 5.5 GPa for a similar beam 100 μm on each side.

Taken together, these and other studies suggest that the mechanical properties of cancellous bone tissue are different from those of cortical bone tissue. The difference between the tension test results of Rho and the bending results of Kuhn and Choi, and the beam size results of the latter group, indicate that the elimination of intact osteons may reduce the modulus, and much more so in bending than in uniaxial tension. When cortical specimens as small as trabeculae are made, their osteons must be cut into fragments not unlike those of trabecular remodeling packets and this probably reduces the modulus, especially in bending. The reduced mineralization of cancellous bone tissue, at least in part caused by the increased rate of remodeling, may also explain some of the modulus reduction.

There may be a hierarchy of average measured tissue moduli in descending order from primary cortical bone, to osteonal cortical bone, to cancellous bone tissue. However, it is also important to bear in mind that there is no "correct" value of elastic modulus for any of these tissues because of normal variations in mineralization, collagen fibril orientation, and other factors. It is important to understand the differences between cortical and trabecular bone tissue properties, but it is probably pointless to pursue increasingly "accurate" values for these properties.

Now let us move back up in scale once more and consider how the architecture of a substantial group of trabeculae determines the mechanical properties of a region of cancellous bone a few millimeters across. There are two key aspects of this architecture: the apparent density and the orientation of the trabeculae.

5.11.1 Apparent Density

As we noted earlier, Carter and Hayes [53] found that, when elastic modulus is plotted against apparent density for cortical and cancellous bone specimens combined, E varies with d^3 (Eq. 5.11). However, Rice and coworkers [54], in an extensive analysis of data from several experiments, found that the exponent was 2 when only cancellous bone was included. Typically, such regression analyses show that apparent density or porosity account for 70–80% of the variability in elastic modulus if one is considering a particular species, type of stress (tension or compression), and direction of loading relative to anatomical axes. The percentage goes down if these factors vary.

Rice and coworkers also confirmed that the compressive strength of cancellous bone varies with d^2, and this correlation accounts for about 73% of the variability. Restrictions regarding species and directionality applied to this case as well. Thus, about three-fourths of the variability in the strength and stiffness of cancellous bone can be accounted for by apparent density. Presumably, the balance depends on other factors, one of which is thought to be the directionality of the trabeculae relative to the loading direction.

5.11.2 Anisotropy of Cancellous Bone Architecture

The equations for the linear theory relating elastic constants to porosity and alignment were tested experimentally [117–119] and accounted for 72–94% of the variability in the elastic constants for cancellous bone. Because the theory assumes that the solid matrix of the bone is homogeneous and isotropic, some of the unexplained variability probably resulted from organizational factors within individual trabeculae. For example, individual trabeculae more closely aligned with average principal tensile stresses may have collagen-mineral structures (and thus modulus values) that are different from those trabeculae closely aligned with average principal compressive stress directions. Models for mechanical properties that use directionality measurements continue to be used and have now been extended to model the nonlinear mechanical properties of cancellous bone [120].

5.12 Predicting Material Properties: Bone as a Composite Material

A number of studies have attempted to mathematically model bone as a composite material. The most general approach has been to assume that bone is composed of two or more phases (e.g., collagen and mineral, or soft and hard tissue) and that these phases contribute to a mechanical property P in direct proportion to the volume fraction F that they occupy (most commonly, P is the elastic modulus). One may simplify the structure of such a model to consist of alternating slabs of the two phases; if two such slabs are arranged so that they have equal strains when they are loaded, one has the Voigt model (Fig. 5.13), in which the composite property is given by

$$P_c = P_1F_1 + P_2F_2 \tag{5.12}$$

Here the subscripts 1 and 2 refer to the two component phases and P_c designates the composite property. If the slabs have equal stresses, the result is known as the Reuss model at right in Fig. 5.13, described by

$$P_c = \frac{P_1P_2}{P_1(1-F_1)+P_2(1-F_2)} \tag{5.13}$$

It is known that the Voigt model provides an upper bound, and the Reuss model a lower bound, for the actual value of the composite property [121, 122]. Unfortunately, these bounds are typically so far apart as to be virtually useless. The Hashin–Shtrickman model [123] gives somewhat narrower bounds. Several authors [124–128] have assumed that compact bone consists of collagen (organic phase) and mineral (inorganic phase) and used these models to derive certain mechanical properties. These efforts have met with limited success, partly because the properties of the component phases are not well known, partly because their interactions are not well understood [129], and perhaps because the void space was ignored. However, Martin [130] had no better success when he assumed that bone consists of calcified matrix and voids and used the same methods to compute the elastic modulus. Others

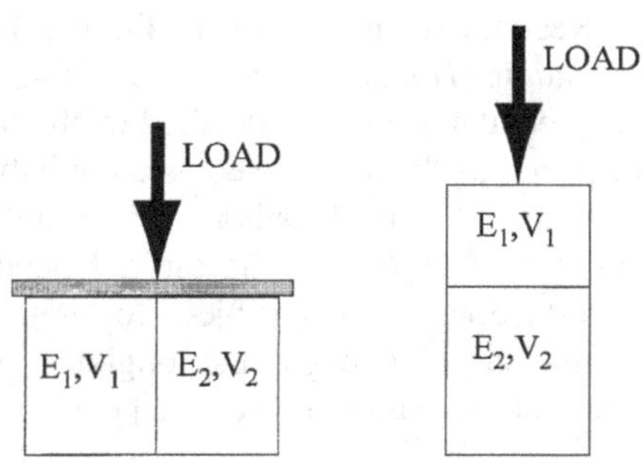

Fig. 5.13 Voigt *(left)* and Reuss *(right)* models for composite material formed from a mixture of two materials. In the Viogt model, each material strains the same amount; in the Reuss model, each material carries the same stress

who have analyzed solid-void two-phase materials include Hirsch [131] and Brown et al. [132]. Stech [133] analyzed the anisotropy of compact bone using a model containing parallel cylindrical voids representing Haversian canals. Bonfield and Clark [60] had some success in combining mixture theories to explain an age-related increase in the elastic modulus of rabbit bone that seemed to be caused by changes in both porosity and specific mineralization. They used a formulation by MacKenzie [134],

$$E = E_0\left(1 - 1.9p + 0.9p^2\right) \tag{5.14}$$

where p is porosity and E_o is the modulus of the calcified matrix, to account for porosity variation, and an equation similar to Eq. (5.9) to account for changes in mineralization. However, the analysis did not account for the entire increase in stiffness, supporting Currey's [58] hypothesis that bone apatite experiences age-related structural changes that increase its elastic modulus.

Many of these models contain constants that are more or less arbitrary and can be manipulated to fit various data. So far, however, no one has developed a theory of bone structure that can satisfactorily explain its mechanical behavior on the basis of its internal structure and composition.

5.13 Summary and Additional Reading

In this chapter, we have seen how the mechanical properties of cortical and cancellous bone are influenced by their structures. Bone behaves as an elastic (and also viscoelastic), anisotropic, composite material. The feature of its histologic structure that dominates its mechanical properties is porosity, also expressed as bone volume fraction (and closely related to apparent density). Mineralization and collagen fibril orientation are also important determinants of mechanical properties. In cortical bone, secondary osteons significantly alter anisotropy and strength, and in cancellous bone the orientations of trabeculae play a similar role. The elastic modulus and strength of cancellous bone are both heavily dependent on apparent density squared. Mixture theories may be used to formulate models for the mechanical behavior of bone, but the results have not been satisfactory.

Recommendations for additional reading include the excellent book, *Bone Mechanics Handbook: Second Edition*, edited by Cowin [135]. This work contains chapters that provide a detailed mathematical presentation of elasticity and strength of materials theory, as well as compilations of data on the mechanical properties of cortical and cancellous bone. The chapter on bone's physical properties in Albright and Brand [136] and the tutorial paper by Turner and Burr [137] are excellent introductions to this subject for non-engineers. For an interesting read on the relationship of strength and toughness in a variety of biological and non-biological materials, we suggest Ritchie's [138] review of this topic. Another resource intended for clinicians is the textbook *Fundamentals of Biomechanics: Second Edition* by Ozkaya and Nordin [139]. Alexander's monograph, *Optima for Animals* [140], contains a section on optimizing the dimensions of tubular bones with respect to strength and weight (pp. 13–23), and presents a succinct overview of optimization theory from the biological perspective.

References

1. Brown TD, Ferguson AB. Mechanical property distributions in the cancellous bone of the human proximal femur. Acta Orthop Scand. 1980;51:429–37.
2. Currey JD. Incompatible mechanical properties in compact bone. J Theor Biol. 2004;231:569–80.
3. Currey JD, Pitchford JW, Baxter PD. Variability of the mechanical properties of bone, and its evolutionary consequences. J Roy Soc Interface. 2007;4:127–35.
4. Currey JD, Butler G. The mechanical properties of bone tissue in children. J Bone Joint Surg. 1975;57A:810–4.
5. Carter DR, Spengler DM. Mechanical properties and composition of cortical bone. Clin Orthop Relat Res. 1978;135:192–217.
6. Ascenzi A, Bonucci E. The ultimate tensile strength of single osteons. Acta Anat. 1964;58: 160–83.
7. Ascenzi A, Bonucci E. The tensile properties of single osteons. Anat Rec. 1967;158:375–86.
8. Ascenzi A, Bonucci E. The compressive properties of single osteons. Anat Rec. 1968;161: 377–91.
9. Ascenzi A, Bonucci E. The shearing properties of single osteons. Anat Rec. 1972;172:499–510.
10. Frost HM, Roth H, Villanueva AR. Physical characteristics of bone. Part III. A semimicro measurement of unit shear stress. Henry Ford Hosp Med J. 1961;9:157–62.
11. Pidaparti R, Burr DB. Collagen fiber orientation and geometry effects on the mechanical properties of secondary osteons. J Biomech. 1992;25:869–80.
12. Katz JL. Composite material models for cortical bone. In: Cowin SC, editor. Mechanical properties of bone. New York: ASME; 1981.
13. Rho JY, Currey JD, Zioupos P, Pharr GM. The anisotropic Young's modulus of equine secondary osteons and interstitial bone determined by nanoindentatiojn. J Exp Biol. 2001;204:1775–81.
14. Moyle DD, Welborn JW, Cooke FW. Work to fracture of canine femoral bone. J Biomech. 1978;11:435–40.
15. Moyle DD, Bowden RW. Fracture of human femoral bone. J Biomech. 1984;17:203–13.
16. Ruggeri A, Raspanti M, Martini D, Guizzardi S. Bone as a composite material: collagen fibers orientation and functional requirements. In: Motta PM, editor. Recent Adv. microscopy of cells, tissue, and organs; 1997. p. 167–1970.
17. Walmsley R, Smith JW. Variation in bone structure and the value of Young's modulus. J Anat (London). 1957;91:603.
18. Currey JD. Differences in the tensile strength of bone of different histological types. J Anat. 1959;98:87–95.
19. Hèrt J, Kucera P, Vavra M, Volenik V. Comparison of the mechanical properties of both the primary and Haversian bone tissue. Acta Anat. 1965;61:412–23.
20. Reilly DT, Burstein AH. The mechanical properties of cortical bone. J Bone Joint Surg. 1974;56A:1001–21.
21. Reilly DT, Burstein AH. The elastic and ultimate properties of compact bone tissue. J Biomech. 1975;8:393–405.

22. Margel-Robertson D, Smith DC. Compressive strength of mandibular bone as a function of microstructure and strain rate. J Biomech. 1978;11:455–71.
23. Carter DR, Hayes WC, Schurman DJ. Fatigue life of compact bone. II. Effects of microstructure and density. J Biomech. 1976;9:211–8.
24. Carter DR, Hayes WC. Compact bone fatigue damage. I. Residual strength and stiffness. J Biomech. 1977b;10:325–37.
25. Burr DB, Schaffler MB, Fredrickson RG. Composition of the cement line and its possible mechanical role as a local interface in human compact bone. J Biomech. 1988;21:939–45.
26. Skedros JG, Holmes JL, Vada EG, et al. Cement lines of secondeary osteons in human bone are not mineral-deficient: new data in a historical perspective. Anat Rec (Part A). 2005;286: 781–803.
27. Burstein AH, Currey JD, Frankel VH, Reilly DT. The ultimate properties of bone tissue: the effects of yielding. J Biomech. 1972;5:35–44.
28. Reilly DT, Burstein AH, Frankel VH. The elastic modulus for bone. J Biomech. 1974;7:271–5.
29. Vincentelli R, Grigorov M. The effect of Haversian remodeling on the tensile properties of human cortical bone. J Biomech. 1985;18:201–7.
30. Evans FG, Vincentelli R. Relation of collagen fiber orientation to some mechanical properties of human cortical bone. J Biomech. 1969;2:63–71.
31. Vincentelli R, Evans FG. Relations among mechanical properties, collagen fibers, and calcification in adult human cortical bone. J Biomech. 1971;4:193–201.
32. Evans FG, Vincentelli R. Relations of the compressive properties of human cortical bone to histological structure and calcification. J Biomech. 1974;7:1–10.
33. Martin RB, Ishida J. The relative effects of collagen fiber orientation, porosity, density, and mineralization on bone strength. J Biomech. 1989;22:419–26.
34. Martin RB, Boardman DL. The effects of collagen fiber orientation, porosity, density, and mineralization on bovine cortical bone bending properties. J Biomech. 1993;26:1047–54.
35. Martin RB, Lau ST, Mathews PV, Gibson VA, Stover SM. Collagen fiber organization is related to mechanical properties and remodeling in equine bone. A comparison of two methods. J Biomech. 1996;29:1515–22.
36. Gebhardt FSMW. Uber funktionell wichtige anordnungsweisen der groberen und feineren bauelemente des wirbeltierknochens. I Allgemeiner Teil Wilhelm Roux Arch fur Entwicklungsmeckanik der Organismen. Archiv für Entwicklungsmechanik der Organismen. 1901;12:167–223.
37. Olivo OM. Rispondenza della funzione meccanica varia degli osteoni con la loro diversa minuta architettura. Boll Soc Ital Biol Sper. 1937;12:400–1.
38. Olivo OM, Maj G, Toajari E. Sul significato della minuta struttura del tessuto osseo compatto. Boll Soc Ital Biol Sper. 1937;109:369–94.
39. Portigliatti-Barbos M, Bianco P, Ascenzi A. Distribution of osteonic and interstitial components in the human femoral shaft with reference to structure, calcification and mechanical properties. Acta Anat. 1983;115:178–86.
40. Portigliatti-Barbos M, Bianco P, Ascenzi A, Boyde A. Collagen orientation in compact bone: II. Distribution of lamellae in the whole of the human femoral shaft with reference to its mechanical properties. Metab Bone Dis Relat Res. 1984;5:309–15.
41. Pauwels F. Biomechanics of the locomotor apparatus: contributions on the functional anatomy of the locomotor apparatus. Berlin: Springer-Verlag; 1980.
42. Boyde A, Riggs CM. The quantitative study of the orientation of collagen in compact bone slices. Bone. 1990;11:35–9.
43. Batson EL, Reilly GC, Currey JD, Balderson DS. Posteexercise and positional variation in mechanical properties of the radius in young horses. Equine Vet J. 2000;32:95–100.
44. Martin RB, Pickett JC, Zinaich S. Studies of skeletal remodeling in aging men. Clin Orthop Relat Res. 1980;149:268–82.

45. Ashman RB, Mayer DC, Cowin SC, Van Buskirk WC. Elastic properties of human and canine femora. Trans Orthop Res Soc. 1983;8:127.
46. Goodwin KJ, Sharkey NA. Material properties of interstitial lamellae reflect local strain environments. J Orthop Res. 2002;20:600–6.
47. Chen HL, Gundjian AA. Determination of the bone-crystallites distribution function by x-ray diffraction. Med Biol Eng. 1974;12:531–5.
48. Hulsen C. Specifisches Gewicht, Elasticitat und Festigkeit des Knochengewebes. Bull Lab Biol St Petersb Bd. 1898;1:7–37.
49. Underwood EE. Quantitative Stereology. Reading: Addison-Wesley Publishing Co.; 1970.
50. Schaffler MB, Burr DB. Stiffness of compact bone: effects of porosity and density. J Biomech. 1988;21:13–6.
51. Currey JD. The effect of porosity and mineral content on the Young's modulus of elasticity of compact bone. J Biomech. 1988;21:131–9.
52. McElhaney J, Alem N, Roberts V. A porous block model for cancellous bones. ASME Paper No. 70-WA/BHF-2, 1–9. New York: American Society of Mechanical Engineers; 1970.
53. Carter DR, Hayes WC. The compressive behavior of bone as a two-phase porous structure. J Bone Joint Surg. 1977a;59A:954–62.
54. Rice JC, Cowin SC, Bowman JA. On the dependency of the elasticity and strength of cancellous bone on apparent density. J Biomech. 1988;21:155–68.
55. Smith JW, Walmsley R. Factors affecting the elasticity of bone. J Anat. 1959;93:503–22.
56. Vose GP, Kubala AL. Bone strength—its relationship to x-ray-determined ash content. Hum Biol. 1959;31:261–70.
57. Currey JD. The mechanical consequences of variation in the mineral content of bone. J Biomech. 1969a;2:1–11.
58. Currey JD. The relationship between the stiffness and the mineral content of bone. J Biomech. 1969b;2:477–80.
59. Wainwright SA, Biggs WD, Currey JD, Gosline JM. Mechanical Design in Organisms. New York: Halsted Press; 1976.
60. Bonfield W, Clark EA. Elastic deformation of compact bone. J Mater Sci. 1973;8:1590–4.
61. Currey JD. Effects of porosity and mineral content on the Young's modulus of bone. Proc Eur Soc Biomech. 1986;5:104.
62. Hernandez CJ, Beaupre GS, Keller TS, Carter DR. The influence of bone volume fraction and ash fraction on bone strength and modulus. Bone. 2001;29:74–8.
63. Bensusan JS, Davy DT, Heiple KG, Verdin PJ. Tensile, compressive, and torsional testing of cancellous bone. Trans Orth Res Soc. 1983;8:132.
64. Ashman RB, Rho JY. Elastic modulus of trabecular bone material. J Biomech. 1988;21(3): 177–81.
65. Enlow DH, Brown SO. A comparative histological study of fossil and recent bone tissues. Part I. Introduction, methods, fish and amphibian bone tissues. Tex J Sci. 1956;8:405–43.
66. Enlow DH, Brown SO. A comparative histological study of fossil and recent bone tissues. Part II. Reptilian and bird bone tissues. Tex J Sci. 1957;9:186–214.
67. Enlow DH, Brown SO. A comparative histological study of fossil and recent bone tissues. Part III. Mammalian bone tissues. General discussion. Tex J Sci. 1958;10:187–230.
68. Frost HM. Intermediary Organization of the Skeleton. Boca Raton: CRC Press; 1986.
69. Schaffler MB, Radin EL, Burr DB. Mechanical and morphological effects of strain rate on fatigue of compact bone. Bone (New York). 1989;10:207–14.
70. Launey ME, Buehler MJ, Ritchie RO. On the mechanistic origins of toughness in bone. Annu Rev Mater Res. 2010;40:25–53.
71. Gray RJ, Korbacher GK. Compressive fatigue behaviour of bovine compact bone. J Biomech. 1974;7:287–92.
72. Carter DR, Hayes WC. Fatigue life of compact bone. I. Effects of stress amplitude, temperature and density. J Biomech. 1976;9:27–34.

73. Saito M, Marumo K. Effects of collagen crosslinking on bone material properties in health and disease. Calcif Tissue Int. 2015;97:242–61.
74. Gautieri A, et al. Advanced glycation end-products: mechanics of aged collagen from molecule to tissue. Matrix Biol. 2017;59:95–108.
75. Willett TL, et al. In vitro non-enzymatic ribation reduces post-yield strain accommodation in cortical bone. Bone. 2013;52(2):611–22.
76. Surowiec RK, Allen MR, Wallace JM. Bone hydration: how we can evaluate it, what can it tell us, and is it an effective therapeutic target? Bone Rep. 2022;16:101161.
77. Wang L. Solute transport in the bone lacunar-canalicular system (LCS). Curr Osteoporos Rep. 2018;16:32–41.
78. Monteiro DA, Dole NS, Campos JL, et al. Fluid shear stress generates a unique signaling response by activating multiple TGFß family type I receptors in osteocytes. FASEB J. 2021;35 (3):e21263.
79. Eberhardsteiner L, Hellmich C, Scheiner S. Layered water in crystal interfaces as source for bone viscoelasticity: arguments from a multiscale approach. Comput Methods Biomech Biomed Engin. 2014;17(1):48–63.
80. Masic A, Bertinetti L, Schuetz R, et al. Osmotic pressure induced tensile forces in tendon collagen. Nat Commun. 2015;6(1):5942.
81. Wang Y, Von Euw S, Fernandes FM, et al. Water-mediated structuring of bone apatite. Nat Mater. 2013;12(12):1144–53.
82. McElhaney JH. Dynamic response of bone and muscle tissue. J Appl Physiol. 1966;21:1231–6.
83. Smith RW, Keiper DA. Dynamic measurement of viscoelastic properties of bone. Am J Med Electron. 1965;4:156–60.
84. Sedlin ED. A rheological model for cortical bone. Acta Orthop Scand. 1965;83:1–78.
85. Burstein AH, Frankel VH. The viscoelastic properties of some biological materials. Ann N Y Acad Sci. 1968;146:158–65.
86. Panjabi MM, White AA, Southwick WO. Mechanical properties of bone as a function of rate of deformation. J Bone Joint Surg. 1973;55A:322–30.
87. Crowninshield RD, Pope MH. The response of compact bone in tension at various strain rates. Ann Biomed Eng. 1974;2:217–25.
88. Bargren JH, Bassett CAL, Gjelsvik A. Mechanical properties of hydrated cortical bone. J Biomech. 1974;7:239–45.
89. Laird GW, Kingsbury HB. Complex viscoelastic moduli of bovine bone. J Biomech. 1973;6 (1):59–67.
90. Ramaekers JGM. The dynamic shear modulus of bone in dependence on the form. Acta Morphol Neerl Scand. 1977;15:185–201.
91. Ramaekers JGM. The rheological behavior of cortical bone and cartilage of Bos taurus. The importance of maximal damping capacity in collagen. Acta Morphol Neerl Scand. 1978;16: 55–67.
92. Ramaekers JGM. The mechanical properties of mallard humeral bone in dependence on the form. Acta Morphol Neerl Scand. 1979a;17:173–80.
93. Ramaekers JGM. The dynamic shear modulus and damping of compact bovine metacarpal bone in dependence on the topography along the bone shaft. Neth J Zool. 1979b;29:151–65.
94. Ramaekers JGM. The rheological behavior of skeletal material originating from several classes of vertebrates. Neth J Zool. 1979c;29:166–76.
95. Lakes RS, Katz JL. Interrelationships among the viscoelastic functions for anisotropic solids: application to calcified tissues and related systems. J Biomech. 1974;7:259–70.
96. Lakes RS, Katz JL. Viscoelastic properties of wet cortical bone. III. A nonlinear constitutive equation. J Biomech. 1979a;12:689–98.
97. Lakes RS, Katz JL. Viscoelestic properties of wet cortical bone. II Relaxation mechanisms. J Biomech. 1979b;12:679–87.

98. Lakes R, Saha S. Long-term torsional creep in compact bone. J Biomech Eng. 1980;102:178–80.
99. Lakes RS, Katz JL, Sternstein SS. Viscoelastic properties of wet cortical bone. I. Torsional and biaxial studies. J Biomech. 1979;12:657–78.
100. Lakes RS, Yoon HS, Katz JL. Ultrasonic wave propagation and attenuation in wet bone. J Biomed Eng. 1986;8:143–8.
101. Johnson M, Katz JL. Some new developments in the rheology of bone. Biorheology (Suppl). 1984;1:169–74.
102. Carter DR, Schwab GH, Spengler DM. Tensile fracture of cancellous bone. Acta Orthop Scand. 1980;51:733–41.
103. Kaplan SJ, Hayes WC, Stone JL, Beaupre GS. Tensile strength of bovine trabecular bone. J Biomech. 1985;18:723–7.
104. Cowin SC. Bone Mechanics. Boca Raton: CRC Press; 1989.
105. Fung YC. A first course in continuum mechanics. Englewood Cliffs: Prentice Hall; 1969.
106. Harrigan TP, Jasty M, Mann RW, Harris WH. Limitations of the continuum assumption in cancellous bone. J Biomech. 1988;21:269–75.
107. Linde F, Hvid I, Madsen F. The effect of specimen geometry on the mechanical behavior of trabecular bone specimens. J Biomech. 1992;25:359–68.
108. Hvid I, Hansen SL. Trabecular bone strength patterns at the proximal tibial epiphysis. J Orthop Res. 1985;3:464–72.
109. Hvid I. Mechanical strength of trabecular bone at the knee. Dan Med Bull. 1988;35:345–65.
110. Wolff J. The Law of Bone Remodeling. (Translation of Wolff's Das Gesetz der Transformation der Knochen by P. Maquet and R. Furlong.). Berlin: Springer-Verlag; 1892.
111. Gong JK, Arnold JS, Cohn SH. Composition of trabecular and cortical bone. Anat Rec. 1964;149:325–32.
112. Dyson ED, Whitehouse WJ. Composition of trabecular bone in children and its relation to radiation dosimetry. Nature (London). 1968;217:576–8.
113. Norrdin RW, Phemister RD, Jaenke RS, LoPresi CA. Density and composition of trabecular and cortical bone in perinatally irradiated beagles with chronic renal failure. Calcif Tissue Res. 1977;24:99–104.
114. Rho J-Y, Ashman RB, Turner CH. Young's modulus of trabecular and cortical bone material: ultrasonic and microtensile specimens. J Biomech. 1993;26:111–9.
115. Kuhn JL, Goldstein SA, Choi KW, London M, Feldkamp LA, Mathews LS. Comparison of the trabecular and cortical tissue moduli from human iliac crests. J Orthop Res. 1989b;7:876–84.
116. Choi K, Kuhn JL, Ciarelli MJ, Goldstein SA. The elastic moduli of human subschodral, trabecular, and cortical bone tissue and the size-dependency of cortical bone modulus. J Biomech. 1990;23:1103–14.
117. Turner CH, Cowin SC. Dependence of elastic constants of an anisotropic porous material upon porosity and fabric. J Mater Sci. 1987;22:3178–84.
118. Cowin SC, Mehrabadi MM. Identification of the elastic symmetry of bone and other materials. J Biomech. 1989;22:503–15.
119. Turner CH, Cowin SC, Rho JY, Ashman RB, Rice JC. The fabric dependence of the orthotropic elastic constants of cancellous bone. J Biomech. 1990;23:549–61.
120. Charlebois M, Pretterklieber M, Zysset PK. The role of fabric in the large strain compressive behavior of human trabecular bone. J Biomech Eng Trans ASME. 2010;132(12):121006.
121. Hill R. The elastic behavior of a crystalline aggregate. Proc Phys Soc Lond. 1952;A65:349.
122. Hill R. Elastic properties of reinforced solids: some theoretical principles. J Mech Phys Solids. 1963;11:357.
123. Hashin Z, Shtrikman S. A variation approach to the theory of the elastic behavior of multiphase materials. J Mech Phys Solids. 1963;11:127.
124. Currey JD. Three analogies to explain the mechanical properties of bone. Biorheology. 1964;2:1–10.

125. Katz JL. Hard tissue as a composite material. I. Bounds on the elastic behavior. J Biomech. 1971;4:455–73.
126. Bonfield W, Li CH. Anistropy of non-elastic flow in bone. J Appl Physiol. 1967;38:2450.
127. Piekarski K. Analysis of bone as a composite material. J Biomech. 1973a;11:557–65.
128. Piekarski K. Analysis of bone as a composite aggregate. Int J Eng Sci. 1973b;11:557.
129. Burstein AH, Zika JM, Heiple KG, Klein L. Contribution of collagen and mineral to the elastic-plastic properties of bone. J Bone Joint Surg. 1975;57A:956–61.
130. Martin RB. Porosity and specific surface of bone. Crit Rev Biomed Eng. 1984;10:179–222.
131. Hirsch TJ. Modulus of elasticity of concrete affected by elastic moduli of cement paste matrix and aggregate. Proc Am Concr Soc. 1962;59:427.
132. Brown SD, Biddulph RB, Wilcox PD. A strength-porosity relation involving different pore geometry and orientation. Am Ceram Soc J. 1964;47:320.
133. Stech EL. A descriptive model of lamellar bone anisotropy. In: Byars EF, editor. Biomechanics monographs. New York: American Society of Mechanical Engineers; 1967. p. 236–45.
134. MacKenzie JK. The elastic constants of a solid containing spherical holes. Proc Phys Soc. 1950;B63:2–11.
135. Cowin SC, editor. Bone mechanics handbook. Boca Raton: CRC Press- second edition; 2001.
136. Albright JA, Brand RA. The scientific basis of orthopaedics. New York: Appleton-Century-Crofts; 1987.
137. Turner CH, Burr DB. Basic biomechanical measurements of bone: a tutorial. Bone (New York). 1993;14:595–608.
138. Ritchie RO. The conflicts between strength and toughness. Nat Mater. 2011;10:817–22.
139. Ozkaya N, Nordin M. Fundamentals of biomechanics: equilibrium, motion and deformation. 3rd ed. New York: Springer Publishing; 2012.
140. Alexander RM. Optima for animals. Princeton: Princeton University Press; 1996.
141. Yamada H. Strength of biological materials. Huntington: Krieger; 1973.
142. Neil JL, Demos TC, Stone JL, Hayes WC. Tensile and compressive properties of vertebral trabecular bone. Trans Orth Res Soc. 1983;8:344.
143. Kuhn JL, Goldstein SA, Ciarelli MJ, Mathews LS. The limitations of canine trabecular bone as a model for human: a biomechanical study. J Biomech. 1989a;22:95–107.
144. Rohl L, Larsen E, Linde F, Odgaard A, Jorgensen J. Tensile and compressive properties of cancellous bone. J Biomech. 1991;24:1143–9.
145. Vahey JW, Lewis JL, Vanderby R. Elastic moduli, yield stress, and ultimate stress of cancellous bone in the canine proximal femur. J Biomech. 1987;20:29–33.
146. Norrdin RW, Histand MB, Sheahan HJ, Carpenter TR. Effects of corticosteroids on mechanical strength of intervertebral joints and vertebrae in dogs. Clin Orthop Relat Res. 1990;259: 268–76.

Chapter 6
Mechanical Adaptability of the Skeleton

Abstract Bone is exquisitely sensitive to its mechanical environment. In Chap. 1, we reviewed the history of Wolff's law and the mechanical adaptability of bone. Here, we dig deeper into the theories regarding the mechanical adaptability of cortical and trabecular bone. The mechanisms by238 which this occurs from the biochemical to the organ level are overviewed through the description of various models that can explain mechanical adaptation.

What You Will Learn

1. Differences between somatic and evolutionary adaptation and how this pertains to mechanical adaptability in bone.
2. Concepts of structure-sensitive and structure-insensitive properties of bone.
3. How bone structure is regulated to find the optimum solution to minimize both fracture risk and bone mass.
4. How local strain thresholds/setpoints define bone's adaptive response to mechanical stimuli.
5. Additional concepts of mechanical adaptability and damage repair.

6.1 Introduction

The idea that bone structure is somehow controlled locally to suit its current mechanical function has become a central tenet of orthopedic medicine and science. The goal of this chapter is to take a deep dive into the idea, which we briefly introduced in earlier chapters as Wolff's law. As a scientific hypothesis, the putative ability of bone to not only repair itself but also to adapt to variations in imposed loading is an extremely appealing idea to engineers and biologists alike. However, developing this idea into a scientific theory supported by appropriate experimental results has proven to be a hefty scientific problem. In the following sections, we take up this problem, trying to organize the principal hypotheses and recent experiments found in the literature into a coherent theory.

M. R. Allen, J. Wallace, *Skeletal Tissue Mechanics*,
https://doi.org/10.1007/978-3-031-90609-1_6

6.2 Mechanical Adaptability and Evolutionary Adaptability

There has been considerable debate among engineers, biologists, and clinicians concerning the balance between genetics and mechanical adaptability in deciding the structures of bones. Some have argued that the anatomy and histology of a bone such as the human femur are determined primarily by genetic information expressed by various cells as they form the skeleton and maintain it throughout life. Others have come to believe that mechanical factors strongly influence the expression of the genetic plan carried by the cells. It is appropriate to consider the relationships between mechanical adaptability and evolutionary change as determinants of skeletal structure. This discussion is based on the ideas of Bateson [1, 2].

6.2.1 Somatic vs Evolutionary Adaptation

The adaptation of bone to alterations in mechanical loading is an example of somatic change, that is, a change in body form or function that does not entail a genetic change and is therefore not passed on to succeeding generations. Somatic change is distinct from evolutionary change, which does involve a modification of the genome of the organism. A simple example of somatic change is obtaining a sun tan during a beach vacation; obviously, this does not alter the genetic code for skin color, and the tan is not passed on to the vacationer's children. Nevertheless, somatic change plays an important role in evolution.

Bateson pointed out that the ability to adjust to environmental changes by somatic modification of body systems is essential to survival because evolutionary change operates relatively slowly, through chance mutations. In addition, somatic adjustments allow the organism to survive when a genetic mutation, which might ultimately have survival value, disrupts one or more body systems. Conversely, genetic adjustments to the somatic system itself allow the homeostatic range to be restored when a mutation has driven it to an extreme. Thus, evolution involves constant interaction between slow, haphazard genetic changes and rapid, predictable somatic changes. In short, somatic flexibility allows time for adaptive modifications of the genome to occur, and also helps the organism tolerate their adverse side effects.

Work in the burgeoning field of epigenetics has demonstrated that interactions between somatic and genetic adaptations are far more complex and interrelated than was once thought. We now know that it is possible for somatic changes to an organism to be passed on to subsequent generations, not through mutations in the genetic blueprint (DNA sequence) but through somatic events such as DNA methylation and histone modifications that change the expression of specific DNA sequences. The processes of epigenetics are well appreciated during growth and development as specialized cells differentiate along countless pathways, depending on morphogens and other factors within their immediate environment. Subsequent daughter cells inherit the characteristics and phenotypes of their differentiated predecessors. This often occurs through an epigenetic process in which an enzyme called methyltransferase adds a methyl group to the carbon 5 position of cytosine (one of the four bases that comprise DNA) converting it to 5-methylcytosine. High levels of 5-methylcytosine alter local transcription patterns and cell phenotype. Going from cells to the whole organism, genetic imprinting occurs when these heavily methylated regions of the genome are passed from parent to offspring through the germline. This tends to silence expression of the allele from that parent in favor of expression of the other parent's allele. Recent work suggests that epigenetics may play an important role in transgenerational adaptation to environmental factors. Cooney and colleagues [3] found that feeding maternal mice methyl-supplemented diets strongly affected the distribution of coat color patterns in their offspring, and correlated with methylation of the sequences controlling its phenotypic expression. A pair of multigenerational studies in humans found associations between early paternal smoking and the body mass index of their sons, and even more striking, between the food supply of paternal grandfathers and the mortality risk ratios of their grandsons [4]. It is not beyond the realm of possibility that epigenetic mechanisms play an important role in bone adaptation to environmental or lifestyle changes (e.g., diet, physical activity) in the individual during development and maybe even in subsequent generations [5].

A Nonskeletal Example

Consider what happens when the vacationer mentioned above goes instead to visit Machu Picchu, the ancient Incan ruins in the Peruvian Andes. After flying to Lima, the tourist's next stop is the city of Cuzco with an elevation of 3700 m (12,000 ft). Picking up her backpack and heading for her hotel, she soon finds her heart pounding and her breath coming in gasps because of the thin air. Clearly, these are responses to the trouble her cells are having in getting enough oxygen. Her increased volume and rate of breathing and pumping blood are somatic changes allowing her to cope with the environmental changes. By the time she leaves Cuzco 2 days later to continue her journey to Machu Picchu itself, carrying her backpack no longer requires her heart and respiration rates to be adjusted upward because her hematopoietic system has produced more red blood cells. This second level of somatic adaptability requires more time to put in place, but much less metabolic energy than pumping blood and air faster.

However, more red blood cells per unit volume of blood (called hematocrit) is not the best physiologic adjustment for living permanently at high altitude. "Lowlanders" who go to high altitudes may suffer severe cardiovascular problems caused by the effects of increased hematocrit on the physical properties of their blood. Some individuals can adjust to living at high altitudes, but many cannot. Groups of people have lived at high altitudes (principally in the Andes and Himalayan mountains) for many generations and appear to be free of altitude sickness. It is possible that these populations have experienced genetic changes, through differential reproductive success, which have adapted them to high altitudes in ways that avoid the problems associated with increased hematocrit [6].

Somatic Change in the Skeleton

Comparing bone's mechanical adaptation to Bateson's altitude adaptation example, the correlate to increased heart and respiration rates is an increased remodeling rate to repair load-induced microdamage. The correlate to increased hematocrit is periosteal and endocortical modeling to decrease stresses in the bone. These skeletal responses can have problematic sequelae, just as the respiratory ones can. The increased rate of remodeling requires metabolic energy, and it introduces remodeling space porosity, which can increase strains and the damage formation rate. The formation of new bone on periosteal and endosteal surfaces also requires metabolic resources, and the additional bone mass must be carried about each day, at a significant energy cost.

6.2.2 Somatic vs Evolutionary Effects

Bateson's point was that all evolutionary change involves endless somatic and genetic adjustments, which together achieve adaptations that would be impossible if only one or the other existed. Thus, it is shortsighted to debate which caused which. It is more important to learn how somatic and genetic changes interact in the skeleton, as they do in other organ systems. This knowledge should ultimately be useful in treating many bone injuries and diseases.

Box 6.1: Teleology and Adaptation, Optimization and Evolution

An engineer who has designed a structure can say, "The purpose of this strut is to reduce bending moments on this beam." We cannot analogously say that the purpose of a strut of trabecular bone is to reduce the bending moment on an adjacent trabecula. The difficulty lies with the word "purpose." The attribution of purpose to physiologic mechanisms is called teleology, and in this text we have tried to avoid it. We might say instead, "a consequence of this trabecular strut is that bending moments on this adjacent trabecula are reduced." This is not just an esoteric point of semantics; it is important in understanding the nature of biological adaptation, whether it is somatic or evolutionary.

In the course of natural selection, a group of organisms produces a great many genetic variants, some of which are better adapted in the sense that they will leave more progeny in the next generation. By this mechanism, species evolve, and adaptations such as bone remodeling come into being and are preserved. The process is a passive and stochastic one: the concept of purpose does not enter into it.

There are two sides to biological adaptation—the organism and its environment. When the environment is noisy and unpredictable, the adaptive process has a more difficult problem to solve (for some examples see: Crummett et al., 2013; Ng et al., 2013; Jansen et al., 2011; Eranen et al.,

(continued)

Box 6.1 (continued)
2009 [7–10]). As we saw in Chap. 2, the mechanical environment of vertebrate bones is provided to a large extent by their own muscles; much of this loading is predictable and controllable. But the outside environment also directly loads bones, and these forces may be less predictable and controllable. If a creature walks on level ground all day, and at the end of the day steps unexpectedly into a hole, which of these is mechanically most significant in driving adaptation: the 20,000 cycles of gait or the one cycle of tripping that generates twice as much skeletal stress? Lanyon [11] has favored the idea that skeletal adaptation is primarily an error-driven process, and it is the unusual loads that are most significant, but this important question has not been the subject of much research. More recent work by Currey and colleagues [12] tends to suggest otherwise.

Organisms that live in water inhabit a mechanical environment that is seemingly more predictable than the terrestrial one, which should allow for more metabolically efficient adaptation of their skeletons relative to land dwellers. But the buoyancy of water works against the adaptive advantage of a lighter skeleton in that a heavy skeleton balances low-density fat, resulting in an organism with an average density close to that of water, thereby facilitating aquatic locomotion. Conversely, birds (that is, flying birds, not terrestrial birds like chickens and ostriches) find exceptional adaptive value in a minimal-weight skeleton, which improves the efficiency of flight. Other factors contribute to the skeleton's mechanical milieu as well, including pregnancy and other seasonal fluctuations in body weight, antler growth, hibernation, migration, and fighting among males. These factors must complicate the optimization of skeletal structure.

This chapter is devoted to discussing the proposition that the skeleton contains adaptive mechanisms that optimize its structure in some sense. Deterministic models [13] suggest that bones do not evolve an optimal structure or properties from any one perspective, because there are competing interests (e.g., mechanical demand vs. mineral homeostasis vs. nutrient transfer vs. immune function), although they may approach optimal. It would therefore be a mistake to hypothesize that evolution is directed toward maximizing the efficiency of structures. Evolution is directed toward maximizing genetic representation in subsequent generations, which in turn is associated with efficiency, but only rather loosely. Indeed, the physiques of many ancient relict species contain features that are hardly paragons of efficiency. Evolution can only modify the basic plan that exists. Take for instance the now well-known example of The Panda's Thumb popularized by Steven Gould [14]. It works quite well and is adapted to its role, but it is certainly not designed in a way an engineer would design it were she to start "from scratch." The human knee joint, and the vertebral column, are two other examples, perhaps more immediate to those using this book.

6.3 What Is the Design Goal—To What Does Bone Adapt?

It is difficult to discuss mechanical adaptation of bone unless we can define the term "adaptation." What does this really mean in the context of purely mechanical skeletal function? For the purposes of our following discussion, we propose that bone structure (its shape and mass) is regulated to find the best (the optimum) solution to minimize both fracture risk and bone mass for the loadings that the bone experiences. The adaptation of bone is a somatic change, the limits of which are determined by the genome of the organism. We will show later that the unknown genetic potential for somatic change severely hampers the formulation of a generally applicable theory for the adaptation of bone to mechanical load.

Minimization of fracture risk is important in most bones, and minimizing bone mass allows the organism to reduce the metabolic energy required for its daily activities, increasing its ability to survive. Following the lead of Currey [15], we speak of optimizing stiffness because bones need to support loads without excessive deformation, but they also need some flexibility for shock absorption and to avoid brittle fracture (consider chalk). The optimal stiffness surely varies with the function of the particular bone. For example, Currey [16] has discussed the divergent functions of three bones. The antlers of the male red deer (which are made of bone as opposed to the keratinous tissue that forms cow horns) require impact strength and energy-absorbing power; stiffness is deemphasized. The femoral cortex of the cow requires more stiffness. The tympanic bulla in the ear of the fin whale requires a high acoustic impedance ($=\sqrt{\rho E}$) and therefore high stiffness. Table 6.1 shows how the elastic modulus and other attributes vary among the three bones. Although most bones would be more like the cow femur than antler or tympanic bulla in their function, the optimal stiffness obviously varies from bone to bone. At this point, it seems clear that fracture risk and bone weight need to be minimized; but we shall see several alternative possibilities for the optimized variable in the following pages.

Table 6.1 Material properties of three different bones

Variable	Red deer antler	Bovine femur	Whale bulla
Work-of-fracture, J/m^2	6190	1710	200
Bending strength, MPa	179	247	33
Elastic modulus, GPa	7.4	13.5	31.3
Density, g/mL	1.86	2.06	2.47
Acoustic impedance, 109 kg/m^2/s	3.71	5.27	8.79
Mineral content, % by weight	59.3	66.7	86.4

Adapted from Ref. [16]

6.3.1 *Optimization, Mechanical Adaptability, and Structure-Sensitive or Insensitive Properties*

There is always a trade-off in nature; improvement in one property may be detrimental to another, and to maximize evolutionary fitness requires balance. In bone, this is reflected by the relationship between stiffness and toughness (or work to fracture), which are typically inversely related. Stiffer bones tend to absorb less energy prior to fracture [17] and therefore optimizing stiffness does not necessarily mean that fracture risk will be reduced. Osteopetrosis provides an illustrative example of this trade-off. This is a condition in which osteoclasts are not working well; so the bone is not resorbed as it should be (which also means that new bone cannot be formed by remodeling processes). The bone as a whole thus becomes more highly mineralized, and overly stiff and strong. However, it has little capacity to absorb energy once it has passed the yield point, and therefore demonstrates brittle behavior and breaks easily, especially under impact loading.

Currey proposed that bone adapts to these competing needs by optimizing toward structure-insensitive mechanical properties [12]. Structure-insensitive properties are those that are not greatly influenced by small flaws or inhomogeneities in the structure [18]. These properties tend to be less variable across bones or species than those that are structure-sensitive. For bone, structure-insensitive properties are those that determine properties in the pre-yield region of the stress–strain curve, such as elastic modulus and bending strength. Structure-sensitive mechanical properties are more greatly affected by structural inhomogeneities. Examples include toughness and work to fracture, which determine the energy expended in the post-yield region, as well as impact strength.

Evolutionarily, Currey argues, bone adapts to the more predictable (i.e., less variable) properties. Thus, stiffness would be deterministic, even at the expense of greater uncertainty in toughness. Mineral content and collagen cross-linking contribute to stiffness and toughness, albeit in an inversely proportional way. In the trade-off between stiffness and toughness, the fittest individuals would be those with the greatest certainty in pre-yield properties, even if this created less predictability in post-yield properties. From a structure-function standpoint, this makes a great deal of sense given that we spend most of our time loading our bones in the pre-yield range. It also makes sense in a self-repairing structure like bone, in which damage caused by unpredictable loads can eventually be repaired so that the animal can survive to reproduce another day.

6.4 How Does Bone Adapt: What Are the Signals and How Are They Transduced?

Mechanotransduction is a word used frequently by researchers, but it is not often defined. It is a general term that can have a variety of meanings, depending on the context in which it is used. Mechanotransduction should be considered a series of inter-related mechanical and biological events, rather than a single process. In bone, it was defined most clearly and thoroughly by Duncan and Turner [19], who divided the sequence of events from the initial mechanical signal to the eventual biological effect into four stages: Mechanocoupling, Biochemical Coupling, Signal Transmission, and Effector Cell Response.

Mechanocoupling refers to the mechanical stimulus that initiates the events leading to a particular cellular response. We treat the potential "actors" in this play extensively in this chapter, concluding that strain may or may not have a direct impact on cellular response, but it certainly is the driving force behind fluid movements in bone that cause fluid shear stresses on osteocyte and canalicular membranes which, in turn, produce (although this is less likely to be significant) electrical potentials (streaming potentials) to which bone cells can respond. Fluid motion also modifies the diffusional mixing of extracellular signaling molecules, which can affect cell behavior, independent of fluid shear stress. Fluid mixing of the molecules is important because the concentration and the rate of presentation of signaling molecules changes cell responses in complex and incompletely understood ways.

Once the mechanical stimulus has been perceived by the cell, it must transduce that mechanical signal into an intracellular biochemical response in the sensor cell (the osteocyte in bone) that has the potential to either transmit the signal to other cells or to initiate the recruitment and eventual maturation process that causes bone formation or resorption. This biochemical coupling may take several forms, and include physical changes to the cell (e.g., integrin–cytoskeleton-nuclear matrix deformations [20, 21], alterations in G-protein pathways, phosphorylation of proteins or regulation of mineral channels (e.g., calcium, phosphorus, or chlorine channels are common), or combinations of physical and biological regulation (e.g., stretch-activated protein channels).

Once sensor cells (now commonly assumed to be osteocytes, bone-lining cells, and pre-osteoblasts) detect the signal and begin to respond, they must transmit this information to other cells that will actually do the work of making more bone, or removing it. Signal transmission may involve a host of autocrine and paracrine factors, but the messages themselves must be transmitted through a dendritic network that is much like the neural network in the brain [22]. This transmission seems to require connexins (in particular connexin-43, but probably also others) that form gap junctions between cells.

Finally, effector cells that actually produce the response must mature and go to work. In bone, these are the osteoclasts and osteoblasts. It is worth emphasizing that they do not work in a vacuum generated only by the information provided by the sensor cells, but are regulated during their activities by hormones, other metabolic products, noncollagenous matrix proteins, and by communication among themselves.

In this chapter, we will concentrate mostly on the first step in the mechanotransduction process, mechanocoupling, and to a lesser degree on signal transmission, as the other steps are largely complex biological processes, outside the scope of this text. If cells adapt bone to mechanical loads by responding to a mechanical stimulus, it is obviously important to know what that stimulus is and to be able to measure it. One attractive suggestion, supported by many experiments, is that bone adapts to strain. Cells attached to bone's surfaces may "feel" the displacements of the surface and respond to them, or the osteocytes within the bone matrix may sense the strain-induced flow of fluid within their network. Strain is also an appealing variable from the point of view of minimizing fracture risk. We saw in Chap. 5 that yield strains are reasonably constant for bone. Therefore, keeping strains near a point well below the yield strain should maintain a margin of safety against monotonic, creep, and fatigue failure. Finally, if stiffness is to be optimized in some sense, the fact that strain is a measure of bone stiffness under a given load (i.e., the stiffer the bone, the less strain that occurs for a given load) may be significant. These features and others make it easy to see why strain is targeted as the prime functional stimulus, and why measurement of bone strains in vivo has been such an important component of the experimental work aimed at bone adaptation.

6.4.1 Measurement of Strain in Living Animals

Until the advent of in vivo strain gauging techniques, the adaptation of bone to mechanical loading could not be effectively analyzed because bone's "mechanical environment" could not be quantitatively assessed at a level commensurate with cellular activity. Lanyon and Smith [23, 24] reported the first measurement of in vivo bone strains in animals when they implanted uniaxial strain gauges on the medial tibial cortex in sheep. They found mean longitudinal surface strains of 260 microstrain during walking and a 16% increase when the animals were trotting.

Subsequent to this, Lanyon [25, 26] applied strain gauges to the thoracic and lumbar vertebrae in sheep. He reported average peak surface strain ranges in thoracic vertebrae during walking of 142 microstrain ($\mu\varepsilon$), increasing 25% to 187 $\mu\varepsilon$ during trotting. Cochran [27, 28] applied strain gauges to the tibia and ulna in dogs, measuring strains in the tibia between 103 and 665 $\mu\varepsilon$, depending on the load and the anatomical position of the strain gage. The mode of the strain, compressive or tensile, was not specified. Shortly after, Barnes and Pinder [29] and Turner et al. [30] added the radius, tibia, metacarpus, and metatarsus of horses to this list. Lanyon et al. [31] demonstrated that strain magnitudes in the human tibia during locomotion were similar to those in bones of other animals.

6.4.2 Principal Stresses, Strains, and Trabecular Orientations

About this time, Lanyon began to use rosette strain gauges, a grouping of three gauges typically oriented at intervals of 45°, which permits the measurement of principal strain directions (i.e., the directions of maximum and minimum strain). Applying these gauges to the calcaneus of a sheep, Lanyon [32, 33] demonstrated that the principal orientations of trabeculae coincided with the principal compressive and principal tensile strain directions. Later, Hayes and Snyder [34] conducted an intensive study of the human patella, which concluded that its trabecular orientations are highly correlated with principal stress directions, as determined by finite element analysis. Although this study did not examine in vivo strains on the patella, both studies suggest that the proposed "trajectory criterion" of Culmann, Meyer, and Wolff described in Chap. 1 is essentially correct and trabeculae do align themselves with the continuum level principal stress (or strain) directions (i.e., the principal directions in a homogenous model structure).

There have been several efforts to mathematically understand the trajectory criterion for cancellous bone by combining properties of the stress (or strain) with properties of cancellous bone structure and/or cancellous bone mechanical properties. Cortical bone [35] and cancellous bone [36] were experimentally demonstrated to be mechanically anisotropic and essentially orthotropic—having three orthogonal principal material axes similar to the three principal axes of stress. If cancellous bone is an orthotropic material, then if one aligns the largest principal modulus (stiffness) with the largest magnitude of principal stress and so on down to the smallest principal modulus and stress, the elastic energy put into the material by an arbitrary stress is minimized [37]. This suggests that trabecular orientation is a result of elastic energy minimization through alignment of the material properties with the applied stress. It was also experimentally demonstrated that the orientations of the surfaces of the trabeculae could be modeled fairly accurately as a second-rank tensor called the mean intercept length (MIL) and that the orientations of the principal axes of the MIL tensor matched the orientation of the orthotropic material properties of the bone [38]. As a result, there are three mathematical objects that one can consider aligned in cancellous bone in order to explain the trajectory criterion: (1) the principal directions of the orthotropic material properties, (2) the principal directions of the mean intercept tensor, and (3) the principal stress (or strain) directions. These alignments can be condensed into the rule that trabecular orientation matches the orientation of the principal stresses [39] and, with additional arguments, that their orientation is driven by the goal (an optimization) to make the principal apparent axial strains the same [40].

All of these ideas for mathematically explaining the orientation of cancellous bone with loading result in solely orthogonal networks of trabeculae because all of the principal axes that are being aligned are orthogonal. Unfortunately, trabeculae do not always meet orthogonally—for example, the two main portions of the trabecular network in the proximal femur meet at an angle different from 90°. This was recognized early [41] but was improperly rejected by arguing that early work on the theory of elasticity predicted orthogonal stresses [42]. (For the philosophers among our readers, this was a triumph of the Platonic Ideal (mathematical theory) over Aristotelian Materialism (anatomical data)). Non-orthogonality of trabeculae was mathematically explained by a superposition of multiple loading cases at both the average [43] and trabecular level [44], but the underlying biological mechanisms for the sensation and correction of trabecular orientation to loading are still not known.

6.4.3 Peak Strains Are Similar Across Species

Subsequently, Rubin and Lanyon [45] surveyed their own work and that of others and published the data shown in Table 6.2. These data show that when an adult animal's musculoskeletal system is functioning at a high physiologic level (e.g., a dog is running or a bird is flying), peak periosteal strains usually lie between 2000 and 3000 $\mu\varepsilon$ in compression. The fact that measured strains were nearly equivalent on different bones and in different animals supported the hypothesis that the skeletons of all animals are adapted to control strain.

There are at least two notable exceptions to the data in Table 6.2. First, strains on the human tibia at a site where stress fractures occur have not been observed to exceed 2000 $\mu\varepsilon$, even during strenuous activity by Special Forces soldiers [46]. More recent experiments, however, show that strains in excess of 3000 $\mu\varepsilon$ can be generated during jumping activities, which may explain why "shin splints" are a relatively common injury among basketball players (see Yang et al. for a comprehensive

Table 6.2 Peak functional strains in various animal bones

Bone	Activity	Peak strain ($\mu\varepsilon$)	Reference
Horse radius	Trotting	−2800	Rubin and Lanyon (1982) [45]
Horse tibia	Galloping	−3200	Rubin and Lanyon (1982) [45]
Horse metacarpal	Acceleration	−3000	Biewener et al. (1983) [151]
Dog radius	Trotting	−2400	Rubin and Lanyon (1982) [45]
Dog tibia	Galloping	−2100	Rubin and Lanyon (1982) [45]
Goose humerus	Flying	−2800	Rubin and Lanyon (1982) [45]
Cockerel ulna	Flapping	−2100	Rubin and Lanyon (1982) [45]
Sheep femur	Trotting	−2200	Rubin and Lanyon (1982) [45]
Sheep humerus	Trotting	−2200	Rubin and Lanyon (1982) [45]
Sheep radius	Galloping	−2300	O'Conner et al. (1982) [105]
Sheep tibia	Trotting	−2100	Lanyon et al. (1982) [49]
Pig radius	Trotting	−2400	Goodship et al. (1979) [152]
Fish hypural	Swimming	−3000	Rubin and Lanyon (1982) [45]
Macaca mandible	Biting	−2200	Hylander (1979) [153]
Turkey tibia	Running	−2350	Rubin and Lanyon (1984) [91]

Adapted from Ref. [45]

review of in vivo bone strain measurements in people [47]). Second, thoroughbred racehorses generate 5000–6000 με on the surface of their metacarpal (cannon) bone when running at racing speeds. However, these horses are special animals bred and trained to run exceptionally fast while they are still juveniles. Strains of this magnitude probably only occur in young animals (or people) in which bone turnover is rapid and many osteons are still mineralizing.

6.4.4 Compressive Strains Are Larger Than Tensile Strains

Because compressive strains in a long bone are usually superimposed on bending strains, the neutral axis is shifted and the peak compressive strains are usually greater than the peak tensile strains (Fig. 6.1). This is not universally true, however—an example would be in the palatomaxillary segment of some lizards, where tensile strains exceed compressive strains [48]. Lanyon et al. [49] demonstrated that the ratio of (the absolute values of) peak compressive strain to peak tensile strain on the sheep radius varies between 1.45 and 2.34, and tensile strain was never greater than compressive strain during normal locomotion. In fact, tensile strain was always less than, and compressive strain was always greater than, 1000 με. Even after osteotomy of the adjacent ulna, the tensile strain on the radius was less than 1000 με. High-fidelity cadaver simulations of human gait suggest that this phenomenon also holds true in the weight-bearing bones of humans. Such experiments enable comprehensive but otherwise prohibitively invasive assessments of bone strain through placement of rosette gauges at multiple sites on a single bone [50, 51]. Results from studies focused on the tibia under loading conditions representative of normal walking nicely illustrate the greater compressive strains experienced in the posterior aspect of the tibial diaphysis relative to the tensile strains present on its anterior aspect (Fig. 6.2).

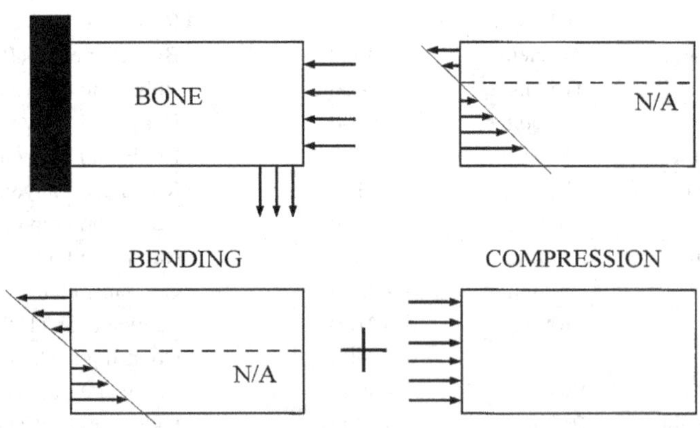

Fig. 6.1 Most long bones are loaded in combined bending and compression, producing an asymmetric stress distribution in which compressive stresses are greater than tensile stresses

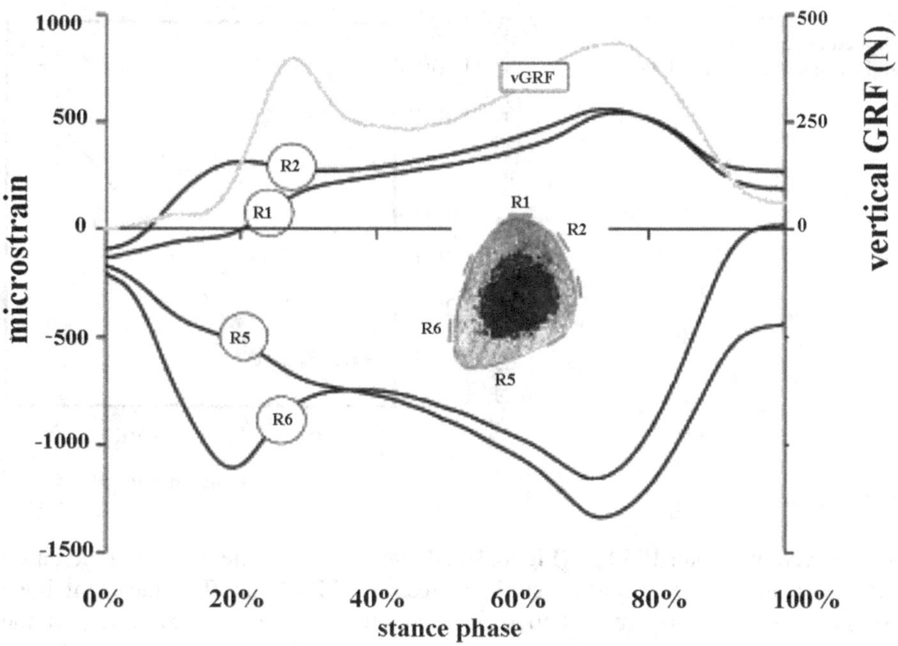

Fig. 6.2 Axial strain profiles at two anterior (R1 and R2) and two posterior (R5 and R6) sites of the tibial shaft, illustrating compressive loading superimposed with bending. The relationship between vertical ground reaction force (vGRF) and tibial strains during stance is also readily apparent. The left ordinate axis has units of microstrain (με) and the right ordinate axis corresponds to the dashed vGRF profile, with units of Newton [51]

6.4.5 What Aspects of Strain Regulate Adaptation?

The pioneering studies by Lanyon and others to measure strains on the surface of bones during activity provide some guidelines about which features in the strain environment may be critical to the adaptive response. The experiments of Lanyon and Rubin, in particular showing that strain magnitude seems constrained within relatively strict boundaries are convincing, that the magnitude of strain is a component of the signal. The uniformity of peak strain across species, and the correspondence between trabecular orientation and principal strain directions—are strong evidence that strain magnitude is one regulator of skeletal adaptation to the mechanical environment.

The *principle of uniformity of peak strain* implies a threshold-driven process. To further identify the relevant magnitudes of strain that could elicit a response, Turner et al. [52] developed a rodent model in which bone formation rate (BFR) could be measured in response to application of different magnitudes of strain generated by the four-point bending of the tibia. Turner et al. [53] loaded a series of rat tibiae at peak compressive strains ranging from 400 to 2000 με and found no difference in BFR in the loaded tibia when compared to the contralateral nonloaded tibia when

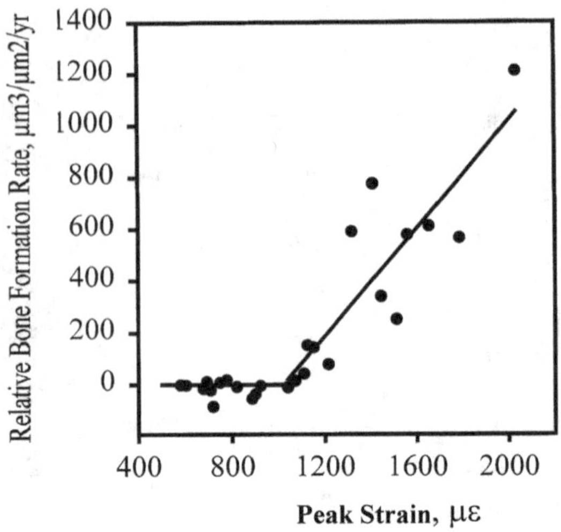

Fig. 6.3 Apposition rate vs applied bending load for the endosteal surface of rat tibia [53]

strains were less than 1050 με (Fig. 6.3). Above 1050 με, bone formation increased monotonically by sevenfold as strain increased to 2000 με. The idea that bone formation was linearly related to strain magnitude was not a new one, but the experimental demonstration that there was a threshold, and about where that threshold lay, was a new observation.

Nine years earlier, Rubin and Lanyon [54] performed a similar experiment in which they applied loads to avian forelimbs from zero (unloaded, and in disuse) to 4000 με and noted a linear increase in cortical bone area related to increasing strain. They did not identify a threshold for response, but instead presented a linear response over the entire range of strains. In more recent work, this group has continued to assert this linearity of response over the range of loads and activities from disuse to rigorous loads involving high strains with adaptation by woven bone [55]. However, when their data are viewed more closely, it becomes clear that there is a threshold for response, and that this threshold occurs slightly above 1000 με, just as in the Turner experiment using rats. Although presented as loads rather than strains, their more recent data also suggest a threshold-driven process because mean values for cortical and cancellous bone areas and volumes do not change substantially when peak loads (in mice) are less than 6 N. These studies appear to validate the idea of a threshold for adaptive response, and imply that this threshold may be similar across species.

The concept of a regulatory threshold is consistent with the view that the osteocyte lacunar–canalicular system performs as a neural [22] or "small world" [56] network that connects osteocytes with osteoblasts and lining cells to enhance mechanical signals. As Gross points out, such networks exhibit threshold-related behavior, and also have the capacity, once the threshold has been reached, to synchronize the signal, which can effectively amplify it. This does not necessarily mean that the strain threshold is exactly the same at all locations on bones. In fact, the idea of a "neural network" involving osteocytes suggests the opposite: the capacity for learning and memory, as in a true neuronal network. Using the rat ulnar-loading model, Hsieh et al. [57] showed that bone formation was initiated at compressive strain thresholds that varied from about 1300 to 3000 $\mu\varepsilon$ along the length of the ulna. These thresholds correlated with regions of the ulna that were normally under higher strain, suggesting that bone tissue habitually receiving higher mechanical strains can become less responsive to loading, i.e., the system "remembers" its local loading history and responds accordingly. In other words, cells accommodate (or habituate) to their routine loading history, and probably their hormonal environment, and only respond when the perceived stimulus is outside what they normally experience. The biological mechanism for such an accommodation has not been determined, but may reside in the osteocyte network [22].

Box 6.2: Cellular and Acellular Bone

Not all bone contains an osteocyte network. It has been known since the middle part of the nineteenth century that bone in some fish is acellular, although cells may be present during skeletal development. It may be that fish have limited need for cellular signaling to repair bone damage. Being in an aquatic environment certainly lowers gravitational forces associated with muscle contraction. Nevertheless, muscle contraction requires calcium, and so fish must use other mechanisms for systemic calcium regulation—and indeed they do. During development, humans develop a structure called the ultimobranchial body, which is eventually incorporated into the thyroid gland. Cells from this structure give rise to C cells in the thyroid gland that produce calcitonin, which helps to regulate plasma calcium levels. Fish also have an ultimobranchial body that produces calcitonin. When calcitonin is administered to fish with osteocytes, plasma calcium levels decrease. However, when calcitonin is administered to fish without osteocytes, plasma calcium does not change. To see reviews of this work, refer to Parenti [58]. This finding suggests that osteocytes are important for the systemic regulation of calcium in the blood, which they must perform by signaling other cells to resorb bone to release calcium stored there, or stop resorbing when circulating Ca^{++} levels become too high. When osteocytes are not present, those signals are not generated and plasma calcium is invariant. Moss [59] performed an experiment to demonstrate differences in fracture repair between fish with cellular vs acellular bone. Moss fractured the jaws of both kinds of fish, and observed the repair process under normal conditions in an aquatic environment with normal levels of calcium. In fish with and without cells, fracture healing occurred via endochondral ossification by the formation of a large cartilage callus that subsequently calcified. When a similar experiment was performed in water that had no calcium salts in it, the fish with cellular bone healed and mineralized normally, but those with acellular bone failed to calcify the cartilage callus. This finding again suggests that fish without cellular bone are not able

(continued)

Box 6.2 (continued)
to internally regulate their calcium balance as do fish with osteocytes. These studies demonstrate the capacity of osteocytes to signal other cells, and the critical role played by osteocytes in metabolic regulation, but do not speak to the exclusivity of osteocytes in transducing mechanical signals. In this regard, there is preliminary evidence that acellular fish bone may have the capacity to form new bone in response to experimentally imposed mechanical loading, suggesting other, perhaps more primordial, mechanisms of mechanotransduction that function independent of osteocytes [60]. Regardless of eventual conclusions, cellular and acellular fish bone provide a natural model to further investigate the regulatory pathways responsible for bone growth and maintenance.

Thus, bone adaptation can be considered a location-specific and error-driven process. This has formed the basis for a number of computational models that attempt to predict bone's modeling and remodeling behavior, including those devised by Cowin [61, 62], Hart and Davy [63], Carter and Beaupre [64, 65], and Frost [66]. For this to be true, of course, the opposite must also be demonstrated: loss of a mechanical signal should result in bone loss.

6.4.6 Canine Disuse Experiments

Uhthoff and Jaworski [67, 68] studied the effects of disuse in growing dogs by casting one of their forelimbs for up to 40 weeks. The metacarpal diaphyseal bone loss in growing dogs was characterized primarily by reduced formation and increased resorption on the periosteal surface, indicating that the normal periosteal bone formation accompanying growth was inhibited or converted to resorption. Relative to the contralateral control bone, bone mass decreased rapidly at first, then more slowly. After 40 weeks of immobilization, it was asymptotically approaching about half the bone mass of the contralateral control bone. When the cast was removed after 32 weeks of immobilization, about 65–70% of the deficit in bone mass was recovered in 28 weeks. In the diaphysis, this occurred both through endosteal apposition and through additions of woven bone to the periosteal surface. When the experiment was repeated in older dogs, bone loss occurred primarily at the endosteal surface and through increased intracortical porosity [69]. There was no effect on periosteal diameter. Older dogs were less able to recover lost bone following remobilization, recovering only 40% of the lost diaphyseal bone. In older dogs, remobilization stopped the endosteal bone loss and caused apposition of small amounts of lamellar bone to the periosteal surface.

6.4.7 The Mechanostat and the Equilibrium Strain Range

Frost [66, 70] attempted to put these ideas about how bone adapts together into a conceptual model, integrating the concepts to define not only when a response would occur, but the nature of the response (resorption or formation) and the location of the response on one of the four bone envelopes. He likened the modulation of bone adaptation to strain to a home thermostat responding to temperature, and therefore termed his conceptual model "*The Mechanostat*."

The mechanostat theory begins with the concept that there is a *minimum effective* strain, which must be exceeded to excite an adaptive response (Fig. 6.3). However, instead of simply postulating that strains below a certain threshold will evoke no response, Frost suggests that there is an *equilibrium range* of strain values, which will evoke no response. Strains above this range evoke a positive adaptive response (increased bone), and strains below this range cause a negative adaptive response (loss of bone). Others have suggested similar physiologic ranges, sometimes called "*dead zones*" or "*lazy zones*." In this framework, the minimum effective strain becomes the upper limit of the equilibrium strain range, or the strain beyond which an augmenting adaptive response is initiated. There would also be a lower strain limit defining the strain below which bone loss would occur. The thresholds separating these zones or ranges are termed "setpoints." These setpoints are determined by many factors in the hormonal and metabolic environment, and so cannot be viewed as fixed at a specific strain level. At very high strains, there is another setpoint, which defines the strain range within which damage is severe enough to elicit a repair response. This can be either a remodeling response to repair internal damage to the bone matrix—microdamage—or rapid deposition of woven bone on external surfaces, such as is seen in a stress fracture.

Frost's mechanostat theory asserts that modeling and remodeling have opposite responses to loading that pushes strains above or below the equilibrium range:

Increased loading—modeling is increased and remodeling is inhibited.
Decreased loading—modeling is inhibited and remodeling is increased.

The logic behind this algorithm is as follows. While the inhibition of remodeling by high strain may seem "backwards" because of the long-held concept that increased loads or strain stimulate "remodeling" to increase bone strength, it must be remembered that we are using the strict definition of remodeling, which does not include modeling. Think first of cortical bone—osteonal remodeling increases porosity as a result of the remodeling space and the new Haversian canals are created. It is, therefore, desirable that osteonal remodeling should be depressed by overload and stimulated by disuse. However, in those cases in which loads are so high that bone is damaged, it is desirable not to reduce remodeling, but to stimulate a repair reaction. We now know that the mechanostat theory is not entirely correct because remodeling is indeed activated at very high loads, when bone is damaged and dying osteocytes produce signals that recruit osteoclasts to remove the damage.

It is also reasonable that mechanical overload should activate modeling to adapt the bone to future, similarly high loads. As we have seen, this has been observed experimentally in many different models.

It is important to understand that an identical mechanical strain can elicit a different response if the setpoint for response is altered. For example, if the setpoint to activate remodeling were raised, the bone could perceive that normal physical activity was decreased when it was not, activating an inappropriate remodeling response causing loss of bone. This may be, in part, what occurs in postmenopausal osteoporosis. With the loss of the estrogen stimulus, the setpoint for remodeling is higher, causing a loss of bone from endocortical, trabecular, and intracortical surfaces even in the presence of normal activity and normal strains. The parathyroid hormone (PTH) is well known to increase osteoblast sensitivity to a mechanical load [71–73]. When given intermittently, as in therapies for osteoporosis, it elicits bone formation directly on endocortical and trabecular surfaces (i.e., a modeling response). Thus, the upper threshold to stimulate modeling is lowered, making the bone "see" usual strains from normal activity as an error signal.

6.4.8 Mechanical Compensation and the Roles of Bone Modeling and Remodeling

The competing architectural demands of remodeling and modeling show that mechanical adaptation occurs at both the structural level and at the tissue level, and these processes must be somehow coordinated. Shape and size changes are effected on external (i.e., periosteal and endosteal) bone surfaces, where there is more modeling than remodeling during growth, and less capacity for change in adults [70]. Changes in material properties are effected on internal (i.e., Haversian and trabecular) bone surfaces and within the lacunar–canalicular network. Internal bone surfaces are governed primarily by remodeling throughout life [70].

The gross anatomy (i.e., the shape and size) of a bone is, in large part, determined during growth. Modeling serves to sculpt the bone to suit both the genetic plan and the demands of current mechanical usage. In adults, modeling is reduced in rate and extent, so that alterations of bone geometry require much more time (or a mechanical stimulus large enough to produce woven bone). This restriction is important with respect to joints as well as bones. An adult who greatly increases the loads on their skeleton is, in principle, more likely to acquire osteoarthritis than a juvenile because his bones have insufficient capacity to alter their metaphyseal size to reduce joint stresses [74].

The mechanical properties of the bone material are matched to mechanical usage by remodeling. Unlike modeling, remodeling rates can be substantially increased in adults by various events, including changes in mechanical loading. But the purpose of remodeling is to maintain bone and to repair material damage, not to change gross architecture; large architectural adaptations to mechanical loading can only be effected by modeling processes. This limitation of remodeling is significant because it is practically impossible to substantially increase the strength or stiffness of cortical bone by remodeling, there being no large voids to be filled. Therefore, in adults it is much easier to reduce cortical bone strength than to increase it.

Because of these competing goals of modeling and remodeling, in both health and disease there are compensatory modifications of structure and material that maintain or adapt the bone to its mechanical environment. For example, postmenopausal osteoporosis involves loss of cancellous bone mass in the spine, which reduces the failure stress of the vertebral bodies. However, in men there is a concurrent increase in vertebral diameter that prevents the structural strength from being reduced to the same extent as the material strength [75]. A similar phenomenon occurs in long-bone diaphyses [76–79] and the neck of the femur [80], although others [81] have shown a more complex redistribution of bone with age in the femoral neck. This form of compensation does not occur as effectively in women, for reasons that are not entirely clear. In part, the difference between men and women may be related to both size differences and material property differences. Cancellous or intracortical bone loss in smaller bones requires greater compensatory periosteal bone formation to maintain a given stiffness than it does in larger bones. Moreover, reduced total bone stiffness during aging, which can be caused by the increased rate of remodeling at menopause in women, requires a much greater periosteal response to maintain stiffness than does a similar percentage loss of bone mass [82]. For example, a 2% reduction in bone apparent modulus might require twice as much apposition on the periosteal surface to maintain stiffness, as does a 2% reduction in bone mass.

These adaptive functions between structure and material occur even during pathologic episodes, resulting in the strongest structure that circumstances allow. For example, human femurs drastically alter their cross-sectional geometry (by modeling of the periosteal and endosteal surfaces) to compensate for the poor material quality and large bending stresses arising from curvatures produced by rickets. Another example is shown in Fig. 6.4, which shows silhouettes of young (left) and old (right) femoral midshaft cross sections. The neutral axis for bending during gait is approximately from the top (anterior) to the bottom (posterior) of the sections. Note that the increased porosity associated with aging is more apparent in endosteal regions and quite striking near the neutral axis, where the resulting reduction in the bone's failure stress has minimal effect on torsional and bending strength.

6.4.9 Cellular Habituation: Do Cells Get Tired?

The idea that bone adaptation is threshold-related (strain magnitude) and error-driven (strain magnitude and distribution) led to the notion that as long as bone is loaded above its threshold, or in an unusual direction, cells will continue to respond until bone adaptation (which, after all, takes some time to achieve) lowers the strain magnitude to below some response "setpoint" (in Harold Frost's terminology). We

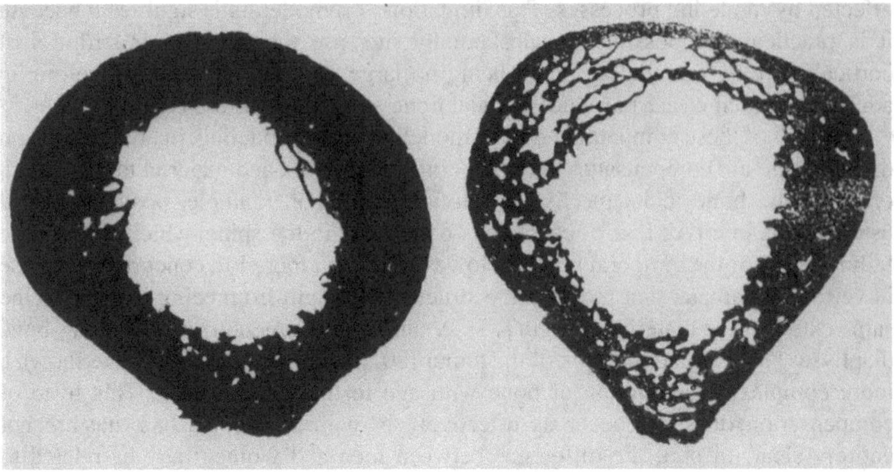

Fig. 6.4 Midshaft cross sections of femur from a woman in her thirties (Left) and from a woman in her nineties (Right)

know that bone responds to a few cycles of loading if the strain magnitude is high enough, and may not respond at all to many cycles of loading if the strain is insufficient to achieve a certain threshold. But what if bone is repeatedly loaded above the strain threshold—will that threshold eventually be insufficient to cause a cellular response? In other words, do cells "adapt" to strain magnitude and become desensitized to it over a period of time?

Experiments suggest that cells do become less sensitive, and that the timeline and/or number of loading cycles in which this occurs can be quite short. The idea that cells may become less responsive was embedded in the early experiments of Rubin and Lanyon [45], but because those experiments were focused on the initiation of bone adaptation, the process of habituation was not recognized from these experiments.

The Rubin and Lanyon experiments showed that, if the strains were sufficiently high, bone loss could be prevented with only four cycles/day of loading, and bone formation (in modeling mode on periosteal surfaces) would be stimulated with only 36 cycles per day. These experiments also showed that bone formation was no greater when bone was loaded 100 cycles/day, or 1800 cycles/day at the same strains, clearly showing that just increasing the number of applied cycles did not stimulate the cells to work harder. Other studies following the early Rubin and Lanyon experiments, using different experimental designs and different animal models, suggested that when bone is loaded to about 1000 $\mu\varepsilon$, the adaptive response saturates within approximately 200–400 loading cycles [83] (Fig. 6.5). These findings led Robling and Turner [84–86] to use the tibial four-point bending model in rats to deliver identical numbers of load cycles each day (i.e., an identical energy input), but to vary the timing of load delivery. They either delivered the load as 360 cycles at one time, or divided those 360 loading cycles into 2, 4, or 6 bouts of

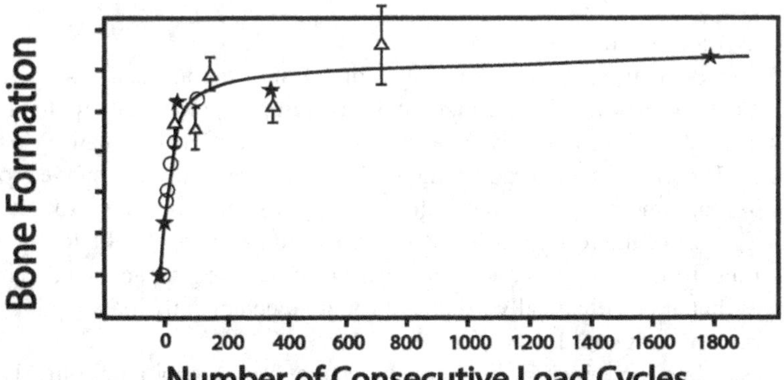

Fig. 6.5 The relationship between bone formation and the number of load cycles per day based on experiments in roosters [91], rats subjected to jumping exercise [83], and four-point tibial bending in rats [53]. In combination, these data demonstrate that a small number of loading cycles creates the signals necessary for bone formation. Additional loading cycles do not increase the amount of bone formation or bone mass

180, 90, or 60 cycles in each bout, separated by different periods of time over the course of a single day. One would expect that, without cell saturation, it would not matter how the 360 cycles were delivered, since the amount of energy imparted to the system in each case was identical. However, these experiments showed that it did matter, and that bone formation rates (BFR/BS) were significantly increased when the load was delivered in bouts of 60 or 90 cycles each over the course of the day, compared to group in which the 360 cycles were delivered at one time.

Box 6.3: Adaptation to Load vs Evolutionary Adaptation

In the theory of evolution, an *adaptation* (or adaptive trait) is a trait of an organism that has a current functional role. In evolutionary theory, *adaptation* refers both to the trait and to the evolutionary process that led to the adaptation. For a theory of how bone responds to load, the word *adaptation* means that the organism is using *phenotypic plasticity* to improve bone function. The two meanings of the word adaptation are similar; however, *evolutionary* adaptation changes the genotype and *load* adaptation is a somatic (phenotypic) adjustment that (as far as we know today) is not heritable. For bone adaptation, the implication is that the bone is improved (in some sense of the word) by the somatic changes. This idea of improvement of the bone in response to loading is a *theory*, albeit one with substantial experimental support. Current experimental results do show that many of the changes in bone in response to loading should reduce the risk of fracture by, for example, increasing the cortical width in a bent bone or by reducing the curvature of an axially loaded long bone. However, this is not to say that all changes of bone in response to load make it

(continued)

Box 6.3 (continued)
better mechanically. For example, a blow on the shin can injure the periosteum and cause formation of a permanent bony lump that appears to have no mechanical function, but may have a *biological* function in repairing tissue damage. Therefore, in general, changes of a bone in response to changed loads seem to make the bones mechanically better, but sometimes they do not. On balance, most of the responses to loading do seem to be "for the better" so it is not wrong to call the process bone *adaptation* as long as any load related changes that are mechanically irrelevant (or are mechanically maladaptive) are not forgotten or ignored.

These data convincingly demonstrate that cells habituate quickly to repeated loading (a relief to those of us who do not care to run 5 miles a day to maintain our bone health). At compressive strain levels of approximately 1500 $\mu\varepsilon$, cell accommodation to load occurs within the first 100–200 load cycles (because the loading cycles were delivered at 2 Hz, this represented less than 2 min of cyclic loading). However, these experiments did not demonstrate the length of the refractory period. How long does it take cells to become resensitized to loads once they have become saturated? A second set of experiments was designed in which all groups were given 4 bouts of 90 cycles per bout, and the time between the bouts varied from 30 min to 8 hours [86]. Bone formation rate was maximized when bouts were separated by 4–8 hours (Fig. 6.6). However, re-sensitization did not occur all at once, but was a gradual process that extended beyond the 8-hour inter-loading period used in these experiments. Inter-loading periods longer than 8 hours were not examined, but extrapolating data from the shorter periods suggests that return to full response may require as long as 12 hours.

Related studies have been performed using low-strain, high-frequency vibrations applied to mesenchymal stem cells (MSC) in culture [87]. In this case, refractory periods of 1–3 hours between loading bouts enhanced the response of the MSCs, more so with the longer refractory period. This effect appeared to be independent of the magnitude or the frequency of the loading stimulus, further suggesting that strain magnitude may not be the predominant stimulus for bone adaptation. Taken as a whole, these results have significant implications for the design of training and exercise programs intended to enhance skeletal health. On the one hand, skeletal health can be maintained or improved by several short periods of exercise over the course of the day. It is also true that, if skeletal health is the goal, injuries such as stress fractures can be avoided by designing training programs that include modest amounts of exercise. However, it should be kept in mind that short bouts of exercise are insufficient for proper maintenance of cardiovascular fitness.

(continued)

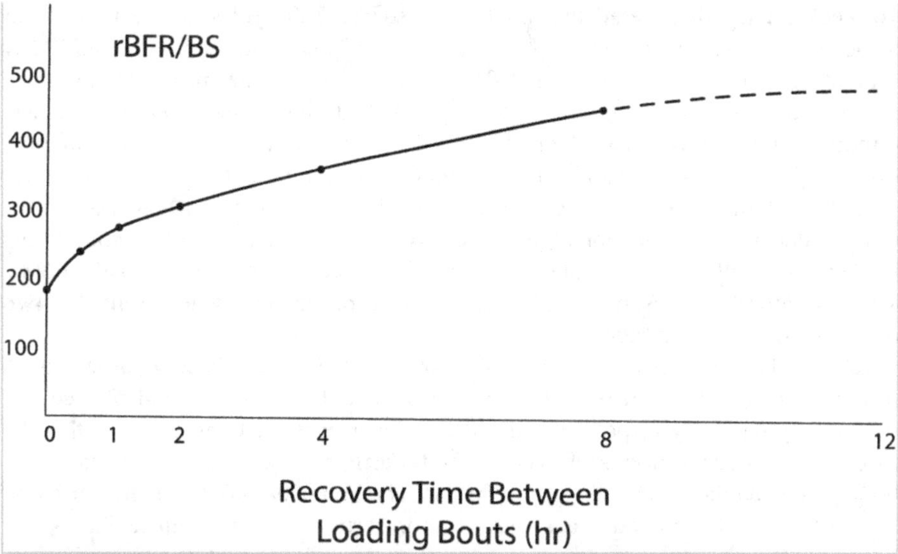

Fig. 6.6 Bone cells have a refractory period following loading, and only gradually recover their full sensitivity to subsequent loads. Osteoblasts recover most of their responsiveness to loads within about 8 hours, but extrapolation from that data indicates that full recovery may not occur for about 12 hours following a loading session. (Adapted from Ref. [86])

6.4.10 Optimization and the Mathematical Basis of Wolff's Law

Our assumption for this discussion is that bone structure is regulated to find the optimum solution to minimize both fracture risk and bone mass for the loadings that the bone experiences. A mathematical difficulty in this formulation is that we do not specify whether this is a "global" or a "local" optimum for the bone. A local optimum is a state of bone mass and geometry that is "better than any close by." That is, small changes in the mass or structure make the bone worse (higher fracture risk or greater bone mass) for the loading experienced. A global optimum is (not to be too specific) the best local optimum in the whole range of the possible variation of the system—as an example, the absolute minimum fracture risk for the given loads. The fact that bone could have more than one local minimum means that a real (as opposed to a mathematical) bone may not reach the "best solution" during adaptation because the changing tissue could easily fall into a local optimum where *any further changes would be stopped* because all perturbations of the tissue from the local optimum make it *worse* with respect to the load to which it is adapting.

Experiments performed by Robling et al. [88] suggest that this could be happening in real bones. Using an experimental approach similar to that previously described, rat ulnae were given 360 axial loading cycles, either delivered at one time or in 90 cycle "bouts" 4 times throughout the day 3 days per week, for

16 weeks. They discovered that bones exposed to 360 cycles at one time never achieved the strength or mass of those that were loaded in shorter bouts. After 4 months, bone formation rate (BFR/BS) was threefold higher in the group which received the loads in bouts with fewer cycles. In terms of the remodeling cycle, 4 months in a rat is about equivalent to 1.5–2 years in a human, certainly sufficient time for bone to fully adapt to the mechanical stimulus. Although stiffness as a specific endpoint was not investigated, bone mineral content, whole bone areal mineral density, cross-sectional area, and rigidity in the direction of bending at the midshaft were all 40–50% greater in animals that received the loads in smaller doses. This is a remarkable result—an identical input of mechanical cycles resulted in two different "optimal" structures.

The explanation for there being two results was that cells become desensitized following relatively short periods of stimulation and require a period of recovery before they can fully respond again [88]. Another possible explanation is that the loading with shorter groups of cycles found a different local optimum because the bone perceived the two loadings as *different*. The bone does not have any particular long-term agenda, but only a series of short-term goals that reduce its risk of breaking. But the experiments raise a series of other important questions: If bone loading continued in short bouts, would the bone continue to respond, and for how long? Which cell populations become desensitized—the sensors (osteocytes) or the effectors (osteoblasts)? These are all exciting questions that remain to be answered by biologists and engineers.

6.4.11 Statics and Dynamics in Bone Adaptation

The experiments explored in this chapter have demonstrated that few load cycles are necessary to maximize bone's adaptive response. However, they also showed that there is a lower limit of loading below which bone is lost. That means that one single loading event is not sufficient to prevent bone loss. Taken to its natural extreme, a single loading event is the same as one statically applied load, suggesting that dynamic loading is required to generate an adaptive response in bone. This, in turn, suggests that bone adaptation is a rate-related phenomenon. We already know that bone is responsive to strain magnitudes that exceed local thresholds or setpoints. Is bone also sensitive to strain rate? If so, what rate is optimal, what are the thresholds (if any), and what is the relationship between strain rate and strain magnitude?

It has been known for over 50 years that bone does not respond to a static load unless, as in the case of orthodontia, a dynamic load is imposed on top of the static load. Static loads, even those that are quite high, do not elicit any adaptive response. Early experiments by Hèrt in 1969 first suggested that dynamic loads were necessary for bone to respond. Later, this was underscored by Lanyon and Rubin [89], who applied static loads to the avian ulna for 2 months and observed bone loss endocortically and increased intracortical porosity, changes identical to those found in immobilized animals. Dynamic loads at equivalent peak strain increased net bone formation periosteally, and prevented loss both endocortically and within the cortex. Mosley and Lanyon [90] followed this work by investigating low, moderate, and high strain rates using the rat ulnar axial-loading model. They found that low strain rate loading suppressed the normal periosteal bone formation response, whereas high strain rate loading significantly increased the amount of new periosteal bone. These were growing animals, however, and other studies have shown that low-to-moderate impact loads can maintain skeletal mass and reduce the amount of loss during aging.

Lanyon and Rubin [89] observed that when no loading was applied over an 8-week period, the ulna suffered endosteal resorption and increased osteonal remodeling. Compressive static loading to 2000 με for the same period of time led to a similar result, but the same strains applied dynamically at 1 Hz produced a 24% increase in cortical area, primarily from woven bone formation on the periosteal surface. Subsequently, Rubin and Lanyon [54] did a dose-response experiment for 1-Hz loading in a bending mode such that peak compressive strains were about twice the magnitude of the tensile strains. When 100 cycles/day were applied at peak strains from 0 to 4000 με, there was a linear relationship between cortical area and peak strain magnitude after 8 weeks (% change in area = $-15.3 + 0.015 \times$ (peak microstrain), with $R^2 = 0.69$). Setting the left side of this equation equal to zero indicates that an equilibrium loading for maintenance of bone mass (zero change in area) would be 100 cycles/day at about 1000 με. This was an important benchmark. These investigators did another frequently cited experiment [91] in which 2050 με was applied at 0.5 Hz in daily loading bouts of 4, 36, 360, and 1800 cycles. It turned out that only 4 cycles/day were sufficient to maintain bone mass at this strain magnitude. After 4 weeks of loading, each of the three higher numbers of cycles produced the same increase in bone mass, approximately 40% ± 10%, indicating that the woven bone response was an all-or-nothing phenomenon.

Subsequently, an experiment performed by Robling using the rat axial ulnar loading model further demonstrated this concept. Bone formation on the periosteal surface in rats loaded statically at 8.5 N and 17 N showed the same amount of bone formation, and was not different than formation on the non-loaded contralateral limb. Even at double the load, a static application did not generate a response; however, when a dynamic load was applied at 17 N for an equivalent length of time, there was a large proliferation of bone on the periosteal surface. Thus, loads of equivalent magnitudes have different effects, depending on whether they are applied statically or cyclically.

6.4.12 How Frequently Should Dynamic Loads Be Applied: The Role of Strain Rate

Dynamic load is required for bone's mechanical adaptation, and the rate of load application can be fast or slow for the same peak load. How fast the load is applied could be an important signal to bone or bone cells, since we already know that a static load (a rate of "zero") does not stimulate them. Exercises that involve loads applied at higher rates, such as running and jumping, are more osteogenic than other forms of exercise in many experiments. Rough estimates have suggested that 68–81% of the variance in bone surface adaptation may be accounted for by strain rate [92]. A recent study [93] reported that in adolescent boys, only high-impact (>4.2 g) loads were related to significant increases in femoral neck bone mineral density and that accelerations lower than that had no effect. High-impact loads were associated with increased periosteal bone formation (measured by hip structure analysis) and reduced endosteal width. Prior studies had similarly discovered that jumping exercises in young boys and girls had a positive effect on hip structural properties, including strength [94, 95].

Experiments using the rat ulnar axial loading model have demonstrated the importance of strain rate to bone's response. When loaded to a high strain, at strain rates corresponding to walking, (0.01/s), running (0.03/s), or jumping (0.05/s and higher), the response is 54% higher at the strain rate for jumping than at the strain rate for running, and even greater compared to the rate generated during walking [90]. Ulnar loading at frequencies that are 5–10 Hz are associated with higher periosteal bone formation rates than those delivered at equivalent strains but only 1 Hz [57]. Thus, given an equivalent strain, bone will respond more strongly at higher rates of load application.

How can this be reconciled with the idea that bone formation initiates in response to strain magnitude? This is because strain rate depends on strain magnitude (ε) and loading frequency (f). For a sinusoidal strain ($\varepsilon(t) = \varepsilon_{\max} \sin(2\pi ft)$):

$$\text{Strain rate} = 2\pi f \varepsilon_{max} \cos(2\pi ft) \tag{6.1}$$

so that as maximum strain and frequency increase, so does maximum strain rate. During our normal gait cycle, strain is (approximately) applied in a sine wave, and any increase in frequency, e.g., when we are running, will be translated into an increased strain rate (Fig. 6.7). Experimentally, rate and frequency can be dissociated by utilizing a triangular waveform for loading and varying the period between cycles to alter either strain rate or frequency independently. Using this concept, Turner et al. [53] showed that given the same peak strain and strain rate, frequencies above 0.2 Hz will generate an adaptive response, whereas those below that will not. Given our natural walking cycle of about 1 Hz, this is encouraging and "makes sense" as a story about the evolution of mammalian physiology. In a second experiment, Turner and colleagues [96] loaded rat tibiae in four-point bending to the same maximum load (54 N). The load was separated into a *constant* component plus a 2-Hz sinusoidal

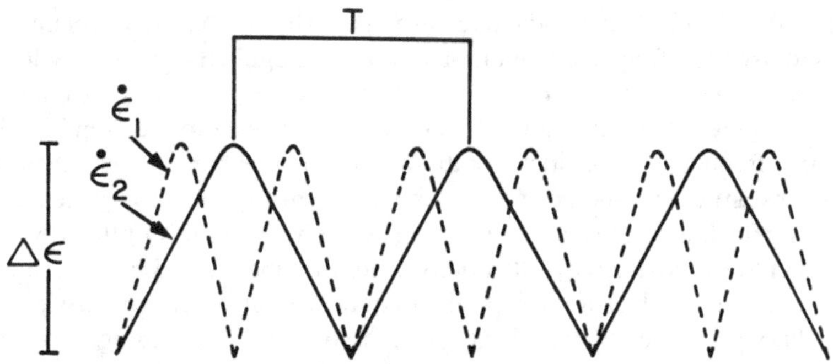

Fig. 6.7 Two strain–time curves showing that as frequency increases, strain rate increases when the peak strain is the same. Frequency = 1/T where T is the interval between peak loads. $\Delta\varepsilon$ = peak strain magnitude; the ε are differing strain rates for low and high frequency

component of varying magnitude (0, 18, 36, or 54 N) so that the constant plus peak sinusoidal load was always 54 N. In this experiment, cycling over a strain range of 0 is equivalent to static loading (loading to 54 N and holding it there), and one would not expect any response. This is what occurred. But over the range of constant plus a sinusoidal component of 18–54 N, there was a dose-response effect with increasing bone formation rate as strain rate (peak sinusoidal component) increased.

Taking these two experiments, one in which frequency was varied, and one in which strain rate was varied, Turner [97] calculated the total *strain stimulus* (S_σ) as

$$S_\sigma = k_i \sum_{i=1}^{n} \varepsilon_{\max,i} f_i \tag{6.2}$$

where "*i*" denotes the number of loading types, k_i is an empirical constant, $\epsilon_{\max_i}$ is the peak strain achieved, and f_i is the frequency of load application. This equation looks very much like Eq. (6.1) for strain rate, indicating that over a series of cycles, the adaptive response (S_σ) will be proportional to strain and frequency. When the strain stimulus is calculated for the two experiments in which either frequency or cyclic strain magnitude are varied, the relationship between strain stimulus and bone formation rate is almost identical. This demonstrates that it is a combination of strain magnitude and frequency—strain rate—that drives the adaptive response.

6.4.13 Effects of High-Frequency Loading: Vibration

Since strain rate is an important factor in skeletal adaptation, skeletal mass and strength could be governed by the animal's most dynamic activities even if those activities are relatively infrequent. The relationship between strain and frequency suggests that using high loading frequencies could minimize the actual strains that are required to elicit a bone-adaptive response. This is the theory behind recent attempts to use high-frequency vibrations, which are applied at extremely low strain magnitudes, to increase bone formation or to reverse the effects of bone loss.

Several variations of vibrational "exercise" have been explored (and marketed). In one, high frequency sinusoidal vibrations that generate higher lateral than vertical accelerations are used. Another form uses vertical displacement to generate higher strains along the longitudinal axis versus transversely [98, 99]. A further division can be made between those devices that deliver high frequency and large accelerations (>1 g) and those that deliver high frequencies at lower intensity (<1 g) [100]. Although these devices have been associated with strong tonic muscle contraction (contractions up to 82% of maximal muscle contraction—enough to cause tomorrow's muscle pain!), and may affect biochemical indices of bone resorption, they have not been shown consistently to have any significant effect on bone density [101, 102].

6.4.14 When Theory Meets Reality

Have you ever heard someone say: "Well, at least in theory this should work"? When an experiment does not conform to theory, either the theory is wrong, or something has been overlooked in the experiment. According to our theory about the importance of strain rate, low strain-high frequency loading should be effective at building bone if the strain rates created are sufficiently high. So far, however, this just does not seem to be correct. Let us explore the reasons why this promising theory has predicted an approach that has not been effective.

One possibility is that there really is a strain magnitude threshold for response of cells, and that very low strains are just not sufficient to cross that threshold, no matter how many of them there are. In animal models, the high-frequency (15–30 Hz) strain signals are more than 20 times smaller in magnitude than low-frequency (<15 Hz) strain components. Cell-culture experiments have demonstrated repeatedly that connective tissue cells (e.g., osteoblasts and fibroblasts) are highly sensitive to frequencies in the 15–30 Hz range. We discussed earlier in the chapter that there appears to be a threshold for the initiation of bone formation at about 1000 με, but some investigators have shown that strains of only 250 με, when delivered at 15 Hz, are sufficient to induce bone formation [103]. However, strain-gauge experiments using dogs walking on a treadmill [96] show that strains within the 15–20 Hz frequency range are <50 με, and even lower at slightly higher frequencies. There is no convincing evidence that strains this low will cause net bone formation. It may simply be that there is insufficient energy in that signal to create a cellular signal in vivo.

A second possibility is that strain and strain rate are not a direct stimulus to the sensor cells, and that downstream events are affected differently by very high frequency loading. The idea that strain and strain rate are a direct stimulus to cells implies that the effector cells (the ones doing the work, osteoblasts and osteoclasts) sitting on bone surfaces sense deformation when the bone is deformed. That is, there is a direct link between the mechanical stimulus and the cell that produces the response, but there is good evidence now that *osteocytes* are the mechanical sensor cells in mammalian bone [104], and that they are the primary cells controlling the induction of bone remodeling. Osteocyte death stimulates bone resorption in cases of estrogen deficiency or disuse, and is directly associated spatially and temporally with bone loss [105–107]. Osteocytes and osteoblasts produce osteoprotegerin (OPG) and RANKL (a ligand for a receptor on osteoclasts) in response to bone damage, eliciting osteoclast recruitment and activation, and subsequently bone resorption to repair damage [108–110]. In mice with RANKL removed only in osteocytes, bone loss does not occur in response to unloading [111]. Moreover, mice in which osteocytes cannot produce RANKL become osteopetrotic [112].

On the other side of the bone remodeling coin, the formation of new bone on periosteal surfaces following mechanical loading is associated specifically with the downregulation of sclerostin production by osteocytes [113]. Transgenic mice engineered to maintain high levels of sclerostin suppressed new bone formation by 70–85% during mechanical loading [114]. Moreover, two osteocyte-specific proteins related to matrix mineralization, DMP1 and MEPE, are upregulated after mechanical loading [115]. Recent evidence has shown that variations in new bone formation within a single bone (using the mouse tibia) are more closely related to the absence of sclerostin production by osteocytes than peak strains [116]. All of these lines of evidence suggest that osteocytes sense mechanical loading in some manner. However, osteocytes sit within a space in the mineralized bone matrix and although they are tethered to the walls of their lacunae by proteins (the glycocalyx) and could potentially "feel" deformation, it is likely that they are protected to some degree from significant amounts of deformation. How then can they sense a mechanical stimulus? Or perhaps the better question is: what mechanical stimuli are they sensing?

6.4.15 Is Strain a Direct Stimulus to Bone Cells? Do Strain Gradients and Distributions Regulate Regional Bone Cell Activity?

Changes in surface curvature induced by bending create stress gradients perpendicular to the surface. Increased surface convexity is associated with tangential strain that become more tensile as one moves toward the surface (Fig. 6.8). Decreased surface convexity (or increased surface concavity) is associated with strain that become less tensile nearer the surface. Because there is not a perfect correspondence between experimentally measured strain magnitude and specific locations of bone formation, and also because that model would lead to results that were inconsistent with observations of bone adaptation, Frost suggested that it is these strain gradients, rather than the magnitude of the strain itself, which regulates the location and magnitude of bone adaptive response. This idea has been tested experimentally

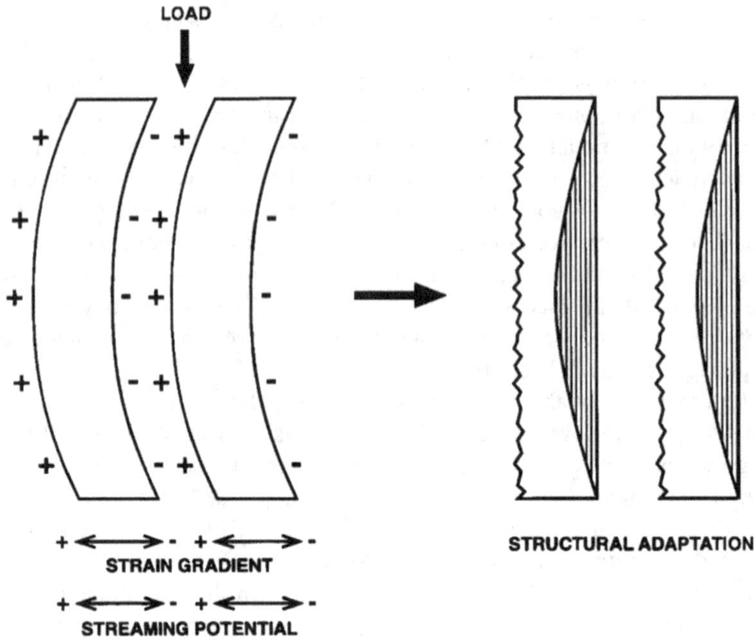

Fig. 6.8 Bending is associated with stress (and strain) gradients. If the polarity of these gradients is as defined in the text, then they would have + and − signs as shown in the longitudinal section at *left*. The streaming potentials produced by these stress gradients would have similar polarity. If positive and negative signals produce resorption and formation, respectively, the bone would straighten, as shown in the right diagram. Computational analyses are consistent with the prediction that bone formation will be enhanced in regions where curvature-induced bending moments are high. Bones will not completely straighten, however, perhaps because some curvature is necessary to control stresses generated by limb acceleration [154]

using several different avian models. Although strains, by necessity given the limitations of current technology, are typically measured on external surfaces of bone, loads create distributions of strain within the bone that result in a strain gradient, which is affected both by the properties of the bone and by its external shape or curvature. This idea was tested by Rubin and Lanyon [54, 91, 117] in a set of classic experiments using turkeys and roosters in which they altered the direction of application of a load in a limb otherwise deprived of any load bearing. They discovered that when bone was loaded in directions in which the bone was not loaded during normal movements (in this case 90° to that for wing flapping), periosteal or endosteal woven bone could be produced at loads below those required to cause bone formation under normal conditions.

Thus, when the strain distribution is usual, the threshold to initiate an adaptive response may be higher than it is for strain distributions produced by unusual or less frequent loading conditions. The implication of this is that bone is adapted to a range of strain situations, and that adaptive responses are more easily generated by a variety of different loading conditions than by increased repetitions of the normal pattern of loading. This may be one reason that exercise physiologists find it so challenging to identify a significant bone response to exercise in adult people, who may be already relatively well adapted to the activities they are asked to perform. The other side of the coin, however, is a cautionary tale—attempting to understand bone's adaptive response to loads that are applied in nonphysiological directions in an animal model may provide solutions to problems that do not exist!

An elegant demonstration of the complexity of strain distribution to bone adaptation [118] used the turkey ulnar osteotomy model to isolate the radius, which then was subjected to a 4-week experimental loading protocol after which periosteal bone formation was measured in 24 different sectors around the circumference of the radius. Interestingly, there was bone formation in every sector, and the amount of new periosteal bone was not well predicted by the levels of strain. New bone in every sector would not be predicted under a model of adaptation that only considered strain magnitude; one would reasonably expect that bone formation close to the neutral axis would be absent or negligible. However, approximately 60% of the new bone formation was predicted by a combination of circumferential, radial, and longitudinal strain gradients. A subsequent study subjected intact roosters that had strain gauges attached to their tarsometatarsal bones to high-speed running for up to a year. Again, it was discovered that peak strains and even peak strain rates were poorly correlated with the amount and location of new bone formation, but that circumferential strain gradients predicted 63% of the response.

These experiments address one question that had perplexed scientists for years (i.e., if bone only adapts to strain magnitude, then why does not the bone along the predominant neutral axis simply disappear over time?). One answer to this, of course, is that we do not always use our limbs in the same ways, and so a specific place in a bone may not always see high or low strains. Another explanation is that bone is not optimally adapted (remember the discussion earlier in this chapter) but only finds biologically acceptable solutions. However, understanding the more complex aspects of strain distribution and gradients in bone, and how it changes with different activities, may provide a more satisfying explanation to this mystery.

The importance of strain gradients suggests that it is less important whether a region of bone is under compressive or tensile strains, but whether that region is becoming more or less compressive or tensile. Consider the flow of fluids in a bone as it is bent. About 23% of the volume of bone is composed of fluid, some of it bound to collagen, but much of it free to flow through the LCN. In general, fluids flow down the fluid-pressure gradient from more compressed regions to less compressed regions. Weinbaum et al. [119] developed a mathematical theory predicting that physiological stress gradients will produce fluid flows in the annular spaces between the osteocyte processes and the canalicular walls. As these fluids flow over the membranes of these processes and interact with the glycocalyx, the resulting fluid shear stresses on the cell wall were predicted to be in the range that causes the release of intracellular calcium ions in osteoblasts and endothelial cells. Similar calculations show that stresses of similarly high magnitude are not created by fluid motion on surfaces adjacent to larger spaces, such as Haversian canals or on marrow cavity surfaces where lining cells (pre-osteoblasts) reside. Partitioning the mechanically relevant signaling to the osteocyte lacunar space suggests that the effect of mechanical loading is not a direct one to the lining cells or osteoblasts, but instead is communicated through the osteocyte network to these cells to activate modeling and remodeling.

6.4.16 Fluid Flow

The idea that fluid flow through canalicular channels is the direct stimulus for cellular activity is consistent with the observation that bone is strain-rate sensitive, because the rate of flow is directly proportional to the strain rate. The more rapidly bone is loaded, the faster fluid is forced through these channels. In the late 1990s, several studies were able to demonstrate that osteoblasts seemed to respond to the flow of fluid over them, independent of the amount of membrane deformation [120, 121]. It had always been a conundrum how osteocytes, in particular, sitting within a mineralized matrix, could sense physical deformations. These experiments suggested that the signal for cellular response was not the deformation itself, but the deformation-induced fluid flow through channels, which could create either a membrane shear stress or chemical transport that would stimulate cells to react. This was predicted earlier by the important modeling studies of Weinbaum and Cowin [119, 122], which also predicted that fluid velocities within the canaliculi are proportional to the loading rate. More recent in vivo work showed that high-frequency oscillatory axial displacement (45 Hz) applied to mouse lower limbs in the absence of weight bearing led to significant increases in trabecular bone volume (17%) and stiffness (38%) without bone deformation [123–125]. Although these experiments demonstrate rather convincingly that bone-matrix deformation is not required for cellular response, they do not preclude the possibility of stimulatory cell signaling from tissues adjacent to the periosteum (muscle and fascia) and endosteum (marrow) rather than from osteocytes exclusively.

It is now generally accepted that both osteoblasts and osteocytes are stimulated by fluid flow [126], and that the specific response is, in part, dictated by whether the flow is continuous, pulsatile, or oscillatory. Flows in the range of 0–20 dynes/cm^2 have been shown to cause osteoblast proliferation, and to upregulate the cellular production of a range of factors important for the regulation of bone formation, including nitric oxide, cyclo-oxygenases (COX 1 and 2) and the prostaglandins (PGE2) that are formed downstream of them, and osteopontin. It also stimulates cellular processes resulting in release of ATP, and the stimulation of purinergic cell-surface receptors (e.g., P2X7).

Fluid motion following bone loading has now also been demonstrated in a more dynamic manner using fluoroscein tracers and photobleaching (FRAP) techniques [127, 128] (Table 6.3). Although prior studies examined tracer perfusion (usually Procion Red) through the LCN with loading [129, 130], they were only able to examine the extent of perfusion in a static manner at defined time points, rather than true dynamic flow. More recent dynamic studies demonstrated that physiologic

Table 6.3 Experimentally measured Lacunar–Canalicular dimensions and average diffusivity

Ca.Le (μm)	Lc.Rd. (μm)	Ca.Ar (μm^2)	Lc.Ar (μm^2)	Diffusivity(10–10 m^2/sec)	Loaded diffusivity(10–10 m^2/sec)
30[a]	3.9–8.7	0.315	16.3	2.72	3.56
26[b]	2.3–11.6	NM	91.3	3.3	NM

Lacunar areas from Price et al. appear to be low, based on the measured radius. NM, not measured
[a]From results in Price et al. (2011) [128] and [b]Wang et al. (2005) [127]

strain magnitudes (400 με) increase diffusion of the fluoroscein tracer by 31% (compared to non-loaded bones), and produce a flow rate of 60 μm/s [128]. This was calculated to impart a fluid shear stress on the membrane of the osteocyte or its processes of about 50 dyne/cm^2, more than double the shear stress that is typically used in *in vitro* laminar flow studies of cellular response. Of course, this calculation is highly dependent on the morphology of the LCN, but these were experimentally measured and validated using a model of transport kinetics that was not dependent on bone's poroelastic material properties (Table 6.3). However, this approach allows one to estimate how variations in canalicular connectivity would affect transport, membrane shear, and ultimately the capacity of cells to sense a mechanical signal. The model predicts that more canaliculi, or presumably greater connectivity, would linearly increase transport rate, whereas a decline in canalicular length would result in a nonlinear decrease in transport, and presumably impaired mechanical signaling. This is an interesting result given experimental observations that osteocyte apoptosis occurs in disuse [105], estrogen deficiency [106], or microdamage formation [131], the last of which significantly reduces the canalicular network, possibly preventing adaptation, or in some cases repair.

The idea that fluid motion provides the mechanocoupling signal that will lead to the cell's biochemical response can explain the observation that very high frequency loading is ineffective at generating a strong cellular response. We know that the voltage and calcium-gated channels that control trafficking of molecules into and out of cells are themselves frequency dependent. We also know that there is a marked decline in the rate of fluid flow within the canalicular system when the frequency of loading is increased from 1 to 30 Hz [56, 132, 133]. If strain-induced flow of fluid through bone is the local mechanical stimulus for the cellular response, then high-frequency, low-magnitude loading could compromise mechanical transduction simply because there is insufficient time for the fluid movement, at least to the degree required for ostoecyte stimulation. This, then, may be the solution to our question about why low magnitude—very high-rate or high-frequency vibration, does not generate massive responses. Fluid motion above a certain magnitude is necessary in addition to the number of cycles in order to elicit a full cellular response.

The nature of the osteocyte (sensor cell) and osteoblast and osteoclast (effector cells) response, however, is likely dependent on the nature of the flow itself. Physiologically, bone cells are probably not exposed to continuous or unidirectional pulsatile flow, but rather to flows that are oscillatory. Cell response to oscillatory flow is heavily modulated by chemotransport, unlike responses to continuous or unidirectional pulsatile flow regimes [134–136]. Oscillatory flow also is a much weaker stimulus to cell response, and requires a longer period of exposure. This is somewhat inconsistent with the rapid saturation of response that is observed from in vivo studies [84].

6.4.17 Summary

The experiments reviewed in this chapter thus far suggest that the mechanical adaptation of bone is governed by a set of controlling mechanisms that can be summarized as follows:

- Trabeculae seem to be aligned with the principal strain directions calculated from a continuous homogenous model.
- Disuse of a growing long-bone diaphysis inhibits periosteal bone formation; disuse of a mature diaphysis results in increased endosteal resorption and intracortical remodeling.
- Mild overloading of a growing long-bone diaphysis inhibits endosteal resorption, and can stimulate periosteal apposition, producing a thicker cortex.
- A set of local strain thresholds or setpoints defines the bone's adaptive response to mechanical stimuli. The upper threshold for an anabolic response is probably above 1000 $\mu\varepsilon$ in typical long bones, and in general this will keep peak periosteal bone strains during even vigorous activities below approximately 2000–3000 $\mu\varepsilon$. This appears to be conserved for the long bones across species.
- Bone adaptation is governed by a combination of strain magnitude and frequency, both of which modulate strain rate. Magnitude and frequency can be varied within limits to create a predictable response.
- Strain is probably an indirect stimulus for bone adaptation; current evidence points to strain-driven motion of fluid within the LCN as the most likely transduction signal. Sensor cells (osteocytes) can detect this flow (by as-yet unknown mechanisms), and create biochemical signals that govern the response of effector cells (osteoblasts and osteoclasts).
- Fluid can only flow over a significant length of time if bone is loaded cyclically. This goes along with the observation that bone only responds to loads that are dynamically applied, and is very responsive even with small numbers of loading cycles. Cell sensitivity saturates quickly, and takes hours to recover. Oscillatory fluid flow regulates cell response by chemotransport mechanisms, whereas continuous fluid flow regulates cell response via membrane shear stress.
- High-frequency vibration does not produce as vigorous an anabolic response as theory might suggest because the fluid flow magnitude is small.
- Experiments investigating bone adaptation emphasize diaphyseal cortical bone responses, as opposed to trabecular bone architecture, which was the historical focus. A major reason for this is that, experimentally, it is difficult to control the mechanical environment to which individual trabeculae are subjected. The

exigencies of experimental science—getting statistically significant results in a few weeks or months—led investigators to conduct experiments that may not be very representative of the sublime, gradual, day-in, day-out way in which mechanical adaptation probably works in healthy individuals. In human beings, most cortical bone adaptations are likely to involve more modest load increases.

6.5 Other Factors to Consider

6.5.1 Experimental Consistency

We may use Frost's mechanostat theory as a basis for comparing the results of several experiments described in the literature, some previously explored. Woo et al. [137] exercised growing pigs and found no effect on periosteal diameter, but endosteal expansion was inhibited. This result is consistent with the mechanostat theory if the endosteal surface in young pigs is remodeling (inhibited by overuse) rather than modeling. The failure of the periosteal surface to show increased bony apposition would make sense if the rate of apposition was high already because the animals were growing. This is a reasonable supposition because pigs form plexiform bone at the periosteal surface during growth, and the function of this kind of primary bone formation seems to be to add bone quickly.

Disuse osteopenia in the casted forelimbs of growing dogs [67, 68] was produced both by the failure of the periosteal surface to expand and by expansion of the medullary cavity. This result is consistent with inhibition of modeling on the periosteal surface and stimulation of remodeling on the endosteal surface. However, bone loss was also seen on the periosteal surface; according to the mechanostat, this should have been produced by remodeling, but this is unclear. Remobilization of the young dogs reversed both these trends: woven bone was opposed to the periosteal surface, while the slow apposition of lamellar bone to the endosteal surface of normal young dogs was also restored. In older dogs, in which modeling is less responsive, disuse enhanced endosteal resorption and increased intracortical porosity, consistent with stimulated remodeling [68, 69]. There was virtually no effect on periosteal diameter, consistent with the observation that this envelope almost always is restricted to modeling in the formation mode. Remobilization [68] caused a small amount of bone to be added to both the endosteal and periosteal surfaces, but only partially restored bone mass in the diaphysis. The endosteal apposition was consistent with refilling of the remodeling space caused by an inhibition of remodeling. Much less bone was opposed periosteally than that in younger dogs, and the new bone was more highly organized, demonstrating the relative slowing of modeling processes in adults.

The isolated avian ulna model (involving adult animals) of Rubin and Lanyon [54] also provides a good test of mechanostat predictions, and demonstrates that bone surfaces respond differently depending on whether the strain stimulus is above or below the equilibrium range. Disuse was associated with increased intracortical remodeling and endosteal resorption, as in immobilized adult dogs. Mechanical overload was associated with insignificant osteonal remodeling, a reduction in endosteal resorption, and abundant periosteal modeling producing woven bone. Similar observations concerning the effects of overload were obtained in the osteotomy studies of Lanyon and colleagues, as we have described. (It is noteworthy that increased osteonal remodeling has been observed in these experiments, but only after the woven bone formed by modeling had been present for some time and, presumably, needed remodeling to repair fatigue damage.)

Another relevant experimental model is that of Jee and coworkers [138], who unloaded one hindlimb of rats by bandaging, thus overloading the opposite limb. The increased resorption and reduced bone formation in the trabecular bone of the tibial metaphysis of the unloaded limb was consistent with increased remodeling and inhibited refilling, resulting in a 50% reduction in bone mass and a new equilibrium state after 26 weeks. Overloading resulted in increased bone mass caused by reduced bone trabecular remodeling and increased periosteal modeling in the formation mode. Schaffler [139] has obtained similar results in dogs with one limb immobilized in a cast.

Thus, while some questions remain, the available experimental data support the essential features of the mechanostat theory. Reduced mechanical loading increases activation of new BMUs, increasing the number of new osteons and producing trabecular and endosteal resorption. Periosteal modeling in the formation mode is inhibited. Conversely, periosteal modeling is commonly activated by overload, which also inhibits activation of new BMUs, depressing formation of new osteons and limiting loss of bone from trabecular and endosteal surfaces. However, there must be an important limitation to the overload aspect of the mechanostat: if the overload produces damage, remodeling is activated to repair it (see Sect. 7.8).

6.5.2 Relationship of Mechanically Adaptive Responses to Other Control Factors

The response of bone to mechanical factors can be expected to depend on the balance that occurs between mechanically prompted physiologic signals and those from systemic or local nonmechanical factors. The nature of competing systemic (or global) and local signals to bone cells has been discussed by both Lanyon [140–142] and Frost [70]. The idea is that various factors within the body compete for the control of bone cells. For example, bone adjacent to marrow, particularly fatty marrow, has a tendency toward net resorption, regardless of the hormonal or mechanical environment [70, 143]. Thus, local "modulators" compete with or modify other signals, which undoubtedly accounts for some of the envelope specificity in bone remodeling (e.g., periosteal vs endosteal).

Conversely, global factors may modify local responses, as when female distance runners lose rather than gain bone from excessive training-induced amenorrhea, a simulacrum of menopause. Exogenously supplied estrogen reverses the bone loss in these runners [144]. Similarly, a study of male basketball players found that in a rigorous training regimen they lost bone mass unless given supplemental dietary calcium; calcium loss through perspiration resulted in a physiologic calcium demand that was met by bone resorption [145]. Skeletal health has been likened to a three-legged stool, the legs being diet, hormonal status (estrogen, vitamin D), and exercise. All three legs must be robust for the stool to stand. It is apparently also possible for local skeletal disturbances to have global effects. When marrow is destroyed within a bone, its replacement involves filling the medullary canal with woven bone, which is subsequently resorbed and replaced by new marrow. Bab et al. [146] found increased bone formation in the mandibular condyle of rats after removal of bone marrow from the tibia. Thus, bone messengers intended to govern local responses may "escape" to the systemic circulation and produce global remodeling responses.

Frost [147] suggested that certain osteoporoses result from an increase in the lower limit of the physiologic strain range, so that normal mechanical loading is perceived as a disuse state and bone is lost. It is possible that estrogen affects this "set point," so that it changes at menopause. If so, postmenopausal bone loss may occur because the normal mechanical stimulus becomes low relative to the cells' altered reference value. Conversely, the anabolic effects of low doses of parathyroid hormone (PTH) suggest it may move the set point in the other direction [148]. Thus, alterations in modeling and remodeling may occur because of changes in strain, changes in the physiologic strain range set points, or combinations of these factors.

6.5.3 Mechanical Adaptability and Damage Repair

We have hypothesized that mechanisms exist to optimize the stiffness of bone while minimizing fracture risk and bone mass. We have examined several possibilities for such mechanisms, such as the ability of osteocytes to sense fluid flow as a consequence of mechanical strain and regulate effector cells accordingly. These cells have the capability of activating modeling or remodeling to change the architecture of a bone in ways that affect its stiffness, fracture risk, and weight. However, as we discussed in Chap. 7, these cells also may activate remodeling to repair fatigue or creep damage. What is the relationship between damage repair and mechanical adaptability?

The answer has to do with the trade-off between weight and fracture risk from creep and fatigue. Let us digress again to Culmann and nineteenth-century engineering for a useful analogy. In bridges, fatigue can be an important element of failure. After some disasters, early American engineers began to build their bridges of elements large and robust enough that stresses were very low and fatigue failure would not occur. Not having the benefit of the mechanical theory being developed by Culmann and others in Europe, they overdesigned, wasting material and effort. As a young engineer, Culmann toured America and recognized this problem [149]. Later, he used theory to design bridges in which stresses were low enough to avoid fatigue failure, but not unnecessarily low. Now, imagine what would happen if a way could be found to repair fatigue damage in a bridge soon after it occurs. Then, the stress level could be allowed to rise, and the bridge could be made even lighter, saving further on materials and expense. That is the advantage that bones have over bridges. Damage repair by remodeling allows bones to be lighter, even though they carry stresses that would otherwise lead to fatigue failure within the organism's lifetime.

Damage repair is not part of mechanical adaptability in the sense that it shapes elements of the skeleton to control stiffness and stresses, but it is an important part of the picture insofar as it allows higher stresses to be tolerated. The same may be said of the elements of bone tissue structure that limit and control microdamage. In fact, all these matters are components of the system that has evolved to optimize bone stiffness while minimizing fracture risk and weight. This system works similarly in the skeletons of many different animals, from hummingbirds to elephants.

6.6 Summary and Further Reading

Wolff's law and the mechanical adaptability of bones to mechanical loading are not only of great interest to basic scientists interested in the skeleton; they are of great clinical utility as well. Orthopedists see the effects of these phenomena in their daily practices. In Chap. 1, we reviewed the history of these concepts in some detail to gain a better understanding of how the current ideas evolved. We saw that there are three key concepts, including optimization of strength with respect to weight, alignment of trabeculae with principal stress directions, and regulation by systems of cells able to respond to a controlling stimulus. We then reviewed some of the more important experiments in this area, and saw that they have emphasized cortical rather than trabecular bone. Although disuse and overload effects are clear, it may be not be appropriate to relate the latter to normal development because woven rather than lamellar bone formation occurred.

We also reviewed some of the theories regarding mechanical adaptability of cortical and trabecular bone, and found that the putative ability of osteocytes to sense strain or strain energy, likely as a consequence of fluid movement down strain gradients, is emerging as an important element of such theories. Finally, we asserted that damage repair is an important adjunct to mechanical adaptability because it allows higher strains to be tolerated, and that damage repair and mechanical adaptability are somatic changes, which extend the organism's ability to cope with environmental changes without committing to an irreversible genetic change.

Chapter 1 provided much of the historical concepts of bone mechanical adaptation prior to the early twentieth century. Frost's perspective remains unique in its attempt to balance biologic, mechanical, and clinical observations. An excellent and more detailed description of events associated with mechanical signaling (mechanotransduction) is given in Duncan and Turner [19]. Finally, Bone Structure and Remodeling, edited by Odgaard and Weinans [150], is an excellent collection of papers by key investigators in this field, including a thought-provoking and entertaining exchange of biologic and engineering views of the subject by Currey and Huiskes.

References

1. Bateson G. The role of somatic change in evolution. Evolution. 1963;17:529–39.
2. Bateson G. Steps to an ecology of mind. Northvale, NJ: Jason Aronson; 1987.
3. Cooney CA, Dave AA, Wolff GL. Maternal methyl supplements in mice affect epigenetic variation and DNA methylation of offspring. J Nutr. 2002;132(8 Suppl):2393S–400S.
4. Pembrey ME, Bygren LO, Kaati G, Edvinsson S, Northstone K, Sjöström M, Golding J. Sex-specific, male-line transgenerational responses in humans. Eur J Hum Genet. 2006;14 (2):159–66.
5. Grossniklaus U, Kelly B, Ferguson-Smith AC, et al. Transgenerational epigenetic inheritance: how important is it? Nat Rev Genet. 2013;14:228–35.
6. Cruz-Coke R. A genetic description of high-altitude populations. In: Baker PT, editor. The biology of high altitude peoples. Cambridge: Cambridge University Press; 1978. p. 47–63.
7. Crummett LT, Sears BF, Lafon DC, et al. Parthenogenetic populations of the freshwater snail Campeloma limum occupy habitats with fewer environmental stressors than their sexual counterparts. Freshw Biol. 2013;58:655–63.
8. Ng CYP, Choi VWY, Lam ACL, et al. The multiple stressor effect in zebrafish embryos from simultaneous exposure to ionising radiation and cadmium. J of Radiol Protection. 2013;33: 113–21.
9. Jansen M, Stoks R, Anja C, et al. Collateral damage: rapid exposure-induced evolution of pesticide resistance leads to increased susceptibility to parasites. Evolution. 2011;65:2681–91.
10. Eranen JK, Nilsen J, Zverev VE, et al. Mountain birch under multiple stressors—heavy metal-resistant populations co-resistant to biotic stress but maladapted to abiotic stress. J Evol Biol. 2009;22:840–51.
11. Lanyon LE. The importance of mechanical adaptation in controlling bone architecture and averting bone fracture. In: Marcelli C, Sebert JL, editors. Architecture et Resistance Mecanique Osseuses. Masson, Paris; 1993. p. 58–67.
12. Currey JD, Pitchford JW, Baxter PD. Variability of the mechanical properties of bone, and its evolutionary consequences. J Roy Soc Interface. 2007;4:127–35.
13. Nowlan NC, Prendergast PJ. Evolution of mechanoregulation of bone growth will lead to non-optimal bone phenotypes. J Theoret Biol. 2005;235:408–18.
14. Gould SJ. The panda's thumb. New York: WW Norton and Co; 1980.
15. Currey JD. The mechanical adaptations of bones. Princeton: University Press; 1984.
16. Currey JD. What is bone for? Property-function relationships in bone. In: Cowin SC, editor. Mechanical properties of bone. New York: American Society of Mechanical Engineers; 1981. p. 13–26.
17. Currey JD. Incompatible mechanical properties in compact bone. J Theoret Biol. 2004;231: 569–80.
18. Ashby MF. Materials selection in mechanical design. Oxford: Butterworth Heinemann; 2004.
19. Duncan RL, Turner CH. Mechanotransduction and the functional response of bone to mechanical strain. Calcif Tiss Int. 1995;57:344–58.
20. Wang N, Butler JP, Ingber DE. Mechanotransduction across the cell surface and through the cytoskeleton. Science. 1993;260:1124–7.
21. Chen CS, Ingber DE. Tensegrity and mechanoregulation: from skeleton to cytoskeleton. Osteoarth Cart. 1999;7:81–94.
22. Turner CH, Robling AG, Duncan RL, Burr DB. Do bone cells behave like a neuronal network? Calcif Tiss Int. 2002;70:435–42.
23. Lanyon LE, Smith RN. Measurements of bone strain in the walking animal. Res Vet Sci. 1969;10:93–4.
24. Lanyon LE, Smith RN. Bone strain in the tibia during normal quadrupedal locomotion. Acta Orthop Scand. 1970;41:238–48.
25. Lanyon LE. Strain in sheep lumbar vertebrae recorded during life. Acta Orthop Scand. 1971;42:102–12.

26. Lanyon LE. In vivo bone strain recorded from thoracic vertebrae of sheep. J Biomech. 1972;5: 277–81.
27. Cochran GBV. Implantation of strain gages on bone in vivo. J Biomech. 1972;5:119–23.
28. Cochran GBV. A method for direct recording of electromechanical data from skeletal bone in living animals. J Biomech. 1974;7:563–5.
29. Barnes GRG, Pinder DN. In vivo tension and bone strain measurement and correlation. J Biomech. 1974;7:35–42.
30. Turner AS, Mills EJ, Gabel AA. In vivo measurement of bone strain in horse. Am J Vet Res. 1975;36:1573–9.
31. Lanyon LE, Hampson WGJ, Goodship AE, Shah JS. Bone deformation recorded in vivo from strain gauges attached to the human tibial shaft. Acta Orthop Scand. 1975;46:256–68.
32. Lanyon LE. Analysis of surface bone in the calcaneus of sheep during normal locomotion. J Biomech. 1973;6:41–9.
33. Lanyon LE. Experimental support for the trajectorial theory of bone structure. J Bone Joint Surg. 1974;56B:160–6.
34. Hayes WC, Snyder B. Toward a quantitative formulation of Wolff's law in trabecular bone. In: Cowin SC, editor. Mechanical properties of bone. New York: American Society of Mechanical Engineers; 1981. p. 43–68.
35. Dempster WT, Liddicoat RT. Compact bone as a non-isotropic material. Am J of Anatomy. 1952;91:331–62.
36. Harrigan TP, Mann RW. Characterization of microstructural anisotropy in orthotropic materials using a 2nd rank tensor. J Mater Sci. 1984;19:761–7.
37. Fyhrie DP, Carter DR. A unifying principle relating stress to trabecular bone morphology. J Orthop Res. 1986;4:304–17.
38. Cowin SC. The relationship between the elasticity tensor and the fabric tensor. Mech Mater. 1985;4:137–47.
39. Cowin SC. Wolff's law of trabecular architecture at remodeling equilibrium. J Biomech Eng. 1986;108:83–8.
40. Turner CH. On Wolff's law of trabecular architecture. J Biomech. 1992;25:1–9.
41. von Meyer GH. Die Architektur der Spongiosa. Archs Anat Physiol Wiss Med. 1867;34:615–28. [Trans by Amundson PK, Skedros J and Brand R. Clin Orthop Rel Res 2011; 469: 3079–3084.]
42. Wolff J. The law of bone remodeling. Berlin: (Translation of Wolff's Das Gesetz der transformation der Knochen by P. Maquet and R. Furlong.) Springer-Verlag; 1892.
43. Carter DR, Orr TE, Fyhrie DP. Relationships between loading history and femoral cancellous bone architecture. J Biomech. 1989;22:231–44.
44. Mullender MG, Huiskes R, Weinans H. A physiologic approach to the simulation of bone remodeling as a self-organizational control process. J Biomech. 1994;27:1389–94.
45. Rubin CT, Lanyon LE. Limb mechanics as a function of speed and gait: a study of functional strains in the radius and tibia of horse and dog. J Exp Biol. 1982;101:187–211.
46. Burr DB, Milgrom C, Fyhrie D, Forwood M, Nyska M, Finestone A, Hoshaw S, Saiag E, Simkin A. In vivo measurement of human tibial strains during vigorous activity. Bone (New York). 1996;18:405–10.
47. Yang PF, Brüggermann GP, Rittweger J. What do we currently know from in vivo bone strain measurements in humans? J Musculoskelet Neuronal Interact. 2011;11:8–20.
48. Smith KK, Hylander WL. Strain gauge measurement of mesokinetic movement in the lizard Varanus exanthematicus. J Exp Biol. 1985;114:53–70.
49. Lanyon LE, Goodship AE, Pye CJ, MacFie JH. Mechanically adaptive bone remodeling. J Biomech. 1982;15:141–54.
50. Goodwin KJ, Sharkey NA. Material properties of interstitial lamellae reflect local strain environments. J Ortho Res. 2002;20:600–6.
51. Peterman MM, Hamel AJ, Cavanagh PR, Piazza SJ, Sharkey NA. In vitro modeling of human tibial strains during exercise in micro-gravity. J Biomech. 2001;34(5):693–8.

52. Turner CH, Akhter MP, Raab DM, Kimmel DB, Recker RR. A non-invasive, in vivo model for studying strain adaptive bone modeling. Bone. 1991;12:73–9.
53. Turner CH, Forwood M, Rho J-Y, Yoshikawa T. Mechanical loading thresholds for lamellar and woven bone formation. J Bone Miner Res. 1994;9:87–97.
54. Rubin CT, Lanyon LE. Regulation of bone mass by mechanical strain magnitude. Calcif Tissue Int. 1985;37:411–7.
55. Sugiyama T, Meakin LB, Browne WJ, Galea GL, Price JS, Lanyon LE. Bones' adaptive response to mechanical loading is essentially linear between the low strains associated with disuse and the high strains associated with the lamellar/woven bone transition. J Bone Miner Res. 2012;27:1784–93.
56. Gross TS, Poliachik SL, Ausk BJ, Sanford DA, Becker BA, Srinivasan S. Why rest stimulates bone formation: a hypothesis based on complex adaptive phenomenon. Exerc Sport Sci Rev. 2004;32:9–13.
57. Hsieh Y-F, Robling AG, Ambrosius WT, Burr DB, Turner CH. Mechanical loading of diaphyseal bone in vivo: the strain threshold for an osteogenic response varies with location. J Bone Miner Res. 2001;16:2291–7.
58. Parenti LR. The phylogenetic significance of bone types in euteleost fishes. Zool J Linnean Soc. 1986;87:37–51.
59. Moss ML. Studies on the acellular bone of teleost fish. II. Response to fractgure under normal and acalcemic conditions. Acta Anat. 1962;48:46–60.
60. Shahar R, Dean MN. The enigmas of bone without osteocytes. BoneKEy Reports. 2013;2: Article 343.
61. Cowin SC, Hegedus DH. Bone remodeling. I: a theory of adaptive elasticity. J Elast. 1976;6: 313–25.
62. Cowin SC, Nachlinger RR. Bone remodeling. III: uniqueness and stability in adaptive elasticity theory. J Elast. 1978;8:285–95.
63. Hart RT, Davy DT. Theories of bone modeling and remodeling. In: Cowin SC, editor. Bone mechanics. Boca Raton: CRC Press; 1989. p. 253–77.
64. Beaupre GS, Orr TE, Carter DR. An approach for time-dependent bone modeling and remodeling—theoretical development. J Orthop Res. 1990a;8:651–61.
65. Beaupre GS, Orr TE, Carter DR. An approach for time-dependent bone modeling and remodeling—application: a preliminary remodeling simulation. J Orthop Res. 1990b;8:662–70.
66. Frost HM. Bone "mass" and the "mechanostat": a proposal. Anat Rec. 1987b;219:1–9.
67. Uhthoff HK, Jaworski ZFG. Bone loss in response to long-term immobilization. J Bone Joint Surg. 1978;60B:420–9.
68. Jaworski ZFG, Uhthoff HK. Reversability of nontraumatic disuse osteoporosis during its active phase. Bone (New York). 1986;7:431–9.
69. Jaworski ZFG, Liskova-Kiar M, Uhthoff HK. Effect of long-term immobilization on the pattern of bone loss in older dogs. J Bone Joint Surg. 1980;62B:104–10.
70. Frost HM. Intermediary organization of the skeleton. Boca Raton: CRC Press; 1986.
71. Ryder KD, Duncan RL. Parathyroid hormone modulates the response of osteoblast-like cells to mechanical stimulation. Calcif Tiss Int. 2001;67:241–6.
72. Li J, Duncan RL, Burr DB, Gattone VH, Turner CH. Parathyroid hormone enhances mechanically induced bone formation, possibly involving L-type voltage-sensitive calcium channels. Endocrinol. 2003;144:1226–33.
73. Kim CH, Takai E, Zhou H, von Stecho D, Muller R, Dempster DW, Guo XE. Trabecular bone response to mechanical and parathyroid hormones stimulation: the role of mechanical micro-environment. J Bone Miner Res. 2003;18:2116–25.
74. Frost HM. Perspectives: a vital biomechanical model of synovial joint design. Anat Rec. 1994;240:1–18.

75. Mosekilde L, Mosekilde L. Sex differences in age-related changes in vertebral body size, density and biomechanical competence in normal individuals. Bone (New York). 1990;11:67–73.
76. Martin RB, Atkinson PJ. Age and sex-related changes in the structure and strength of the human femoral shaft. J Biomech. 1977;10:223–31.
77. Martin RB, Pickett JC, Zinaich S. Studies of skeletal remodeling in aging men. Clin Orthop Relat Res. 1980;149:268–82.
78. Ruff CB, Hayes WC. Sex differences in age-related remodeling of the femur and tibia. J Orthop Res. 1988;6:886–96.
79. Ahlborg HG, Johnell O, Turner CH, Rannevik G, Karlsson MK. Bone loss and bone size after menopause. N Engl J Med. 2003;349:327–34.
80. Beck TJ, Ruff CB, Bissessur K. Age-related changes in female femoral neck geometry: implications for bone strength. Calcif Tissue Int. 1993;53:S41–6.
81. Poole KES, Mayhew PM, Rose CM, Brown JK, Bearcroft PJ, Loveridge N, Reeve J. Changing structure of the femoral neck across the adult female lifespan. J Bone Miner Res. 2010;25:482–91.
82. Jepsen KJ, Andarawis-Puri N. The amount of periosteal apposition required to maintain bone strength during aging depends on adult bone morphology and tissue-modulus degradation rate. J Bone Miner Res. 2012;27:1916–26.
83. Umemura Y, Ishiko T, Yamauchi T, Kurono M, Mashiko S. Five jumps per day increase bone mass and breaking force in rats. J Bone Miner Res. 1997;12:1480–5.
84. Robling AG, Burr DB, Turner CH. Partitioning a daily mechanical stimulus into discrete loading bouts improves the osteogenic response to loading. J Bone Miner Res. 2000;15:1596–602.
85. Robling AG, Burr DB, Turner CH. Skeletal loading in animals. J Musculoskelet Neuronal Interact. 2001a;1:249–62.
86. Robling AG, Burr DB, Turner CH. Recovery periods restore mechanosensitivity to dynamically loaded bone. J Exper Biol. 2001b;204:3389–99. Skeletal loading in animals
87. Sen B, Xie Z, Case N, Styner M, Rubin CT, Rubin J. Mechanical signal influence on mesenchymal stem cell fate is enhanced by incorporation of refractory periods into the loading regimen. J Biomech. 2011;44:593–9.
88. Robling AG, Hinant FM, Burr DB, Turner CH. Improved bone structure and strength after long-term mechanical loading is greatest if loading is separated into short bouts. J Bone Miner Res. 2002;17:1545–54.
89. Lanyon LE, Rubin CT. Static vs. dynamic loads as an influence on bone remodeling. J Biomech. 1984;17:897–905.
90. Mosley JR, Lanyon LE. Strain rate as a controlling influence on adaptive modeling in response to dynamic loading of the ulna in growing male rats. Bone. 1998;23:313–8.
91. Rubin CT, Lanyon LE. Regulation of bone formation by applied dynamic loads. J Bone Joint Surg. 1984;66A:397–402.
92. O'Connor JA, Lanyon LE, MacFie H. The influence of strain rate on adaptive bone remodeling. J Biomech. 1982;15:767–81.
93. Deere K, Sayers A, Rittweger J, Tobias JH. Habitual levels of high, but not moderate or low, impact activity are positively related to hip BMD and geometry: results form a population-based study of adolescents. J Bone Miner Res. 2012;27:1887–95.
94. Fuchs RK, Bauer JJ, Snow CM. Jumping improves hip and lumbar spine bone mass in prepubescent children: a randomized controlled trial. J Bone Miner Res. 2001;16:148–56.
95. Kato T, et al. Effect of low-repetition jump training on bone mineral density in young women. J Appl Physiol. 2006;100:839–43.
96. Turner CH, Chandran A, Pidaparti RMV. Anisotropy of cortical bone and its constituents. Transactions of the Orthopaedic Research Society. 1995;20:155.
97. Turner CH. Three rules for bone adaptation to mechanical stimuli. Bone. 1998;23:399–407.

98. Rittweger J, Schiesel H, Felsenberg D. Oxygen uptake during whole-body vibration exercise: comparison with squatting as a slow voluntary movement. Eur J Appl Physiol. 2001;86:169–73.
99. Gusi N, Raimundo A, Leal A. Low-frequency vibratory exercise reduces the risk of bone fracture more than walking: a randomized controlled trial. BMD Musculoskelet Disord. 2006;7:1–8.
100. Wysocki A, Butler M, Shamliyan T, Kane RL. Whole body vibration therapy for osteoporosis. State of the science. Ann Intern Med. 2011;155:680–6.
101. Elmentaser M, McMillan M, Smith K, Khanna S, Chantler D, Panarelli M, Ahmed SF. A comparison of the effect of two types of vibration exercise on the endocrine and musculoskeletal system. J Musculoskelet Neuronal Interact. 2012;12:144–54.
102. Iwamoto J, Sato Y, Takeda T, Matsumoto H. Whole body vibration exercise improves body balance and walking velocity in postmenopausal osteoporotic women treated with alendronate: Galileo and Alendronate Intervention Trail [sic] (GAIT). J Musculoskelet Neurobnal Interact. 2012;12:136–43.
103. McLeod KJ, Bain SD, Rubin CT. Dependence of bone adaptation on the frequency of induced dynamic strains. Trans Orthop Res Soc. 1990;15:103.
104. Bonewald LF. The amazing osteocyte. J Bone Miner Res. 2011;26:229–38.
105. Tomkinson A, Reeve J, Shaw RW, Noble BS. The death of osteocytes via apoptosis accompanies estrogen withdrawal in human bone. J Clin Endocrinol Metab. 1997;82:3128–35.
106. Noble BS, Stevens H, Loveridge N, Reeve J. Identification of apoptotic changes in osteocytes in normal and pathological human bone. Bone. 1997;20:273–82.
107. Divieti Pajevic P. Regulation of bone resorption and mineral hoemostasis by osteocytes. IBMS BoneKey. 2009;6:63–70.
108. Noble BS, Peet N, Stevens HY, et al. Mechanical loading: biphasic osteocyte survival and targeting of osteoclasts for bone destruction in rat cortical bone. Am J Physiol Cell Physiol. 2002;284:C934–43.
109. Cardoso L, Herman BC, Verborgt O, Laudier D, Majeska RJ, Schaffler MB. Osteocyte apoptosis controls activation of intracortical resorption in response to bone fatigue. J Bone Miner Res. 2009;24:597–605.
110. Kennedy OD, Herman BC, Laudier DM, Majeska RJ, Sun HB, Schaffler MB. Activation of resorption in fatigue-loaded bone involves both apoptosis and active pro-osteoclastogenic signaling by distinct osteocyte populations. Bone. 2012;50:1115–22.
111. Xiong J, Onal M, Jilka RL, et al. Matrix-embedded cells control osteoclast formation. Nat Med. 2011;17:1235–41.
112. Nakashima T, Hayashi M, Fukunaga T, et al. Evidence for osteocyte regulation of bone homeostasis through RANKL expression. Nat Med. 2011;17:1231–4.
113. Robling AG, Niziolek PJ, Baldridge LA, et al. Mechanical stimulation of bone in vivo reduces osteocyte expression of Sost/Sclerostin. J Biol Chem. 2008;283:5866–75.
114. Tu X, Rhee Y, Condon KW, et al. Sost downregulation and local Wnt signaling are required for the osteogenic response to mechanical loading. Bone. 2012;50:209–17.
115. Harris SE, Gluhak-Heinrich J, Harris MA, et al. CMP1 and MEPE expression are elevated in osteocytes after mechanical loading in vivo: theoretical role in controlling mineral quality in the perilacunar matrix. J Musculoskelet Neuronal Interact. 2007;7:313–5.
116. Moustafa A, Sugiyama T, Prasad J, Saman G, Gross TS, Lanyon LE, Price JS. Mechanical loading-related changes in osteocyte sclerostin expression in mice are more closely associated with the subsequent osteogenic response than the peak strains engendered. Osteoporos Inbt. 2012;23:1225–34.
117. Rubin C, Gross T, Qin Y-X, Fritton S, Guilak F, McLeod K. Differentiation of the bone-tissue remodeling response to axial and torsional loading in the Turkey ulna. J Bone Joint Surg. 1996;78A:1523–34.
118. Gross TS, Edwards JL, McLeod KJ, Rubin CT. Strain gradients correlate with sites of periosteal bone formation. J Bone Miner Res. 1997;12:982–8.

119. Weinbaum S, Cowin SC, Zeng Y. A model for the excitation of osteocytes by mechanical loading-induced bone fluid shear stresses. J Biomech. 1994;27:339–60.
120. Owan I, Burr DB, Turner CH, Qui J, Tu Y, Onyia JE, Duncan RL. Mechanotransduction in bone: osteoblasts are more responsive to fluid forces than mechanical strain. Am J Physiol. 1997;273:C810–5.
121. You J, Yellowley CE, Donahue HJ, Zhang Y, Chemn Q, Jacobs CR. Substrate deformation levels associated with routine physical activity are less stimulatory to bone cells relative to loading-induced oscillatory fluid flow. J Biomech Eng. 2000;122:387–93.
122. Wang L, Cowin SC, Weinbaum S, Fritton SP. Modeling tracer transport in an osteon under cyclic loading. Ann Biomed Engng. 2000;28:1200–9.
123. Garman R, Rubin C, Judex S. Small oscillatory accelerations, independent of matrix deformations, increase osteoblast activity and enhance bone morphology. PLoS One. 2007;2:e653.
124. Ozcivici E, Garman R, Judex S. High frequency oscillatory motions enhance the simulated mechanical properties of non-weight bearing trabecular bone. J Biomech. 2007;40:3403–11.
125. Ozcivici E, Luu YK, Adler B, Qin Y-X, Rubin J, Judex S, Rubin CT. Mechanical signals as anabolic agents in bone. Nat Rev Rheum. 2010;6:50–9.
126. Riddle RC, Donahue HJ. From streaming potentials to shear stress: 25 years of bone cell mechanotransduction. J Orthop Res. 2009;27:143–9.
127. Wang L, Wang Y, Han Y, Henderson SC, Majeska RJ, Weinbaum S, Schaffler MB. In situ measurement of solute transport in the bone lacunar-canalicular system. Proc Nat Acad Sci. 2005;102:11911–6.
128. Price C, Zhou X, Li W, Wang L. Real-time measurement of solute transport within the lacunar-canalicular system of mechanically loaded bone: direct evidence for load-induced fluid flow. J Bone Miner Res. 2011;26:277–85.
129. Knothe-Tate ML, Knothe U, Niederer P. Experimental elucidation of mechanical load-induced fluid flow and its potential role in bone metabolism and functional adaptation. Am J Med Sci. 1998;316:189–95.
130. Knothe-Tate ML, Steck R, Forwood MR, Niederer P. In vivo demonstration of load-induced fluid flow in the rat tibia and its potential implications for processes associated with functional adaptation. J Exp Biol. 2000;203:2737–45.
131. Mori S, Harruff R, Ambrosius W, Burr DB. Trabecular bone volume and microdamage accumulation in the femoral heads of women with and without femoral neck fractures. Bone (New York). 1997;21:521–6.
132. Srinivasan S, Weimer DA, Agans SC, Bain SD, Gross TS. Low-magnitude mechanical loading becomes osteogenic when rest is inserted between each load cycle. J Bone Miner Res. 2002;17:1613–20.
133. LaMothe JM, Zernicke RF. Rest insertion combined with high-frequency loading enhances osteogenesis. J Appl Physiol. 2004;96:1788–93.
134. Haut Donahue TL, Haut TR, Yellowley CE, et al. Mechanosensitivity of bone cells to oscillating fluid flow induced shear stress may be modulated by chemotransport. J Biomech. 2003;36:1363–71.
135. Riddle RC, Hippe KR, Donahue HJ. Chemotransport contributes to the effect of oscillatory fluid flow on human bone marrow stromal cell proliferation. J Orthop Res. 2008;26:918–24.
136. Bakker AD, Soejima K, Klein-Nulemnd J, et al. The production of nitric oxide and prostaglandin E2 by primary bone cells is shear stress dependent. J Biomech. 2001;34:671–7.
137. Woo SL, Kuei SC, Amiel D, Gomez MA, Hayes WC, White FC, Akeson WH. The effect of prolonged physical training on the properties of long bone: a study of Wolff's law. J Bone Joint Surg. 1981b;63A:780–7.
138. Jee WSS, Li XJ, Schaffler MB. Adaptation of diaphyseal structure with aging and increased mechanical usage in the adult rat: a histomorphometrical and biomechanical study. Anat Rec. 1991;230:332–8.
139. Schaffler MB. Immobilization induced bone loss: quantitative histological studies of cortical bone resorption. Transactions of the Orthopaedic Research Society. 1990;15:187.

140. Lanyon LE. Osteoporosis and mechanically related bone modeling. In: Menczel J, Robin GC, Makin M, Steinberg R, editors. Osteoporosis. Proceedings of the International Symposium, Jerusalem Osteoporosis Center. Chichester: Wiley; 1981.
141. Lanyon LE. Functional strain as a determinant of bone remodeling. Calcif Tissue Int. 1984;36: S56–61.
142. Lanyon LE, Rubin CT, Baust G. Modulation of bone loss during calcium insufficiency by controlled dynamic loading. Calcif Tissue Int. 1986;38:209–16.
143. Johnson LC. Morphologic analysis in pathology: the kinetics of disease and general biology in bone. In: Frost HM, editor. Bone biodynamics. Boston: Little, Brown; 1964.
144. Cumming DC. Exercise-associated amenorrhea, low bone density, and estrogen replacement therapy. Arch Intern Med. 1996;156:2193–5.
145. Klesges RC, Ward KD, Shelton ML, Applegate WB, Cantler ED, Palmieri GM, Harmon K, Davis J. Changes in bone mineral content in male athletes. Mechanisms of action and intervention effects. JAMA J Am Med Assoc. 1996;276:226–30.
146. Bab I, Gazit D, Massarawa A, Sela J. Removal of the tibial marrow induces increased formation of bone and cartilage in rat mandibular condyle. Calcif Tissue Int. 1985;37:551–5.
147. Frost HM. The pathomechanics of osteoporoses. Clin Orthop Relat Res. 1985;200:198–225.
148. Gunness M, Hock JM. Anabolic effect of parathyroid hormone on cancellous and cortical bone histology. Bone (New York). 1993;14:277–81.
149. Timoshenko SP. History of strength of materials. New York: McGraw-Hill; 1953.
150. Odgaard A, Weinans H. Bone structure and remodeling. Singapore: World Scientific; 1995.
151. Biewener AA, Thomason J, Goodship A, Lanyon LE. Bone stress in the horse forelimb during locomotion at different gaits: a comparison of two experimental methods. J Biomech. 1983;16: 565–76.
152. Goodship AE, Lanyon LE, McFie H. Functional adaption of bone to increased stress. J Bone Joint Surg. 1979;61A:539–46.
153. Hylander WL. The functional significance of primate mandibular form. J Morphol. 1979;160: 223–40.
154. Dodge T, Wanis M, Ayoub R, et al. Mechanical loading, damping and load-driven bone formation in mouse tibiae. Bone. 2012;51:810–8.

Chapter 7
Fatigue and Fracture Resistance of Bone

Abstract In Chap. 5, we learned to apply general engineering principles to normalize away the impacts of bone mass, and measure estimated material-level mechanical properties. This allowed us to examine the mechanical properties of bone when it is fractured by a load sufficient to exceed the failure stress of the material. This type of analysis is useful and important, but there are two principal ways in which structures may fail more gradually: creep and fatigue. In these situations, stresses are applied at levels below the ultimate stress/strength of the material, and the accrual of damage ultimately leads to failure. This chapter will cover the theory behind these loading modalities and their applications in bone.

What You Will Learn

1. Basics of fracture mechanics, starting with linear elastic fracture mechanics but then moving into more complicated models of cracking.
2. How and why crack propagation occurs in bone and how this influences bone toughness.
3. The properties of bone that affect crack propagation and fracture mechanics.
4. How creep and fatigue damage of bone are related.
5. Mathematical modeling of stress fractures.

7.1 Introduction

In Chap. 5, we examined the mechanical properties of bone when it is exposed to a load sufficient to exceed the failure stress of the material. Structures may also fail more gradually at subcritical loading levels. There are two principal ways in which this can happen: creep and fatigue. In both, a stress less than the ultimate stress is applied, and damage from this stress develops and accumulates until failure occurs. In creep, the stress is applied and held; in fatigue, the stress is applied periodically. Obviously, these phenomena may occur simultaneously in the components of an engineering structure like a bridge, where the weight of the structure provides creep loading and the periodic passage of vehicles produces fatigue. Similarly, in the

M. R. Allen, J. Wallace, *Skeletal Tissue Mechanics*,
https://doi.org/10.1007/978-3-031-90609-1_7

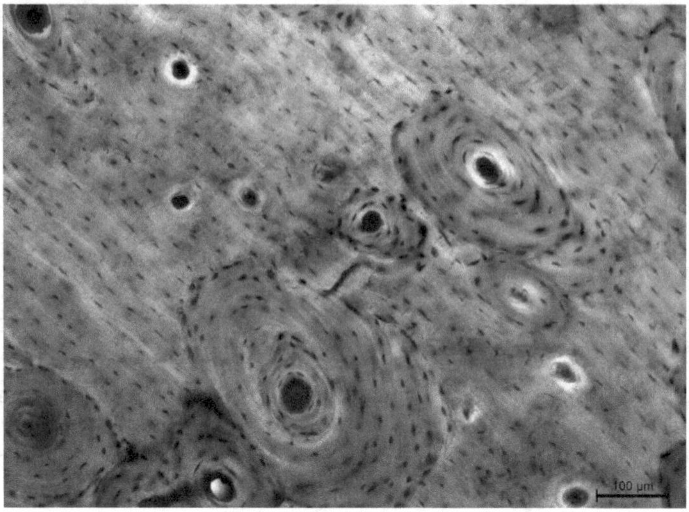

Fig. 7.1 A typical microcrack as seen in a 100-µm-thick histologic cross section of human cortical bone. The specimen was stained in basic fuchsin en bloc to mark the surfaces of cracks existing before the section was cut. Artifactual cracks produced by the sectioning process are not stained [25]. This crack is about 80 µm long in the plane of the section, and it ends about the cement lines of two secondary osteons. Most microcracks in cortical bone are of this type

skeleton, bones often support more or less constant loads for prolonged periods of time (such as the vertebral bodies in your spine as you sit reading this book), and equally often carry cyclic loads (such as when you walk to class). In this chapter, we shall see that creep and fatigue are closely related, and will explore the ways in which bone limits and repairs the damage that they produce. We also discuss what seems to happen when the limits of this capacity for damage control are exceeded and the bone fails by what is called a *stress fracture*.

At least some of the damage caused by creep and fatigue can be seen in histologic sections of normal, healthy bones (Fig. 7.1). These *microcracks* can, under continued loading, multiply and grow into "macrocracks," eventually resulting in a complete fracture. Ordinarily, microcracks (or other kinds of microdamage) are removed by targeted remodeling. If fatigue damage in bone accumulates more rapidly than remodeling can remove it, fatigue failure or a "stress fracture" can occur. These fractures can be partial or complete, and tend to occur in soldiers and joggers, as well as ballet dancers and other elite athletes, including racehorses. In principle, inhibition of normal bone remodeling could cause an accumulation of unrepaired fatigue damage that, over time, could lead to fracture.

It is important to understand, however, that the prevention of fatigue failure is not entirely dependent on repair by remodeling. The growth of microdamage is also controlled by the lamellar structure of bone. In ways completely analogous to the crack deflection and blunting effects achieved in man-made composite materials, bone's internal structure limits the development of fatigue damage. Much of the information contained in this chapter has been extrapolated from the extensive experimental data on the fatigue behavior of nonbiologic composite laminates (e.g., graphite-reinforced epoxies, or polyester laminates). Knowledge of their behavior allows us to understand how various features of lamellar bone structure function to improve its fatigue properties. Before considering such crack-resistant materials, however, we should look at what is known about cracking in simpler materials.

7.2 Basic Fracture Mechanics

Fracture mechanics considers the conditions under which an existing flaw or crack in a load-bearing structure grows and causes catastrophic failure. In conventional engineering applications, the flaws of concern may be produced as the structure is manufactured or by overloading or repetitive loading of the structure thereafter. In the case of skeletal structures, "manufacturing flaws" are usually (i.e., in healthy people) of less concern than those arising from heavy usage. However, genetic errors can result in abnormal bone tissue (usually because of a collagen defect, such as in the disease Osteogenesis Imperfecta) that becomes damaged under normal loads. Sometimes damage is introduced by surgeons as they try to correct other difficulties. Whatever the cause of the cracking, bones have an exceptional ability to control it. The purpose of this section is to provide a brief summary of fracture mechanics to help us understand how cracks are propagated generally, and how they are controlled in bone.

7.2.1 *Linear Elastic Fracture Mechanics*

The simplest way to approach this topic is to consider it in the context of stress concentrations. Consider the stress concentration on the edge of an elliptical hole in an infinite 2D plate (Fig. 7.2a). The plate bears a tensile stress, s, normal to the long axis of the ellipse. The stress at each end of this axis is defined as

$$\sigma_{\mathrm{A}} = s(1 + 2a/b) \tag{7.1}$$

where b and a are the minor and major radii of the ellipse, respectively. For a circular hole, this equation reduces to 3 s since $a = b$. This equation loses its validity if $a >> b$ and the ellipse becomes crack-like (Fig. 7.2b). In that case, the stresses at a point given by polar coordinates (r, θ) relative to the end of the b-axis are

$$\sigma_{\mathrm{x}} = s(a/2r)^{1/2}\cos(\theta)/[2(1 - \sin(\theta)/2\sin(3\theta/2))] \tag{7.2a}$$

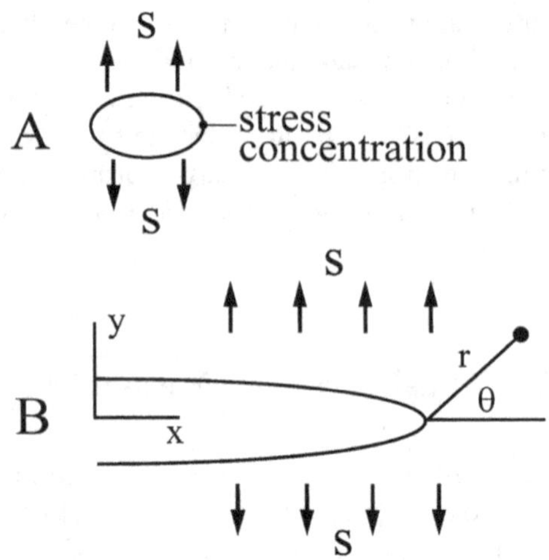

Fig. 7.2 (**a**) An elliptical hole in a plate or solid structure bearing a tension stress s has a stress concentration several times *s* at the point shown. (**b**) A crack may be thought of as a very eccentric elliptical hole. The stress at a point near the crack tip *(dot)* is a function of its distance *r* and angular position θ

$$\sigma_y = s(a/2r)^{1/2}\cos(\theta)/[2(1-\ \sin(\theta)/2\sin(3\theta/2))] \tag{7.2b}$$

$$\sigma_{xy} = s(a/2r)^{1/2}\cos(\theta)/[2\sin(\theta)/2\sin(3\theta/2)] \tag{7.2c}$$

The stress in the y direction along the $\theta = 0$ line is

$$\sigma_y = s(a/2r)^{1/2} \tag{7.3}$$

As one approaches the crack tip, $\boldsymbol{r}$ becomes small relative to the crack length, a, and the stress concentration becomes very large. At the crack tip itself, $r = 0$ and $\boldsymbol{\sigma}_y$ is a singularity (i.e., it is infinite). To avoid this difficulty, one may move the $2r$ term to the other side of the equation, obtaining $\boldsymbol{\sigma}_y(2r)^{1/2}$, which is not singular at the crack tip but is still proportional to the "intensity" of the stress at the crack tip. One can then define:

$$\text{Stress Intensity} = \sigma_y(2r)^{1/2} = sa^{1/2} \tag{7.4}$$

That is, the stress intensity in the vicinity of the crack tip is proportional to the nominal stress, s, and the square root of the crack length, a. This stress intensity is what drives the crack forward.

G.R. Irwin [1] used linear elasticity theory to obtain equations for the stresses near the tips of cracks with several different geometries and loading situations. He found that in general

$$\sigma = k(2\pi r)^{-1/2} + \text{other terms} \tag{7.5}$$

where k is a coefficient specific to the particular problem. As one approaches the crack tip, the "other terms" became less important and can be ignored. Because of the π in these equations, a formal *stress intensity factor*, K, was defined so as to include π as well:

$$K = Cs(\pi a)^{1/2} \tag{7.6}$$

K has units of Pa-m$^{1/2}$; compare Eq. (7.6) with Eq. (7.4). C is a dimensionless constant that depends on the size and shape of the flaw and the size, shape, and mode of loading of the specimen, and s is a pertinent stress component in the specimen beyond the influence of the crack, for example, tensile stress perpendicular to the plane of the crack. K characterizes the state of stress near the crack tip; therefore, it is known as a *characterizing parameter*. Its utility stems from the fact that if K exceeds a critical value, the crack will propagate; otherwise, it will not. There is no need to make calculations of the local stresses to gain this information.

To reiterate, the fundamental concept in linear elastic fracture mechanics is that K determines whether a crack will propagate or not, and if so, how fast. In practice, K bears a subscript denoting the type of cracking that is being considered.

7.2.2 Kinds of Cracks

Of the various ways in which a specimen may be loaded to extend cracks, three are commonly discussed in fracture mechanics (Fig. 7.3). Mode I cracks are propagated perpendicular to the line of action of a tensile load, as when a wedge splits a log lengthwise. Mode II cracks result when shearing forces are applied parallel to the plane of the crack and parallel to the direction of its propagation. Mode III cracks are a "tearing" phenomenon; in this case, the shearing is perpendicular to the crack propagation direction. Mode I is the most commonly discussed type of cracking.

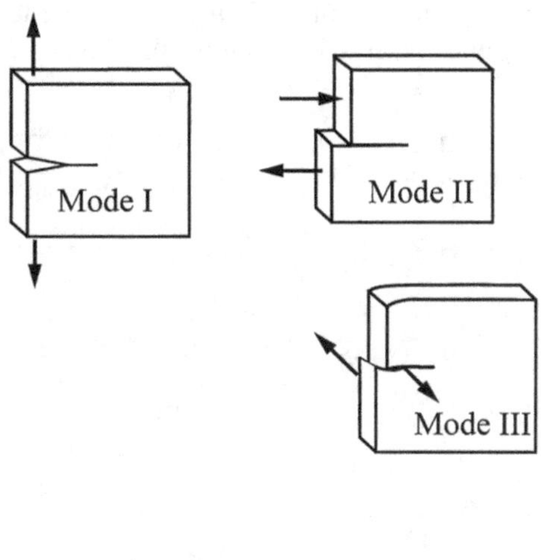

Fig. 7.3 Modes of cracking

7.2.3 Modes of Cracking

As an example, consider the mode I cracking that occurs when a plate of width b, containing a crack of length a on one edge, is loaded in tension. In this case, the C in Eq. (7.6) turns out to be

$$C = [(\tan q)/q]^{1/2} \frac{0.752 + 2.02(a/b) + 0.37(1 - \sin q)}{\cos q} \tag{7.7}$$

where $q = \pi a/2b$. As shown in Eq. (7.6), when multiplied by $(\pi a)^{1/2}$ and the tensile stress in the plate, this C gives the stress intensity factor, K_{I}, for this situation. If this K_{I} is greater than a critical stress intensity factor, K_{Ic}, the crack propagates rapidly through the structure. K_{Ic} is also called the *fracture toughness* and is a material property similar to yield stress. Table 7.1 gives some typical values of K_{Ic} for metals, ceramics, polymers, and cortical bone [2]. By conducting many experiments and tabulating the results, the means for calculating K_{I} in various situations and comparing them to the critical values have been amassed for many conventional engineering applications. Figure 7.4 gives formulae for C in a few simple cases.

The measurement of K is beyond the scope of this book because it is complicated by the need to ensure that all assumptions underlying its calculation are fulfilled. The basic process is simple, however, as illustrated in Fig. 7.5 for mode I cracking. A specimen with a standardized crack is manufactured, and the crack is propagated across the specimen by loading it in tension. The applied load and crack opening displacement are measured and plotted. A failure load is identified on the load–displacement curve using several rules. K_{Ic} is then calculated by multiplying the failure load by a formula derived from the length of the crack and the shape and dimensions of the specimen.

Several investigators have measured K_{Ic} for cortical bone (Table 7.2). Values for bone are an order of magnitude less than those for structural metals but similar to a tough ceramic like Al_2O_3. In making these comparisons, it needs to be emphasized that the values for bone are for crack propagation parallel to the long axis of the bone, which is to say "with the grain." If one tries to propagate a crack transversely, or perpendicular to the primary and secondary lamellar structure, the crack turns and goes "with the grain." Thus, the fracture toughness of bone is decidedly anisotropic, and the values shown are for the direction offering the least resistance to cracking.

Table 7.1 Fracture toughness for various materials

Material	K_{Ic}, MPa-m$^{1/2}$
2024 aluminum	20–40
4330 V steel	86–110
Ti-6Al-4 V	106–123
Concrete	0.23–1.43
Al_2O_3 ceramic	3–5.3
SiC ceramic	3.4
PMMA polymer	0.8–1.75
Polycarbonate polymer	2.75–3.3
Cortical bone	2.2–6.3

Crack Geometry	C
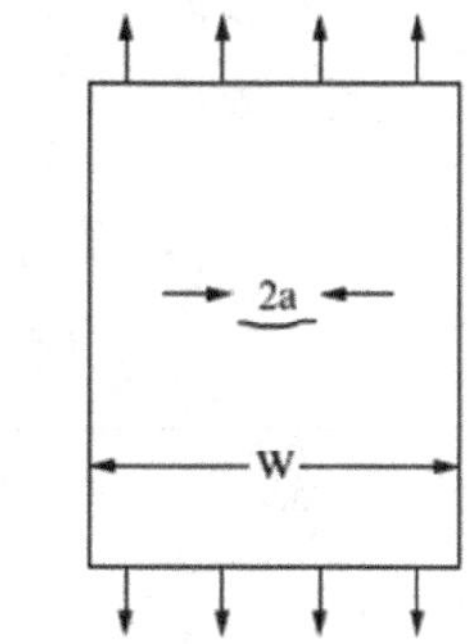	Plate of width W in tension, Mode I crack of length $2a$ in center. $C = [1 - 0.1(a/W)^2 + 0.96(a/W)^4]\,[\sec(\pi a/W)]^{1/2}$
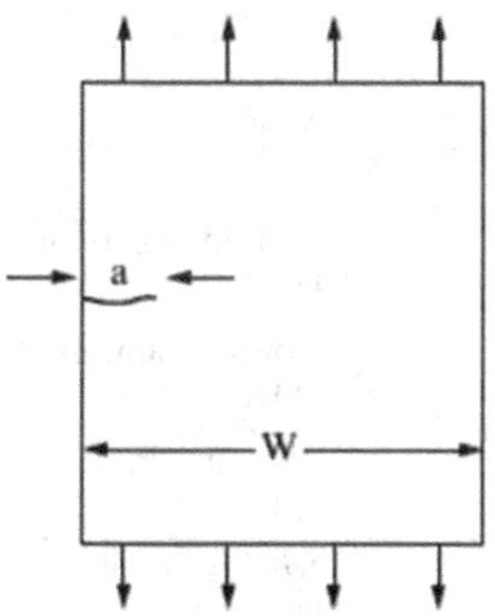	Plate of width W in tension, Mode I crack of length a on one edge. $C = [2W/\pi a)\tan(\pi a/2W)]^{1/2} \cdot \dfrac{0.752 + 2.02(a/W) + 0.37[1 - \sin(\pi a/2W)]}{\cos(\pi a/2W)]}$
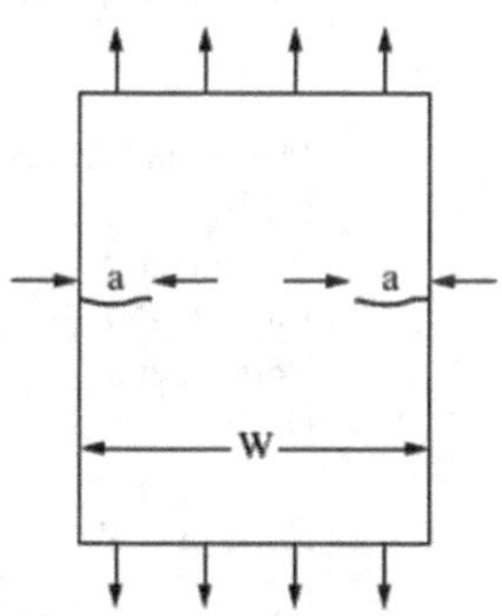	Plate of width W in tension, Mode I crack of length a on each edge. $C = [1 + 0.122\cos^4(\pi a/W)]\,[(W/\pi a)\tan(\pi a/W)]^{1/2}$

Fig. 7.4 Equations for the constant C in three simple cases

7.2.4 Maximum Tolerable Flaw Size

By rearranging Eq. (7.6) and setting the stress intensity factor equal to the critical value, one may solve for the flaw size that will precipitate unstable crack propagation. For human tibial compact bone, Norman et al. [3] found that $K_{Ic} \approx 4\,\text{MPa-m}^{1/2}$. For a semicircular crack of radius a at the surface of a semi-infinite body loaded in

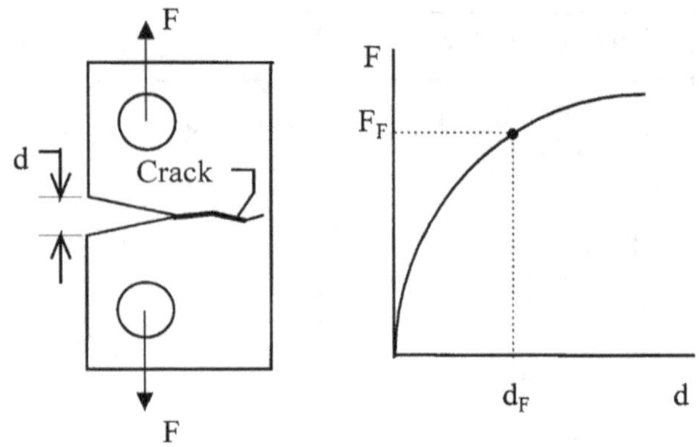

Fig. 7.5 The basics of measuring K_{Ic}. A specimen with a notch is loaded in tension to drive a crack to the right as shown. The tensile force F is plotted against the crack opening displacement, d. A failure point, P(FF, dF), on the curve is defined by certain rules, and K_{Ic} is calculated from FF and dF using a formula

Table 7.2 Experimental values of K_c and G_c for bone

Bone type	K_c, MPa-m$^{1/2}$	G_c, Jm^{-2}	Source
Mode I, transverse fracture			
Bovine femur	5.49	3100–5500	Melvin and Evans [115]
Bovine tibia	2.2–4.6	780–1120	Bonfield and Datta [116]
Equine metacarpus	7.5	2340–2680	Alto and Pope [117]
Human tibia Human femur (young and old)	2.2–5.7 5.1 and 4.3	350–900 2262 and 1712 (derived from K_c and E)	Norman et al. [114] Wang et al. [118]
Mode I, longitudinal fracture			
Bovine femur	3.21	1388–2557	Melvin and Evans [115]
Bovine tibia	2.8–6.3	630–2238	Behiri and Bonfield [56]
Human femur	2.2–5.7	350–900	Norman et al. [114]
Mode II			
Human tibia	2.2–2.7	365	Norman et al. [114]

tension (e.g., on the tension side of a large bone loaded in bending), the stress intensity factor is [2]

$$K_I = 2(a/\pi)^{1/2} s\left[1.211 - 0.186(\cos\theta)^{1/2}\right] \tag{7.8}$$

Here, $\boldsymbol{\theta}$ is as defined in Fig. 7.2 and s is the stress elsewhere in the body, far from the crack. Setting $\boldsymbol{\theta} = 0$, the maximum tolerable crack radius would be

$$a_c = 0.238\pi K_{Ic}{}^2/s^2 \tag{7.9}$$

Assuming that $E = 20\,\text{GPa}$ and the strain ($\boldsymbol{\varepsilon}$) is a physiologic 2000 με, one has $s = E\boldsymbol{\varepsilon} = (20 \times 10^9)\,(2 \times 10^{-3}) = 40\,\text{MPa}$ and

$$a_c = 0.238\pi\left(4^2/40^2\right) = 0.0075\ \text{m} = 7.5\ \text{mm} \tag{7.10}$$

which is a rather large flaw. Remember also that the K_{Ic} used here is probably considerably smaller than the one for a crack propagating across a bone.

7.2.5 It Takes Energy to Propagate a Crack

Before Irwin and others developed the concept of a stress intensity factor, A.A. Griffith [4] formulated the problem of crack propagation in brittle solids in terms of energy. He reasoned that energy was required to create the surfaces of the extended crack because all surfaces have "surface energy" associated with them. He conducted many experiments with glass rods to develop his idea into a theory.

Suppose it takes energy or work, dW_{c}, to propagate a crack for a distance, da. This energy usually comes from the work done in deforming the specimen via the strain energy, U, stored throughout its interior. Conservation of energy requires that

$$dW_{\text{c}}/da = dU/da \tag{7.11}$$

The right-hand side of this equation is called the *strain energy release rate* and is denoted by G, after Griffith, who showed that for mode I propagation of a thin ellipsoidal crack, the strain energy release rate is

$$G = dU/da = \pi a s^2/E \tag{7.12}$$

where E is the elastic modulus. By setting up an experiment in which a crack can be slowly propagated in a material of modulus E, and its length (a) measured as a function of the applied stress (s), the strain energy release rate can be determined from Eq. (7.12).

The left-hand side of Eq. (7.11) is called the *crack growth resistance, R*:

$$R = dW_{\text{c}}/da \tag{7.13}$$

For a crack to grow, strain energy must be produced and released at least as fast as it is consumed by breaking molecular bonds in the material to extend the crack. For the crack to grow precipitously, G must be greater than a critical value, which is R, that is, $G \geq R$.

By comparing Eq. (7.6) and Eq. (7.12), it is apparent that G is proportional to K^2. For mode I cracking in a plane stress situation,

$$G_\mathrm{I} = K_\mathrm{I}^2/E \tag{7.14}$$

It is important to understand that a propagating crack is "nourished" by the elastic energy of the mass under stress. Finkel [5] pointed out that although cracks are driven by the strain energy stored in the mass under stress, they are themselves without inertia, and thus capable of turning instantaneously in any direction that promises more elastic energy.

7.2.6 Beyond the Linear Theory: Real Cracks Have Ears

In most materials, and certainly in bone, linear elasticity theory does not hold when stresses become high, and that includes the situation at the tip of a crack. There is a zone of plastic deformation around the crack tip that is generally shaped as depicted in Fig. 7.6. Note the "Mickey Mouse ears" shape of the plastic zone. Finkel has called the plastic zone the "ears" of the crack, and refers to computations showing that these ears may "wiggle" as the crack propagates. The plastic zone at the leading edge of a propagating crack has an ear-like characteristic shape.

For mode I cracking, the distance directly ahead of the crack that is plastically deformed is given by

$$r_\mathrm{p} = \frac{1}{k\pi}\left(\frac{K_I}{\sigma_\mathrm{y}}\right)^2 \tag{7.15}$$

where $\boldsymbol{\sigma}_\mathrm{y}$ is the yield stress; $k = 2$ for plane stress or 6 for plane strain. In a plane strain situation, the strains are all within a plane (i.e., deformations perpendicular to

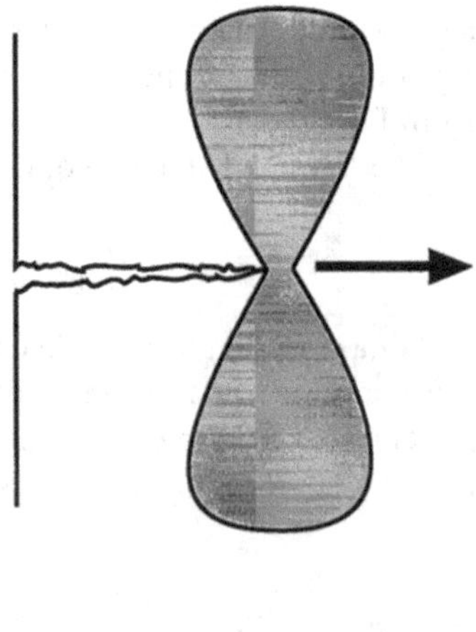

Fig. 7.6 The plastic zone at the leading edge of a propagating crack has an ear-like characteristic shape

the plane may be assumed zero). Plane stress is similarly defined in terms of stress components perpendicular to the plane. Thanks to Poisson's ratio, the two situations are not the same.

One way to accommodate the presence of a plastic zone in linear elastic fracture mechanics theory is to add this r_p (or some portion of it) to the crack length in the equations for K and G. Equivalently, when using the energy approach, another term must be added to Eq. (7.14): the rate at which energy is consumed by plastic deformation per unit length of crack extension.

Text Box 7.1 Dynamic Fracture Mechanics

Things break differently when they break extremely rapidly. Dynamic fracture mechanics is the study of cracking under these conditions [6]. When a bone is broken very rapidly in a situation involving a great deal of kinetic energy, the fracture is highly comminuted (i.e., the bone breaks into many pieces). The reasons for this are complex, but stem from two important phenomena. When a crack propagates slowly through a structure, it causes the adjacent material to be unloaded. This load change prevents nucleation of, or arrests, other cracks in the vicinity. When a structure is impulsively loaded, a stress wave passes through it at the speed of sound (about 3000 m/s for bone). This results in crack initiation and propagation at many sites simultaneously, before any one crack can arrest others. When these many cracks grow to intersect one another, many fracture fragments are produced. Emergency room physicians see this result in victims of high-energy trauma.

The second phenomenon promoting fragmentation at high fracture rates is bone's viscoelastic nature. The more rapidly bone is loaded, the stiffer it is. The velocity of the stress wave increases with the square root of the elastic modulus, so that the stress field more rapidly fills the bone, accentuating the effects just described. On the other hand, bone strength increases with loading rate. This effect is seen in other materials, too. The reasons for it are not clear.

In summary, when a bone is rapidly loaded, it becomes more fracture-resistant, but if it breaks, the fracture becomes more fragmented. Also, logically enough, the severity of damage to the soft tissues tends to parallel the degree of damage to the bone; thus, the seriousness of severe fractures is compounded by impairment of the associated soft tissues, on which fracture healing depends.

7.2.7 Beyond the Linear Theory: Real Cracks Are Complicated

Propagation of a crack in bone is slowed (or accelerated!) by changes in the material ahead of or behind the tip of the propagating crack (Fig. 7.7). Crack toughening mechanisms ahead of the crack tip are called "intrinsic mechanisms" and those that occur behind the crack tip are called "extrinsic mechanisms" [7]. Intrinsic

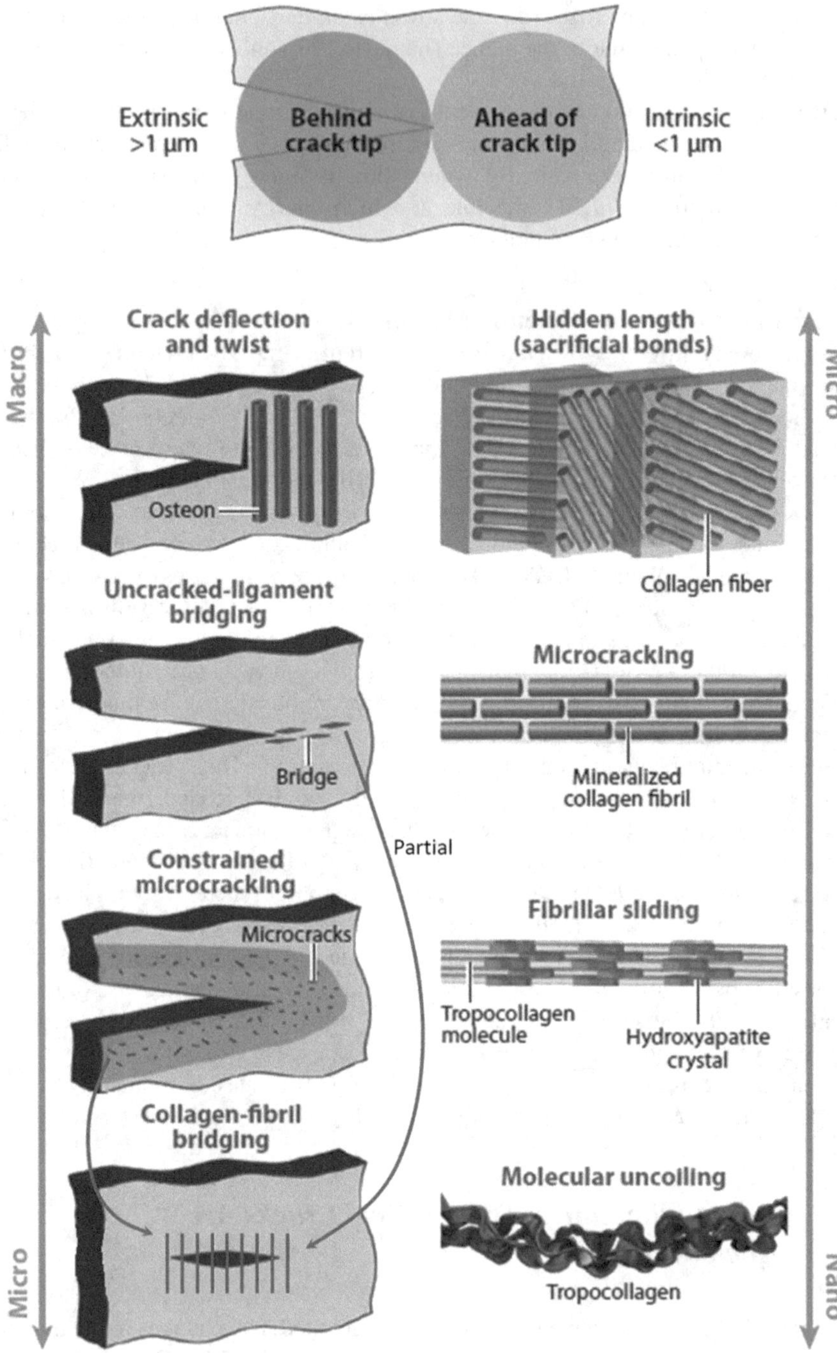

Fig. 7.7 The extrinsic and intrinsic toughening mechanisms of bone. Crack bridging by uncracked ligaments (large cracks) and by protein/collagenous bone matrix (small cracks) are very important

mechanisms involve changes in the material properties, for example, in a metal the material properties change from elastic to plastic in front of the crack tip. Extrinsic mechanisms act at the location of crack propagation (not in front of the crack) and, in bone, are strongly related to bridging of the crack by "ligaments" of unbroken tissue. For relatively large cracks (the kind in a fracture toughness specimen), a great deal of toughening comes from unbroken "ligaments" of bone tissue that bridge across the faces of the propagating crack. The ligaments hold the faces of the crack together creating a "closure stress." Examination of the crack faces of a piece of bone after failure finds the remains of numerous ligaments that had to be broken in order to finally fracture the tissue.

There continues to be some debate about the role of microcrack formation in the toughening of bone tissue, but it is clear that younger/tougher bone forms much greater numbers of microcracks during failure than does older more brittle bone [8]. This is likely due to a more functional collagen matrix, a very important contributor to bone fracture toughness [9–12]. The fracture toughness of bone tissue with the collagen matrix removed is extremely low and almost no ligaments are observed in the wake of the crack tip. Ligament bridging between the faces of microcracks by mineralized collagen fibrils and bundles allows the microcrack to open, changing the shape of the surrounding tissue, which consumes energy and thereby blunts propagation. One effect is that arrays of collagen-bridged cracks at the micro level provide enough controlled deformation to enable tissue bridging across larger cracks at the macro level. This is an example of a hierarchical dependency in bone tissue where a macroscopic toughening mechanism (i.e., ligament bridging of large macroscopic cracks) depends on a more microscopic toughening mechanism (collagen matrix bridging of microcracks).

As a crack is propagated in a fracture specimen, the stress intensity required to propagate the crack could decrease, stay the same, or increase. Recall for linear elastic fracture mechanics that: $K = Cs(\pi a)^{1/2}$. Experimentally, the process is to determine "s" and "a" as the crack propagates and then calculate the stress intensity factor "K." Clearly, the best situation is for the stress intensity factor to increase with crack length so that there is a tendency for the crack to stop propagating. Healthy bone has the property that K increases with crack length resulting in what is called a "rising R-curve," where "R" is for "Resistance" (Fig. 7.8). The name "R-curve" is probably the most obscure in all of bone mechanics. Fortunately, the concept is simple: An R-curve plots the changing fracture propagation toughness of the tissue as a function of the extension of the crack.

Fig. 7.7 (continued) to bone toughness. Removal of, or damage to, the bone matrix leaves an extremely brittle remnant mineral matrix. The energy absorbing mechanisms of the bone matrix include uncoiling of hidden length, fibrillar sliding, and molecular uncoiling, all of which depend on the animal's genetic background, age, and matrix cross-linking [119]

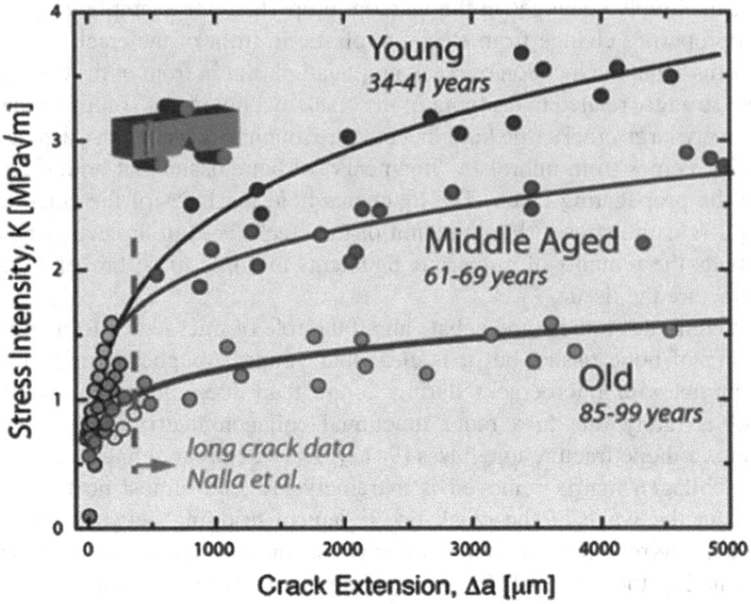

Fig. 7.8 The increase in bending fracture toughness with crack length is strongly affected by age [119]

Both aging and sequential degradation of the bone matrix using radiation correlate with a reduction of the propagation resistance (Fig. 7.9). The radiation experiments demonstrate the importance of the collagen matrix to fracture toughness. The implication of the age-related decrease in the propagation resistance is that the collagen matrix degrades with age in humans. There is a great deal of research currently being done to characterize how age and other conditions in humans affect both the bone matrix and fracture toughness. It is clear that a significant portion of age-related bone fragility in humans is caused by changes in the collagen matrix. What remains to be determined is what is changing in the matrix and whether it can be prevented or treated in vivo.

7.2.8 Crack Growth and Fatigue

Paris [13] proposed that the number of cycles required for fatigue failure in a specimen with a crack could be related to the crack's stress intensity factor. His idea was that the crack length increase per load cycle, da/dN, would be proportional to the range of K values experienced during the loading: $\Delta K = K_{\max} - K_{\min}$. A commonly suggested form for such a relationship is

$$da/dN = \xi(\Delta K)^{\zeta} \tag{7.16}$$

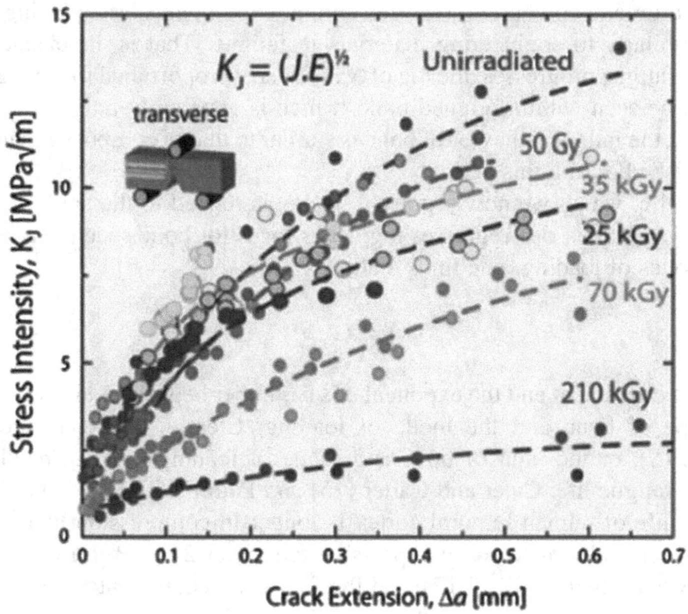

Fig. 7.9 The increase in bending fracture toughness with crack length is strongly reduced by radiation damage to the collagen matrix [119]

where ξ and ζ are constants. Unfortunately, the Paris equation has not proven to be a consistently valid relationship. It describes the results at intermediate crack growth rates better than at low or high values of *da*/*dN*. Within that range, if crack propagation data are presented as a log–log plot of *da*/*dN* vs ΔK, the slope and intercept should yield the coefficients ζ and ξ, respectively.

7.3 Fatigue Behavior of Bone

Depending on the fracture toughness of a material and the magnitude of the applied loads, cracks start and grow more or less rapidly under repetitive loading. In this section, we examine the basics of fatigue and fatigue damage behavior of bone.

7.3.1 The S–N Curve

One basic means of presenting fatigue data is an S–N curve: a plot of the applied stress or strain as a function of the number of cycles required for failure at that loading level. A fatigue S–N curve was first measured for bone by King and Evans in 1967 [14]. Subsequently, in the 1970s and 1980s, several other investigators (notably Carter and coworkers) reported similar experiments, establishing that bone behaves similarly to engineering materials in fatigue. That is, its elastic modulus declines as fatigue progresses, the log of N is linearly proportional to the log of S, and cracks can be seen within fatigued bone, which is apparently fatigue damage. The concept that the fatigue behavior of bone is similar to that of composite materials was also established during this period.

Fatigue life, which we now represent by N_F, is related to the applied stress in a nonlinear way. When one regresses log N_F vs log S for bone specimens fatigued in various modes of loading, one finds that

$$N_F = c/S^q \tag{7.17}$$

where c is a coefficient and the exponent q is a number between 5 and 15, depending on the type of bone and the mode of loading. Clearly, small changes in load magnitude (S), or the kind of bone and mode of loading (q), can result in large changes in fatigue life. Caler and Carter [15] and Pattin et al. [16] have shown that the fatigue life of human femoral cortex is longer in compression than in tension. Testing in uniaxial tensile and compressive fatigue at 2 Hz, Pattin et al. produced data that, when fitted to Eq. (7.17), had the following coefficients:

$$N_F = 1.445 \times 10^{53}/S^{14.1} \text{ (tension)} \tag{7.18}$$

$$N_F = 9.333 \times 10^{40}/S^{10.3} \text{(compression)} \tag{7.19}$$

where the stress, S, ranges from zero to a peak stress normalized by the initial elastic modulus (i.e., S is also the initial strain in units of microstrain). These data were obtained at superphysiologic strains (2600–6600 **με**). If one extrapolates the results to a physiologic peak strain of 2000 με, the fatigue life would be 4.1 million cycles in tension and 9.3 million cycles in compression.

7.3.2 Fatigue Damage in Bone

Although we typically think of microdamage in bone as linear cracks, damage caused by cyclic loading of bone comes in different flavors. Three distinct forms of damage can be identified in bone, distinguished by their morphology, their mechanical and physiological roles, and by the manner in which they are repaired. *Linear microcracks* are typically 40–100 μm "long" in cross section but may run for hundreds of microns or more than a millimeter longitudinally [17, 18]. They are less than 2 μm "wide" in cross section (Fig. 7.1) and are typically repaired by normal bone remodeling processes in which the crack is resorbed by osteoclasts, and new bone is laid down in its place. If many linear microcracks accumulate, either because of high and repeated stresses or because remodeling is suppressed through disease or treatments for disease (e.g., bisphosphonates used to mitigate osteoporosis), they can degrade the elastic modulus of the bone to the extent that a stress fracture, or a complete fracture, can occur. A second form of microdamage, *diffuse damage,* is composed of a collection of small cracks, most less than 10 μm long in cross section. The cracks themselves can be difficult to detect with standard light microscopy, even on stained sections, but can be identified as individual cracks using confocal microscopy [19, 20]. These cracks probably cause less mechanical degradation than linear microcracks, and may be important in stress relief when bone is repeatedly loaded. They do not create the signals necessary to stimulate whole bone remodeling [21], so why do they not accumulate? Recent observational and experimental data suggest that they may heal through an intrinsic matrix repair process in which the defect is eliminated by filling with mineral [22], or in which non-collagenous proteins (perhaps osteopontin) act like a glue to bring the surfaces of bone back into contact [23]. *Microfractures* are a third type of microdamage in bone. These are not cracks at all, but full fractures of trabecular bone, which heal through callus formation and subsequent callus remodeling, just as a full fracture of a long bone would heal. Microfractures have a deleterious effect on the mechanical properties of whole bone as they disrupt the connectivity of the trabecular lattice.

Thus, fatigue damage is not a narrowly defined quantity in either bone or other materials. In the engineering literature, it has usually been defined theoretically as a parameter that varies from zero in virgin structures to unity in failed structures, based on changes in mechanical properties, but without any specific reference to cracks or other evidence of physical damage, which may be difficult to detect. For example, damage is often defined as the fractional reduction in elastic modulus. Alternatively, when cracks are apparent, as in certain kinds of composite materials, damage may be defined in terms of the numbers or lengths of microscopic cracks (often defined per unit area) that appear after loading.

Cracks are relatively easy to find in healthy bone. Frost [24] first described such damage, and linked it to fatigue. Burr and Stafford [25] counted 0.014 such cracks/mm^2 in the ribs of a 60-year-old man. The number of such cracks increases with age in the human femoral diaphysis and neck cortex [26], and in the trabecular bone of the femoral head [27]. Microcracks accumulate exponentially after age 40 [26]. Although the data suggest increased skeletal fragility caused by microcrack damage in aging women, they do not show why some older women fracture and others do not. Mori et al. [27] found that older women have significantly more microcracks than younger women, but older women who fracture did not have more microcracks than older women who did not fracture. Microdamage burdens may be higher in trabecular than cortical bone. Wenzel et al. [28] found approximately 5 microcracks/mm^2 in trabecular bone from the spine, without a significant increase with age. In the spine, disk pathology may contribute to microdamage [29], either because loads are improperly attenuated or because stress concentrations are created in the vertebral body.

If linear microcracks and diffuse damage both perform the function of stress relief or energy dissipation, then why are two different kinds of damage created? This probably has to do with the nature of the stresses that create the damage, and the quality of the bone matrix in which the damage forms. Linear microcracks are preferentially found in areas of higher compressive stress [30], are associated with greater matrix deformation before the cracking occurs [31], and may be associated with shear strains/stresses in the matrix caused by the compression [32]. About 80–90% of all linear microcracks in cortical bone are found in the interstitial matrix between osteons where mineralization is higher [26, 33, 34]. Because bone from older individuals tends to contain more nonenzymatic cross-links and may be more fully mineralized, it may be stiffer and less able to deform to accommodate loading. As a consequence, linear microcracks are more commonly found in older people than in younger people, and tend to be on average three times longer (136 μm in older people vs 63 μm in younger folks) [31]. This reduces the fatigue life of bone by about 75% between the ages of 40 and 80 years. Diffuse damage, on the other hand, is more likely to be found in areas of high tensile stress and is more often seen than linear microcracks in bones from younger people [30, 31]. Although stress is dissipated by linear microcracks that are attracted to and trapped by lamellar interfaces, the stress reduction associated with larger areas of smaller cracks initially increases the toughness of bone [35], and is likely to be more effective at resisting a catastrophic fracture [8]. When linear cracks form, on the contrary, residual toughness declines. Although some have suggested that linear cracks are simply the natural sequelae to diffuse damage [35, 36], the specific age-, location-, and stress-associated patterns of these two types of damage [37] make it more likely that they are distinct forms of damage that are not related by any temporal sequence.

7.4 Creep Behavior of Bone and Its Relationship to Fatigue

As we noted in Chap. 5, bone is viscoelastic and therefore creeps or deforms under a constant load. In contrast, a perfectly elastic material would have no deformation if a load below the elastic limit were applied and held. If a bone specimen is loaded to a constant stress well below the ultimate stress and held, the material non-elastically deforms and could eventually fracture. Figure 7.10a shows this behavior for cortical bone from a human femur. The initial strain response to an applied stress is immediate and therefore elastic. The strain–time curve is similar to the damage–cycle number curve for fatigue. After a period of rapid creep there is a plateau, and the rate of damage then increases again just before failure. If groups of specimens are loaded to various stress levels, $\boldsymbol{\sigma}$, and the time to failure, $\boldsymbol{t}_{\mathrm{F}}$, is recorded, the log–log graph of $\boldsymbol{\sigma}$ vs t_{F} is linear (Fig. 7.10b) and described by an equation similar to Eq. (7.17):

$$t_{\mathrm{F}} = A/\sigma^{B} \tag{7.20}$$

where A and B are coefficients specific to the kind of bone and the mode of loading.

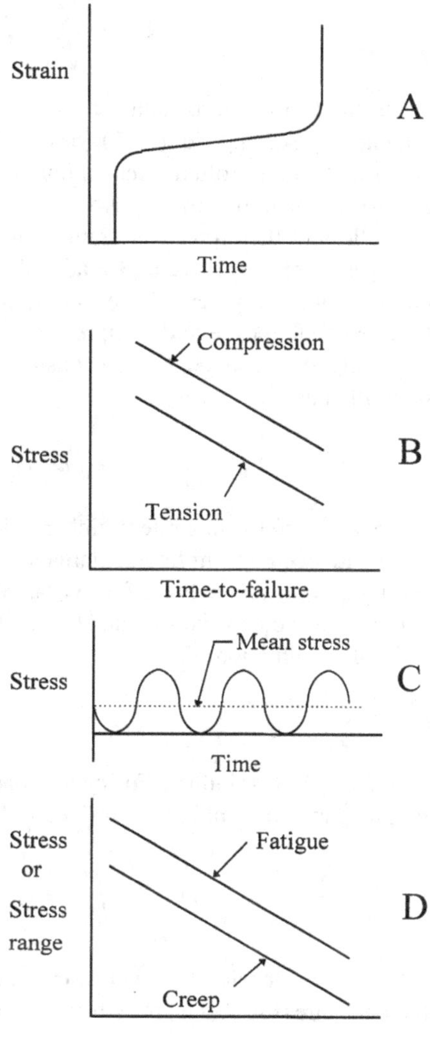

Fig. 7.10 (**a**) Typical plot of strain vs time for a cortical bone creep test to failure. The initial elastic response is followed by relatively constant strain during slow damage accumulation, eventually leading to rapid damage accumulation and failure. (**b**) Log–log plot of applied stress vs creep time-to-failure is linear with different behaviors for compression and tension. (**c**) Cyclic loading from zero to a tensile or compressive stress produces a nonzero mean stress that superimposes creep effects on fatigue behavior. (**d**) Comparison of creep and fatigue responses for human femoral bone

Caler and Carter [15] found that the exponent B was about 17 and similar for tensile and compressive creep. On the other hand, the coefficient A was greater for compression than for tension.

If the creep damage (D_c) occurs at a constant rate, its formation rate would be

$$dD_c/dt = \sigma^B/A \tag{7.21}$$

If the stress varies with time, the creep damage at any time, t, is

$$D_c = \int_0^t \left[\sigma^B/A\right] dt \tag{7.22}$$

In this formulation, failure occurs when $D_c = 1$, and the time required can be obtained by solving Eq. (7.22) for this condition. Carter and Caler pointed out that creep and fatigue often occur simultaneously. If a specimen is loaded in tensile fatigue, for example, the stress varies from zero to a peak tensile value. The loading is cyclic and therefore it is "fatigue loading," but it also has a creep component because there is an average load value that is not zero (Fig. 7.10c). The creep component is only zero if the fatigue loading is "fully reversed," that is, it cycles between peak tensile and compressive stresses that are equal in magnitude.

If fatigue loading occurs at a constant stress range, S, then the fatigue damage may be written as

$$D_f(t) = R_L t S^\beta/\alpha \tag{7.23}$$

where R_L is the loading rate (number of load cycles/day) and $\boldsymbol{\alpha}$ and $\boldsymbol{\beta}$ are coefficients. If fracture occurs from fatigue damage alone, then the time to fracture is obtained by setting $D_F = 1$ and solving for $t = t_F$. When damage is accumulating in response to simultaneous creep and fatigue effects, it might be possible to formulate the result as a simple summation:

$$D_{total} = D_c + D_f \tag{7.24}$$

with $D_{total} = 1$ at failure. For cyclic loading with a nonzero mean stress, creep and fatigue damage combine to produce failure after N_F cycles and at time t_F, and

$$D_{total} = \int_0^{t_F} \left[\sigma(t)^B/A\right] dt + R_L t_F S^\beta/\alpha = 1 \tag{7.25}$$

For a zero-tension fatigue test in which stress is a haversine function of time and the peak stress is S, the stress history may be written as

$$\sigma(t) = (1 + \sin 2\pi R_L t) S/2 \tag{7.26}$$

Substituting this function into Eq. (7.22) gives, at failure,

$$1 = (1/A)(S/2)^B \int_0^{t_F} (1 + \sin 2\pi R_L t)^B dt \tag{7.27}$$

which with two changes of variables can be solved to give

$$t_F = KA/S^B \tag{7.28}$$

for the time to failure in the zero-tension fatigue test. Here,

$$K = \frac{2^B}{\int_0^1 (1 + \sin 2\pi(t/t_F))^B d(t/t_F)} \tag{7.29}$$

Now compare Eq. (7.20), for time-to-failure in a creep test, to Eq. (7.28), for time-to-failure in a zero-tension fatigue test. The only difference is the coefficient K in Eq. (7.28). If we rewrite these equations in log–log form so as to place stress on the y-axis and time on the x-axis, as is customary, we have

$$\log S = B \log A - B \log t_F \text{ (creep)} \tag{7.30}$$

and

$$\log S = B \log KA - B \log t_F (\text{zero} - \text{tension fatigue}) \tag{7.31}$$

The slopes of the two equations are the same but their intercepts are different. Figure 7.10d shows the results of creep and zero-tension fatigue experiments on human femoral bone performed by Carter and Caler [38]. Note that the fatigue data fall along a line with the same slope as the creep data, but displaced up and to the right. The specimens loaded in fatigue required longer to fail at a given stress level because the stress noted on the y-axis was only applied part of the time. In the case of the creep specimens, it was applied continuously.

In later papers, Carter and Caler [15, 39] showed that when fatigue stresses are high, so that N_F and t_F are low, creep damage dominates, and vice versa. This is significant because most fatigue experiments are done with supraphysiologic stresses to keep the time necessary to bring the specimen to failure reasonable. Damage in these experimental specimens is mostly caused by creep. Conversely, bones are fatigued in vivo at relatively low stresses, so N_F and t_F are large; damage then is mostly caused by fatigue. Insofar as the mechanisms of damage caused by creep and by fatigue are different, the results of laboratory experiments cannot be extrapolated to the in vivo situation. On the other hand, differences between creep and fatigue damage have not been identified. These are important caveats to bear in mind as you read the remainder of this chapter and work through the exercises at the end.

7.5 Fatigue Behavior of Fiber-Reinforced Composite Laminates

Engineers have learned to manufacture materials with exceptional creep and fatigue resistance. These materials limit damage through the presence of lamellae or fibers in their structure. Because bone is a lamellar structure in which both ultrastructural (i.e., collagen fibrils) and microstructural (i.e., osteons) fibers figure prominently, we can gain some insight into bone's damage resistance by examining the behavior of these man-made materials.

7.5.1 The Birth and Growth of Cracks

The resistance of any material to fatigue failure is a function of its resistance to both the *initiation* and *propagation* of cracks. These two factors are, in turn, highly dependent on the microstructural characteristics of the material. Materials that resist the initiation of microcracks often cannot effectively resist their propagation. On the other hand, those materials in which cracks form easily are often resistant to the propagation of a crack through the material. Because the fatigue life of a material is a function of both crack nucleation and growth, a material in which cracks are easily started, but find it difficult to grow, can often provide greater resistance to fatigue failure than one in which cracks are difficult to initiate. In fact, many highly fatigue-resistant materials derive this property primarily from their resistance to crack growth rather than initiation. Bone is such a material.

7.5.2 Material Strength and Fiber Diameter

There are two primary microstructural factors that affect crack initiation and growth: *material strength* and *fiber diameter*. In general, decreased yield strength *increases* the potential for crack initiation but also *increases* the resistance to crack growth (Fig. 7.11). This is because plastic flow at the leading edge of the advancing crack blunts the tip, reducing the stress concentration and the strain energy to be found at this critical location. As discussed more fully later in this chapter, crack initiation and growth also depend on the size and number of fibers running through the material. In cortical bone, osteons serve as large fibers and collagen fibrils as small fibers. The interface between a fiber and the surrounding matrix is a source of weakness that can initiate cracks. On the other hand, the cracks tend to remain small and follow the fibers instead of propagating across the structure. Therefore, it is important that the fibers be oriented so as to lead the cracks in harmless directions, for example, along rather than across a beam.

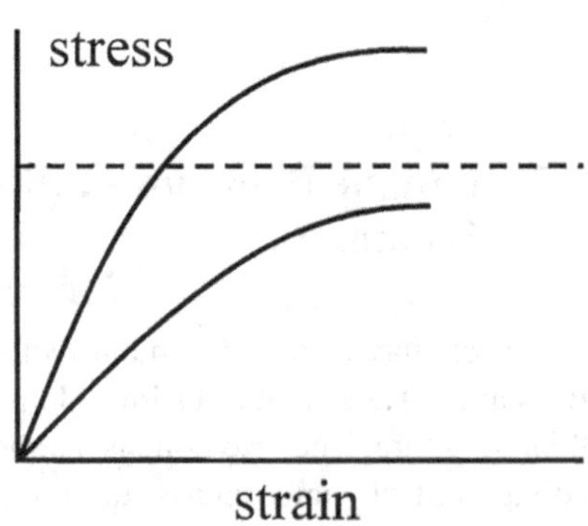

Fig. 7.11 Stress–strain curves for materials having high and low yield stress. If the latter material's yield strength is less than the stress required to propagate a crack *(dashed line)*, it may suffer from plastic deformation, but cracks cannot grow because the stresses never become high enough

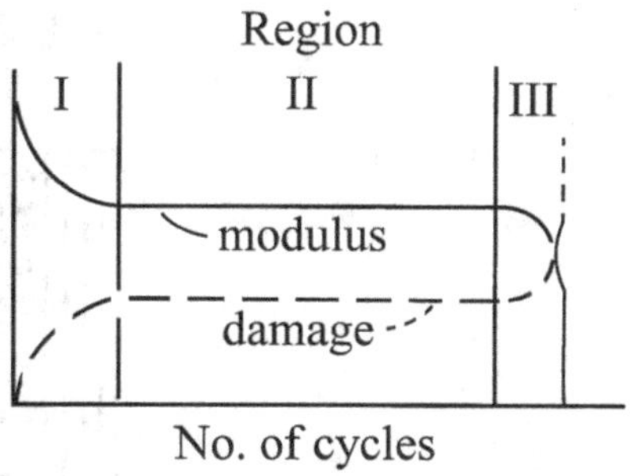

Fig. 7.12 The road to fatigue failure for a typical composite material. In *Region I*, elastic modulus declines as damage begins under the initial loading. In *Region II*, the rates of change of damage and modulus are much lower. In *Region III*, damage builds up rapidly to failure, and the modulus plummets

7.5.3 The Road to Failure

The ability of fibrous, lamellar composites to keep cracks small and running in harmless directions gives them a characteristic, three-phase damage accumulation history as they fail. Consequently, the long-term fatigue behavior of fibrous composite laminates is often divided into three regions, as shown in Fig. 7.12 [40]. We may think of the path to failure as a road running through these three regions.

Region I: Initiation

The first stage of fatigue is characterized by a relatively rapid but limited loss of stiffness coincident with the initiation of microdamage. The kind and amount of microdamage is determined by the properties of the composite structure. As illustrated in Fig. 7.13, in a cross-ply lamellar structure, cracks begin at structural discontinuities within a lamella and propagate to the interface with an adjacent lamella, where they are apt to stop, particularly if the fibers in the adjacent laminae are aligned in a different direction. These self-limiting cracks redistribute the stresses in the material, reducing the stress alongside the cracks and increasing the stresses near the crack tips, at lamellar interfaces.

Region II: Benign Accumulation

The rate of damage accumulation and the stiffness loss quickly stabilize, generally within the first 25% of the fatigue life, and this marks the transition to a period of benign damage accumulation. Stabilization occurs for two reasons. First, the crack tips are stopped at lamellar interfaces, so their progress across the specimen is halted. Second, as the cracks become more numerous, and the distance between their sides

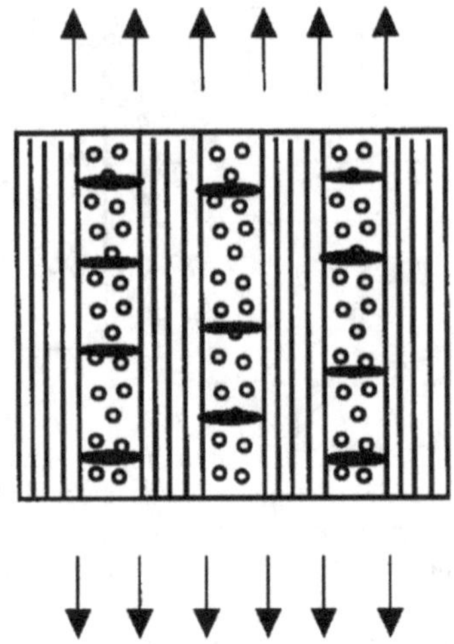

Fig. 7.13 Typical cracking in a cross-ply laminate loaded in tension. Cracks open in the lamina having fibers perpendicular to the load direction, and propagate to the lamellar interface, where they stop. The material above and below each crack is unloaded because of the crack's free surfaces. When the cracks become so numerous that the peak stress between them is sufficiently low, no more cracks can form; a saturation damage state has been reached

diminishes, the peak stresses between them become so low that additional cracks can no longer be initiated in the intact material. The level of damage at which this stabilization occurs is called the *characteristic damage state*. Because this intralamellar damage is now shut off, damage accumulation in region II shifts to interlamellar debonding or delamination. These cracks travel parallel to the lamellar interfaces, which, if the design is right, are aligned with the principal stress directions. Consequently, these cracks cannot directly cause catastrophic failure and do not even change the elastic modulus much. The surface area of the resulting secondary cracks is large, serving as an energy sink, and preventing the growth of a single, large, and more dangerous crack. Region I damage tends to be the same regardless of whether the material is loaded cyclically or monotonically. The kind of lamellar debonding damage typical of region II is characteristic of cyclic rather than monotonic loading.

Region III: Failure

Near the end of a fibrous laminate fatigue life, stiffness declines rapidly. Eventually, debonding at lamellar interfaces, combined with matrix degeneration, increases the stress in the load-aligned fibers or lamellae, and they begin to fracture. Although the damage formed in region I is self-limiting because cracks stop at fiber or lamellar interfaces, the damage predominating in region III is self-feeding because as fibers or load-aligned lamellae break, they increase the stress in those elements that remain intact. This process usually occurs rapidly, constituting the final 10% or so of the total fatigue life, and ends in ultimate failure of the structure.

7.6 Osteonal Bone as a Fibrous Lamellar Composite Material

We saw in Chap. 5 that Haversian bone is mechanically weaker than primary bone in monotonic bending and tension, in part because of its greater porosity, and probably because of the presence of cement lines. However, there is evidence to suggest that the unique microstructural arrangement of the various components of Haversian bone is adapted to control crack growth and thereby extend fatigue life. Bone tissue contains collagen fibrils organized into lamellae; this is a fibrous laminate on a microscopic scale. In secondary bone, these fibrous lamellae are in turn organized into larger fibers (osteons) running through a matrix (primary bone or osteonal fragments). An important hypothesis of bone mechanics is that this structure is responsible for the fact that osteonal bone subjected to cyclic loading behaves much like an engineered fibrous-lamellar composite.

7.6.1 Osteonal Bone's Road to Failure

In the upper part of Fig. 7.14, the tensile elastic modulus of human femoral bone declines over its fatigue life [16]. There is an early, rapid diminishment of stiffness, followed by a prolonged, much more gradual loss until, very suddenly, the modulus plummets, as failure occurs. This sort of modulus degradation curve is characteristic of osteonal bone loaded in tension or bending. The lower part of Fig. 7.14 shows that this behavior is substantially different for compression: the modulus falls slowly at first, but at an increasing rate until failure. Later in this chapter, we attempt to explain this difference based on the kind of damage associated with each mode of loading.

There is evidence that if the imposed loads are physiologic in magnitude, region II of the road to failure may persist for many millions of cycles. Schaffler [41] tested secondary and primary bovine cortical bone in tension at a strain range of 0–1200 microstrain and a frequency of 4 Hz. In both cases, there was an initial rapid stiffness loss over the first one to three million loading cycles (region I), after which no further stiffness loss occurred (region II), up to 45 million loading cycles. Thus, the results looked just like Fig. 7.12, but region III was not reached. It is believed that bone's region I is associated with multiple crack initiation events, just as in other composite materials. Much of this damage is apparently in the interstitial bone outside of the cement lines of *intact* osteons (i.e., those osteons not overlapped by more recent osteons). However, at physiologic strains and strain rates, region III and failure were beyond the capacity of Schaffler's experimental apparatus to keep cycling. Bone has excellent fatigue properties at physiologic strains and strain rates.

This fatigue behavior may be contrasted with that seen by Carter et al. [42] when bovine bone was fatigued at superphysiologic strains and strain rates (Fig. 7.15). Under these conditions, remodeled bone had less fatigue resistance than primary bone, and both had fatigue life of fewer than a million cycles at 3300–5500 microstrain.

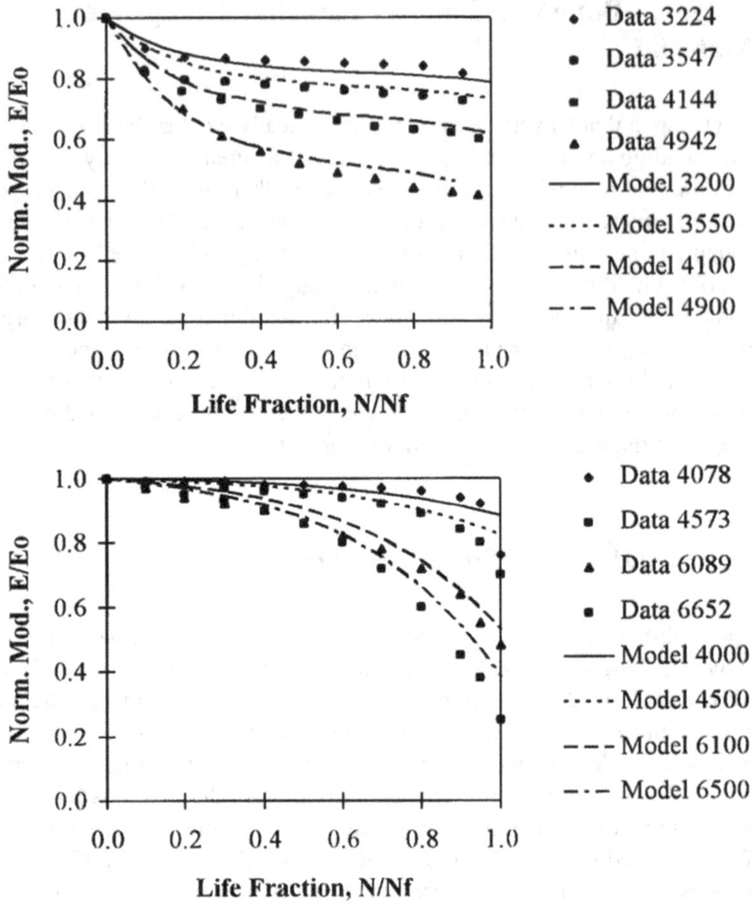

Fig. 7.14 Degradation of elastic modulus during the fatigue life for human femoral bone specimens loaded in tension (upper graph) and compression (lower graph). Symbols are experimental data from Pattin et al. [16]; curves are from the theory described previously. In both cases, numbers shown in the legend represent initial strain values in $\mu\varepsilon$

7.6.2 Bone Toughness

We have defined toughness in two ways: as the resistance to crack growth and as the amount of energy required to produce complete fracture. Clearly, if more effort is required to propagate cracks through a specimen, more energy is required to produce complete fracture, too. Moyle and Bowden [43] found that bone's work-to-failure was positively correlated ($r^2 = 0.31$) to the fractional area occupied by osteons. Haversian bone is similar in its osteonal arrangement to tough composite materials reinforced with discontinuous fibers. Fiber-reinforced composites are often highly anisotropic and have reduced transverse strength because of the directionality of the

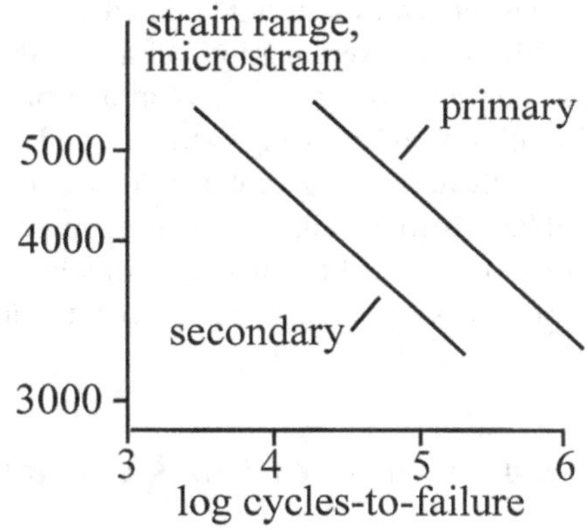

Fig. 7.15 Effects of remodeling on fatigue life of bovine bone. For a given strain range, primary bone specimens had a longer fatigue life than those composed of secondary bone

fibers. Aligned fibers provide high strength and stiffness in the predominant load direction at the expense of properties in the circumferential and radial directions. Secondary osteons, each only 5–10 mm long [44] and oriented nearly in the direction of principal strain [45], behave like oriented, discontinuous fibers within a matrix of more brittle material. The osteonal "fibers" each have a cylindrically lamellar structure inside and a relatively weak osteon–matrix interface. Failure at these weak interfaces by interlamellar debonding inside the osteon, or delamination at the cement lines, stops or deflects crack propagation. Crack deflection serves the dual purposes of energy absorption and reorientation so that the applied stresses are no longer great enough to spread the crack tip apart. Both mechanisms increase fracture toughness while at the same time minimizing damage to vascular and neural structures within the Haversian canal.

7.6.3 The Conflict Between Strength and Toughness in Bone

Bone's structure has evolved under the demand to be strong and also to be tough in order to reduce the chances of catastrophic failure. Strength and toughness in materials are generally mutually exclusive [7] and, as a result, bone is a "lower-strength" material that also has sufficient toughness for its function. High strength is often associated with high material hardness, but hard materials tend to be brittle. For high toughness, a material must be able to deform and reduce the stress concentrations that would cause fracture. The deformation mechanisms adding toughness to bone include the opening of bridged microcracks, sliding of mineralized collagen fibrils, motion along osteonal cement lines, uncoiling of collagen fibrils, breaking of sacrificial non-covalent bonds, and others as discussed above. Many of these toughening mechanisms reduce the *strength* of bone, but are essential to its proper function.

The general motif in bone and some manmade materials that are both strong and tough is the combination of a nano/microstructure of a hard phase to provide stiffness that is incorporated in a strong yet softer matrix that has (1) mechanisms for alleviating local high stresses and (2) mechanisms to oppose crack opening. In bone, there is a tough collagen matrix that is impregnated with, and surrounded by, a stiff and strong reinforcing mineral phase. This arrangement is akin to reinforced concrete where the concrete particulate provides compressive stiffness and strength while the steel rebar "fibers" carry the tensile load.

7.6.4 Controlled Crack Propagation Studies

Controlled crack propagation may be accomplished by machining a wedge-shaped specimen (Fig. 7.16a). This specimen is loaded slowly in three-point bending so that the apex of the central wedge is in tension, and the area under the load–deformation curve is taken as the work-to-failure, or toughness, of the specimen. The fact that the width of the crack front increases as it moves across the wedge helps keep the crack from accelerating. The first such experiment on bone was reported by Piekarski [46], using bovine femurs. He showed that the microscopic appearance of the fracture surface depended a great deal on the rate at which the crack propagated across the wedge. Typically, cracks traveled slowly at first, and then suddenly accelerated to produce a catastrophic failure. The initial portion of the fracture surface was always rough and exhibited many pulled-out osteons; the surface produced by the subsequent rapid failure was consistently smooth, with osteons breaking across rather than pulling out. When the crack was advancing slowly, it tended to enter and follow cement lines, thereby avoiding Haversian canals. Work-to-failure was dependent

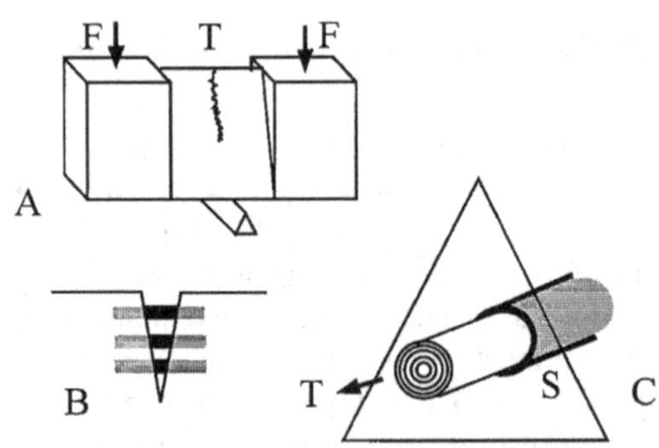

Fig. 7.16 (a) Specimen geometry for controlled crack propagation studies. The central region is wedge shaped and loaded in bending. As the crack propagates from the tensile edge *(T)*, it encounters more bone, and this helps control the crack velocity. (**b**) As the crack propagates through the bone, it tends to leave osteons in its path intact, bridging the opening. (**c**) An osteon bridging the crack experiences a tensile load *(T)* in the gap as well as a shear load *(S)* on the embedded portion

upon the rate of crack propagation. The slow, controlled cracks required almost 60 kJ/m^2 of crack surface to propel them, while the catastrophic cracks required less than 1 kJ/m^2.

7.6.5 Mathematical Analysis of Osteonal Pullout

Applying techniques used to study reinforced plastics, Pope and Murphy [47] mathematically analyzed the phenomenon of osteon pullout. Their analysis was directly applicable to the sort of experiment done by Piekarski and raised important questions about the mechanisms by which histologic structure may affect crack propagation. Consider a region of osteonal bone loaded in longitudinal tension as shown in Fig. 7.16b. A crack is propagating transversely through the cortex. As the leading edge of the crack envelops individual osteons, they experience two forces. First, they continue to be stretched by the tensile stress in the cortex. The tensile force from the left side of the crack attempts to pull the osteon out of the bone matrix on the right side of the crack, and vice versa. The tensile pullout forces are resisted by the shear forces that the surrounding bone exerts on the walls of the embedded portions of the osteon. If the tensile strength of the portion of the osteon bridging the crack is great enough, these shear forces may exceed the shear strength of the cement line and the osteon will begin to pull out of the bone matrix. Subsequently, one would expect the osteon to fail not within the primary, transverse crack, but away from the primary crack, at the stress concentration at the end of the debonded region. If, on the other hand, the osteon is relatively weak in tension, it will break within the primary crack, without pulling out. Let us consider a simplified version of Pope and Murphy's mathematical formulation of this problem.

The tensile force acting on the osteon in the primary crack is

$$T = \pi R_0^2 \sigma_t \tag{7.32}$$

where $\boldsymbol{\sigma}_t$ is the tensile stress and R_o is the radius of the osteon. The shear force on the walls of the embedded portion of the osteon on, say, the right side is given by

$$S = 2\pi R_0 L \sigma_s \tag{7.33}$$

where $\boldsymbol{\sigma}_s$ is the shear stress in the cement line and L is the distance over which the surrounding bone grips the osteon. The osteon will pull out when the tensile force exceeds the resisting shear force, that is, when $T > S$. By substituting the foregoing equations into this inequality, one finds that pullout occurs when

$$\sigma_{tf} > 2(L/R_0)\sigma_{sf} \tag{7.34}$$

where σ_{sf} is the shear failure stress for the cement line and σ_{tf} is the tensile failure stress of the osteon. Note that it is easier for pullout to occur if R_0 is relatively large. Conversely, smaller osteons are more likely to break off when forced to bridge a transverse crack.

7.6.6 Comparing Theory with Data

Moyle and coworkers [48] performed an experiment similar to Piekarski's on canine femoral bone. They confirmed the correlation between slow cracking and osteon pullout, and they also found that the work-to-failure for slow cracking was significantly higher in the lateral aspect of the dog's femur than on the medial side (Table 7.3). Because the bending caused by eccentric loading at the hip joint produces tensile stress on the lateral side of femurs and compression on the medial side, this difference in resistance to tensile cracking may be a manifestation of mechanical adaptation. Like Piekarski, Moyle's group found that most specimens failed catastrophically, but some cracked slowly all the way across the wedge. These specimens turned out to have significantly *smaller* osteons than did the catastrophic failures (Table 7.4), in direct contradiction to the analytical result obtained above.

Osteon size and control of crack propagation are also important factors in fatigue resistance. In flexural fatigue experiments, Gibson et al. [49] found that the dorsal region of the racehorse "cannon" (third metacarpal) bone had a longer fatigue life than the lateral region, which was monotonically stronger and stiffer. Fatigue resistance was strongly associated with signs of osteon pullout on the fracture surfaces; there were no signs of pullout in the lateral specimens. The dorsal region had smaller osteons than the lateral region: 156 ± 19 vs 179 ± 13 μm. The reason for this repeated disagreement with theory may be that osteon pullout depends as much on the number of osteons as it does on their diameter. In Chap. 4, it was shown that as secondary osteons accumulate in a section, their population asymptotically approaches a limit that is the reciprocal of their mean individual area. Thus, if the osteons are smaller, more of them can be packed into a section, and there are more to pull out.

Table 7.3 Work-to-failure in different regions of canine femur [48]

Lateral side ($N = 10$)	11,800 ± 3000 J/m^2
Medial side ($N = 13$)	6960 ± 1350 J/m^2

Table 7.4 Effects of osteon diameter and density on crack propagation [48]

Type failure	Mean osteon diameter, μm	Osteon density, mm^{-2}
Slow cracking ($N = 23$)	148 ± 22	20 ± 7
Fast cracking ($N = 45$)	202 ± 29	14 ± 5

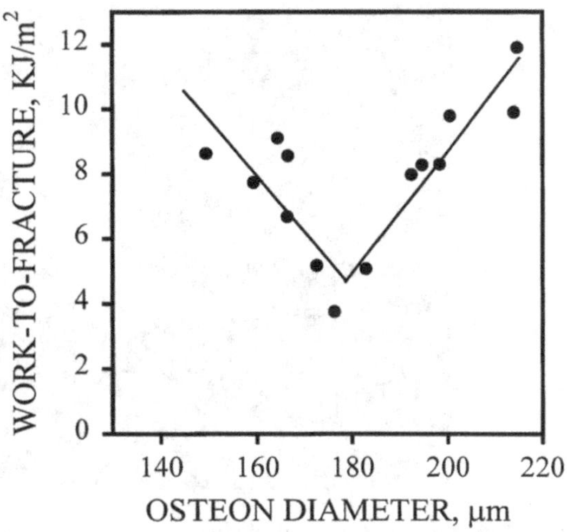

Fig. 7.17 Relationship between work-to-failure and osteon diameter [43]

Thus, the energy dissipated by osteon pullout should depend on both the likelihood of pullout (which increases with osteon diameter) and the number of osteons available to be pulled out (which decreases with diameter). Figure 7.17 shows data supporting such an argument [43, 48]. The left side of the graph shows that greater energy absorption can be achieved when there are many small osteons in the area of a developing crack. As osteon diameter increases, energy absorption declines. However, as shown on the right side of the graph, work of fracture again increases when osteon diameters are greater than 200 μm. The reason might be that these larger osteons are more likely to pull out rather than break off, as predicted by Eq. (7.34). Corondan and Haworth [50] presented additional data which further demonstrate that fracture is inhibited by increased numbers of osteons as well as by larger osteons. Comparing the morphology of fractured surfaces of bone with that of sections adjacent to the fracture, they found that bone broke where there are fewer *and* smaller osteons. Bone *did not* break in adjacent regions where osteon packing *and* size were greater.

7.6.7 Crack Initiation

Persistent slip bands are "pre-failure" or plastic slip lines representing partial failure within a material, generally along maximum shear planes. These bands may represent the starting point for microcracks within a material. There are features in cyclically loaded bone that look like irreversible microscopic slip bands, which occur in metals from plastic deformation (Fig. 7.18). Frost first demonstrated these in 1960. As in metals, slip bands in bone may form a focus for subsequent crack formation.

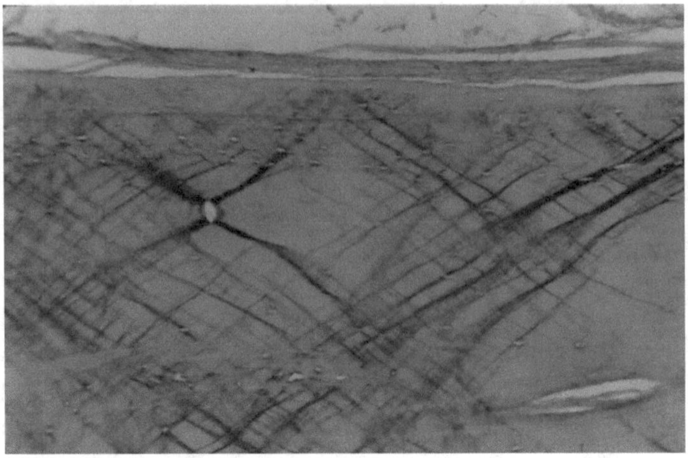

Fig. 7.18 Slip bands in cortical bone

The relatively weak cement line interface between osteons and interstitial bone matrix may reduce the shear strength of osteonal bone [51, 52]. At the same time, however, this interface may, by slipping, relax shear stresses, reduce strain energy, and slow crack propagation. Lakes and Saha [53] suggested that cement line displacement may be a manifestation of shear failure in bone. Their observations of displacement at lamellar interfaces in bovine plexiform bone supported the notion that lamellar interfaces in bone reduce stress concentrations by reducing shear stress. In studying the viscoelastic creep behavior of human bone, Park and Lakes [54] observed displacement at the cement line, but it was one-third of what would be expected if it were the only source of the creep. This observation suggests that the osteonal cement line fails only after other damage (e.g., slip bands) has begun to grow in the matrix.

7.6.8 Crack Stopping

The fatigue life of Haversian bone is governed by both crack initiation and propagation. Because cracks can be removed through bone remodeling, it can be postulated that crack initiation is not so deleterious as the rapid propagation of cracks across the bone material. From this point of view, the most useful microstructural characteristics of bone tissue would be those able to slow or alter the direction of crack propagation, rather than those that inhibit crack nucleation. In fact, Haversian bone appears to be constructed in just this manner. The cement lines between osteons and the interstitial matrix are weaker than the bone matrix and, when less mineralized, are probably more compliant. On reaching a more compliant region, a crack will accelerate and then decelerate, and finally be arrested when it reaches another stiff region [55]. Even if crack initiation occurs repeatedly during cyclic loading events, Haversian bone is postulated to stop the transverse growth of these cracks, diverting them along the bone axis and allowing them to expend the strain energy at the crack tip without seriously reducing structural integrity.

7.6.9 Strain Rate

The fracture toughness of bone may be reduced at high strain rates because crack velocity is generally increased under these conditions. When crack velocities exceed approximately 1 mm/s, the likelihood of osteon pullout is reduced and brittle failure is more common [56]. Increased crack velocity requires a higher stiffness ratio between bone matrix and the cement line to halt crack propagation. A change in mechanical properties of the cement line material from viscous to elastic at high strain rates would be expected to compound this problem [57, 58]. Even within the relatively narrow physiologic range of strain, Schaffler [41] found that a strain rate of 0.03/s caused significantly more microscopic damage and greater loss of stiffness than loading at 0.01/s.

Simkin and Robin [59] found, in monotonic tests at tensile strain rates of only 0.004/s, that nearly two-thirds (59%) of the microcracks were trapped either at cement lines or circumferential lamellae within an osteon; only 41% cut across these boundaries. At higher strain rates (0.4/s), osteons were capable of stopping fewer than 45% of the microcracks. Maximum strain rates during moderate activity range between 0.0034 and 0.015/s [60, 61]. Thus, at strain rates experienced by sedentary individuals, osteons might be expected to stop close to two-thirds of the propagating microcracks coming their way. These observations are in concert with those of Norman and Wang [33], who found that 67% of all microcracks in cortical bone from human femurs were interstitial rather than intraosteonal, and most interstitial cracks abutted a cement line. The bones of more active individuals would be expected to experience higher strain rates and thus more intraosteonal cracking.

One must be careful to distinguish between strain rate and frequency effects, as discussed in Chap. 6. Ordinarily, strain rate increases as the frequency of loading increases, so that one might expect fatigue life to diminish as frequency increases. However, Caler and Carter [15] found that the fatigue life of human femoral bone increased rather than decreased as the loading frequency increased from 0.02 to 2 Hz. This change is apparently related to creep effects. Even though the bone is more resistant to crack propagation at the low strain rate associated with a lower frequency, the amount of time that the bone is loaded for each cycle is longer at a low frequency. Thus, the number of cycles required for failure (i.e., the fatigue life) is less at a low frequency. Schaffler [41] was careful to keep frequency constant (at 4 Hz) in his tests of the effects of strain rate; he did this by using a triangular waveform and varying the "off period" between "spikes" to vary frequency.

7.6.10 *Effect of Remodeling*

The effects of remodeling on fatigue life are unclear. Carter et al. [42, 62, 63] tested bovine cortical bone using a rotating technique for fatigue testing at stress amplitudes of 65–108 MPa. Specimens with the most remodeling were significantly less fatigue resistant than primary bone specimens of equal density. Evans and Riolo [64] found the opposite effect in human tibial specimens loaded in bending to 34 MPa. The source of this discrepancy could be the species tested, or it could be the different stress amplitudes. Assuming a modulus of 20 GPa, the Carter experiments produced superphysiologic strains (3250–5400 με); the Evans–Riolo experiment did not (1700 με). We have seen that much of the damage to bone at high strain magnitudes is caused by creep rather than fatigue, and osteons may affect fatigue and creep damage mechanisms differently. Strain rate may have also entered the picture: although Evans and Riolo did not specify their frequency, Carter et al. tested at 125 Hz. At the high strain rate associated with such a high frequency (and strain magnitude), the crack-stopping effects of osteons may disappear.

Schaffler's [41] tensile fatigue tests at physiologic strains and strain rates revealed no differences between primary and secondary bone for either percent stiffness loss or microdamage accumulation. However, the modulus of the primary bone was greater than that of remodeled bone, both initially and just before failure. In addition, the primary bone's modulus held up better over the fatigue life. These experiments suggest that remodeling does not substantially reduce the fatigue life of bone *at physiologic strain magnitudes*. There may be a strain rate at which crack resistance and fatigue life are optimal. At low strain rates, creep-like conditions occur and damage is more severe. At high strain rates, the effectiveness of cement lines and other tissue interfaces as crack stoppers is diminished. Perhaps the bone structure is adapted to crack stopping best at physiologic strain rates.

7.7 Modeling Fatigue Damage Effects in Osteonal Bone

Krajcinovic et al. [65] developed a continuum damage theory for osteonal bone, which assumed that when a crack propagates transversely through the cortex, osteons become debonded, break, and pull out as described previously. He calculated the modulus degradation and residual strength of the specimen as this process continued. He was able to predict both bending and tensile elastic moduli, and the relationships between elastic modulus and tensile failure properties.

Griffin and coworkers [66] proposed a more detailed model for fatigue of osteonal bone, based on the observation that cortical bone fatigue behaviors are different in tension and compression. Fatigue lives are longer in compression than in tension, and the shapes of the modulus degradation curves are different (Fig. 7.14). In tension, there are generally three distinct regimens of behavior with increasing numbers of cycles: the first is an initial reduction in modulus, the second is a plateau,

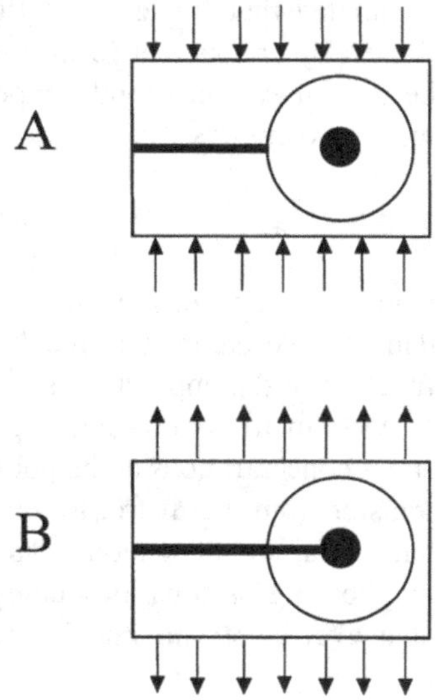

Fig. 7.19 Hypothetical basis for cracking differences in tensile and compressive fatigue. (**a**) In tensile loading, Poisson stress is compressive, closing cracks and making them more likely to stop at cement lines. (**b**) In compressive loading, Poisson stress is tensile, opening cracks and making them less likely to stop at cement lines

and the third is a rapid modulus reduction before failure. In compression, on the other hand, the modulus diminishes slowly at first, but at an increasing rate until failure.

Carter and Hayes [67] also observed differences in the kind of microdamage associated with fatigue in tension and compression. In tension, most cracks were found in the interstitial matrix or in osteonal cement lines. In compression, the cracks tended to run from one vascular channel to another. Figure 7.19 shows similar behavior in the tensile and compressive regions of human femur specimens fatigued in four-point bending. Is it possible to mechanistically account for the differences in modulus degradation in tension and compression on the basis of the differences in the nature of the damage?

The theory of Griffin et al. [66] models cortical bone as a fiber-matrix composite, the fibers being osteons and the matrix consisting of primary bone and remodeled osteon fragments. Damage is partitioned into components that independently affect the fibers and the matrix, and the "damaged modulus" is obtained from the isostrain rule of mixtures given in Eq. (7.12). Here, we write this equation in the form

$$E = (1 - D_{\rm f})E_{\rm f}V_{\rm f} + (1 - D_{\rm m})E_{\rm m}V_{\rm m} \tag{7.35}$$

where E is the elastic modulus of the bone, V is volume fraction, and D represents damage. The subscripts m and f refer to matrix and fibers, respectively. $D_{\rm f}$ and $D_{\rm m}$ have values that range from zero, in an undamaged state, to one, in a failed state. When there is no damage, this equation is reduced to Eq. (7.12).

For ceramic matrix composites loaded in tension, the initial damage is caused by microcracks in the matrix. The number of cracks does not increase indefinitely, but approaches a limit known as the saturation or characteristic damage state. It is postulated that for tensile fatigue of cortical bone, the stiffness loss is primarily caused by cracks that form in the interstitial matrix and are self-limiting in length because they consistently appear as shown in Fig. 7.1. Therefore, damage to the matrix evolves as

$$dD_{\mathrm{m}}/dN = K_1(D_{\mathrm{s}} - D_{\mathrm{m}}) \tag{7.36}$$

where D_{m} represents the matrix damage, N is the current cycle number, K_1 is a matrix damage rate parameter, and D_{s} represents the saturation level of matrix damage, a function of the applied stress.

Most matrix cracks intercept osteon cement lines and can be expected to produce stress concentrations at the point of impingement. Matrix damage also results in load transfer from the matrix to the osteons. These factors should eventually lead to the tensile failure of osteons. As osteons break, the load is further redistributed to neighboring osteons, producing a positive feedback and resulting in osteon failure at an ever-increasing rate. Therefore, the fiber damage, D_{f}, is assumed to accumulate as

$$dD_{\mathrm{f}}/dN = K_2 D_{\mathrm{s}} + K_3 D_{\mathrm{m}} \tag{7.37}$$

where K_2 is an osteonal damage rate coefficient and K_3 is an osteon–matrix damage interaction parameter. Equations (7.36) and (7.37) can be solved simultaneously to obtain

$$D_{\mathrm{m}} = D_{\mathrm{s}}\left(1 - e^{-AN'}\right) \tag{7.38}$$

$$D_{\mathrm{f}} = D_{\mathrm{s}} K_3 [K_2 \exp(-K_1 N') + K_1 \exp(K_2 N') - (K_1 + K_2)]/(K_1 + K_2) \tag{7.39}$$

where $N' = N/N_{\mathrm{F}}$ is the normalized fatigue life in tension. Substitution of these two equations into Eq. (7.35) provides an expression for the reduction in modulus as a function of the normalized fatigue life.

In compression, microcracks appear less likely to stop at the osteon–matrix interfaces than they do in tensile loading. If so, then in bone that is fatigued under compression, cracks may tend to grow in length and number without bound, degrading the material in a positive feedback manner, and damage is assumed to evolve as

$$dD_{\mathrm{c}}/dN = K_4(D_{\mathrm{I}} + D_{\mathrm{c}}) \tag{7.40}$$

where K_4 is a compression damage rate parameter and D_{I} represents incipient damage in the form of stress concentrators, such as Haversian or Volkmann's canals, which initiate or extend D_{c}. Eq. (7.40) is readily solved to obtain

$$D_c = D_I[\exp(K_4 N'') - 1] \tag{7.41}$$

where N'' is the normalized fatigue life in compression. For the compression case, D_c is used for both damage terms in Eq. (7.35) because there do not appear to be distinct damage modes in osteons and interstitial matrix in this mode of loading.

By fitting the resulting modulus degradation equations to Pattin's [16] data, the constants K_1, K_2, K_3, and K_4 were determined. Return to Fig. 7.14 to see examples of how well these results fit data for typical compression and tension specimens. These results support the hypothesis that the different shapes of these curves are caused by different damage modes in tension and compression. In particular, it appears that matrix damage under tensile loading is self-limiting, with cracks being stopped by osteonal cement lines, but this may not be the case in compressive loading.

Griffin et al. [66] went on to demonstrate that a mathematical model for the fatigue of a beam loaded in four-point bending, which incorporated the foregoing equations to describe the development of damage on the tensile and compressive sides, was able to predict modulus degradation and fatigue life reasonably well.

7.7.1 *Why Would Cracks Be Stopped in Tension but Not in Compression?*

What could account for the observation that osteons stop cracks better in uniaxial tensile fatigue than in compression? Figure 7.19 offers a possible explanation. When a specimen is loaded in uniaxial tension, the stress acting across the fibers due to Poisson's effect (see Chap. 5) is compressive and tends to close a matrix crack of the kind shown in Fig. 7.1. Thus, it is more difficult to propagate the crack and easier for a cement line to stop it. On the other hand, when uniaxial compression is applied, the Poisson stress is tensile and tends to open the crack. This should make it more difficult for cement lines to stop the crack, consistent with the observation that such cracks travel from Haversian canal to Haversian canal, passing through cement lines.

7.7.2 *If Cracks Are Self-Limiting in Tensile Fatigue but Not in Compressive Fatigue, Why Is Fatigue Life Longer in Compression than in Tension?*

Although we have argued that the toughness and fatigue resistance of bone are based on resistance to crack propagation rather than crack initiation, perhaps this is an oversimplification. Consider the following possibility. When cortical bone is loaded in uniaxial tension, cracks are easily initiated but difficult to propagate substantial distances. When loaded in uniaxial compression, the opposite situation may occur: difficult initiation but then easier propagation. This hypothesis is consistent with the possibility that the Poisson effect drives crack propagation in compressive loading. It is also in concert with the modulus degradation curves of Pattin and coworkers (Fig. 7.14). In tensile fatigue, the modulus drops rapidly at first, then plateaus until just before failure. This finding is consistent with a form of damage that is easily initiated but self-limiting, followed eventually by accelerating osteon breakage. In compression, the early, self-limiting form of damage may not occur because crack initiation is more difficult in this mode of loading. However, if continued loading eventually breaks down the tissue's resistance to crack initiation, the cracks may be of a kind that grow relentlessly, and the modulus may then diminish rapidly. To test this hypothesis, more work is needed to elucidate the relative ease of crack initiation and crack propagation in different modes of loading oriented in various ways to the lamellar structure.

7.8 The Role of Fatigue in Activating Bone Remodeling

We have asserted that one of the primary functions of remodeling is to repair fatigue damage. In this section, we discuss the evidence for that. To begin, it is useful to compare the fatigue life of bone to the rate at which bone is renewed or turned over by remodeling. We have seen Schaffler's data indicating that bovine cortical bone can endure 30–45 million load cycles at 1200 microstrain without fatigue failure. Pattin's data for the fatigue life of human femoral bone suggest that the fatigue life at a peak strain range of 2000 $\mu\varepsilon$ is four to ten million cycles. At 1200 $\mu\varepsilon$, this might be substantially extended. How long would it take an average person to accumulate 30 million load cycles in, say, the femur? If a person's daily activities were the equivalent of walking 5 miles/day, each femur would receive about 5 miles × 5280 ft./mile ÷ 6 ft./stride = 4400 load cycles/day, assuming each step covers 3 ft. This works out to be roughly 1.5 million load cycles/year, so that 20 years would be required to accumulate 30 million cycles. The turnover rate for the cortex of a bone like the femur is about 3% per year in adults [68], but this rate should be higher in children and the elderly. Assuming an average value of 5%, 20+ years is just about what would be required to replace all the bone in any given region. Thus, there is a rough correspondence between the fatigue life of cortical bone and the rate at which it is replaced.

7.8.1 Random or Directed Repair

Osteonal remodeling could replace microdamage in one of two ways: by directing the BMUs to the location of damage, or by simply randomly remodeling the cortex at a rate designed to keep up with the overall rate of damage accumulation. Random remodeling would not be ideal. Some regions can be expected to habitually have higher stresses than others (e.g., near points of muscle attachment). The process might be more efficient if osteonal activation occurred at sites where microdamage has actually developed.

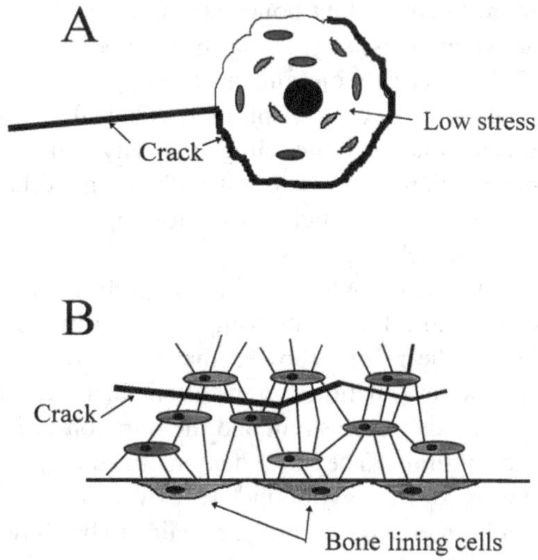

Fig. 7.20 Alternative hypotheses regarding the mechanism for activation of remodeling by fatigue damage. (**a**) Debonding crack in cement line isolates osteon from stress, creating local disuse state that produces activation signal. (**b**) Damage to osteocytic network produces activation signal

It has been proposed that osteonal cement lines not only stop cracks, but that this event initiates the reparative remodeling process [69]. This could happen because debonding of the osteon during crack trapping produces a local disuse state, which activates a new BMU. This new remodeling unit could then resorb the damaged region and replace it with new bone (Fig. 7.20a). Alternatively, Frost [70] postulated that damage triggers a response through disruption of the lacunar–canalicular network. There is evidence that stress or strain provoke responses in osteocytes that are transmitted from cell to cell through their processes [71–73]. The interruption of normal osteocyte signaling to bone lining cells could activate remodeling if these signals are inhibitory (Fig. 7.20b). These two theories provide distinctly different mechanisms for activation of the repair of microdamage in bone and a contrast that should allow for experimental testing of their validity. They are similar, however, in that they both use the same pathway; in one case, the signal is not generated because cement line debonding relieves stress, and in the other, the signal is blocked by intraosteonal damage.

7.8.2 The Physiological Basis for Targeted Remodeling

The idea of targeted remodeling of bone damage has appeared in many guises over the years. Tschantz and Rutishauser [74] suggested that microcracks in the compressed ulnar cortex of dogs stimulate remodeling. Chamay and Tschantz [75] proposed that loading stimulates bone remodeling. Rahn et al. [76] showed that overload zones adjacent to fractures in the process of repair are also remodeled, resulting in the resorption of damage and the formation of more secondary osteons. Comparative primate morphology studies [77] show that osteonal remodeling is more intense when bones are impulsively and repetitively loaded, that is, in very active animals, suggesting that remodeling may be initiated by such loads. Frost's [24] earlier data on human ribs support the same concept: the ribs are loaded many times each day by respiration. Typically, ribs have more microcracks and also a greater rate of remodeling than any other cortical bone site in the body. These observations are consistent with the hypothesis that bone remodeling is activated by microdamage, but they do not implicate microdamage itself as the direct cause of the remodeling process.

Burr and coworkers [78, 79] performed in vivo fatigue experiments to verify a cause-and-effect relationship between microcracks and BMU activation in cortical bone. The results showed that microcracks were associated with bone resorption spaces several times more often than expected by chance alone, even when the magnitude of the strain and the duration of loading were within reasonable physiological limits. They also demonstrated that cracks caused the resorption spaces, and not the other way around. However, the mechanisms behind this association have not been elucidated, and it remains to be shown how long remodeling takes to repair a "batch" of fatigue damage.

Over the past 20–25 years, experimental evidence has shown that microdamage has clear physiological significance in signaling through osteocytes for targeted bone remodeling [80–82]. Bentolila and coworkers [83] demonstrated that remodeling can be activated in association with fatigue damage in a bone that ordinarily lacks osteons and therefore has no cement lines. By cyclically loading the ulnas of rats, which do not ordinarily remodel their cortical bone, they were able to demonstrate microcracks in association with BMU resorption cavities. This experiment was important for two reasons. First, it suggested that cement line debonding, as proposed by Martin and Burr [69], need not play a role in activation of BMUs by fatigue damage because the rat's ulnar cortex initially has no osteons. Instead, it supports Frost's proposal that damage acts directly on osteocytes and perhaps through disruption of osteocyte canaliculi. Second, it supports the hypothesis that the control of fatigue damage is an important function of remodeling, because in the normal rat ulna, there is neither damage nor remodeling to be seen, but when damage is introduced, remodeling is activated within days. (Remodeling may also be activated in these animals by calcium deficiency, however [84].)

We now know that osteocyte signaling is indeed responsible for the initiation of the repair process [85], and that this occurs through alterations in the RANKL–OPG signaling axis. Verborgt et al. [86] showed in the rat overload model that osteocyte apoptosis occurred following the induction of microdamage by loading, and that those osteocytes that became apoptotic were those that were closest to the damage. Moreover, osteocyte death occurred prior to the formation of resorption spaces, suggesting that the apoptosis was also temporally associated with the subsequent remodeling response. When osteocyte apoptosis is suppressed, no remodeling occurs even though damage is present [87]. These experiments clearly showed that osteocyte apoptosis is required for the targeted repair of damage. At this time, the signals for the response had not been identified. Moreover, it was not clear whether osteocyte death was permissive (i.e., healthy osteocytes produce signals that prevent remodeling, but when the cells die, those signals are not present and allow the remodeling reaction), or whether the process of osteocyte death actively produced signals that enticed cells to begin the remodeling process. This question also now seems to have been resolved. Again using the same model, the RANKL/OPG ratio was increased significantly within 3 days following the induction of bone damage and continued to increase over the following week [88]. Apoptotic cells close to the damage produced less OPG, whereas healthy osteocytes away from the damage upregulated their release of RANKL. Thus, in damaged bone, there is more RANKL, which promotes the full differentiation of osteoclasts and is essential for the activation of resorption, and less OPG, which neutralizes the RANKL. Consequently, the initiation of the remodeling sequence and the activation of resorption is stimulated.

Several key experiments [89, 90] have since been done to demonstrate that osteocytes are a key source of RANKL. Indeed, these studies showed that regulation of bone remodeling in adults is controlled by the osteocytes, and that the RANKL produced by osteoblasts probably does not contribute to the signaling for bone remodeling at all. There may be a dose–response element to the apoptotic response. Investigators attempting to simulate the effects of microcracks in vitro and studying osteocytic responses to larger cracks demonstrate that large cracks are associated with a greater increase in RANKL and a greater decrease in OPG than smaller cracks [91]. This would lead to a more robust remodeling and repair response.

All of these experiments provide compelling evidence that osteocytes help to regulate bone remodeling, and that the locations of the osteocytes undergoing apoptosis provide direction signals that permit specific areas of bone, those that have been damaged, for instance, to be targeted for remodeling and/or repair (Fig. 7.21). This provides a significant new dimension in our understanding of the role that microdamage plays in normal bone remodeling. It may be productive to begin considering skeletal microdamage not only in mechanical terms, but very much in physiological terms.

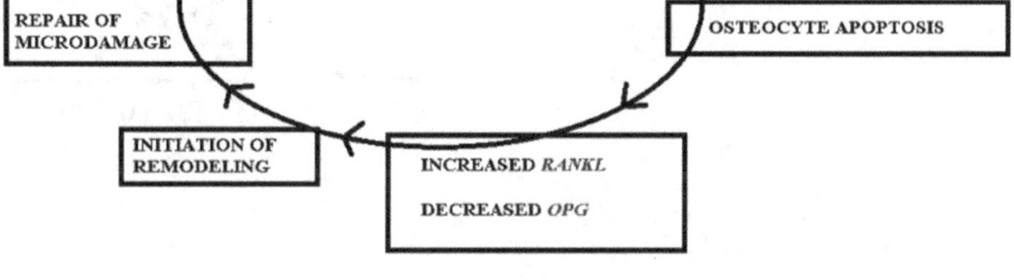

Fig. 7.21 Steps in the targeted repair of microdamage

7.8.3 Microdamage and Bone Fragility in the Elderly

The idea that fatigue damage may be an important component of bone fragility in the elderly has gained credence among bone researchers [92]. This idea is supported by data showing that microdamage increases exponentially with age in human femoral bone [26, 27] (Fig. 7.22). This increase could occur because reduced bone mass increases bone strains, and fatigue damage is probably sensitive to strain magnitude. It follows from this view that many nontraumatic fractures in the elderly (i.e., osteoporotic fractures) could be essentially stress fractures. This hypothesis is supported by the fact that callus-like woven bone formations have been observed on periosteal surfaces adjacent to femoral neck fractures [93]. It is clear that bone does indeed become more fragile with age, even apart from the loss of bone mass itself. The tissue of bone becomes less tough and more brittle with each ensuing decade [94–96]. Bone from older individuals can sustain less plastic deformation, consequently the amount of energy it absorbs in the post-yield region of the stress–strain curve is significantly reduced and fracture occurs more quickly. There may be several reasons underlying this reduced energy absorption capacity beyond the well-documented accumulation of microdamage. These may include alterations in mineralization, changes in collagen cross-linking, increases in bone turnover that create resorption pits that serve as stress concentrations for damage, changes in tissue hydration state, or some combination of these factors. Whether the reduced toughness that occurs with age really contributes to fractures in older people is still not known.

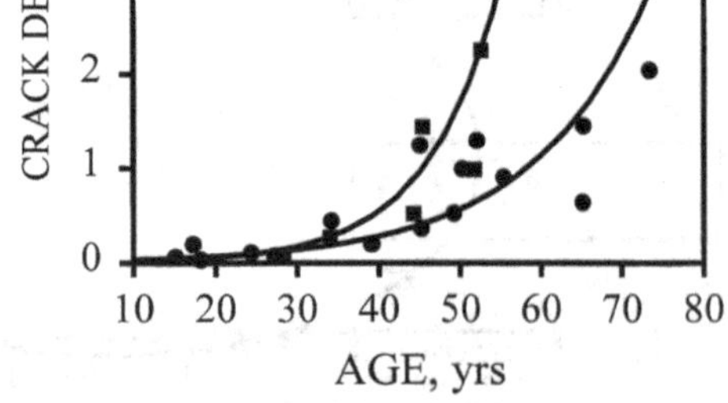

Fig. 7.22 Microcrack density vs age for men *(circles)* and women *(squares)* [26]

7.9 Modeling Stress Fractures

Stress fractures are the clinical manifestation of the accumulation of fatigue damage in bone. When cracks are not stopped and repaired, they can grow to become larger cracks visible radiographically. These "stress fractures" represent the imbalance between cyclic loading and bone adaptive responses to the loading that allow the formation and growth of cracks [97]. Stress fractures in humans occur most commonly in the mid-distal tibia, metatarsals of the foot, the navicular, and occasionally the femur, although the specific incidence and location vary by age and activity [98]. They are an important medical problem among military recruits, ballet dancers, race horses, and other high-performance athletes and in recreational athletes. Their diagnosis and treatment is complicated by their capricious nature and the difficulty in predicting when bone pain is indicative of an incipient fracture. The earliest sign of a stress fracture is often not the fracture itself, but the bony reaction to it. The earliest radiographic evidence is generally the formation of a small periosteal callus, sometimes also referred to as a periosteal stress reaction, although the latter can also refer to localized soft tissue growth from the bone surface that is similar to the bridging callus formed during the repair of a complete bone fracture. The healing response to stress fracture increases the local metabolism of the tissue. Increased bone metabolism is measured using a radioactive bisphosphonate that collects at sites of bone formation. The localized radiation is detected using scintigraphy or positron emission tomography (PET). Stress fractures do not generally progress to full fracture of the bone if activity levels are reduced, although full fractures can develop in instances of metabolic disease (e.g., hypophosphatasia, which is represented by low levels of alkaline phosphatase that indicates insufficient bone formation) or in some cases in which repair is suppressed pharmacologically (as in atypical femoral fractures in those taking potent remodeling–suppressors for osteoporosis) [99].

The development of a stress fracture depends on the remodeling (repair) rate, but also on the material properties of the bone tissue, the magnitude (and perhaps direction) of strain, and the number of loading cycles. In racehorses, in which strains may be as high as 5000 $\mu\varepsilon$, stress fractures can occur within 50,000 loading cycles. In military recruits and athletes in which lower strains are prevalent, stress fractures can occur with 100,000 or so cycles of loading. Therefore, it is important to minimize even brief periods of high strain or strain rate. There is some evidence that in athletic and military populations, stress fractures are associated with muscular fatigue, as muscles are now known to control strain distributions in bone. When muscles fatigue, strains on bone can increase significantly [100, 101], and perhaps as importantly, strain distribution can be altered [102, 103]. Thus, the prevention of stress fractures in our bones is a delicate balance between the loading environment, the properties of the bone, the adaptive response of bone tissue to loading, and the ability of our muscles to maintain the loading environment within proper physiological limits.

We have argued for the importance of remodeling in repairing fatigue damage in bone to prevent stress fracture. We now attempt to learn, by analysis, more about how this repair system may work.

Nash [104] made an early attempt to mathematically analyze the accumulation of fatigue damage in a self-repairing structure. Prendergast and Taylor [105] used damage as a stimulus for adaptive remodeling on bone's external surfaces, but did not consider the problem of damage repair itself. Later, Martin [106] studied the consequences of assuming that damage itself activates remodeling, and found that there could be negative as well as positive repercussions. When a bone experiences a "fatigue challenge," its mechanical properties may be changed not only by damage but also by increased remodeling space (resorption cavities and refilling osteons) produced by the induced remodeling. It has been postulated that this may actually accelerate progression to a stress fracture [107]. That is, increased remodeling introduces more porosity, which decreases the elastic modulus, elevates the strains produced by the applied loads, and further increases damage formation. Thus, internal remodeling is required to repair fatigue damage but may, in the process, make the bone more susceptible to damage. Modeling responses on the periosteal surfaces of a fatigued bone can offset the negative effects of internal remodeling. Sufficient fatigue damage can stimulate the proliferation of woven bone from the periosteal surface [108, 109]. Presumably, this external modeling response reduces strains in the weakened structure by increasing its cross-sectional area. It is of great interest to know what determines whether the remodeling and modeling responses to fatigue ultimately have a positive or negative effect.

Let us use some of the methods in the approach of Martin [106] to analyze this problem. Consider the fatigue damage and remodeling in a representative cross section of a long bone diaphysis uniformly loaded in simple axial compression. A sequence of calculations simulates daily mechanical and remodeling changes in the bone. For convenience, fatigue damage is defined as millimeters of observable crack length per square millimeter of bone cross section. The damage formation rate is assumed to be a function of the loading rate (R_L in cycles/day) and the resulting strain range (S). Based on Eqs. (7.16 and 7.23), the damage formation rate may be assumed to be proportional to the strain range raised to a power, q:

$$dD_F/dt = k_D S^q R_L \tag{7.42}$$

where k_D is a damage rate coefficient that depends on loading conditions and bone structure.

The BMU activation frequency within the section is *Ac.f.* As each BMU passes through the section, its resorption cavity expands to the diameter of an osteonal cement line, removing a portion of the cortical area (which is subsequently refilled) and any damage it contains. Assuming that fatigue damage can only be removed by this process, and that the rate of removal is proportional to the amount of existing damage (because the more damage there is, the greater the probability that a BMU will remove it), the rate of damage removal (or repair) is

$$dD_R/dt = D(Ac.f)\pi(Cm.Ln.Rd)^2 F_S \tag{7.43}$$

where D is the existing damage and *Cm.Ln.Rd* is the osteonal cement line radius. (Note that $Ac.f\,\pi\; Cm.Ln.Rd^2$ is the amount of cross section removed by all the BMUs created in 1 day.) F_S is a *damage repair specificity factor*, which accounts for any mechanisms that serve to direct remodeling to fatigue-damaged sites, as opposed to strictly random remodeling. The net rate of accretion of fatigue damage is $dD/dt = dD_F/dt - dD_R/dt$. In the equilibrium state, $dD_R/dt = dD_F/dt$ and the corresponding burden of damage waiting to be repaired is

$$D_0 = (k_D S^q R_L)/\left(\pi(Cm.Ln.Rd)^2 Ac.f_0 F_S\right) \tag{7.44}$$

where the subscripts "0" indicate equilibrium values.

The "dose–response" relationship between damage and activation frequency is unknown, but many such curves in biology are sigmoidal, so this is a reasonable assumption. Mathematically, this may be expressed as

$$Ac.f_a = \frac{Ac.f_0 Ac.f_{\max}}{Ac.f_0 - (Ac.f_{\max} - Ac.f_0)\exp[k_R Ac.f_{\max}(D - D_0)/D_0]} \tag{7.45}$$

where $Ac.f_0$ and $Ac.f_{max}$ are the minimum and maximum allowable activation frequencies and k_R is a coefficient that determines the steepness of the response curve. The next part of the analysis is a bit involved. How does the bone's porosity change as a result of the changes in *Ac.f*? This depends on the current numbers of resorbing (*N.Rs*) and refilling (*N.F*) BMUs in the section. The net change in bone volume fraction can be obtained by calculating

$$Q = Q_B N.F - Q_C N.Rs \tag{7.46}$$

where Q_B and Q_C are the rates (mm^2/day) at which bone is added and removed by refilling and resorbing BMUs, respectively. Because the time rate of change of porosity is just $-Q$, porosity can be tracked by constantly recalculating *N.Rs*, $N.F_F$, and Q as Ac.f changes.

The elastic modulus, E, is computed from porosity, *Po*, using a relationship similar to those seen in Chap. 5

$$E = E_0(1 - Po)^3 \tag{7.47}$$

where E_0 is the modulus of the bone tissue. It is recognized that E may also be diminished by fatigue damage, but this effect is ignored for the moment.

Damage is also assumed to stimulate periosteal bone formation, increasing the area of the cross section. No dose–response data are available for this effect, but Turner et al. [110] found that when rat tibias were cyclically loaded, surface bone formation was lamellar at lower loads and converted to woven bone at higher loads.

Lamellar bone apposition rates were linearly proportional to load or strain magnitude, but woven bone production was an "all or nothing" response, that is, the rate of bone formation was not correlated with load. Mathematically, this may be represented by

$$MAR_P = k_p(D - D_O)/D_O \text{ for } D < D_C \quad (7.48)$$

$$MAR_P = MAR_W \text{ for } D < D_C \quad (7.49)$$

where k_P is a coefficient and MAR_W is the apposition rate for woven bone formation. If $MAR_P{}^*$ is the maximum apposition rate for lamellar bone formation, achieved when the damage reaches a critical value, D_C, then

$$k_p = MAR_P^*/[(D_C/D_O) - 1] \quad (7.50)$$

The daily change in periosteal radius is calculated as ΔPs.Rd. = $MAR_P\Delta t$, and cortical area is calculated assuming the endosteal radius is constant. Strain is computed by dividing the applied load by the model's cross-sectional area and elastic modulus.

All of these calculations may be incorporated in a computer model and the behavior of the system under various conditions can be explored. The model's parameters can be partially calibrated against the results of the Burr experiments [78, 79], in which remodeling was activated by fatiguing canine bones in vivo. The model can then be used to examine the general behavior of the system and its sensitivity to its various parameters. Suppose that R_{LE} cycles per day (CPD) of additional loading, initially producing 2500 με, is superimposed on the equilibrium loading. When the periosteal response is disabled and R_{LE} is 90 CPD, the additional damage increases the remodeling rate and porosity, and the bone becomes less stiff, but a new equilibrium is reached. However, when R_{LE} surpasses a critical value, the remodeling never catches up with the increased damage formation rate. Porosity, strain, and damage increase at an ever-increasing rate, without limit. It is reasonable to consider this instability to be fatigue failure in the model (i.e., a complete "stress fracture"). If the loading is only slightly supercritical, this instability can be quite insidious. Damage levels off at what seems to be a new equilibrium level. Then, suddenly, after what may be a long time, the damage "explodes," rising rapidly at a very high rate.

Thus, without a periosteal response, the model can withstand only relatively small fatigue challenges because of the increased strain caused by the remodeling space. When a periosteal callus is included in the model, the results are quite different (Fig. 7.23). Now, the increases in porosity from remodeling are offset by periosteal diameter changes, and the porosity comes back down as the bone's cross-sectional area increases. The addition of external bone acts to reduce strain, and the system returns toward a new equilibrium level—after a transient response lasting about a year, increasing the loading causes larger transients, which end more quickly. Eventually, however, another critical load level is reached, and instability implying

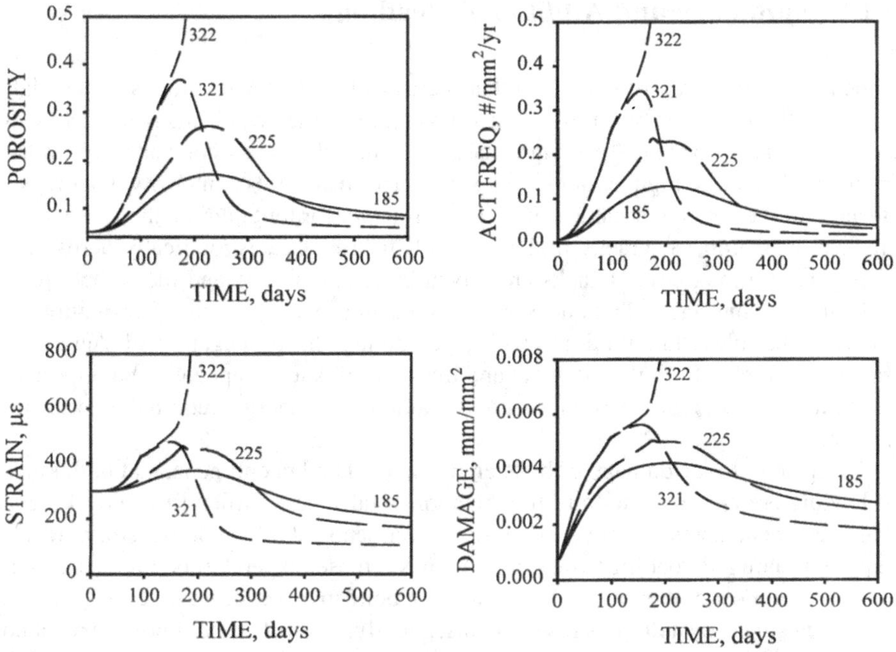

Fig. 7.23 Stress fracture responses for model with periosteal modeling are shown for loading at four different loading levels: $R_{LE} = 185, 225, 321$, and 322 cycles per day (CPD). The first of these rapidly produced a *stress fracture* in the model lacking a periosteal response. Now the model survives this loading easily, showing only transient responses in damage, activation frequency, porosity, and strain. As the loading increases, the responses increase in magnitude but recovery is quicker, until suddenly the system "blows up" [106]

failure ensues once more. Again, the inherent instability of the system is remarkable. As loading increases, remodeling responds ever more vigorously, and the challenge is overcome more and more quickly. Then, when just one more cycle per day of loading is added, instead of damage simply not decreasing, the system precipitously fails.

Such instability is typical of feedback control systems that are nonlinear or contain delays in their responses. Bone remodeling in response to fatigue would have both of these characteristics. It is clear that damage is a nonlinear function of strain, and modulus is a nonlinear function of porosity. The activation of remodeling may be a fairly linear function of damage, but there are delays in the development of porosity changes after f_a changes. These traits are fully consistent with the response of the model, particularly the fact that small changes in the loading conditions or the system parameters can produce large changes in the outcome if the system is near a critical state. This kind of behavior in the model, and the difficulty in predicting when someone will suffer a stress fracture and when they will not, are in concert with one another. However, much remains to be learned about the intriguing interactions between remodeling and mechanical behavior in the skeleton.

7.10 Summary and Additional Reading

Bone is a natural composite material that shares with engineered composite materials many mechanisms for resisting crack propagation. Fracture mechanics teaches us that cracks are propagated through materials when the stress concentration at the crack tip becomes high enough to exceed a critical stress intensity factor, K_c. Because of the lamellar structure of bone, K_c is lower for longitudinal than transverse crack propagation, so cracks find it difficult to travel catastrophically across the diaphysis of a long bone. Cracks are driven by strain energy, and the cement lines and lamellar interfaces in bone serve to consume energy in the propagation of relatively harmless longitudinal cracks, preventing the propagation of dangerous transverse cracks. Both the numbers and the sizes of osteons appear to be important in maximizing fatigue resistance and the amount of energy required to fracture a bone.

Creep and fatigue damage of bone are closely related phenomena, and both kinds of loading are common in daily life. Experimentally, it is difficult to study fatigue damage mechanisms in bone because experiments at physiologic strain levels require prolonged loading times which make these experiments impractical. It appears that most of the fatigue failure data on bone involve primarily creep damage caused by superphysiologic stresses. Consequently, we still do not know very much about the differences between creep and fatigue damage in bone. At relatively high stress levels, crack damage appears to be different for cyclic tensile and compressive loading. During most of the tensile fatigue life, cracks are self-limiting, occurring primarily in interstitial regions between osteons. In compression, cracks seem to form later, running between Haversian canals and crossing cement lines. The differences in the shapes of the modulus degradation curves seen in tensile and compressive fatigue seem to be attributable to these differences in crack propagation characteristics.

Cyclic loading at physiologic stress levels produces microcracks in osteonal bone and activates increased intracortical remodeling; the mechanism behind this is unclear. If one assumes that bone remodeling is activated in proportion to fatigue damage, and that the increased porosity produced by this remodeling reduces the elastic modulus of the bone, increasing strains and therefore increasing damage, it becomes clear that limits exist on the ability of remodeling to control fatigue damage under increased loading. Modeling this system mathematically suggests that periosteal callus formation is essential to maintaining its stability. Even so, the nonlinearity of the system seems to make it critically stable, so that small differences in skeletal physiology may greatly alter the ability of individuals to avoid a stress fracture under conditions of increased physical activity.

Additional reading suggestions include the early clinical perspective of Devas [111] on stress fractures and the more recent book edited by Burr and Milgrom [97] that covers many aspects of stress fracture epidemiology and pathogenesis. A survey of the work of D.R. Carter between 1975 and 1990 will produce much of the literature on the mechanics of fatigue and creep in bone. A review by Donahue and Galley [112] provides an excellent overview of microdamage and its implication for skeletal fragility and remodeling. Some information can also be found in the more recent paper "Why bones bend but do not break" [113].

References

1. Irwin GR. Analysis of stresses and strains near the end of a crack traversing a plate. J Appl Mech. 1957;24:361–4.
2. Pilkey WD. Formulas for stress, strain, and structural matrices. New York: Wiley; 1994.
3. Norman TL, Vashishth D, Burr DB. Fracture toughness of human bone under tension. J Biomech. 1995;28:309–20.
4. Griffith AA. The phenomenon of rupture and flow in solids. Philos Trans R Soc Lond. 1920; A221:163–98.
5. Finkel V. Portrait of a crack (translated from the Russian by Y. Nadler). Moscow: Mir Publications; 1985.
6. Freund LB. Dynamic fracture mechanics. Cambridge: Cambridge University Press; 1990.
7. Ritchie RO. The conflicts between strength and toughness. Nat Mat. 2011;10:817–22.
8. Green JO, Wang J, Diab T, Vidakovic B, Guldberg RE. Age-related differences in the morphology of microdamage propagation in trabecular bone. J Biomech. 2011;44:2659–66.
9. Barth HD, Zimmermann EA, Schaible E, Tang SY, Alliston T, Ritchie RO. Characterization of the effects of x-ray irradiation on the hierarchical structure and mechanical properties of human cortical bone. Biomaterials. 2011;32:8892–904.
10. Russell NA, Pelletier MH, Bruce WJ, Walsh WR. The effect of gamma irradiation on the anisotropy of bovine cortical bone. Med Eng Phys. 2012;34:1117–22.
11. Nyman JS, Reyes M, Wang XD. Effect of ultrastructural changes on the toughness of bone. Micron. 2005;36:566–82.
12. Nalla RK, Kinney JH, Ritchie RO. Mechanistic fracture criteria for the failure of human cortical bone. Nat Mat. 2003;2:164–8.
13. Paris PC. The fracture mechanics approach to fatigue. In: Fatigue—an interdisciplinary approach. Proceedings, tenth Sagamore Army materials research conference. Syracuse, NY: Syracuse University Press; 1964.
14. King AI, Evans FG. Analysis of fatigue strength of human compact bone by the Weibull method. In: Jacobson B, editor. Digest of the 7th International Conference on Medical and Biological Engineering. Stockholm: The Royal Academy of Engineering Sciences; 1967.
15. Caler WE, Carter DR. Bone creep-fatigue damage accumulation. J Biomech. 1989;22:625–35.
16. Pattin CA, Caler WE, Carter DR. Cyclic mechanical property degradation during fatigue loading of cortical bone. J Biomech. 1996;29:69–80.
17. Burr DB, Martin RB. Calculating the probability that microcracks initiate resorption spaces. J Biomech. 1993;26:613–6.
18. Taylor D, Lee TC. Measuring the shape and size of microcracks in bone. J Biomech. 1998;31: 1177–80.
19. Fazzalari NL, Forwood MR, Mathey BA, Smith K, Kolesik P. Three-dimensional confocal images of microdamage in cancellous bone. Bone. 1998;23:373–8.
20. Vashishth D, Koontz J, Qiu SJ, Lundin-Cannon D, Yeni YN, Schaffler MB, Fyhrie DP. In vivo diffuse damage in human vertebral trabecular bone. Bone. 2000;26:147–52.
21. Herman BC, Cardoso L, Majeska RJ, Jepsen KJ, Schaffler MB. Activation of bone remodeling after fatigue: differential response to linear microcracks and diffuse damage. Bone. 2010;47: 766–72.
22. Boyde A. The real response of bone to exercise. J Anat. 2003;203:173–89.

23. Seref-Ferlengez Z, Kennedy O, Schaffler MB. Diffuse microdamage induced in cortical bone in vivo decreases with time in the absence of bone remodeling. In: 34th Annual Meeting of the American Society for Bone and Mineral Research; 2012.
24. Frost HM. Presence of microscopic cracks in vivo in bone. Henry Ford Hosp Med Bull. 1960a;8:25–35.
25. Burr DB, Stafford T. Validity of the bulk-staining technique to separate artifactual from in vivo bone microdamage. Clin Orthop Relat Res. 1990;260:305–8.
26. Schaffler MB, Choi K, Milgrom C. Aging and matrix microdamage accumulation in human compact bone. Bone (New York). 1995;17:521–5.
27. Mori S, Harruff R, Ambrosius W, Burr DB. Trabecular bone volume and microdamage accumulation in the femoral heads of women with and without femoral neck fractures. Bone (New York). 1997;21:521–6.
28. Wenzel TE, Schaffler MB, Fyhrie DP. In vivo trabecular microcracks in human vertebral bone. Bone (New York). 1996;19:89–95.
29. Hasegawa K, Turner CH, Chen J, Burr DB. Effect of disc lesion on microdamage accumulation in lumbar vertebrae under cyclic compression loading. Clin Orthop. 1995;311:190–8.
30. Diab T, Vashishth D. Effects of damage morphology on cortical bone fragility. Bone. 2005;37: 96–102.
31. Diab T, Condon KW, Burr DB, Vashishth D. Age-related change in the damage morphology of human cortical bone and its role in bone fragility. Bone. 2006;38:427–31.
32. Yeni YN, Hou FJ, Ciarelli T, et al. Trabecular shear stresses predict in vivo linear microcrack density but not diffuse damage in human vertebral cancellous bone. Ann Biomed Engng. 2003;31:726–32.
33. Norman TL, Wang Z. Microdamage of human cortical bone: incidence and morphology in long bones. Bone (New York). 1997;20:375–9.
34. Wasserman N, Yerramshetty J, Akkus O. Microcracks colocalize within highly mineralized regions of cortical bone. Europ J Morphol. 2005;42:43–51.
35. Parsamian GP, Norman TL. Diffuse damage accumulation in the fracture process zone of human cortical bone specimens and its influence on fracture toughness. J Mat Sci Mat Med. 2001;12:779–83.
36. Vashishth D, Behiri JC, et al. Crack growth resistance in cortical bone: concept of microcrack toughening. J Biomech. 1997;30:763–9.
37. Boyce TM, Fyhrie DP, Glotkowski MC, et al. Damage type and strain mode associations in human compact bone bending fatigue. J Orthop Res. 1998;16:322–9.
38. Carter DR, Caler WE. Cycle-dependent and time-dependent bone fracture with repeated loading. J Biomech Eng. 1983;105:166–70.
39. Carter DR, Caler WE. A cumulative damage model for bone fracture. J Orthop Res. 1985;3: 84–90.
40. Reifsnider KL. Damage and damage mechanics. In: Reifsnider KL, editor. Fatigue of composite materials. New York: Elsevier; 1990. p. 11–77.
41. Schaffler MB. Stiffness and fatigue of compact bone at physiological strains and strain rates. Doctoral dissertation. Morgantown, WV: West Virginia University; 1985.
42. Carter DR, Hayes WC, Schurman DJ. Fatigue life of compact bone. II. Effects of microstructure and density. J Biomech. 1976;9:211–8.
43. Moyle DD, Bowden RW. Fracture of human femoral bone. J Biomech. 1984;17:203–13.
44. Cohen J, Harris WH. The three-dimensional anatomy of haversian systems. J Bone Joint Surg. 1958;40A:419–34.
45. Lanyon LE, Bourn S. The influence of mechanical function on the development and remodeling of the tibia. J Bone Joint Surg. 1979;62A:263–73.
46. Piekarski K. Fracture of bone. J Appl Phys. 1970;41:215–23.
47. Pope MH, Murphy MC. Fracture energy of bone in a shear mode. Med Biol Eng. 1974;12: 763–7.

48. Moyle DD, Welborn JW, Cooke FW. Work to fracture of canine femoral bone. J Biomech. 1978;11:435–40.
49. Gibson VA, Stover SM, Martin RB, Gibeling JC, Willits NH, Gustafson MB, Griffin LV. Fatigue behavior of the equine third metacarpus: mechanical property analysis. J Orthop Res. 1996;13:861–8.
50. Corondan G, Haworth WL. A fractographic study of human long bone. J Biomech. 1986;19: 207–18.
51. Saha S. Longitudinal shear properties of human compact bone and its constituents, and the associated failure mechanisms. J Mater Sci. 1977;12:1798–806.
52. Frasca P. Scanning electron microscopy studies of ground substance in the cement lines, resting lines, hypercalcified rings and reversal lines of human cortical bone. Acta Anat. 1981;109:115–21.
53. Lakes R, Saha S. Cement line motion in bone. Science. 1979;204:501–3.
54. Park HC, Lakes RS. Cosserat micromechanics of human bone: strain redistribution by a hydration sensitive constituent. J Biomech. 1986;19:385–97.
55. Kendall K. Control of cracks by interfaces in composites. Proc R Soc Lond. 1975;341:409–28.
56. Behiri JC, Bonfield W. Fracture mechanics of bone—the effects of density, specimen thickness and crack velocity on longitudinal fracture. J Biomech. 1984;17:25–34.
57. Balazs EA, Gibbs DA. The rheological properties and biological function of hyaluronic acid. In: Balazs EA, editor. Chemistry and molecular biology of the intercellular matrix. New York: Academic Press; 1970.
58. Gottesman T, Hashin Z. Analysis of viscoelastic behavior of bones on the basis of microstructure. J Biomech. 1980;13:89–96.
59. Simkin A, Robin G. Fracture formation in differing collagen fiber pattern of compact bone. J Biomech. 1974;7:183–8.
60. Lanyon LE, Baggott DG. Mechanical function as an influence on the structure and form of bone. J Bone Joint Surg. 1976;58B:436–43.
61. Rubin CT, Lanyon LE. Regulation of bone formation by applied dynamic loads. J Bone Joint Surg. 1984;66A:397–402.
62. Carter DR, Caler WE, Spengler DM, Frankel VH. Uniaxial fatigue of human cortical bone. The influence of tissue physical characteristics. J Biomech. 1981;14:461–70.
63. Carter DR, Hayes WC. Compact bone fatigue damage. I. Residual strength and stiffness. J Biomech. 1977;10:325–37.
64. Evans FG, Riolo ML. Relations between the fatigue life and histology of adult human cortical bone. J Bone Joint Surg. 1970;52A:1579–86.
65. Krajcinovic D, Trafimow J, Sumarac D. Simple constitutive model for cortical bone. J Biomech. 1987;20:779–84.
66. Griffin LV, Gibeling JC, Martin RB, Gibson VA, Stover SM. A model of flexural fatigue damage accumulation for cortical bone. J Orthop Res. 1997;15:607–14.
67. Carter DR, Hayes WC. Compact bone fatigue damage: a microscopic examination. Clin Orthop Relat Res. 1977;127:265–74.
68. Parfitt AM. The physiologic and clinical significance of bone histomorphometric data. In: Recker RR, editor. Bone histomorphometry techniques and interpretation. Boca Raton: CRC Press; 1983. p. 143–223.
69. Martin RB, Burr DB. A hypothetical mechanism for the stimulation of osteonal remodeling by fatigue damage. J Biomech. 1982;15:137–9.
70. Frost HM. Bone remodeling and its relationship to metabolic bone disease. Springfield: Thomas; 1973b.
71. Weinbaum S, Cowin SC, Zeng Y. A model for the excitation of osteocytes by mechanical loading-induced bone fluid shear stresses. J Biomech. 1994;27:339–60.
72. Cowin SC, Weinbaum S, Zeng Y. A case for bone canaliculi as the anatomic site of strain generated potentials. J Biomech. 1995;28:1281–97.

73. Owan I, Burr DB, Turner CH, Qui J, Tu Y, Onyia JE, Duncan RL. Mechanotransduction in bone: osteoblasts are more responsive to fluid forces than mechanical strain. Am J Physiol. 1997;273:C810–5.
74. Tschantz P, La Rutishauser E. surcharge mechanique de l'os vivant. Ann Anat Pathol. 1967;12:223–48.
75. Chamay A, Tschantz P. Mechanical influences in bone remodeling. Experimental research on Wolff's law. J Biomech. 1972;5:173–80.
76. Rahn BA, Gallinaro P, Schenck RK. Compression interfragmentaire et surcharge locale de l'os. In: Boitzy A, editor. Interarthrite de l'Epaule. Osteogenese et Compression. Bern: Huber; 1972.
77. Schaffler MB, Burr DB. Primate cortical bone microstructure: relationship to locomotion. Am J Phys Anthropol. 1984;65:191–7.
78. Burr DB, Martin RB, Schaffler MB, Radin EL. Bone remodeling in response to in vivo fatigue microdamage. J Biomech. 1985;18:189–200.
79. Mori S, Burr DB. Increased intracortical remodeling following fatigue damage. Bone (New York). 1993;14:103–9.
80. Parfitt AM. Targeted and nontargeted bone remodeling: relationship to basic multicellular unit origination and progression. Bone. 2002;30:5–7.
81. Burr DB. Targeted and nontargeted remodeling. Bone. 2002;30:2–4.
82. Martin RB. Is all cortical bone remodeling initiated by microdamage? Bone. 2002;30:8–13.
83. Bentolila V, Hillam RA, Skerry TM, Boyce TM, Fyhrie DP, Schaffler MB. Activation of intracortical remodeling in adult rat long bones by fatigue loading. Trans Orthop Res Soc. 1997;22:578.
84. Ross RD, Sumner RD. Bone matrix maturation in a rat model of intra-cortical bone remodeling. Calcif Tissue Int. 2017;101:193–203.
85. Schaffler MB, Kennedy OD. Osteocyte signaling in bone. Curr Osteop Rep. 2012;10:118–25.
86. Verborgt O, Gibson GJ, Schaffler MB. Loss of osteocyte integrity in association with microdamage and bone remodeling after fatigue in vivo. J Bone Miner Res. 2000;15:60–7.
87. Cardoso L, Herman BC, Verborgt O, Laudier D, Majeska RJ, Schaffler MB. Osteocyte apoptosis controls activation of intracortical resorption in response to bone fatigue. J Bone Miner Res. 2009;24:597–605.
88. Kennedy OD, Herman BC, Laudier DM, Majeska RJ, Sun HB, Schaffler MB. Activation of resorption in fatigue-loaded bone involves both apoptosis and active pro-osteoclastogenic signaling by distinct osteocyte populations. Bone. 2012;50:1115–22.
89. Nakashima T, Hayashi M, Fukunaga T, et al. Evidence for osteocyte regulation of bone homeostasis through RANKL expression. Nat Med. 2011;17:1231–4.
90. Xiong J, Onal M, Jilka RL, et al. Matrix-embedded cells control osteoclast formation. Nat Med. 2011;17:1235–41.
91. Mulcahy LE, Taylor D, Lee TC, Duffy GP. RankL and OPG activity is regulated by injury size in networks of osteocyte-like cells. Bone. 2011;48:182–8.
92. Heaney RP. The natural history of vertebral osteoporosis. Is low bone mass an epiphenomenon? Bone. 1992;13(Suppl 2):S23–6.
93. Rutishauser E, Majno G. Lesions osseuses par sur charge dans le squelette normal et pathologique. Bull Schweiz Akad Med Wiss. 1950;6:333–41.
94. Burstein AH, Reilly DT, Martens M. Aging of bone tissue: mechanical properties. J Bone Joint Surg. 1976;58A:82–6.
95. Currey JD, Brear K, Zioupos P. The effects of ageing and changes in mineral content in degrading the toughness of human femora. J Biomech. 1996;29:257–60.
96. Burr DB, Turner CH. Biomechanical measurements in age-related bone loss. In: Rosen CJ, Glowacki J, Bilezikian JP, editors. The aging skeleton. London: Academic Press; 1999.; Chapter 20. p. 301–11.
97. Burr DB, Milgrom C, editors. Musculoskeletal fatigue and stress fractures. Boca Raton: CRC Series in Exercise Physiology; 2001.

98. Hulkko A, Orava S. The role of age in the development of stress and fatigue fractures. In: Burr DB, Milgrom C, editors. Musculoskeletal fatigue and stress fractures. Boca Raton: CRC Series in Exercise Physiology; 2001.; Chapter 4. p. 35–54.
99. Shane E, Burr DB, Ebeling P, Abrahamsen B, Adler RA, Brown TD, et al. Atypical subtrochanteric and diaphyseal femoral fractures: report of a task force of the American Society for Bone and Mineral Research. J Bone Miner Res. 2010;25:1–28.
100. Fyhrie DP, Milgrom C, Hoshaw SJ, Simkin A, Dar S, Drumb D, Burr DB. Effect of fatiguing exercise on longitudinal bone strain as related to stress fracture in humans. Ann Biomed Eng. 1998;26:660–5.
101. Milgrom C, Radeva-Petrova DR, Finestone A, Nyska M, Mendelson S, Benjuya N, Simkin A, Burr D. The effect of muscle fatigue on in vivo tibial bone strains. J Biomech. 2007;40:845–50.
102. Yoshikawa T, Mori S, Santiesteban AJ, Sun TC, Hofstad E, Chen J, Burr DB. The effects of muscle fatigue on bone strain. J Exp Biol. 1994;188:217–33.
103. Donahue SW, Sharkey NA, Modanlou KA, Sequeira LN, Martin RB. Bone strain and microcracks at stress fracture sites in human metatarsals. Bone. 2000;27:827–33.
104. Nash CD. Fatigue of a self-healing structure: a generalized theory of fatigue failure. ASME Publication #66-WA/BHF-3. New York: American Society of Mechanical Engineers; 1966.
105. Prendergast PJ, Taylor D. Prediction of bone adaptation using damage accumulation. J Biomech. 1994;27:1067–76.
106. Martin RB. A mathematical model for fatigue damage repair and stress fracture in osteonal bone. J Orthop Res. 1995;13:309–16.
107. Scully TJ, Besterman G. Stress fracture—a preventable training injury. Mil Med. 1982;147: 285–7.
108. Johnson LC, Stradford HT, Geis RW, Dineen JR, Kerley E. Histogenesis of stress fractures. J Bone Joint Surg. 1963;45A:1542.
109. Uhthoff HK, Jaworski ZFG. Periosteal stress-induced reactions resembling stress fractures. Clin Orthop Relat Res. 1985;199:284–91.
110. Turner CH, Forwood M, Rho J-Y, Yoshikawa T. Mechanical loading thresholds for lamellar and woven bone formation. J Bone Miner Res. 1993a;9:87–97.
111. Devas MB. Stress fractures. London: Churchill Livingstone; 1975.
112. Donahue SW, Galley SA. Microdamage in bone: implications for fracture, repair, remodeling, and adaptation. Crit Rev Biomed Eng. 2006;34:215–71.
113. Burr DB. Why bones bend but don't break. J Musculoskelet Neuronal Interact. 2011;11:270–85.
114. Norman TL, Vashishth D, Burr DB. Effect of groove on bone fracture toughness. J Biomech. 1992;25:1489–92.
115. Melvin JW, Evans FG. Crack propagation in bone. In: Fung YC, Brighton JA, editors. Biomaterials symposium, AMD2. New York: American Society of Mechanical Engineers; 1973.
116. Bonfield W, Datta PK. Fracture toughness of compact bone. J Biomech. 1976;9:131–4.
117. Alto A, Pope MH. On the fracture toughness of equine metacarpi. J Biomech. 1979;12:415–21.
118. Wang X, Shen X, Li X, et al. Age-related changes in the collagen network and toughness of bone. Bone. 2002;31:1–7.
119. Zimmerman EA, Barth HD, Ritchie RO. The multiscale origins of fracture resistance in human bone and its biological degradation. JOM. 2012;64:486–93.

Chapter 8
Mechanical Properties of Ligament and Tendon

Abstract Ligament and tendon serve very specialized functions in the musculoskeletal system. These functions require unique mechanical properties. The structural and material-level properties that determine mechanical competency are covered, as well as how these properties are affected by various factors and when damage occurs. The mechanical properties can be described through mathematical modeling, which can then help elucidate novel theories about functional optimization.

What You Will Learn

1. Structural and material-level mechanical properties of tendon/ligament
2. Mechanical properties at tendon/muscle and tendon/ligament/bone interfaces
3. Mathematical modeling of ligament/tendon mechanical properties
4. How mechanical properties of ligament/tendon adapt with age, exercise, and disuse
5. Properties of ligament/tendon during healing

8.1 Introduction

In Chap. 3, we explored the architecture composition of ligaments and tendons, noncalcified but nonetheless extremely mechanically competent connective tissue entities responsible for binding bones together, transmitting forces from muscles, and constraining motion within normal limits. Recall that ligaments and tendons are composed of linearly arranged collagen molecules assembled in a hierarchal fashion into subfibrils, fibrils, and fibers. Collagen fibers within ligaments are architecturally orientated to effectively control and constrain joint motion; in tendons, they are grouped into distinct but parallel fascicles. In this chapter, we will take an in-depth look at the mechanical behavior of these skeletal soft tissues and some of the specialized laboratory and in vivo approaches used to determine their behaviors. We will also briefly examine some common sites of injury, repair, and assessment.

M. R. Allen, J. Wallace, *Skeletal Tissue Mechanics*,
https://doi.org/10.1007/978-3-031-90609-1_8

8.2 Quasi-Static Tensile Properties

When loaded in the laboratory, ligaments and tendons produce characteristic load–deformation patterns that reflect their structural architecture. Materials such as steel and aluminum display linear stress–strain characteristics because of their crystalline nature. In these materials, the modulus of elasticity is constant throughout the elastic region and is representative of the atomic forces at work within their crystalline architecture. If the elastic limit is exceeded, crystalline materials will yield and eventually fail. In contrast, noncrystalline materials display what is termed *rubbery elasticity*. Tensile tests of these materials produce continuous upwardly concave load–displacement and stress–strain curves. In the noncrystalline materials, the elastic modulus is dependent on load magnitude because of intra- and intermolecular forces and molecular cross-linking. Failure of these materials is often abrupt and can occur without a well-defined yield point. Tensile tests of tendon and ligament produce curves that display both crystalline and rubbery characteristics, reflecting their complex structural hierarchy and molecular interactions.

A typical load–deformation (or stress–strain) curve for tendons or ligaments (Fig. 8.1) can be divided into three regions. In the *toe region*, collagen crimps are removed by elongation. Initially, minimal force is required for this elongation, but as the collagen fibers straighten, more and more force is required and the curve swings upward. The toe region typically ends at about 1.5–3.0% strain. It is a matter of some controversy whether the toe region is physiologically or functionally important in vivo. It may actually be artifactual and the result of assuming that the normal resting length of the tendon or ligament in vivo occurs at zero load. Gillis et al. showed that in the forelimb of the standing horse, the superficial flexor tendon has a

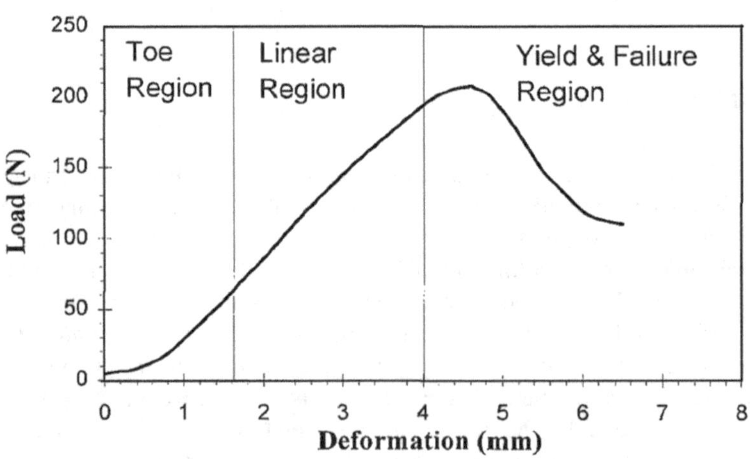

Fig. 8.1 Typical load–deformation curve from a tendon or ligament showing the toe, linear, and yield and failure regions. The stress–strain curve would have a similar shape [2]

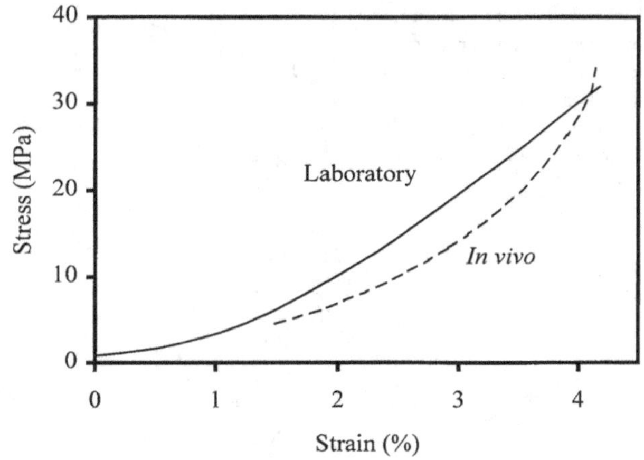

Fig. 8.2 Different stress–strain curves for equine superficial flexor tendon are obtained when measured in vivo *(dashed line)* and on specimens removed from the animal and loaded in the laboratory *(solid line)* [1]

strain of about 1.5% compared to the extracted tendon under zero load [1] (Fig. 8.2). Therefore, during standing, a resting position for the horse, the structure is pre-strained beyond the toe region seen in laboratory tests of extracted tendons.

Further loading produces a *linear region* in which the molecular cross-links of the collagen are stressed (see Fig. 8.1). As with bone, the slope of this region of the load–deformation curve represents the stiffness of the tendon or ligament, and the slope of the stress–strain curve is a measure of the tissue's modulus of elasticity. In practice, the slopes of these curves are usually determined at a standardized load or stress, close to the limits of the most linear portion of these curves. Elastic modulus values reported for tendon and ligament range between 1.0 and 2.0 GPa; the modulus of highly specialized tissues (such as the *ligamentum nuchae*, described later in this chapter) can be substantially less.

In the final region of the curve, a reduction in slope marks the *yield point* or onset of cross-link or fiber damage. If loading is continued, the tendon or ligament will eventually fail. Tendons and ligaments can be strained to between 5% and 7% without damage. The maximum strain that a ligament or tendon can endure before failure is generally in the neighborhood of 12–15%; however, substantial damage to the collagen network can occur well before complete rupture. Ligaments with unusually high elastin content can be strained to 30% or more without damage. Ultimate tensile strengths for tendons and ligaments range from 50 to 150 MPa (for comparison, recall that the tensile strength of bone is about 150 MPa).

8.3 Viscoelastic Properties

The mechanical behavior of ligament and tendon is *viscoelastic*; thus, their load–displacement and stress–strain relationships are rate- and history-dependent. Testing parameters must always be carefully considered when establishing or comparing

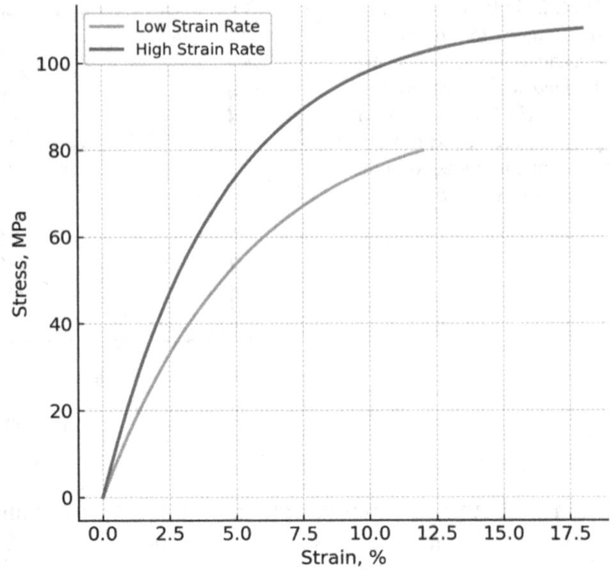

Fig. 8.3 Effect of strain rate on stress–strain curve for tendon. At high strain rates, the linear portion of the curve is extended. (Adapted from concepts presented in Ref. [3])

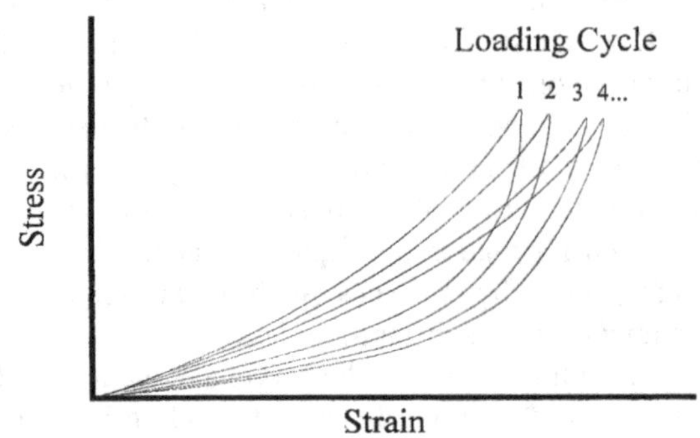

Fig. 8.4 Energy is lost each time a tendon is repeatedly loaded and unloaded, a process called hysteresis. The consequence of this over time is reduced stiffness and greater deformation for a given load or stress. Generally, a steady state develops after 10–20 cycles

material properties such as ultimate strength, ultimate strain, and modulus of elasticity. Two representative stress–strain curves for tendon are plotted in Fig. 8.3. As can be seen, the higher strain rate results in higher stress and strain at both failure and yield, and a slightly increased elastic modulus. This contrasts with the behavior of bone. Note that the ductility (energy absorption) of bone is decreased at higher strain rates, but that of tendon is increased.

Tendons and ligaments display *hysteresis*, another behavior caused, at least in part, by viscoelasticity. Hysteresis refers to loss of energy during a load–deformation cycle (Fig. 8.4). If a purely elastic material is loaded in its elastic range and then immediately unloaded, the load–deformation curve for the loading phase is identical to the load–deformation curve for the unloading phase. The energy absorbed during loading and returned to the system during unloading will be the same, and the system will be 100% efficient. The loading–unloading profiles of ligaments and tendons are not identical, however. The area under the loading curve is greater than the area under the unloading curve, and the difference between the two represents the energy lost during the load cycle, which is referred to as hysteresis. The amount of hysteresis observed depends on the particular ligament or tendon and on the testing conditions. Tendons involved in gait tend to be efficient storers of elastic energy (i.e., have minimal hysteresis). For example, the digital flexor tendons of horses studied by Gillis et al. lost only about 5% of their strain energy during a loading–unloading cycle [1].

Where the lost energy goes is difficult to determine. Frictional forces at work within the sample cause some energy to be converted to heat, some of which flows to the surroundings while the remainder remains within the sample, raising its temperature and altering its mechanical properties. When the load is removed, the accumulated thermal energy slowly dissipates and the tendon or ligament may shorten further with time. However, *residual strain* may remain in the tissue for some time or indefinitely. Some of the lost strain energy has probably been used to break molecular cross-links or otherwise disrupt the molecular structure of the tissue.

Figure 8.4 demonstrates the effect of loading history. The mechanical behavior in cycle four is clearly different than that in cycle one. The shift to the right reflects a decrease in tendon or ligament stiffness and may be accompanied by a decrease in ultimate strength. For most tendons and ligaments, a steady state is achieved after 10–20 cycles of loading. However, if the peak load is increased on a subsequent cycle, the steady state is disrupted and additional cycles of loading will be required to reestablish stability. Conditioning effects have been measured in vivo and are an inherent time-dependent property reflective of actual physiologic behavior and not just an artifact of laboratory testing [4]. Tendons and ligaments are also subjected to constant loads or deformations, and, like bone and cartilage, they exhibit *creep* under constant load and *stress relaxation* under constant deformation (Fig. 8.5).

8.4 Mathematical Modeling of Mechanical Properties

When modeling musculotendon dynamics, many investigators have chosen to simulate tendon as a purely elastic component for simplicity's sake. Zajac assumed pure tendon elasticity in developing a generic musculotendon force–strain curve using two parameters, the *tendon slack length*, L_S, and the *peak isometric muscle force*, F_0^M. L_S is the length at which the tendon just begins to exhibit some resistance to lengthening [5]. Tendon strain is defined as $\varepsilon = (L - L_S)/L_S$. Suppose that the force–strain curve for a tendon has the shape shown in Fig. 8.6. The peak isometric force is measured for the tendon's muscle and used to normalize the tendon force values (as shown on the right-hand axis). By dividing the force values by tendon cross-sectional area, the normalized force–strain curve can be turned into a stress–strain curve, which defines the response of a generic tendon. This relationship can then be scaled to obtain the force–length response of a musculotendon unit having a different area, F_0^M, and L_S.

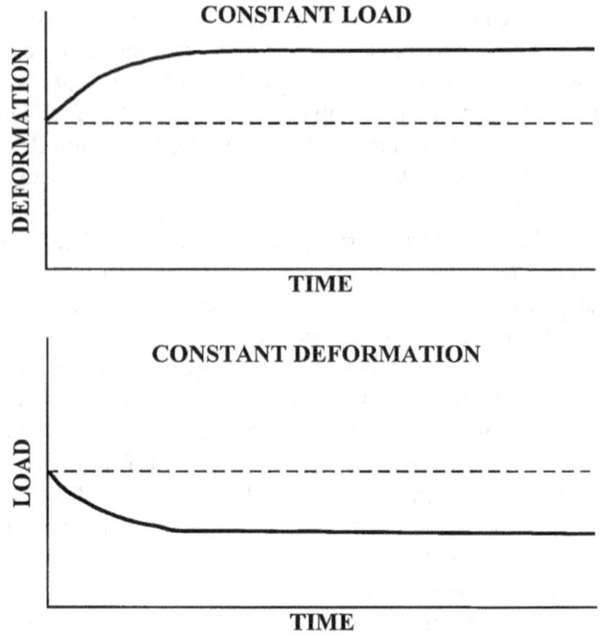

Fig. 8.5 (**a**) When a constant load is applied to a tendon or ligament, its deformation gradually "creeps" to a stable value. (**b**) When a constant deformation is applied, the load (material stress) "relaxes" to a stable value. The times required for these processes are of the order of minutes or hours

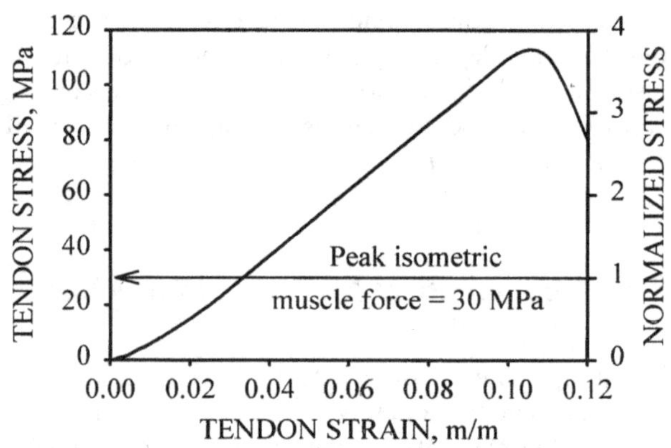

Fig. 8.6 The stress–strain curve for a tendon may be normalized by the peak isometric muscle force, F_0^M (expressed per unit cross-sectional area of the muscle), to obtain a generic tendon response shown by the scale on the right side of the graph [5]

However, tendon tissue is viscoelastic and nonlinear. Lieber and associates developed an empirical model for muscle–tendon units by isolating and measuring the individual mechanical behavior of the frog semitendinosis muscle and its tendon and *aponeurosis* (the connection between the two; see following) [6]. They obtained the following regression equations for tendon and aponeurosis material behavior:

$$F_T = 10^{(\varepsilon + 0.633)/1.35} \tag{8.1}$$

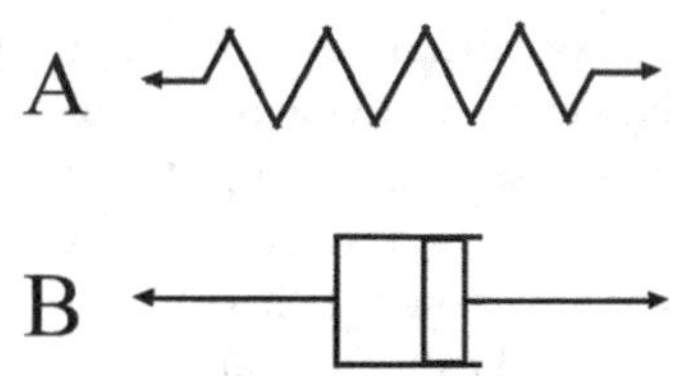

Fig. 8.7 Symbols used for the two basic rheological model components: a spring (**a**) and a dashpot (**b**)

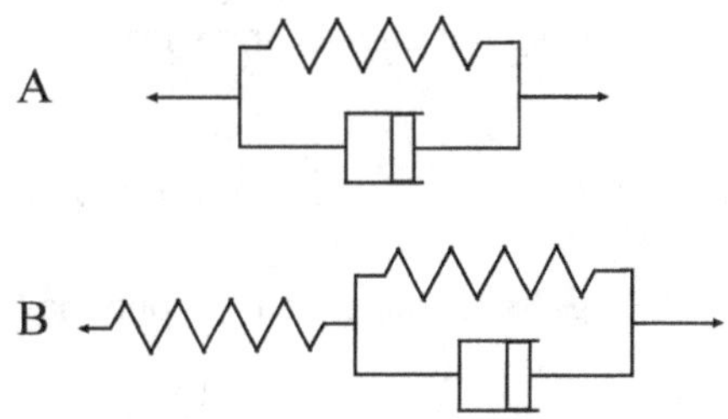

Fig. 8.8 Two rheological models: a Kelvin solid (**a**) and a three-parameter solid (**b**)

$$F_{\mathrm{A}} = 10^{(\varepsilon\ +\ 3.63)/5.66} \tag{8.2}$$

where ε is percent tendon or aponeurosis strain and F_{T} and F_{A} are the corresponding tensions in the tendon or aponeurosis, respectively, expressed as percent of the maximum tetanic muscle tension.

The viscoelastic properties of tendons or ligaments are often represented using rheological models composed of two basic components: springs and dashpots (Fig. 8.7). The elastic component of their behavior is represented by the spring element, in which strain is linearly proportional to the applied stress. The dashpot represents the viscous behavior, in which the strain rate is linearly proportional to the applied stress. (The dashpot is a "leaky" piston in a cylinder; under load, the fluid slowly flows past the piston, allowing it to move at a velocity proportional to the load.) In practice, combinations of elements are connected to simulate various behaviors. Two common models are the Kelvin solid and the three-parameter solid (Fig. 8.8).

It is interesting to compare the behaviors of these models to those seen in Fig. 8.5. For the Kelvin solid, the strain, ε, is the same in each element, and the total stress supported by the combination is

$$\boldsymbol{\sigma} = E\boldsymbol{\varepsilon} + \eta\dot{\boldsymbol{\varepsilon}} \tag{8.3}$$

where $\dot{\varepsilon}$ is the strain rate, E is the elastic modulus of the spring, and η is the viscosity of the dashpot. In a creep test, one has a constant load, σ_0, and Eq. (8.3) can be integrated to obtain

$$\boldsymbol{\varepsilon} = (\sigma_0/E)\left[1 - e^{-Et/\eta}\right] \tag{8.4}$$

As t becomes very large, the strain asymptotically approaches σ_0/E. Initially, however, $t = 0$ and $\varepsilon = 0$; contrary to Fig. 8.5, there is no instantaneous elastic response in a creep test. When a constant strain, ε_0, is applied, the second term in Eq. (8.3) is zero, and the stress is constant at $E\varepsilon_0$. Thus, there is no relaxation effect for this model, either.

For the three-parameter model, a single spring is in series with a Kelvin solid. The equation for the series spring is $\sigma = E'\varepsilon'$ and that for the Kelvin element is $\sigma = E''\varepsilon'' + \eta\dot{\varepsilon}''$. Using the Laplace transform method [7], these equations may be combined to obtain

$$(E' + E'')\ \boldsymbol{\sigma} + \boldsymbol{\eta\sigma} = E'E''\boldsymbol{\varepsilon} + E'\boldsymbol{\eta\dot{\varepsilon}} \tag{8.5}$$

Using the Laplace method, one may obtain for the creep solution

$$\boldsymbol{\varepsilon} = (\boldsymbol{\sigma}_0/E')\ \left[E^*\left(1 - e^{-\lambda t}\right) + e^{-\lambda t}\right] \tag{8.6}$$

where $E^* = (E' + E'')/E''$ and $\lambda = E''/\eta$. In this case, at $t = 0$ there is an instantaneously produced strain σ_0/E', and the strain subsequently approaches $\sigma_0 E^*/E'$. For the relaxation solution, one obtains

$$\boldsymbol{\sigma} = E'\boldsymbol{\varepsilon}_0\ \left[\left(1 - e^{-\lambda t}\right)/E^* + e^{-\lambda t}\right] \tag{8.7}$$

where ε_0 is the applied strain and $\gamma = (E' + E'')/\eta$. Again, there is a stress $E'\varepsilon_0$ at $t = 0$, and this subsequently relaxes asymptotically to $E'\varepsilon_0/E^*$. Thus, this model exhibits the general behavior observed in tendons and ligaments. By fitting its equations to experimental data, values for E', E'', and η can be obtained. These results may be useful for predictive purposes, but the model is still phenomenological rather than mechanistic. That is, we have not learned much about how the viscoelastic behavior is caused by the tissue structure. What parts of the tendon have the attributes associated with E' and E''? Presumably η is associated with a flow of water, but where, exactly? These questions remain to be addressed.

A more elaborate and popular nonlinear viscoelastic theory was described by Fung [8]. This quasi-linear viscoelastic (QVL) model mathematically describes the stress relaxation of collagenous tissues as the product of independent functions of time and strain:

$$\boldsymbol{\sigma}[\boldsymbol{\varepsilon}(t)\ ;\ t] = G(t)^*\ \boldsymbol{\sigma}^{\mathrm{e}}(\boldsymbol{\varepsilon}) \tag{8.8}$$

where σ^{e} is the *elastic response*, a function of strain (ε) only, and G is the *reduced relaxation function*, which is dependent on time (t) only. Versions of this theory have been successfully applied by other investigators as well [9–11]. However, the form of this model assumes time and strain to be separable functions, i.e., time-dependent relaxation behavior (*G(t)*) is the same at any given strain (ε). Experimental evidence demonstrates that this is not the case [12, 13] indicating that even more elaborate nonlinear models are required to accurately describe the strain-dependent relaxation behavior of these tissues. A specialized transversely isotropic viscoelastic constitutive model developed by Davis and De Vita considers finite strain, tissue anisotropy, and the strain-dependent nature of stress relaxation [14]. The model was able to represent the increase in stress relaxation with strain for rat tail tendon fascicles strained to 0.98%, 2.00%, and 3.28%. Their phenomenological model remains in development, but is an improvement for tendons over the QLV model, where the relaxation kernel is a function of strain.

8.5 Measuring Structural and Material Properties

A load–deformation curve, such as that shown in Fig. 8.1, is obtained by securely mounting each end of a sample in a material testing apparatus and then elongating the sample until it ruptures. Stress and strain behavior can then be generated by normalizing load to cross-sectional area and elongation to original length. However, unlike metals, plastics, or even bone, tendons and ligaments cannot easily be machined into standardized testing specimens. To further compound matters, they are difficult to grip, perhaps even more sensitive to moisture and temperature than bone, and have complex internal structures that make data interpretation difficult. For these reasons and others, accurate and reproducible laboratory measurement of tendon and ligament tensile properties is notoriously difficult.

8.5.1 Storage and Testing Environments

The treatment and storage of ligament and tendon specimens and the environment in which testing is conducted are important practical considerations. Because it is not always possible to test tissue immediately after death, the issue of storage by freezing is particularly crucial. Investigations as early as 1847 reported the effects of various storage methods on the mechanical properties of tendons and ligaments, and although the data are somewhat conflicting, it would appear that freezing at −20 °C or below has little effect on their mechanical behavior [15–17]. Later, Woo et al. evaluated the effects of freezing rabbit medial collateral ligaments for up to 3 months [18]. The only parameter significantly changed by freezing was the area of hysteresis in the first few cycles; by cycle 10, this difference became insignificant. Cyclic stress relaxation, load–deformation characteristics, and energy-absorbing capacity were unchanged. Jung et al. subsequently examined the consequences of up to eight freeze–thaw (−20 to 22 °C) cycles on human bone–patellar tendon–bone preparations and found little measurable effect on the structural properties of these constructs [19].

Studies have shown that storage of specimens at room temperature has time-dependent effects that are amplified in small specimens. Davison noted that the breaking strength of rat tail tendon decreased with time [20]. Because this could be

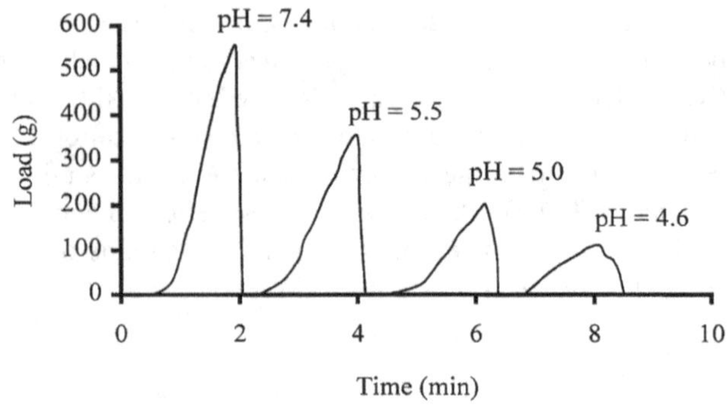

Fig. 8.9 Rat tail tendon strength declines with diminishing pH [20]

prevented by immediately washing the harvested rat tail tendons with enzyme inhibitors and keeping them on ice, Davison concluded that the collagen molecules were being subjected to gradual enzymatic degradation. An acidic environment (pH below 6) can also induce changes in the collagen molecule and lead to decreased material strength. Figure 8.9 shows the breaking strengths of rat tail tendons after being soaked in buffers of varying pH for 30 min at 0 °C.

As we have learned, biologic tissues are viscoelastic with mechanical behavior dependent on fluid pressures and flow. Not surprisingly, numerous studies have shown the mechanical behavior of these tissues to vary as a function of hydration [21–24]. It is therefore important to keep test specimens immersed, or at least saturated, in physiologic testing fluids throughout experimentation. The fluids most frequently used are physiologic saline, phosphate buffered saline and Ringer's solution. The former is a 0.9% (g of solute/ml of solvent) solution of sodium chloride. This ionic concentration of salt is termed isotonic because its osmotic pressure is the same as that within the intracellular fluid of most cells. Cells remain intact and normally shaped when bathed in this fluid. Cells bathed in hypotonic solutions of lower ionic concentration take on water, swell, and eventually burst, releasing hydrolases and other potentially harmful enzymes into the extracellular matrix. Ringer's is a more elaborate isotonic electrolyte solution that closely matches the electrolyte concentration of the blood. It contains small amounts of potassium and calcium in addition to sodium and chloride.

Because the viscoelastic behavior of tendons and ligaments is temperature dependent, good experimental technique dictates careful control of temperature as well as hydration during laboratory testing. Body temperature best simulates the in vivo situation. Substantial heat can be generated in cyclically loaded tendon. In the laboratory situation, the amount of energy stored as strain energy or released to the environment as heat is dependent on the heat conduction characteristics and temperature of the test system. It is interesting to note that Wilson and Goodship measured temperatures in the superficial flexor tendons of racehorses greater than 44 °C and postulated heat-induced cell death as a possible cause of tendon degeneration, a frequent problem in the equine athlete [25].

In should be clear from the foregoing discussion that testing conditions are important and should be closely examined when conduction (and reviewing) work. Discrepancies between experimental results can often be explained by the different conditions under which the materials were evaluated. Obviously, conditions closely matched to the biologic environment are best able to reproduce the normal behavior of these tissues within the living body.

8.5.2 Measuring Cross-Sectional Area

Derivation of stress–strain behavior from load-elongation data requires knowledge of specimen cross-sectional area (CSA) and resting length (L_o). In testing rigid materials these parameters are easily determined and controlled, but the viscoelastic nature of tendon and ligament makes accurate measurement of CSA and L_o difficult. A variety of techniques have been developed to establish the cross-sectional area of ligaments or tendons, including measurement of specimen weight or volume per unit length, physical measurement using a micrometer, shadow amplitude contour reconstruction, laser reflectance, ultrasonographic measurement, and computer digitization of cross sections, photomicrographs, or inkblots. Significant systematic differences between these methodologies have been demonstrated [26–28]. These systematic differences are often responsible for the considerable variation in absolute tissue stress and modulus values reported across studies.

If the density of the tissue is known, one can estimate the specimen's mean CSA by dividing its weight by its length and density. Alternatively, the specimen's volume can be determined by water displacement and divided by length to obtain mean cross-sectional area. These gravitational methods can be very reproducible but are generally useful only with long specimens of uniform dimensions. If a circular or elliptical cross section can be assumed, an alternative approach is to simply measure its diameter(s) with calipers at the site of interest and calculate the area. As might be expected, this technique produces more variable results. A more reproducible approach deforms the cross-sectional shape of the sample into a rectangle by pushing the segment of interest into a slot of known width and closing the slot with a plunger conforming to the dimensions of the slot. A standardized force is applied through the plunger and over time the specimen conforms to the rectangular geometry of the slot, after which height is measured and the area calculated as height × slot width. This method of deriving the cross-sectional area of a specimen yields reproducible values but may introduce substantial systematic error. Clearly, the result will depend on the force used to deform the ligament or tendon into a rectangular cross-section, the tissue's normal architecture, and the time between force application and height measurement. If these variables are carefully controlled, the method is convenient and can produce excellent results.

Shadowing amplitude contour reconstruction, a fairly involved technique that requires a collimated light source, a detector, and a mechanical apparatus to rotate the sample, may give a more precise measure of specimen CSA; however, it is less reproducible than other techniques, and the apparatus is expensive to fabricate. Laser micrometer systems [27] and newer laser reflectance systems [28, 29] offer the advantage of nondestructive measurement on intact specimens and can also be used to measure CSA while the structure is loaded. Ultrasonography should be included in the list of reliable cross-sectional area measurement techniques and can be used for both in vitro and in vivo measurement [30]. And finally, computer digitization of histologic cross-sections taken from the test specimen after testing is accurate and reproducible but labor intensive and incompatible with destructive testing.

8.5.3 Gripping Tissues

The mechanical testing of tendons or ligaments requires that they be grasped at the ends for mounting in the testing apparatus; this is particularly difficult because these tissues are slippery. In addition, the viscoelastic nature of these tissues causes the specimen to deform over time and loosen within clamps, leading to slippage at the interface when loads are applied. Adhesive compounds, cyanoacrylates, have been found to work well for small slips of rat tendon but do not work for larger samples. A variety of methods has been employed for tensile testing larger tendons and ligaments. These include looping the specimen around a post with or without suture augmentation, use of grips with roughened or serrated surfaces, use of bone–tendon–muscle–tendon–bone preparations, and freezing the tendon into the clamps.

Bone–tendon-muscle–tendon–bone (or bone–ligament–bone) preparations are an excellent choice if sub-failure characteristics are to be determined. The bone affords a reliable grip site that can accommodate high loads. Much has been learned about ligamentous behavior using the rabbit medial collateral ligament (MCL) and methods developed by Woo et al. [31]; see the upper panel of Fig. 8.10. However, the method is limited to bone–ligament–bone preparations and failures often occur near the insertions rather than in the mid-substance of ligament. When testing muscle–tendon–bone preparations, the muscle fibers will fail at tendon strains greater than 4% or 5%, which precludes complete recordings of the tendon stress–strain relationship. In either case, appropriate orientation of the bony ends of the specimen is critical to achieve a physiologic application of stress.

Riemersma and Schamhardt were the first to use cryogenic clamps to grip these tissues during testing [33]. In a modification of this method (lower panel of Fig. 8.10), liquid nitrogen flows through hollow chambers in the clamp, causing the specimen to freeze thereby forming a rigid interlock with the serrations of the clamp [32]. Because the mechanical properties of tendons and ligaments are sensitive to temperature, it is important to have a long specimen so that the test region is well away from the clamps and the nitrogen. Investigators have employed other versions of this technique with success [34, 35].

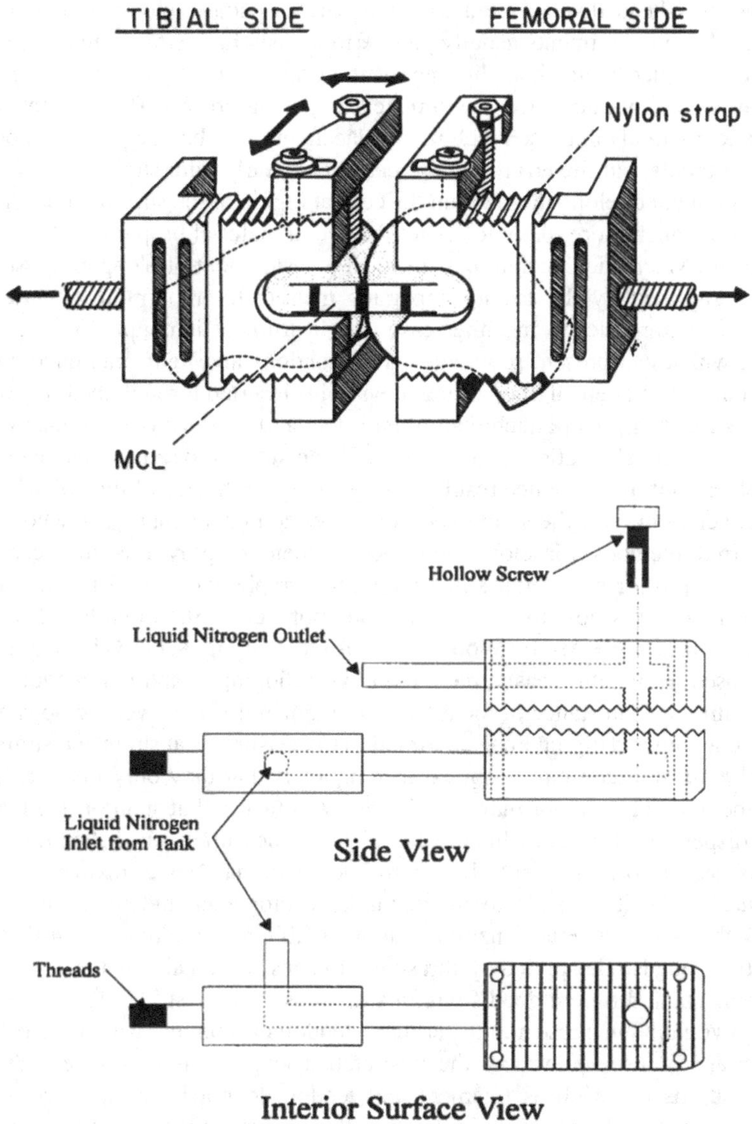

Fig. 8.10 Top: Grips for holding tibia and femur to test the medial collateral ligament (MCL) [31]. Bottom: Apparatus for gripping tendon or ligament tissue by freezing it within the clamp [32]

8.5.4 Assessing Elongation

Methodologies to measure sample elongation can be grouped into two broad categories: measuring the displacement of the testing machine crosshead and measuring changes in a defined gauge length on the test specimen itself. The latter can be done mechanically or by using optical, radiographic, or other imaging techniques.

Using crosshead displacement as a measure of sample elongation is the least preferred. Such experiments usually yield erroneously high strain values because of extraneous displacement (i.e., slipping) at the grips. Bone–ligament–bone preparations can reduce but not always eliminate this testing artifact. Because the error is systematic, methods that use machine displacement may be adequate for comparative experiments, but the error may increase nonlinearly with strain.

Measurement of elongation within the central region of the specimen, far from the grip sites, is much more likely to reflect the actual material properties of the tendon or ligament. Mechanical techniques for local measurement of elongation use *extensometers* supplied by the testing apparatus manufacturer or more generic linear displacement transducers. In either case, the instrument is coupled to the sample directly, which can present problems. Many tendons and ligaments have complex fiber geometry or a helical fiber arrangement; pins inserted through the body of these structures for instrument attachment often rotate and produce out-of-plane motions that hamper axial elongation measurement. Using sutures for attachment can reduce this problem but may produce results that reflect the behavior of the collagen fibers in the direct vicinity of the suture and not the tendon or ligament as a whole.

Preferred methods for elongation measurement employ imaging techniques. Cinematography or sequential still frame photographs were used in the past, but digitized imaging is now the more common approach [36]. An early schematic of this technique as presented by Woo et al. is provided in Fig. 8.11 [31]. An alternative method used for in situ measurement employs radiography and radiopaque spheres placed within the substance of the tendon or ligament [37]. Newer methods employ speckle tracking and digital image correlation to measure local strain and strain fields [14, 38]. One shortcoming of optical techniques is that they only measure surface strain; the internal behavior may be different. Another is that stain or dye markings tend to disperse under load, which makes distance measurements difficult. Recorded behavior may also be influenced by surrounding fascia. Newer texture correlation techniques make it possible to avoid markers altogether and measure strain by tracking the displacements of natural features within the structure [39, 40]. Texture correlation can also be applied to ultrasound images, which allows for in vivo strain measurements at different depths within structures of interest [41, 42].

When you read in a research paper that a tendon or ligament was stretched 5%, it is important to also know how the unstretched length was designated. Extracted collagenous tissues such as ligament and tendon do not have an obvious gauge length, i.e., resting length prior to loading that is subsequently used to determine strain. The reason is that their length depends not only on current loads but also on past loading history. Several conventions are used to handle this problem. One is to apply a small load to the structure for a fixed period of time and to arbitrarily call the resulting length the gauge length. An alternative method is to extrapolate from the linear portion of the load–deformation curve to the point at which the load goes to zero and refer to the length intercept at that point as the gauge length.

A closely related problem is the determination of tendon or ligament lengths in vivo and in situ. For both ligaments and tendons, this depends on body position, joint angles, and muscle activity. For example, strains in the ulnar collateral ligament

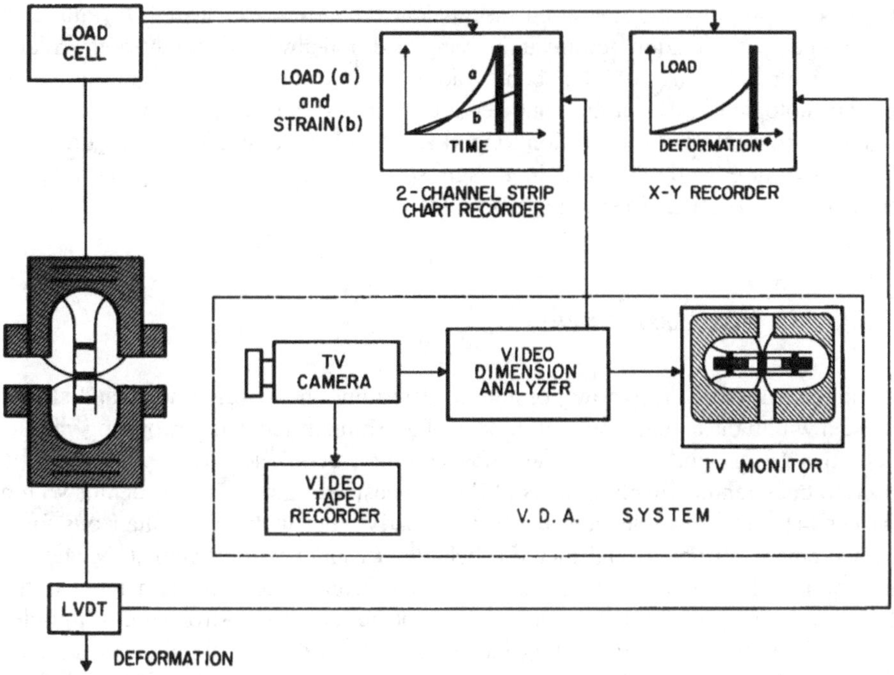

Fig. 8.11 A video dimension analyzer (VDA) system used to videographically record tendon deformations and plot load–deformation curves [31]

of the elbow joint are diminished when the joint is near full extension if the flexor muscles are active. In tendons, the relationships between body position, muscle activity, tendon compliance, and the architecture of the muscle–tendon construct are extremely complex. This is particularly true for tendons that traverse two or more joints. In addition, because of the viscoelastic features of these structures, cyclic loading alters resting length. Finally, we must keep in mind that there is considerable individual variability in the innate tightness of ligaments; some individuals have very lax ligaments and hypermobile joints, whereas others have very tight ligaments.

8.5.5 Assessing Architecture

Although the collagen fibers in tendons typically are parallel, those in ligaments often are not. The pathways from origin to insertion of particular fibers are such that, at any given angle of joint rotation, certain fibers are stressed more than others. In testing a bone–ligament–bone preparation, orientation of the bones or the bone plugs to which the ligament is attached becomes critical. If only one particular joint angle is simulated, the characteristic behavior of the ligament at other angles of the joint will not be seen. Moreover, introduction of any extraneous twist or malalignment of the tissue may alter the measured mechanical properties. For testing to failure of either ligament or tendon, failure at the bony insertion always raises the concern that slight malpositioning of the bony block in the apparatus has resulted in nonphysiologic stresses at the bone–tendon or bone–ligament junction. Because of ligament nonuniformity in relation to joint angles, elongation in loaded ligaments is best in two or even three dimensions using speckle techniques or image correlation so that local variations in strain can be assessed.

8.5.6 In Vivo Assessment

As indicated above, it is now possible to determine the displacements and strains present within the collagenous soft tissues of live human subjects performing various activities. Ultrasound, the safest and most common technique, has been successfully used to track tendon displacements [42] and measure strain at various depths within tissues [41]. These studies are indeed informative, but without simultaneous force measurement, structural and material behaviors cannot be determined. Maganaris overcame this limitation by measuring strain from sonographic images of the tendons (and aponeuroses) of the tibialis posterior and gastrocnemius muscles taken while maximally stimulating the respective muscle and simultaneously measuring ankle joint moments with a dynamometer. Force was then indirectly derived by dividing the measured joint moments through by the tendon moment arms measured on magnetic resonance images. Sonography was used to measure tendon cross-sectional area. Force–displacement and stress–strain relationships were thus derived [43, 44]. The in vivo results calculated by Maganaris and colleagues are in close agreement with data obtained in benchtop experiments, with functional strains for tendon in the 3–5% range, Young's modulus values of 0.9–1.4 GPa, and hysteresis values of 14–22%.

Several different in vivo transducer designs have been employed to measure the forces transmitted by the skeletal soft tissues during normal activities. All designs work on a similar principle in that they deform as axial load is applied to the tendon or ligament to which they are attached. Buckle transducers, implantable force transducers, and fiberoptic cables have all been used to measure tendon and ligament forces in humans. All of these devices require surgical implantation with attendant risk to subjects. And once implanted, the transducers require calibration, which is usually achieved indirectly by measuring joint moments and calibrating the transducer to the theoretic forces derived by dividing the moments obtained by the moment arm through which the tendon works. Once calibrated, forces within the structure during normal activities can be measured.

Forces in the anterior cruciate ligament [45, 46] and patellar tendon [47] have been successfully measured with implanted transducers, but functional forces in the Achilles tendon have been the primary focus of in vivo work because of its important functional role in locomotion and relatively easy accessibility [48–51]. Although not without technical challenges, the fiberoptic transducers have been used in several such studies with good success; they are minimally invasive, easy to place, and inexpensive. Finni and colleagues used fiberoptics to study eight normal subjects and reported peak Achilles tendon forces during walking of 750–2360 N [52].

8.6 Functions at Junctions

Interfaces between tissues with distinctly different properties are complicated and require specialized properties. The musculoskeletal system has two examples of such junctions, from muscle to tendon and from tendon to bone.

8.6.1 The Myotendinous Junction

The transition from muscle to tendon occurs at the *aponeurosis*, a flat, thin sheet of collagenous material that gradually narrows and thickens to form tendon. Individual muscle fibers attach to these tendinous extensions both proximally and distally. At each junction, the force developed by the intracellular contractile apparatus of the fiber is transferred to the extracellular matrix of the tendon. The junction site is characterized by extensive membrane folding, which increases the surface area available for force transmission and thus decreases the junctional stress. The interdigitation also enables more effective load transmission because the interface is subjected primarily to shear stress rather than tensile stress. In fact, the angles at the junctions between muscle fibers and tendon collagen are usually close to zero, which greatly enhances junctional strength. Increases in the junctional angle may be associated with disuse atrophy and junction failure [53].

Binding of the muscle fiber to the tendon collagen apparently occurs at discrete sites along the interface [54, 55]. Subsarcolemmal "densities" have been identified within skeletal muscle cells at these sites. The cytoskeletal proteins *alpha-actinin*, *vinculin*, and *talin* and the adhesive proteins *integrin* and *vitronectin* have all been implicated in the load transmission process, but the fields' understanding of myotendinous junctions is far from complete.

Several studies of musculotendinous preparations have shown that muscle strain injuries usually occur at or near the junction of muscle and its tendon of origin. Lieber and associates, using frog semitendinosis muscle–tendon–bone preparations, demonstrated that during passive stretch the aponeurosis was significantly more compliant than both the tendon and the bone–tendon junction [56]. Maganaris measured very similar behavior in the tibialis anterior aponeurosis of live human subjects [43] (Fig. 8.12).

In another study using frog semitendinosis units, Tidball et al. reported large increases in failure load (30%) and energy absorption (112%) in stimulated (actively contracting) preparations as compared to nonstimulated controls [57]. If the muscle–tendon unit is strained to failure with the muscle stimulated, most fibers fail at the

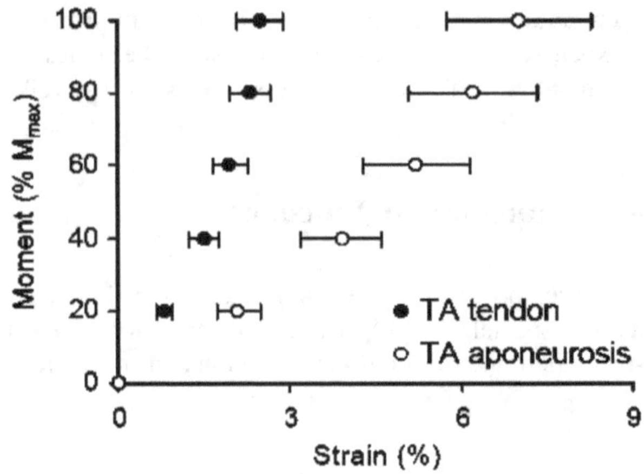

Fig. 8.12 Compliance of the tibialis anterior (TA) tendon and aponeurosis measured in live human subjects [44]

myotendinous junction, whereas failure in nonstimulated muscle–tendon preparations occurs within the inactive fibers a short distance from the junction.

Most studies of aponeurosis mechanics, whether under passive, active, in vivo or in vitro conditions, have only characterized strains in the longitudinal direction, i.e., along the trajectory of its emanating tendon. However, it has been shown that the aponeurosis also expands in the transverse directions, which makes sense given that muscle fibers surround and attach on opposite sides of these tendinous sheets and muscles bulge upon contraction [58–60]. Azizi and Roberts used high-speed fluoroscopy to measure biaxial strains within the aponeuroses of actively contracting gastrocnemius muscles extracted from turkeys [61]. Transverse strain within the aponeurosis was found to be 4× greater than longitudinal strains. More important in terms of functional significance was the fact that longitudinal stiffness varied as a function of transverse strain, thereby providing a mechanism to modulate overall stiffness of the muscle–tendon unit, which in turn modulates energy storage and recovery capacity, factors we briefly explore in forthcoming sections.

8.6.2 Ligament and Tendon Insertions to Bone

Tendons and ligaments attach to bone at *insertions* or *entheses*. Insertions are generally classified as *direct* or *indirect,* but they are anatomically and functionally diverse, constituting a continuum of morphology with each site uniquely suited to its location and function. Direct insertions, also referred to as *fibrocartilaginous entheses*, often meet the adjoining bone at right angles, and have four microscopic transitional zones of varying thicknesses [62]. Zone 1 is represented by the normal collagen and fibroblasts of tendon or ligament; zone 2 is made of fibrocartilage and contains cells with morphology similar to chondrocytes; zone 3 is mineralized fibrocartilage; and zone 4 is bone. Between zones 2 and 3 is a mineralization

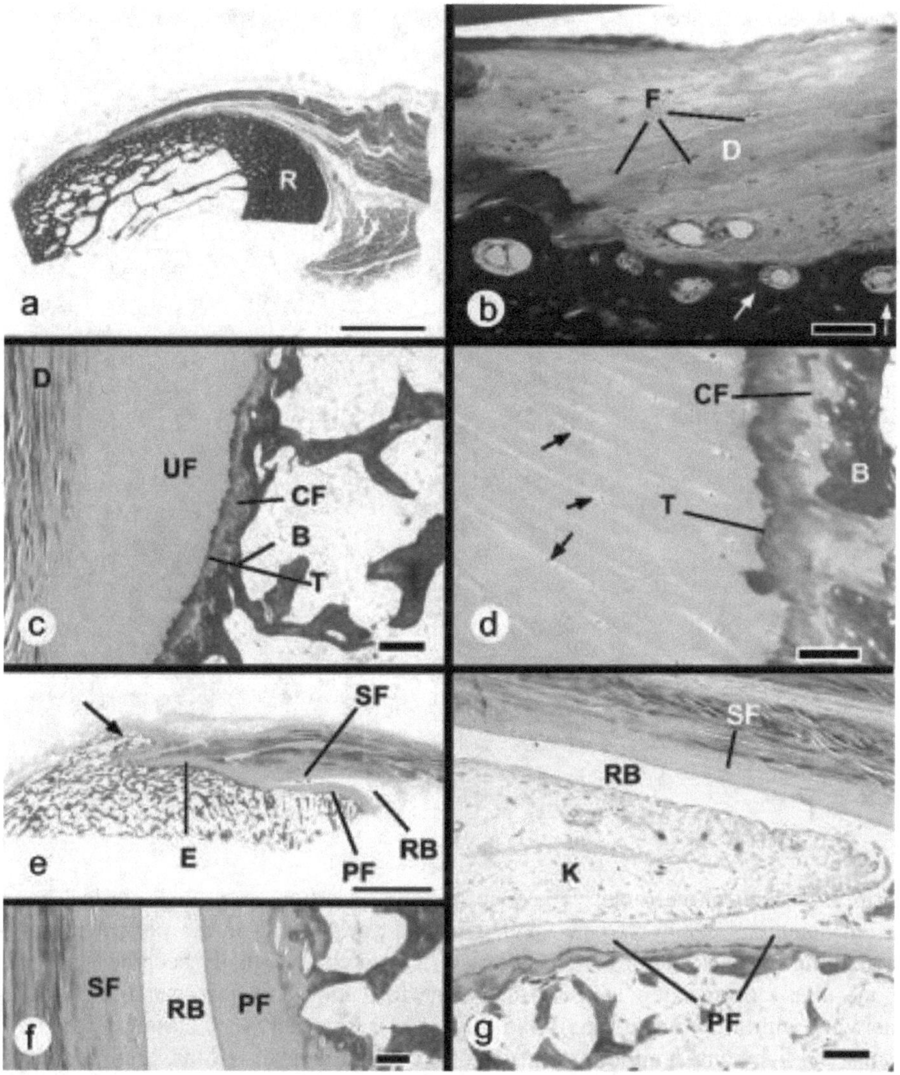

Fig. 8.13 Low [Left] and high [Right] magnification photomicrographs of a direct (fibrocartilaginous) insertion for the Achilles tendon. The four zones of tissue include dense fibrous connective tissue (D), uncalcified fibrocartilage (UF), calcified fibrocartilage (CF), and bone (B). The irregular mineralization tidemark (T) is shown [62]

tidemark (Fig. 8.13). The transition from tendon or ligament to bone typically ranges from 300 to 700 μm across.

Collagen fibers at indirect or *fibrous* insertion sites approach the bone at an acute angle and do not have a fibrocartilaginous zone of transition. However, there is a mineralization tidemark or demarcation between the tendon or ligament and the bone. At indirect insertions, more of the fibers blend with the periosteum, which is

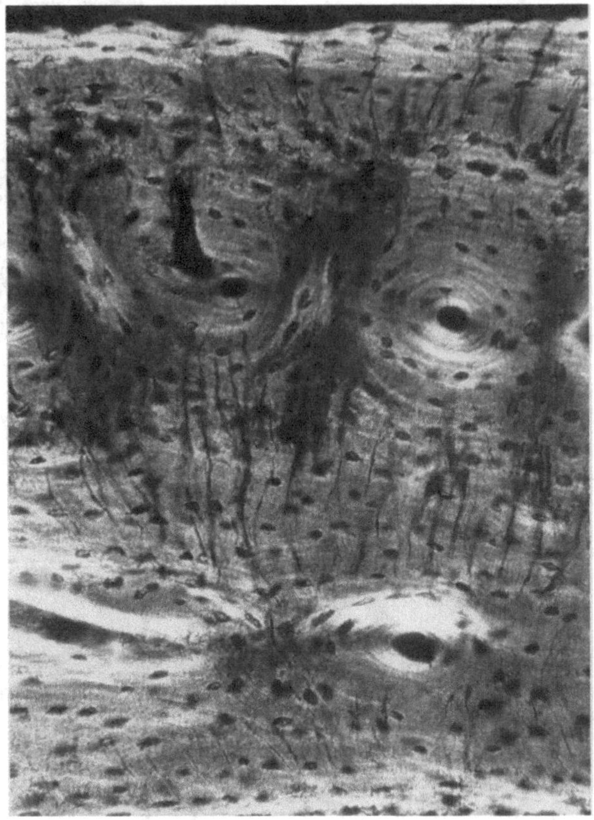

Fig. 8.14 Sharpey's fibers appear as *dark lines* radiating inward from a tendonous attachment at top of this photomicrograph of a canine radius. These fibers have become embedded in the bone matrix as a result of ongoing periosteal apposition during growth. Note how they have also been interrupted by the formation of secondary osteons. Basic Fuchsin-stained, undemineralized cross section; field width, approximately 0.5 mm

firmly attached to the underlying bone. Fibrous insertions into thick cortical bone often display *Sharpey's fibers*—collagen fibers originating in the periosteum and extending into peripheral lamellar bone. These fibers eventually become buried in bone matrix during new subperiosteal bone formation and form a radial collagen network within the underlying bone (Fig. 8.14). Irrespective of classification, tendons and ligaments tend to broaden and flare out at their insertions, often blending with other tendons, ligaments, and collagenous structures, which serves to distribute load, reduce stress, and increase overall capacity. It is also worth noting that tendon insertions are nonuniform and a single insertion (e.g., Achilles) can in fact be purely fibrous in one region and fibrocartilaginous in another. Regional differences in tissue structure and composition are strongly indicative of regional differences in the local stresses.

Failures near insertion sites due to mechanical overload are common clinical injuries. More susceptible sites of injuries include the combined *entheses* of the extensor carpi radialis brevis muscle and the lateral collateral ligament of the elbow (producing a condition known as tennis elbow), the anterior talofibular ligament (the most common site of ankle sprain), and the Achilles tendon. Stouffer et al., in a study

Table 8.1 Failure modes of the medial collateral ligament as a function of age

			Failure modes, %		
	Age, months	No.	Tibial avulsion	Mid-substance	Femur fracture
Growing animals:	1.5	11	100	0	0
	4–5	20	100	0	0
Adult animals:	6–7	18	22	67	11
	12–15	16	0	56	44

Data from [18]

using human patellar tendon–bone units, concluded that bone–tendon strains were much larger than those measured within the substance of the tendon [63]. Butler and colleagues also reported high tissue strains near the insertion sites of human patellar tendon and anterior and posterior cruciate ligaments [64]. Damage at insertion sites usually occurs in the underlying bone (*avulsion fractures*) or in the tendon tissue adjacent to the junction rather than at the junction itself. Woo et al., using rabbit femur—medial collateral ligament—tibia complexes, showed that the specific failure location within these sites is age-dependent [18]. Young growing animals were much more susceptible to avulsion of the ligament from the tibia, whereas older animals were more likely to fail by mid-substance tearing of the ligament (Table 8.1). In still older animals, femur fractures often occurred before any type of observable ligamentous failure. One thought is that this can be explained in part by the age-related increases in cross-linking of collagen within connective tissues.

It is often asserted, and some studies have confirmed, that failure occurs within the substance of the ligament at high strain rates, whereas at lower strain rates bone avulsion at the attachment site is the usual mode of failure. However, other studies have failed to confirm these findings. Several studies have shown that joint angle and loading direction have profound influences on failure mechanics at insertion sites. This experimental consideration is important when evaluating the biomechanical properties of tendons and ligaments in the laboratory.

8.7 Functional Adaptation and Specialization

Ligaments and tendons demonstrate tremendous variation in mechanical properties that reflect the functional demands placed upon them. For example, the ligamentum flavum connecting adjacent vertebra of the human spine functions to protect the spinal cord and keep it from protruding dorsally. This structure functions over a large range of intervertebral angles and must be highly extensible to effectively stretch during intervertebral flexion while not bulging and impinging on the spinal cord during extension. Nachemson and Evans showed that the ligament can withstand strains exceeding 60% without rupture [65]. Figure 8.15 presents load–deformation curves for human anterior cruciate (top) and ligamentum flavum (bottom) ligaments. The ligamentum flavum has a very low elastic modulus (0.1 GPa) compared to most

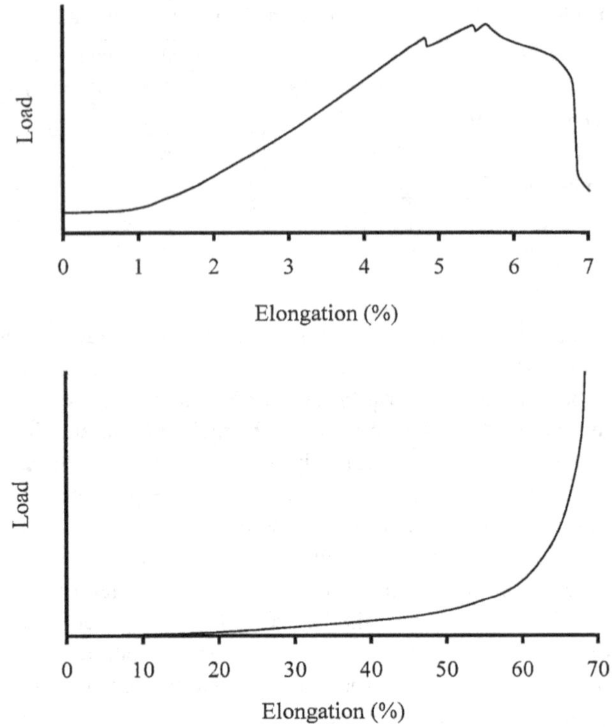

Fig. 8.15 Load–deformation curves for human anterior cruciate [Top] and ligamentum flavum [Bottom] ligaments. Note the tenfold difference in elongation scales

other ligamentous structures. At the other extreme, the elastic modulus of calcified turkey leg tendon approaches that of bone (10 GPa), presumably because it is under constant high stress from the continually flexed position of the turkey's lower limbs.

Specialized function is achieved by physiologic optimization of matrix composition, collagen fiber orientation, and myotendinous and insertional arrangements. Genetic coding undoubtedly controls tissue development; however, it is clear that the expression of this information is affected by local mechanical conditions. Fibroblasts have been shown to be responsive to strain in vitro, and many studies have demonstrated enhanced healing in tendons subjected to active or passive motion. It has been postulated that, in addition to stimulating fibroblastic activity, mechanical factors may also play a role in regulating the extracellular assembly and orientation of collagen fibrils in the matrix.

The mechanical characteristics of tendons and ligaments are dictated by the relative concentrations of collagen, elastin, proteoglycan, and, occasionally, hydroxyapatite (mineral), which may not be uniform along the structure's length. By way of example, the elastin content in the ligamentum flavum described earlier approaches 70%, compared to only 10% in most ligaments, and even less in tendons. As shown in Fig. 8.15, the ratio of collagen to elastin is a strong determinate of mechanical behavior. Calcification also correlates strongly with increases in structural stiffness, as shown by Leonard et al. in the turkey tendon [66].

The importance of achieving the right functional balance of compositional constituents cannot be overemphasized. Short-term energy storage and return within the Achilles during running is perhaps the most widely cited example of tendon's important mechanical contributions to efficient musculoskeletal function. Nevertheless, given their passive nature, it is easy to overlook other critical, albeit more subtle, contributions that tendons make to efficient muscle function during everyday activities. Two examples are the tendons of flexor hallucis longus and flexor digitorum longus, both of which originate from muscles deep in the posterior shank, traverse several joints along their path and function to flex the toes and counterbalance ground reaction forces during the push-off phase gait. Experimental simulations of normal walking strongly suggest that the material properties of these two tendons are precisely tuned to transmit appropriate time-varying resistive forces to the toes while under constant highly efficient isometric contraction, thereby minimizing energy expenditure [67, 68] (Fig. 8.16). In this case, force development in the tendon under constant muscle length is purely a consequence of the rotations in the joints these tendons traverse, making proper force transmission to the toes entirely dependent on tendon stiffness.

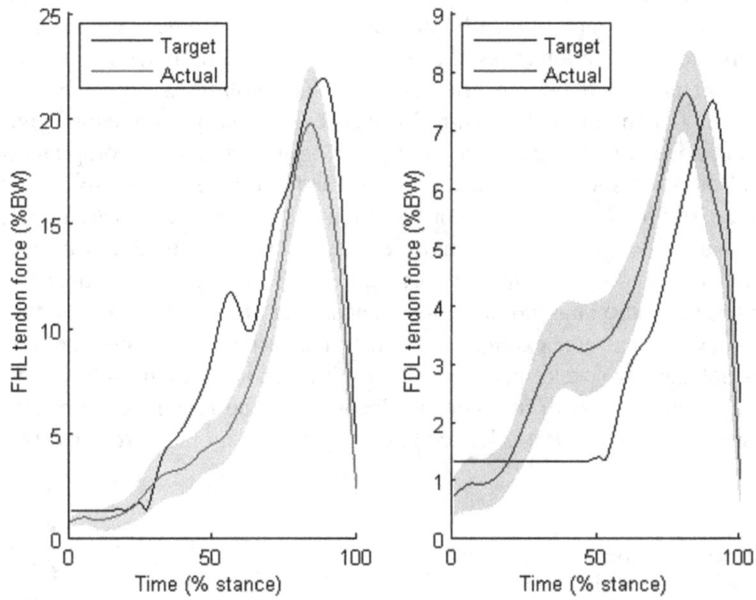

Fig. 8.16 Data from cadaver simulations of normal gait showing time varying forces in the tendons of flexor hallucis longus (red/left) and flexor digitorum longus (blue/right) under constant isometric muscle contraction. Muscle forces estimated from electromyographic recordings of normal subjects (black lines) are included for comparison (BW = body weight) [67]

8.7.1 Determinants of Tendon Architecture

Tendon architecture is determined by considerations of stiffness as well as strength. In thinking about the various factors that govern tendon architecture, it is useful to consider a simple model of tendon action: a loop consisting of a bone, a hinge joint, a second bone, a muscle attached to the first bone, and the tendon of the muscle attached to the second bone so that contraction of the muscle flexes the joint (Fig. 8.17).

Contraction of the muscle exerts a force (F_1) on the tendon that causes rotation of the joint and, in turn, enables the effector end of the second bone to exert a force of some kind on the environment (F_2). The relationship of the two forces, F_1 and F_2, depends in part on the relative moment arms of the tendon insertion and the effector point (with respect to the joint center of rotation). The relationship of the two forces also depends on the compliance of the tendon as a structure, which is, in turn, determined by its elastic modulus, cross-sectional area, and length. If we assume for the moment that the tendon is perfectly elastic, then if it were twice as thick it would stretch only half as much at a given force, resulting in greater motion at the joint for a given level of muscle contraction. Similarly, if it were half as long, a given level of muscle motion would result in greater motion at the joint. From the perspective of an organism that must adapt efficiently, the variables which can be manipulated include muscle architecture, tendon cross-sectional area, tendon length, tendon moment arm, and moment arm of the force effector. From an evolutionary perspective, these variables may change together, with different evolutionary "problems" giving rise to different adaptive solutions. For example, if you have a long, compliant tendon, you could make it shorter or thicker (if greater forcefulness were sought), or you could increase fiber length (excursion) of the muscle if precision were more important, or you could increase both muscle fiber length and tendon stiffness while decreasing the tendon insertion moment arm if speed were the critical factor. In some cases, adaptation has led to tendons that are very robust compared to the size of their muscles. An extreme example is the massive suspensory ligament of the horse, which is not, strictly speaking, a ligament, but the tendon for a muscle so tiny that it appears as a reddish fuzz of rudimentary fibers. This tendon functions principally for energy storage during gait; it is loaded primarily by gravity rather than by its muscle.

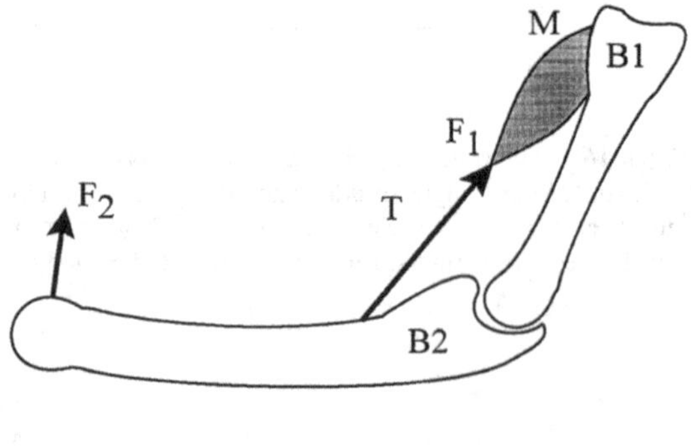

Fig. 8.17 Sketch of a simple model for tendon mechanics. Bones B1 and B2 are articulated at a hinge joint; muscle M pulls on tendon T with force F1, flexing the joint. The distal end of bone B2 produces force F2 against an object in the environment

At the other extreme, some of the human finger flexor muscles have extremely long and compliant tendons; these are adapted to maximize precision of finger motion at the expense of force or elastic energy storage (which is not a factor in the fingers).

8.7.2 Sites of Tendon Compression

Proteoglycan is another component of tendons and ligaments that increases under special conditions. High proteoglycan content in tendons or ligaments produces a fibrocartilaginous tissue. Localized areas of *fibrocartilage* are found in regions of tendon that wrap around and press against bone. Recall that proteoglycan is a major constituent of cartilage and is the molecule most responsible for this tissue's ability to withstand compression. Fibrocartilaginous tissues contain specialized cells that show a gradient of phenotypic expression, ranging between fully differentiated chondrocytes and fully differentiated fibroblasts. These cells produce large amounts of proteoglycan and collagen. (The mix of different collagen types is also varied.) Cell shape is reflective of extracellular matrix composition, and a "*fibrochondrocyte*" is more spindle-shaped in areas of low proteoglycan (such as tendons or ligaments) and more rounded in areas containing large amounts of proteoglycan (like cartilage).

An interesting mechanical theory of tissue differentiation based on localized stress and strain conditions has been proposed by Giori and coauthors [69]. It is postulated that cells produce an extracellular matrix that varies according to the balance between distortional and hydrostatic stresses imposed on them. In this case, however, the theory addresses the balance between cartilage and tendon rather than the calcification of cartilage. A graphical representation of the theory is presented in Fig. 8.18. The horizontal axis represents the amount of compressive or tensile hydrostatic stress, and the vertical axis denotes the amount of distortional strain, at a point in the tendon. It is postulated that cartilaginous tissue is produced in an environment dominated by compressive hydrostatic stress and that fibrous tissue occurs in an environment dominated by tensile hydrostatic stress. Fibrocartilage, such as that seen in tendons that pass around and are forced against a bone, is produced in an environment with high compressive hydrostatic stress combined with high distortional strain. Fibrous tissue is maintained in a situation that combines high distortional strain with tensile hydrostatic stress.

To find out which of these situations occurs in regions of a tendon, the authors developed a two-dimensional finite element model. They assumed tendon tissue to be a two-phase, linear, elastic fiber-reinforced composite consisting of longitudinally oriented collagen fibers and an extrafibrillar matrix. Their model did not iteratively vary the tissue properties according to the local stress history, as the Carter–Wong models did, but simply studied the distribution of the stresses in a tendon wrapped around and forced against a cylindrical bone. Figure 8.19 shows typical distributions of hydrostatic stress (left) and distortional strain (right) in the finite element model. The distortional strain is relatively high throughout the tendon and greatest along the

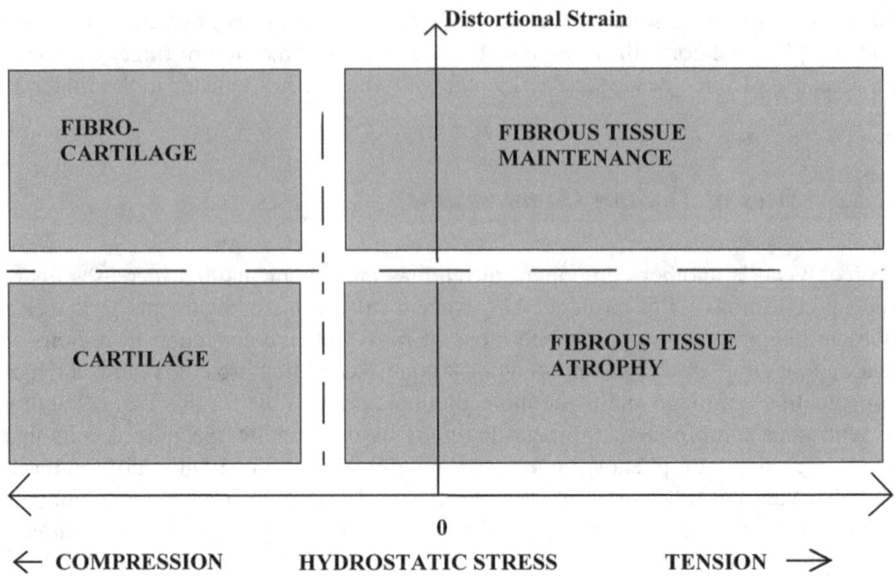

Fig. 8.18 Hypothesis for the effects of stress and strain on tissues produced by cells in a tendon. The hydrostatic portion of the stress is plotted on the abscissa, with compression to the left and tension to the right. Distortional (or shear) strain is plotted on the ordinate. Depending on the balance between these hydrostatic and distortional influences, the cells are postulated to produce the kinds of tissue shown [69]

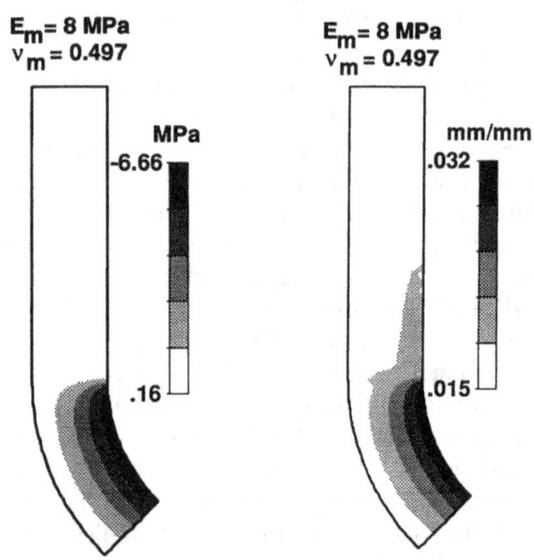

Fig. 8.19 Hydrostatic stress [Left] and distortional strain [Right] in a finite element model of a tendon passing around a cylindrical bone. The elastic modulus and Poisson's ratio values used in the model are shown *above* and the scales of shading used to depict stress results to the *right* of each figure. The bone, which would be adjacent to the *dark region* in the tendon, is not shown, and only the upper half of the tendon appears because the model utilizes the symmetry of the situation [69]

surface contacting the bone. Thus, the tendon's mechanical situation is entirely in the upper half of Fig. 8.18. The hydrostatic stress is high and compressive (negative) where the tendon presses against the bone; it is slightly positive elsewhere. The region in contact with the bone is in the "fibrocartilage" zone because its hydrostatic stress is negative. The other portions of the tendon are considered to be in the "fibrous tissue maintenance" zone because their hydrostatic stresses are slightly positive. Therefore, the mechanical situation in the tendon and the distribution of tissue types correspond to the biologic effects postulated in Fig. 8.18.

Matyas et al. proposed a mechanical hypothesis for tissue differentiation at direct ligamentous insertions [70]. (Recall that direct insertions contain a band of fibrocartilage.) They examined histologic sections of the femoral insertions of rabbit medial collateral ligaments and quantified cell shape as an index of tissue phenotype. They assumed that rounded cells produce a more cartilaginous material and spindle-shaped cells a more fibrous material. To perform their study, the investigators used automated image processing and analysis techniques that differentiated the margins of dark-staining cell nuclei. A one-to-one correspondence of nuclear shape to cytoplasmic shape was assumed based on the authors' preliminary work. High-magnification images of histologic slides were captured and digitized. The software automatically measured the location, area, and perimeter of each nucleus and calculated a roundness coefficient for each cell as follows:

$$\text{Roundness} = (4\pi \times \text{area})/\text{perimeter}^2$$

Using this equation, a perfect circle would have a roundness coefficient of 1. Finite element techniques were used to characterize the local mechanical environments in and about the insertion site, and the results were correlated to the histologic maps of cell roundness. Corresponding gradients of cell shape and principal stress were noted in the fibrocartilaginous zones at insertion sites. Several mechanical properties correlated with tissue phenotype (cell roundness) (Table 8.2), providing further evidence that functional adaptation is controlled, at least in part, by the mechanical environment.

8.7.3 Flexors and Extensors

The long flexor and extensor tendons that contribute to finger motion in humans and quadrupedal gait in animals are also highly specialized structures. The flexors usually have different mechanical properties than the extensors. Shadwick showed that porcine flexor tendons were consistently stronger, stiffer, and more energy conserving than extensor tendons [71]. Thus, flexor tendons were more effective biologic springs than the extensors, which makes good sense biologically in terms of reducing the amount of muscular energy and the metabolic cost of locomotion. As the body decelerates when the foot contacts the ground, kinetic and potential energy

Table 8.2 Mechanical correlations to cell shape

Mean nuclear roundness vs.:	Pearson correlation
Principal stresses	
Compression	−0.68
Tension	−0.53
Derivatives of principal stress	
Hydrostatic stress	−0.67
von Mises stress	−0.31
Strain energy density	−0.41
Principal strains	
ε_1	−0.40
ε_2	+0.69
ε_3	−0.20
Derivatives of principal strain	
Volumetric strain	−0.67
Distortional strain	−0.31

Data from Ref. [70]

are converted to strain energy stored in the flexor tendons of the limb. Elastic recoil of the stretched tendons converts this momentarily stored energy back to kinetic and potential energy at push-off. When the tendons are more efficient at storing energy, less work is required for locomotion. Extensor tendons function differently and are not elongated during deceleration; thus, efficient energy management is less crucial in these structures. Interestingly, Shadwick also showed that the two types of tendon possess identical mechanical characteristics at birth, providing further evidence that functional demand regulates structure and mechanical behavior [71].

Tendon stiffness has been shown to increase with increasing load, largely because of the straightening of crimps in collagen fibers. Stiffness increases dramatically as the toe region of the load–deformation curve is surpassed (see Fig. 8.1); when crimps are removed, stiffness remains reasonably constant. Typically, relative muscle force at the point of transition is less than 25% of maximum. This behavior is amplified in human finger flexors because they are so long. In these structures, high compliance when muscle force is low serves to damp contractile variations, producing smooth and finely controlled motion at the fingers. When higher muscle force is applied, this damping effect is diminished, but high-force tasks usually do not require fine motor control. The greater tendon stiffness during high-load tasks (such as tight gripping) provides more efficient force transfer.

The synovial sheath that surrounds the long flexors and extensors is another important functional adaptation of these tendons. It produces and retains synovial fluid that bathes the tendon and facilitates smooth excursion with minimal friction. Without the synovial membrane, these tendons would be subjected to substantial drag because of their length.

8.8 Effects of Age on Mechanical Behavior

A number of preclinical studies have shown that the mechanical properties of tendons and ligaments change with age. During maturation, the size of the toe region of the stress–strain curve decreases and the elastic modulus and ultimate strength increase [27, 72]. Evidence suggests that three processes contribute to these changes in behavior. The first is an increase in collagen content per unit area, as reflected by a steady decline in water content during maturation. The second is a gradual decrease in collagen crimp, which would account for the reduced toe region. The third is a gradual increase in collagen cross-links, which renders the molecules stiffer. The reduced crimping may simply be a manifestation of the altered conformation of the collagen molecules caused by their additional cross-linking, and the straightening of the collagen may promote additional cross-linking. Increasing tensile strength and stiffness plateau after maturity and deteriorate during old age [17, 73].

Rats demonstrate increasing strength and modulus of the rat tail tendon during maturation [74]. Figure 8.20 shows how the elastic modulus of equine flexor tendon changes with age [1]. Note that, in the horse, tendon stiffness gradually increases and then begins to decline in older animals. Rodents may not live long enough for this to happen to their tendons or they may not use the tendons to the same degree as other species.

Data from human studies also show declining stiffness and modulus of elasticity with age [75–77]. Using ultrasonography, Onambele et al. measured the stiffness and modulus of the gastrocnemius tendon in three cohorts of healthy individuals tightly grouped about mean ages of 24, 46, and 68 years [78]. The Young's modulus of tendon in the older group was found to be 28% lower than in the younger cohort. Similarly, Stenroth and colleagues measured a 17% decline in stiffness and 32% decline in Young's modulus when comparing subjects in their 70s to those in their mid-30s [79]. The cross-sectional area of the tendon in the older subjects was 16% larger than that in the young, which partially compensated for the reduction in modulus. Interestingly, tendon stiffness was highly correlated with maximum

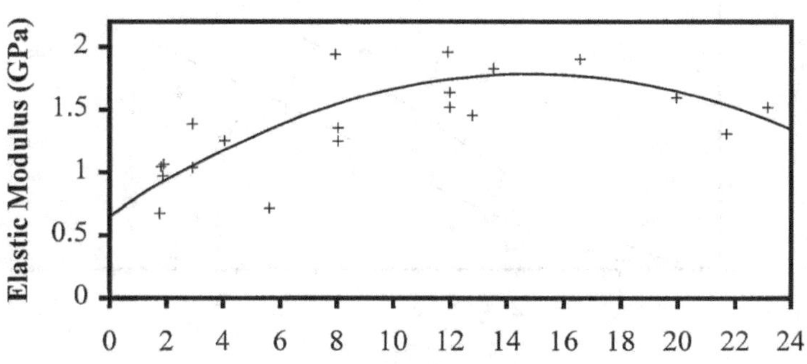

Fig. 8.20 Effects of age on the elastic modulus of the equine flexor tendon [1]

voluntary contraction of the triceps surae and there was no difference in stiffness or modulus when comparing old and young subjects having similar muscle strength. These findings suggest that the Achilles tendon adapts to the mechanical performance of the triceps surae, which was significantly smaller and weaker in the older subjects. Thus, it could be that sarcopenia is the prime driver of the mechanical and compositional changes noted in tendons and ligaments as a function of aging.

8.9 Effects of Exercise and Disuse

Skeletal tissues, including bone, tendon, and ligament, respond in predictable ways to the mechanical demands placed upon them. Specifically, it has been demonstrated that the mass and strength of bone rise and fall according to mechanical demand. To some degree, tendons and ligaments behave similarly. Woo and associates have shown that exercise increases the tensile strength of extensor tendons and flexor tendon insertions but not flexor tendons themselves [80]. Figure 8.21 compares the mechanical behavior of digital extensor tendons extracted from exercised (40 km/week for 1 year) and non-exercised swine. The improved performance of the exercised tendons is obvious. The lack of similar strength changes in the flexor tendons may be because tendons and ligaments that are regularly subjected to high loads by normal activities of daily living are already optimally designed. Perhaps the only way to increase the load-bearing capacity of these structures is to increase the cross-sectional area, which, unlike bone, would be naturally self-limiting because of geometric and kinematic constraints. This reasoning may also explain why bone–tendon and bone–ligament complexes respond to increased demand by improving

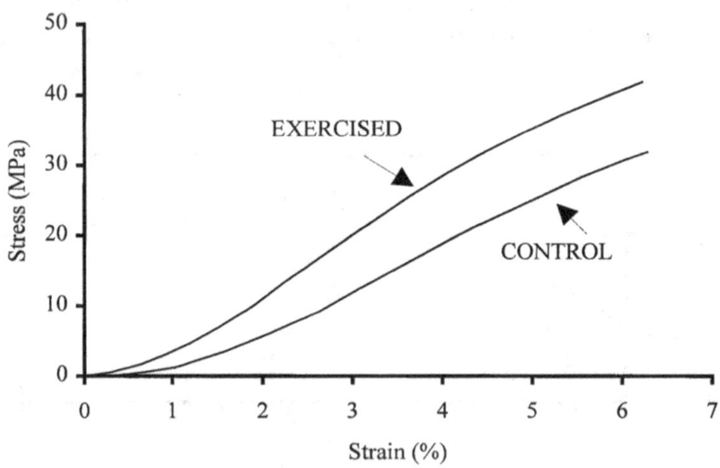

Fig. 8.21 Extensor tendon load–deformation curves for exercised and control swine. (Adapted from Ref. [80])

Table 8.3 Mechanical properties of normal, immobilized, and reconditioned rhesus monkey anterior cruciate ligaments

Parameter	Normal ($n = 30$)	Immobilized ($n = 18$)	Five-month reconditioning ($n = 22$)	Twelve-month reconditioning ($n = 20$)
Maximum load (kg)	87.33[a]	53.14[b]	68.68[c]	79.23[a,c]
	(19.40)	(12.79)	(18.35)	(17.01)
Strain at maximum load (%)	46.66[a]	43.88[a,b]	38.53[b]	42.91[a,b]
	(8.63)	(9.08)	(7.80)	(6.22)
Energy at failure (cm-kg)	34.02[a]	23.16[b]	26.53[b,c]	31.42[a,c]
	(10.72)	(5.27)	(7.19)	(8.16)
Stiffness (kg/mm)	18.49[a]	12.76[b]	17.13[b]	18.10[b]
	(3.01)	(3.29)	(3.01)	(2.77)

SD in parentheses. Values without any superscripts in common are significantly different ($p < 0.05$). Data from Refs. [82, 84]

strength at insertion sites rather than by architectural modifications of the tendon or ligament itself [80, 81]. Conversely, Laros et al. [81] and Noyes et al. [82] demonstrated in dogs and primates, respectively, that the mechanical strength at insertion sites quickly diminishes in response to disuse (produced by immobilization). In the primate study, recovery was incomplete after 5 months of reconditioning exercise (Table 8.3). Woo and coworkers also have shown inferior mechanical properties in the rabbit medial collateral ligament after disuse [83]. These studies indicate that the insertion sites may be more prone to disuse atrophy, and take longer to regain their original mechanical properties, than the ligaments themselves.

Several human studies employing ultrasound and/or MRI imaging to measure tendon elongation and cross-sectional area have confirmed the anabolic effects of training on the structural stiffness and elastic modulus of the patellar tendon [85–87] and Achilles tendon [88, 89] . In most studies, changes in cross-sectional area remain small or nonexistent while strain for a given load is reduced, suggesting that internal remodeling of the tissue rather than gross addition of material is responsible for the observed mechanical adaptations. These same in vivo methodologies also have also been used to show that human tendon stiffness and modulus decline with disuse [75–77].

8.10 Healing

8.10.1 Ligament

Ligament failure typically occurs by mid-substance tear or bony avulsion. Mid-substance failure is more common in adults and avulsions are seen more frequently in children. Ligament *sprains*, stretching the tissue beyond its elastic limit, elicit an inflammatory response that will eventually repair the compromised tissue. Ligament healing, following injury, is divided into the same three phases as bone fracture healing, but requires a much longer time. Phase I (48–72 h), *acute inflammation*, is characterized by pain, swelling, redness, and warmth from vasodilation. During this phase cells are recruited and a hematoma forms. Phase II (up to 6 weeks post injury), *proliferation*, is characterized by increased numbers of fibroblasts, collagen synthesis, and scar tissue formation. Phase III (12 months or longer) involves *remodeling* of the scar. Collagen synthesis decreases, water content drops, and the extracellular matrix becomes more organized (more longitudinal arrangement of fibers). Scar contraction coincides with increasing tensile strength. Unlike bone, the injured ligament may never regain full strength; primate studies indicate tensile strength to be 80% of the intact ligament at 1 year [90]. The reason for the lack of complete property recovery is unclear.

8.10.2 Tendon

Tendon can fail in mid-substance, at or near the myotendinous junction or at its bony insertion. Avulsion is less common for tendons than ligaments but does occur in young athletes. Tendon lacerations, particularly of the finger flexors and extensors, are common injuries. Overloading of tendon can cause damage to the collagen network, as can repetitive motion and abrasion. Diseases that compromise blood flow, such as diabetes, can cause ischemia-induced degeneration, particularly in areas with little collateral blood supply. Steroid injections have also been associated with tendon degeneration and rupture. Tendon healing follows the same pathway as ligaments but is more challenging. The length and quality of the repair process can vary considerably as the result of major differences in blood supply from site to site. In addition, *adhesions* after surgical repair are a significant clinical problem, particularly in the digital flexors/extensors. Scar formation extends into surrounding tissues, tacks the tendon down, and prevents sliding. Rehabilitation protocols employing both active and passive motions, judiciously applied after surgery, can reduce this and improve clinical outcomes.

8.10.3 Role of Motion in Healing

Injuries to mechanically regulated tissues can present difficult clinical problems. Too little loading after surgical repair of a tendon or ligament leads to inadequate tissue remodeling, low tensile strength at the repair site, and diminished overall structural strength. Too much loading will pull the repair apart or elongate the structure, resulting in dysfunction. Absence of motion can permit runaway scar formation, leading to adhesions and poor functional outcomes.

Active motion, whereby loading is gradually increased to pre-injury levels, is perhaps the best postoperative treatment after ligament reconstruction or repair of partial tendon laceration. Vailas and associates demonstrated increased collagen synthesis and strength in rat medial collateral ligaments subjected to progressive exercise during healing as compared to non-exercised controls [91]. In cases of complete tendon laceration, many studies have advocated repair techniques that place more strands of suture across the laceration site. These newer techniques increase the postoperative tensile strength of the repair so that active motion rehabilitation can begin soon after surgery. Nevertheless, overload of the repair site is a significant concern, making *passive motion* of the involved joint a popular alternative that subjects the healing tendon to modest stress and lowers the risk of gap formation at the repair site. Freehan and Beauchene, Takai et al., and others have shown that passive motion improves healing and increases wound strength compared to immobilized controls [92, 93]. Passive motion has also proven useful in healing cartilaginous injuries in experimental animal models, suggesting that the treatment might be useful for joint injuries affecting ligament and cartilage as well as tendon [94]. Although its clinical utility in humans remains controversial, a systematic review of 19 studies examining the ability of continuous passive motion to promote and accelerate healing in the rabbit model pioneered by Salter yielded convincing evidence of improved outcomes [95]. Regardless of tissue type, most animal studies and some human studies indicate improved tissue generation and organization with reduced inflammation and adhesion formation.

Flexor tendon adhesions and joint contractures continue to be problematic despite advances in surgical techniques and postoperative rehabilitation. In many instances, *tenolysis*, surgically freeing the tendon from surrounding tissue, is required to achieve acceptable function. Patients not committed to extensive postoperative rehabilitation may never regain adequate function.

8.11 Summary and Further Reading

Tendons transmit muscle forces to bones, storing and releasing energy in the process; ligaments are passive joint stabilizers. Both these structures provide proprioceptive information that is vital to all kinds of motor activity. Both are composed largely of type I collagen, with ligaments containing somewhat more elastin. Both tendons and ligaments readily exhibit disuse atrophy, and their compositions and mechanical properties seem to vary in accordance with their function; for example, flexor tendons are generally stiffer than their corresponding extensor tendons. It is hypothesized that this is controlled by cellular responses to mechanical signals. Viscoelasticity is an important feature of the mechanical properties of these tissues, so that they are sensitive to strain rate and load history. They also creep, relax, and exhibit hysteresis. Mechanical testing is complicated by their soft, slippery, wet, and viscoelastic nature, making gripping, dimensional analysis, and strain measurement problematic. When lacerated or torn, ligaments and tendons will heal, but the healed tissue may never regain full strength and tendon motion is frequently constrained by adhesions to adjacent tissues.

For further reading about structure relationships in tendons, ligaments, and their insertions we recommend two review articles by Benjamin and colleagues: *Structure–function relationships in tendons: a review* [96] and *Where tendons and ligaments meet bone: attachment sites ('entheses') in relation to exercise and/or mechanical load* [62]. An informative review article by Wang entitled expands upon the limited amount of tendon biology presented here [97]. To further explore biological aspects of tendon and ligament development and repair, we refer readers to an article by Yang and coauthors entitled: *Tendon and ligament regeneration and repair: clinical relevance and developmental paradigm* [98]. A more recent tendon review also focuses on healing [99]. And although somewhat dated, the book *Injury and Repair of the Musculoskeletal Soft Tissues*, edited by Woo and Buckwalter still provides an excellent overview of ligament and tendon tissue biology and mechanics [100]. An excellent recent review provides insight into the computational modeling of ligament mechanics [101].

References

1. Gillis C, Sharkey NA, Stover SM, et al. The effect of maturation and aging on the material and ultrasonographic properties of equine superficial digital flexor tendon. Am J Vet Res. 1995;56: 1345–50.
2. Woo SL-Y, An K-N, Arnoczky SP, et al. Anatomy, biology, and biomechanics of tendon, ligaments, and meniscus. In: Simon S, editor. Orthopaedic basic science. Park Ridge: American Academy of Orthopaedic Surgeons; 1994. p. 45–87.
3. Haut RC. Age-dependent influence of strain rate on the tensile failure of rat-tail tendon. J Biomech Eng. 1983;105:296–9.
4. Maganaris CN. Tendon conditioning: artefact or property? Proc Biol Sci. 2003;270(Suppl 1): S39–42.
5. Zajac FE. Muscle and tendon: properties, models, scaling, and application to biomechanics and motor control. Crit Rev Biomed Eng. 1989;17:359–411.
6. Lieber RL, Brown CG, Trestik CL. Model of muscle-tendon interaction during frog semitendinosus fixed-end contractions. J Biomech. 1992;25:421–8.
7. Flugge W. Viscoelasticity. Waltham: Blaisdell; 1967.
8. Fung YCB. Stress-strain-history relations of soft tissues in simple elongation. In: Fung Y, Perrone N, Anliker M, editors. Biomechanics: its foundations and objectives. Englewood Cliffs: Prentice-Hall; 1972.
9. Haut RC, Little RW. A constitutive equation for collagen. J Biomech. 1972;5:423–30.
10. Jenkins RB, Little RW. A constitutive equation for parallel-fibered elastic tissue. J Biomech. 1974;7:397–402.
11. Woo SL-Y. Mechanical properties of tendons and ligaments: quasistatic and nonlinear properties. Biorheology. 1982;19:385–96.
12. Hingorani RV, Provenzano PP, Lakes RS, et al. Nonlinear viscoelasticity in rabbit medial collateral ligament. Ann Biomed Eng. 2004;32:306–12.
13. Duenwald SE, Vanderby R, Lakes RS. Stress relaxation and recovery in tendon and ligament: experiment and modeling. Biorheology. 2010;47:1–14.

14. Davis FM, Vita R. A nonlinear constitutive model for stress relaxation in ligaments and tendons. Ann Biomed Eng. 2012;40:2541–50.
15. Wertheim MG. Memoirs sur l'elasticite et la cohesion des principaux tissues des corps humain. Anales de Quimica y Physica. 1847;21:385–414.
16. Barad S, Cabaud HE, Rodrigo JJ. Effects of storage at 80 C as compared to 4 C on the strength of rhesus monkey anteriocruciate ligaments. Trans Orthop Res. 1982;7:378.
17. Noyes FR, Grood ES. The strength of the anterior cruciate ligament in humans and rhesus monkeys. Age-related and species-related changes. J Bone Joint Surg. 1976;58:1074–82.
18. Woo SLY, Orlando CA, Gomez MA, et al. Tensile properties of the medial collateral ligament as a function of age. J Orthop Res. 1986;4:133–41.
19. Jung HJ, Vangipuram G, Fisher MB, et al. The effects of multiple freeze-thaw cycles on the biomechanical properties of the human bone-patellar tendon-bone allograft. J Orthop Res. 2011;29:1193–8.
20. Davison PF. The contribution of labile crosslinks to the tensile behavior of tendons. Connect Tissue Res. 1989;18:293–305.
21. Chimich D, Shrive N, Frank C, et al. Water content alters viscoelastic behavior of the normal adolescent rabbit medial collateral ligament. J Biomech. 1992;25:831–7.
22. Haut TL, Haut RC. The state of tissue hydration determines the strain rate sensitive stiffness of human patellar tendon. J Biomech. 1997;30:79–81.
23. Thornton GM, Shrive NG, Frank CB. Altering ligament water content affects ligament pre-stress and creep behavior. J Orthop Res. 2001;19:845–51.
24. Hoffman AH, Robichaud DR, Duquette JJ, Grigg P. Determining the effect of hydration upon the properties of ligaments using pseudo Gaussian stress stimuli. J Biomech. 2005;38:1636–42.
25. Wilson AM, Goodship AE. Exercise-induced hyperthermia as a possible mechanism for tendon degeneration. J Biomech. 1994;27:899–905.
26. Ellis DG. Cross-sectional area measurements for tendon specimens: a comparison of several methods. J Biomech. 1969;2:175–86.
27. Woo SLY, Peterson RH, Ohland KJ, et al. The effects of strain rate on the properties of the medial collateral ligament in skeletally immature and mature rabbits: a biomechanical and histological study. J Orthop Res. 1990;8:712–21.
28. Moon DK, Abramowitch SD, Woo SL. The development and validation of a charge coupled device laser reflectance system to measure the complex cross-sectional shape and area of soft tissues. J Biomech. 2006;39:3071–5.
29. Pokhai GG, Oliver ML, Gordon KD. A new laser reflectance system capable of measuring changing cross-sectional area of soft tissues during tensile testing. J Biomech Eng. 2009;131.
30. Gellhorn AC, Carlson MJ. Inter-rater, intra-rater, and inter-machine reliability of quantitative ultrasound measurements of the patellar tendon. Ultrasound Med Biol. 2013;39:791–6.
31. Woo SLY, Gomez MA, Seguchi Y, et al. Measurement of mechanical properties of ligament substance from a bone-ligament-bone preparation. J Orthop Res. 1983;1:22–9.
32. Sharkey NA, Smith TS, Lundmark DC. Freeze clamping musculo-tendinous junctions for in-vitro simulation of joint mechanics. J Biomech. 1995;28:631–5.
33. Riemersma DJ, Schamhardt HC. The cryo-jaw, a clamp designed for in vitro rheology studies of horse digital flexor tendons. J Biomech. 1982;15:619–20.
34. Matthews GL, Keegan KG, Graham HL. Effects of tendon grip technique (frozen versus unfrozen) on in vitro surface strain measurements of the equine deep digital flexor tendon. Am J Vet Res. 1996;57:111–5.
35. Bowser JE, Elder SH, Rashmir-Raven AM, Swiderski CE. A cryogenic clamping technique that facilitates ultimate tensile strength determinations in tendons and ligaments. Vet Comp Orthop Traumatol. 2011;24:370–3.
36. Holmes G, Sharkey NA, Syftestad G, Reiser K. Changes in collagen crosslinks, fibril diameter, and mechanical strength in aging chicken tendons. Trans Orthop Res Soc. 1991;16:620.

37. Bay BK, Sharkey NA, Szabo RM. Biomechanical behavior of the median nerve in and about the carpal tunnel. J Hand Surg. 1997;22A:621–7.
38. Andarawis-Puri N, Kuntz AF, Kim SY, Soslowsky LJ. Effect of anterior supraspinatus tendon partial-thickness tears on infraspinatus tendon strainthrough a range of joint rotation angles. J Shoulder Elb Surg. 2010;19:617–23.
39. Bey MJ, Song HK, Wehrli FW, Soslowsky LJ. A noncontact, nondestructive methodfor quantifying intratissue deformations and strains. J Biomech Eng. 1997;22A:621–7.
40. Doehring TC, Kahelin M, Vesely I. Direct measurement of nonuniform large deformations in soft tissues during uniaxial extension. J Biomech Eng. 2009;131.
41. Kim YS, Kim JM, Bigliani LU, et al. In vivo strain analysis of the intact supraspinatus tendon by ultrasound speckles tracking imaging. J Orthop Res. 2011;29:1931–7.
42. Doesburg M, Yoshii Y, Henderson J, et al. Speckle-tracking sonographic assessment of longitudinal motion of the flexor tendon and subsynovial tissue in carpal tunnel syndrome. J Ultrasound Med. 2012;31:1091–8.
43. Maganaris CN. Tensile properties of in-vivo human tendinous tissue. J Biomech. 2002;35: 1019–27.
44. Maganaris CN, Paul JP. Tensile properties of the in vivo human gastrocnemius tendon. J Biomech. 2002;35:1639–46.
45. Roberts J, Cummings J, Grood E, Noyes F. In-vivo measurement of human anterior cruciate ligament forces during knee extension exercises. In: 40th annual meeting, orthopaedic research society. New Orleans, LA; 1984. p. 84.
46. Howe JG, Wertheimer C, Johnson RJ, et al. Arthroscopic strain gauge measurement of the normal anterior cruciate ligament. Arthroscopy. 1990;6:198–204.
47. Finni T, Komi PV, Lukkariniema J. In vivo human triceps surae and quadriceps femoris muscle function in a squat jump and counter movement jump. Eur J Appl Physiol. 2000;83: 416–26.
48. Gregor RJ, Komi PV, Järvinen M. Achilles tendon forces during cycling. Int J Sports Med. 1987;8:9–14.
49. Komi PV, Belli A, Huttunen V, et al. Optic as a transducer of tendomuscular forces. Eur J Appl Physiol. 1996;72:278–80.
50. Fukashiro S, Komi PV, JaE' M, Miyashita M. In vivo achilles tendon loading during jumping in humans. Eur J Appl Physiol. 1995;71:453–8.
51. Komi PV, Fukashiro S, JaE' M. Biomechanical loading of Achilles tendon during normal locomotion. Clin Sports Med. 1992;11:521–31.
52. Finni T, Komi PV, Lukkariniemi J. Achilles tendon loading during walking: application of a novel optic technique. Eur J Appl Physiol. 1998;77:289–91.
53. Tidball JG. The geometry of actin filament-membrane associations can modify adhesive strength of the myotendinosis junction. Cell Motil. 1983;3:439–47.
54. Tidball JG, Daniel TL. Myotendinous junctions of tonic muscle cells: structure and loading. Cell Tissue Res. 1986;245:315–22.
55. Trotter JA, Corbett K, Avner BP. Structure and function of the murine muscle-tendon junction. Anat Rec. 1981;201:293–302.
56. Lieber RL, Leonard ME, Brown CG, Trestik CL. Frog semitendinosus tendon loadstrain and stress-strain properties during passive loading. Am J Phys. 1991;261:C86–92.
57. Tidball JG, Salem G, Zernicke R. Site and mechanical conditions for failure of skeletal muscle in experimental strain injuries. J Appl Physiol. 1993;74:1280–6.
58. Scott SH, Loeb GE. Mechanical properties of aponeurosis and tendon of the cat soleus muscle during whole muscle isometric contractions. J Morphol. 1995;224:73–86.
59. Donkelaar CCV, Willems PJB, Muijtjens AMM, Drost MR. Skeletal muscle transverse strain during isometric contraction at different lengths. J Biomech. 1999;32(8):755–62.
60. Maganaris DN, Kawakami Y, Fukunaga T. Changes in aponeurotic dimensions upon muscle shortening: in vivo observations in man. J Anat. 2001;199:449–56.

61. Azizi E, Roberts TJ. Biaxial strain and variable stiffness in aponeuroses. J Physiol. 2009;587: 4309–18.
62. Benjamin M, Toumi H, Ralphs J, et al. Where tendons and ligaments meet bone: attachment sites ('entheses') in relation to exercise and/or mechanical load. J Anat. 2006;208:471–90.
63. Stouffer DC, Butler DL, Hosney D. The relationship between crimp pattern and mechanical response of human patellar tendon-bone units. J Biomech Eng. 1985;107:158–65.
64. Butler DL, Sheh MY, Stouffer DC, et al. Surface strain variation in human patellar tendon and knee cruciate ligaments. J Biomech Eng. 1990;112:38–45.
65. Nachemson AL, Evans JH. Some mechanical properties of the third human lumbar interlaminar ligament (ligamentum flavum). J Biomech. 1968;1:211–20.
66. Leonard F, Moscovitz P, Hodge JW, Adams JP. Age-related Ca-Mg content and strength in Turkey tendon. Calcif Tissue Res. 1976;19:331–6.
67. Hofmann CL, Okita N, Sharkey NA. Experimental evidence supporting isometric functioning of the extrinsic toe flexors during gait. Clin Biomech. 2013;28:686–91.
68. Kirane YM, Michdelson JD, Sharkey NA. Evidence of isometric function of the flexor hallucis longus muscle in normal gait. J Biomech. 2008;41:1919–28.
69. Giori NJ, Beaupre GS, Carter DR. Cellular shape and pressure may mediate mechanical control of tissue composition in tendons. J Orthop Res. 1993;11:581–91.
70. Matyas J, Anton MG, Shrive NG, Frank CB. Stress governs tissue phenotype at the femoral insertion of the rabbit MCL. J Biomech. 1995;28:147–9.
71. Shadwick RE. Elastic energy storage in tendons: mechanical differences related to function and age. J Appl Physiol. 1990;68:1033–40.
72. Tipton CM, Matthes RD, Martin RK. Influence of age and sex on the strength of bone-ligament junctions in knee joints of rats. J Bone Joint Surg. 1978;60A:230–4.
73. Neumann P, Ekstrom LA, Kellar TS, et al. Aging, vertebral density, and disk degeneration alter the stress-strain characteristics of the human anterior longitudinal ligament. J Orthop Res. 1994;12:103–12.
74. Kastelic J, Baer E. Deformation in tendon and collagen. In: Vincent J, Currey J, editors. The mechanical properties of biological materials. Cambridge: Cambridge University Press; 1980. p. 397–435.
75. Karamanidis K, Arampatzis A. Mechanical and morphological properties of different muscle-tendon units in the lower extremity and running mechanics: effect of aging and physical activity. J Exp Biol. 2005;208:3907–23.
76. Reeves ND, Maganaris CN, Narici MV. Plasticity of dynamic muscle performance with strength training in elderly humans. Muscle Nerve. 2005;31:355–64.
77. Maganaris CN, Reeves ND, Rittweger J, et al. Adaptive response of human tendon to paralysis. Muscle Nerve. 2006;33:85–92.
78. Onambele GL, Narici MV, Maganaris CN. Calf muscle-tendon properties and postural balance in old age. J Appl Physiol (1985). 2006;100(6):2048–56.
79. Stenroth L, Peltonen J, Cronin N, et al. Age-related differences in Achilles tendon properties and triceps surae muscle architecture in vivo. J Appl Physiol. 2012;113:1537–44.
80. Woo SL-Y, Gomez MA, Amiel D, et al. The effects of exercise on the biomechanical and biochemical properties of swine digital flexor tendons. J Biomech Eng. 1981;103:51–6. https://doi.org/10.1115/1.3138246.
81. Laros GS, Tipton CM, Cooper RR. Influence of physical activity on ligament insertions in the knees of dogs. J Bone Joint Surg. 1971;53A:275–86.
82. Noyes FR, Delucas JL, Torvik PJ. Biomechanics of anterior cruciate ligament failure: an analysis of strain-rate sensitivity and mechanisms of failure in primates. J Bone Joint Surg. 1974;56:236–53.
83. Woo SL, Gomez MA, Sites TJ, et al. The biomechanical and morphological changes in the medial collateral ligament of the rabbit after immobilization and remobilization. J Bone Joint Surg. 1987;69A:1200–11.

84. Noyes FR. Functional properties of knee ligaments and alterations induced by immobilization. A correlative biomechanical and histological study in primates. Clin Orthop Rel Res. 1977;123:210–42.
85. Reeves ND, Narici MV, Maganaris CN. Strength training alters the viscoelastic properties of tendons in elderly humans. Muscle Nerve. 2003;28:74–81.
86. Malliaras P, Kamal B, Nowell A, et al. Patellar tendon adaptation in relation to load-intensity and contraction type. J Biomech. 2013;46:1893–9.
87. Kubo K, Ohgo K, Takeishi R, et al. Effects of isometric training at different knee angles on the muscle-tendon complex in vivo. Scand J Med Sci Sports. 2005;16:159–67.
88. Kubo K, Morimoto M, Komuro T, et al. Effects of plyometric and weight training on muscle-tendon complex and jump performance. Med Sci Sports Exerc. 2007;39:1801–10.
89. Arampatzis A, Peper A, Bierbaum S, Albracht K. Plasticity of human Achilles tendon mechanical and morphological properties in response to cyclic strain. J Biomech. 2010;43: 3073–9.
90. Clancy WG, Narechania RG, Rosenberg TD. Anterior and posterior cruciate ligament reconstruction in rhesus monkeys: a histological, microangiographic, and biomechanical analysis. J Bone Joint Surg. 1981;63A:1270–84.
91. Vailas AC, Tipton CM, Mattes RD, Gart M. Physical activity and its influence on the repair process of medial collateral ligaments. Connect Tissue Res. 1981;9:25–31.
92. Freehan LM, Beauchene JG. Early tensile properties of healing chicken flexor tendons: early controlled passive motion versus postoperative immobilization. J Hand Surg. 1990;15:63–8.
93. Takai S, Woo SL, Horibe S, et al. The effects of frequency and duration of controlled passive mobilization on tendon healing. J Orthop Res. 1991;9:705–13.
94. Salter RB, Simmonds DF, Malcolm BW, et al. The biological effect of continuous passive motion on the healing of full thickness defects in articular cartilage. An experimental investigation in the rabbit. J Bone Joint Surg Am. 1980;62:1232–51.
95. Knapik DM, Harris JD, Pangrazzi G, et al. The basic science of continuous passive motion in promoting knee health: a systematic review of studies in a rabbit model. Arthroscopy. 2013;29:1722–31.
96. Benjamin M, Kaiswr E, Milz S. Structure-function relationships in tendons: a review. J Anat. 2008;212:211–28.
97. Wang JHC. Mechanobiology of tendon. J Biomech. 2006;39:1563–82.
98. Yang G, Rothrauff BB, Tuan RS. Tendon and ligament regeneration and repair: clinical relevance and developmental paradigm. Birth Defects Res C Embryo Today. 2013;99:203–22.
99. Nourissat G, Berenbaum F, Duprez D. Tendon injury: from biology to tendon repair. Nat Rev Rheumatol. 2015;11:223–33. https://doi.org/10.1038/nrrheum.2015.26.
100. Woo SL-Y, Buckwalter JA. Injury and repair of the musculoskeletal soft tissues. Park Ridge: American Academy of Orthopaedic Surgeons; 1988.
101. Weiss JA, Gardiner JC. Computational modeling of ligament mechanics. Crit Rev Biomed Eng. 2001;29:303–71. https://doi.org/10.1615/critrevbiomedeng.v29.i3.20.

Chapter 9
Mechanical Properties of Cartilage and Synovial Joints

Abstract Most synovial joints have very low friction under dynamic loading, due to their structural features including cartilage and synovial fluid. The viscoelastic properties, fluid flow through, and surface decorating molecules of articular cartilage are keys to the efficient transmission of load through this soft tissue.

What You Will Learn

1. The structural components of synovial joints and how they contribute to the mechanical function of the joint.
2. How the theoretical models of cartage can be used to describe its mechanical behavior.
3. The role of synovial fluid in joint mechanical properties.

9.1 Introduction

Now that we have explored the mechanical behaviors of tendons and ligaments, we will turn our attention to the tissue that directly transmits forces from one bone to another while at the same time allowing for smooth, almost frictionless, motion between them. These motions occur within our synovial joints, with the main players being cartilage and synovial fluid.

9.2 Mechanically Relevant Structural Components of Cartilage

The general structure and composition of cartilage were introduced in Chap. 3. There are three important points to recall about the structure to help understand the mechanical role that cartilage plays in joint mechanics.

M. R. Allen, J. Wallace, *Skeletal Tissue Mechanics*,
https://doi.org/10.1007/978-3-031-90609-1_9

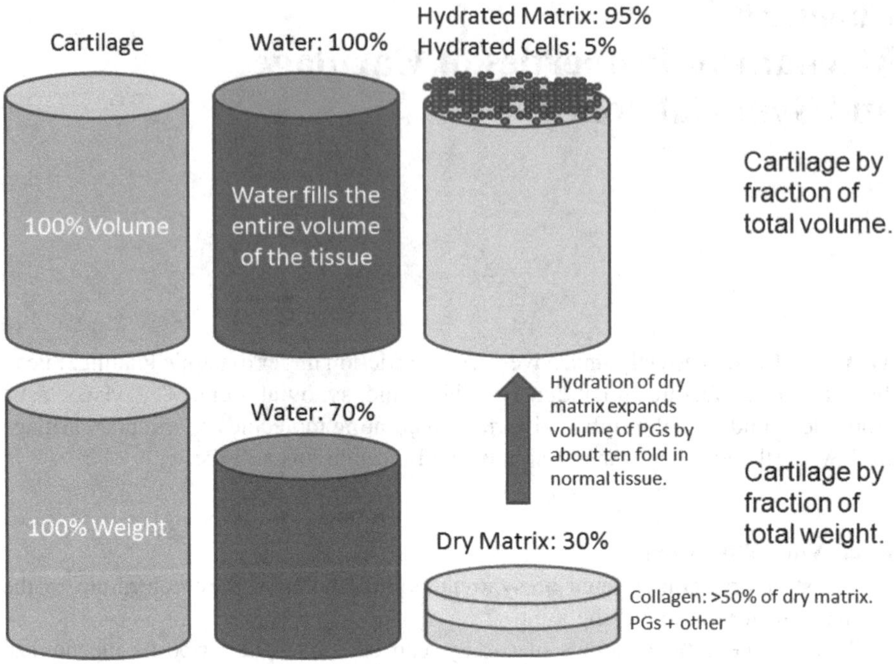

Fig. 9.1 The three components of cartilage are water, extracellular matrix, and chondrocytes

Cartilage is an extracellular matrix inflated with water that is held in place by strong electrochemical interactions with the proteoglycans. The void dimensions of cartilage ECM are about 50 Å, much smaller than the 50–300 μm void size of bone. The small distances between molecular components in cartilage strongly affect its mechanical properties. Cartilage ECM is about 95% of the total tissue volume and the remaining 5% is chondrocytes. The chondrocytes create and assemble the matrix initially and maintain the framework of collagen and proteoglycans. Fibrillar type II collagen and proteoglycans are the primary macromolecules of adult articular cartilage ECM; however, water is 70% (by weight) of the tissue (Fig. 9.1). The water is very tightly bound to the ECM by the physicochemical properties of the macromolecules. Collagen is more than 50% of the dry weight (or 10–20% of the wet weight) of articular cartilage. In adults, approximately 90% of cartilage collagen is of type II, which is found in only a few other tissues such as the vitreous body of the eye. Type II is a fibril-forming collagen and the molecules are made of three identical alpha$_1$(II) polypeptide chains. Each of the alpha$_1$(II) molecules is itself a 300 nm-long triple helix initially formed with short telopeptides at each end that are enzymatically removed before assembly. Once the collagen molecules leave the cell, they assemble into fibrils extracellularly. The resulting network of type II collagen fibrils gives the ECM tensile strength and is essential to maintaining the tissue's volume and shape. The tensile strength of the collagenous matrix is increased by enzymatically produced covalent, intermolecular cross-links.

Proteoglycans (PGs) are highly abundant in the ECM of all connective tissues but especially in cartilages. In articular cartilage, these highly hydrophilic macromolecules are embedded within the fibrillary network in an underhydrated (compressed) form. They give articular cartilage the ability to undergo large, non-damaging, reversible compression. If anything interferes with the proteoglycan content (e.g., the release of destructive enzymes by the chondrocytes), or with the ability of chondrocytes to make normal ECM (e.g., a genetic defect), the mechanical functioning of the cartilage will deteriorate.

The predominant PG of articular cartilage is aggrecan. It contains a protein core (215 kDa) to which glycosaminoglycans and oligosaccharides are covalently attached. Ninety percent of the total mass of aggrecan is from the glycosaminoglycans keratan sulfate and chondroitin sulfate. Aggrecan attaches noncovalently to the very high molecular weight biopolymer hyaluronan. A link protein acts to stabilize this interaction between aggrecan and hyaluronan, making it nearly as stable as a covalent bond. The combination of aggrecan and hyaluronan forms an extraordinarily large space-filling molecule called aggregate. In free solution, aggrecan swells to about 50 times its dry volume. However, in healthy cartilage, the collagen network constrains swelling to about ten times the dry volume (i.e., 20% of free swelling). This permits the accumulation of high concentrations of compressed aggrecan in the tissue. Aggrecan is a highly negatively charged molecule. The highly concentrated negative charges repel each other and strongly attract water molecules creating a swelling pressure. Under compressive load, water is pushed out of aggrecan and the whole tissue. Compression of the proteoglycans increases the negative charge density in the tissue, which in turn increases electrical repulsion between the negative charges and the attraction of water into the tissue. The process of water expression continues until there is equilibrium between loading forces and the electrical and osmotic swelling stresses. Removing the load permits water to return to the tissue and it expands until the expansion force of the proteoglycans is balanced by tension in the fibrillary network. The numbers of negatively charged groups in the proteoglycans will rise and fall in proportion to the pH of the synovial fluid. Cations in the synovial fluid (e.g., Na+ and Ca++) also are attracted to and shield the negative proteoglycan charges reducing the effective charge density, divalent cations having effects much more than twice as strong as monovalent cations [1]. For these and other reasons, the chemistry of cartilage has substantial effects on its mechanical properties.

In addition to the swelling properties of the PGs, their high negative charge concentration has strong effects on the diffusion of solutes. The negative charges of aggrecan are fixed to the molecule and do not diffuse independently. As a result, they create a high density of fixed negative charge that can retard the diffusion of negatively charged (anionic) macromolecules into the tissue while attracting and concentrating positively charged (cationic) molecules [2]. The effect of the fixed negative charges on the diffusion of neutral macromolecules with a mixture of positive and negative regions (zwitterions) is more complex.

Because cartilage consists of an organic biopolymeric matrix and water, it can be useful to think of it as a multiphasic system. The simplest multiphasic system is biphasic, with water being one phase and the collagen–proteoglycan biopolymer

Table 9.1 Properties of lateral femoral condyle cartilage in several different species

Variable	Human	Bovine	Canine	Monkey	Rabbit
Aggregate modulus, Mpa	0.70	0.89	0.60	0.78	0.54
Poisson's ratio	0.10	0.40	0.30	0.24	0.34
Permeability, 10–15 m^4*/N-s*	*1.18*	*0.43*	*0.77*	*4.19*	*1.81*

From Mow et al. [3]

matrix the other. The volume fraction of the water phase can be thought of as the porosity of the cartilage. Although the water molecules are restrained to a significant degree by the matrix molecules, they may nevertheless slowly migrate through the tissue under the influence of a pressure gradient. This may be modeled using Darcy's law:

$$Q = -k(dP/dx) \tag{9.1}$$

where Q is the flow rate per unit cross-sectional area, dP/dx is the pressure gradient, and k is the hydraulic permeability coefficient (sometimes called the apparent permeability). Typical values of k are shown in Table 9.1. Permeability diminishes in the deepest layers of the articular cartilage and also varies with the composition of the liquid phase. Clearly, the resistance of the tissue to deformation depends to some extent on its permeability.

9.3 Mechanical Properties of Cartilage

Cartilage is a water-filled biopolymer that has strong interactions between its constituent components. As a result, its mechanical properties are viscoelastic due to the flow of water in the matrix and time-dependent reconfiguration of the biopolymers under mechanical loading. The elastic modulus of the tissue varies greatly with the rate of loading, demonstrating a broad spectrum of interactions between water and the biopolymers comprising the tissue. At very slow rates of loading, the resulting modulus is only a few Mpa, a characteristic that can be reasonably modeled as the slow percolation of water through an elastic matrix. At physiologic rates of dynamic loading, the modulus of the cartilage is much larger, approaching 500 Mpa [4–7]. The mechanical behavior of cartilage under these more typical dynamic conditions cannot be explained using simple fluid flow models; additional interactions between the constituent biopolymers are needed to explain this more extreme viscoelastic behavior. This range of behaviors should be borne in mind when reading the rest of this chapter, because many of the descriptions of cartilage stiffness measurements refer to behavior at very slow rates of loading. For example, indentation tests are usually quasistatic (extremely slow loading rate) and the "equilibrium" or "aggregate" modulus used in many models refers to the stiffness associated with an infinitely slow rate of loading.

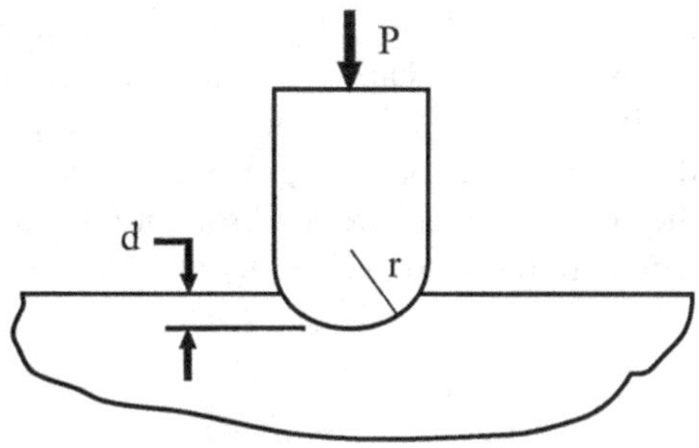

Fig. 9.2 Diagram of indentation experiment. An indenter having a hemispherical end of radius *r* is pressed into the surface of the tissue for a distance d by a load *P*

Finally, while it is much less stiff than cortical bone, cartilage does not serve to "cushion" or reduce impulsive forces in joints. It cannot do this because it is so thin that its capacity to absorb energy is insignificant compared to (a) eccentric contractions of muscles and (b) energy absorption in the bones on either side of the joint. Instead, the role of cartilage is to provide a self-renewing, well-lubricated, compliant, load-bearing surface.

It is difficult to form mechanical test specimens of cartilage because it is only found in thin (2–3 mm) layers. Therefore, many of the early mechanical tests involved indentation of the surface as a means of gaining compressive stiffness data. Such experiments were based on the equation derived by Boussinesq for a hemispherical indenter applied to an elastic half-space [8] (Fig. 9.2). If an indenter of radius r penetrates to a depth d under load P, the elastic modulus is given by

$$E = P(1 - \nu^2)/2dr \tag{9.2}$$

where ν is Poisson's ratio of the indented material. Assuming that the cartilage was indeed of infinite depth and that it was incompressible (so that $\nu = 0.5$), Sokoloff [13] obtained $E = 2.3$ Mpa for the elastic modulus of the articular cartilage. Later, when a mathematical solution was obtained for indentation of an elastic layer on a rigid foundation (i.e., cartilage on relatively rigid bone) by a plane-ended cylinder, Hayes et al. found that E was only 1.6 Mpa [9]. As more work was done, it became clear that this value, too, was probably inaccurate because cartilage is not truly incompressible and fails to meet the assumptions of the tests in other ways as well (e.g., it is anisotropic, inhomogeneous, and viscoelastic). Table 9.1 shows data on the compressive modulus and Poisson's ratio of articular cartilage for various species. These data were obtained from compression tests on machined specimens using techniques similar to those described later in this section.

Woo and coworkers pioneered the tensile testing of cartilage [10]. They sliced articular cartilage into miniature "dumbbells" having a thickness of 0.25 mm and clamped them into a tensile test apparatus. Strain was measured by videotaping the specimen as it was pulled apart and measuring, in each video frame, the distance between two lines drawn across the undeformed specimen.

When cartilage is loaded, the way the force is shared between the fluid and solid phases is highly dependent on the nature of the loading. In tension, it is primarily the solid phase of the cartilage that carries the load; in particular, it is the collagen fibers. As explored in Chap. 8, tendon and ligament tissues exhibit a "toe" region with an initially low modulus, followed by a region of increasing stiffness. Cartilage produces the same sort of curve, except that its upper portion remains nonlinear because the collagen fibers are not brought into full alignment. Its stress–strain relationship can therefore be described by an equation of the form

$$\sigma = A\left(e^{B\varepsilon} - 1\right) \tag{9.3}$$

where σ is stress, ε is strain, and A and B are coefficients. To find the modulus at any point on the curve, one merely differentiates to obtain

$$E = d\sigma/d\varepsilon = B(\sigma + A) \tag{9.4}$$

Figure 9.3 shows the shapes of such stress–strain curves for bovine articular cartilage as a function of depth below the surface and orientation relative to the split line direction. Table 9.2 shows the elastic modulus at a stress of 2.5 Mpa for each

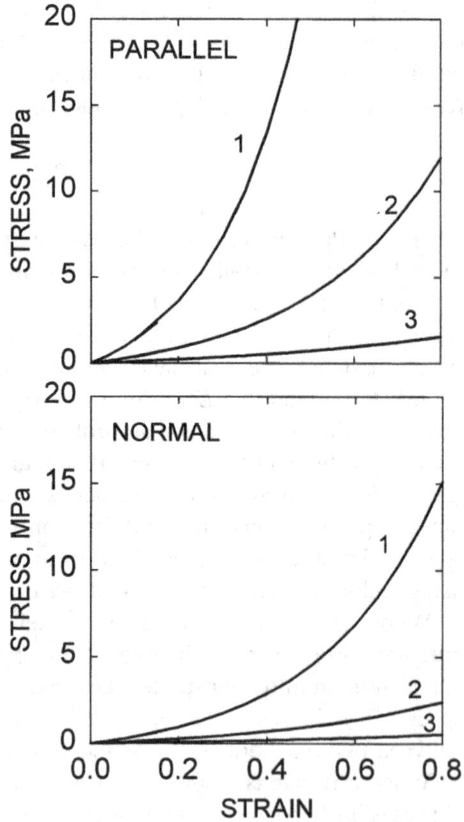

Fig. 9.3 Stress–strain curves for bovine cartilage specimens as functions of orientation and depth. Loading rate is approximately 1.2% per s [10]

Table 9.2 Data for bovine cartilage. Loading rate approximately 1.2% per s

Direction	Slice # (relative depth)	Tensile modulus, Mpa
Parallel to split lines	1	23.0
	2	11.2
	3	5.0
Normal to split lines	1	2.2
	2	6.6
	3	3.6

From Woo et al. [10]

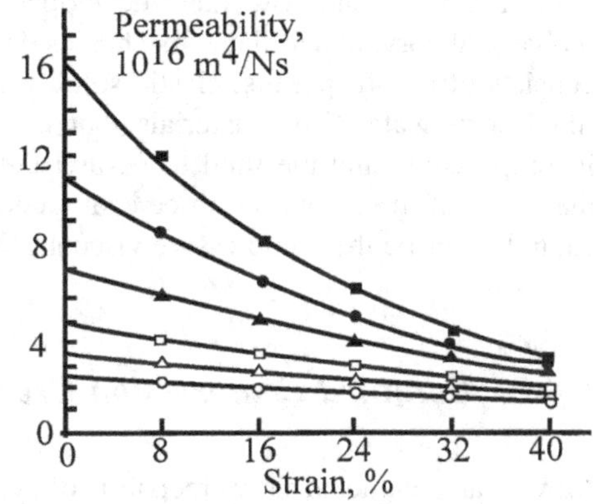

Fig. 9.4 Permeability vs compressive strain for cartilage. The family of curves shows how permeability decreases as increasing pressure is applied. From top to bottom, the pressures were 0.17, 0.34, 0.69, 1.03, and 1.72 Mpa [14]

position and orientation. Note that the tensile modulus is greater parallel to the split lines. Near the articular surface, fluid pressure caused by loading and the tensile stresses in the matrix at the edge of a loaded region have components that are perpendicular to the load (parallel to the surface) and tend to tear the surface of the tissue. As a result, the tangential orientation of the collagen at the tissue surface is a useful direction for resisting the tensile stress. Ultimate tensile stresses for articular cartilage range from 9 to 18 Mpa and ultimate tensile strains from 60% to 120% [11].

9.4 Poroelastic Theory of Cartilage

When a cartilage specimen is loaded in tension or compression, drops of fluid can be seen to exude and collect on its surfaces (Fig. 9.4). This is a clear sign that cartilage liquid flows under the influence of pressure gradients produced by mechanical deformation of the solid phase. Movement of the fluid phase produces some of the viscoelastic behavior of the tissue, especially for slow loading. Additionally, the permeability of the tissue changes when the solid phase is deformed. Mansour and Mow studied this behavior in 200-μm-thick specimens of cartilage clamped between two porous metal platens [12]. Through these porous platens, a pressure gradient was imposed across the specimen: the applied pressure, P_A. The resulting flow was measured and the permeability coefficient k calculated. By compressing the specimen to different strain levels (ε), the relationship between permeability and tissue strain could be studied. Typical results show that an interaction occurs between the deformations of the fluid and solid phases when cartilage is loaded. The curves can be fit to the equation

$$k = Ae^{\alpha\varepsilon} \tag{9.5}$$

where A and α both decrease as the applied pressure increases. To understand this interplay, Mow and coworkers developed what has become known as the KLM biphasic theory of cartilage mechanics [13]. This theory assumes that cartilage consists of a soft, porous, elastic solid, the pores of which communicate and are filled with water. The materials that compose both phases are assumed to be incompressible and the fluid to be inviscid. The details of this theory are beyond the scope of this book, but three examples are presented to show where it can and cannot describe the compressive viscoelastic behavior of cartilage.

9.4.1 Confined Compression Creep Test

Mow et al. conducted slow creep tests of cylindrical specimens of cartilage using the scheme shown in Fig. 9.5 [13]. A cylindrical specimen of cross-sectional area A is cored from a joint surface and placed in a cylindrical metal well with solid walls and bottom. A porous metal loading platen is placed on top of the specimen. A constant load, F, is applied at time $t = 0$ and the specimen deformation is recorded until an equilibrium deformation is reached. In this situation, creep is controlled by the exudation of fluid through the porous loading platen against the specimen's top surface. The movement of fluid to this interface is modeled to understand exudation. In the biphasic theory, movement of the fluid in the tissue is determined by the permeability, k, and the fluid pressure gradients of the specimen.

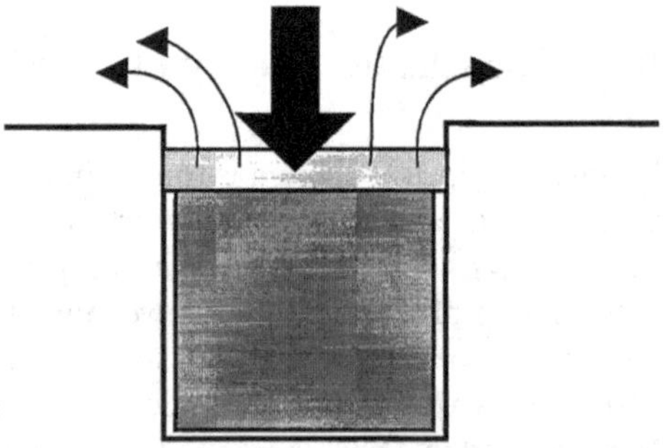

Fig. 9.5 Schematic diagram of the experimental setup used for creep compression tests of cartilage. Fluid is allowed to flow through the porous loading platen

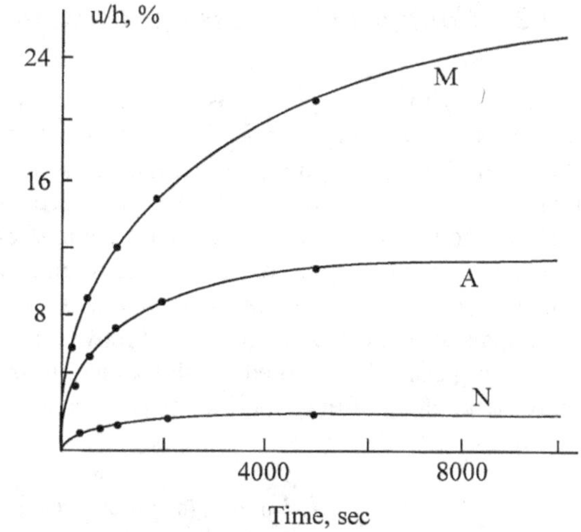

Fig. 9.6 Strain *(u/h)* vs. time for articular cartilage *(A)*, meniscus *(M)*, and nasal cartilage *(N)* for slow loading [14]

Table 9.3 Compressive properties of three types of bovine cartilage

Tissue	Aggregate modulus (HA), Mpa	Permeability (k),10–15 m^4/N-s	H_2O by weight, %
Nasal cartilage	5.64	0.49	75.6
Articular cartilage	0.85	4.67	80.9
Meniscus	0.41	0.81	73.9

From Woo et al. [15]

The equilibrium deformation is determined by the equilibrium, or aggregate, modulus of the specimen, H_A, and the applied loading. If the initial height of the specimen is h, and its displacement under stress $\sigma = F/A$ is $u(t)$, the biphasic theory solution for the ratio of the deformation to the initial specimen height as a function of time is

$$u/h = (\sigma/H_A)\left\{1 - 2\sum_{n=0}^{\infty}\left[\exp\left(-\Lambda_n H_A kt/h^2\right)/\Lambda_n\right]\right\} \tag{9.6}$$

where t is time and $\Lambda_n = (n + ½)^2\pi^2$ [15]. Notice that as $t \to \infty$, this equation simplifies to $\sigma = H_A\ (u/h)$. Figure 9.6 and Table 9.3 show the results of fitting this equation to slow creep data for articular cartilage and two other kinds of cartilaginous tissue: nasal cartilage and knee meniscus. The distinct differences in the properties of these tissues are striking. Notice the inverse relationship between the aggregate modulus and the equilibrium deformation. Also, the cartilage with the greatest water content has the greatest permeability.

9.4.2 Confined Compression Relaxation Test

In this experiment, the apparatus is the same as that depicted in Fig. 9.5, except that the instantaneous load application is replaced by a ramp displacement like that shown in Fig. 9.7. After a specified maximum displacement is reached, it is maintained while the reactive stress produced in the specimen is monitored. As Fig. 9.7 shows, there is a nonlinear increase in stress until maximum displacement is reached; then the stress "relaxes" to an equilibrium value. Fluid flows from the specimen early in the test (e.g., at A and B), but later the strain redistribution is accompanied entirely by internal fluid movements (e.g., at C and D).

The biphasic theory predicts the stress during both the compression and the relaxation phases of the experiment. The equation for the latter is

$$\sigma = -(H_A V_0/h)\left\{ t_0 - 2\sum_{n=1}^{\infty} \left[K_n{}^{-1} \exp(-K_n t) - 1 \right] \right\} \tag{9.7}$$

where V_0 is the displacement rate, t_0 is the time at which the displacement is stopped (B in Fig. 9.7), and $K_n = n^2\pi^2 k H_A/h^2$ [15]. The other symbols retain their former definitions. Note again that as $t \rightarrow \infty$,

$$\sigma = H_A V_0 t_0/h = H_A \varepsilon \tag{9.8}$$

because $V_0\, t_0/h$ is the strain at the end of the slow loading period, ε. These equations fit experimental relaxation data quite well when t_0 is large enough (on the order of a hundred seconds), and the values of H_A and k that achieve these fits agree well with those obtained from creep experiments. There are advantages to both methods of obtaining the aggregate modulus. The relaxation test takes less time to reach equilibrium, but it requires an apparatus to achieve a displacement curve like that shown in Fig. 9.7.

In the course of testing various kinds of specimens, it has been learned that the aggregate modulus of the cartilage is inversely proportional to the water content but increases in proportion to the hyaluronic acid content. For example, diminishing the water content from 85% to 75% increases the modulus from about 0.3 to 1.0 Mpa, and a 30% increase in hyaluronic acid content can double the modulus. There is also a reduction in the modulus with age, but this is not a predictive relationship because the individual variability at every age is large.

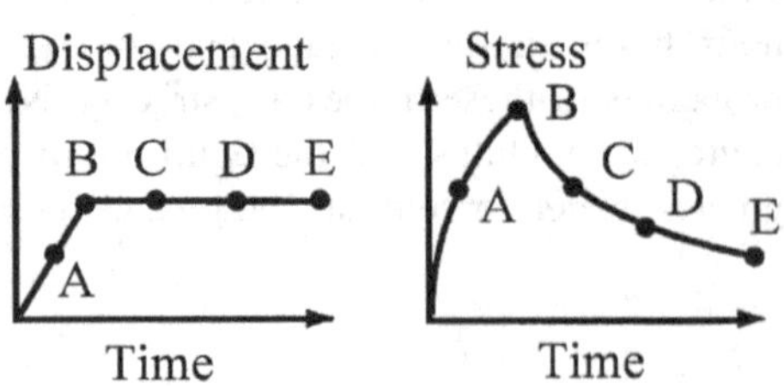

Fig. 9.7 [Left] Displacement applied to cartilage specimen in compression relaxation test. [Right] Stress–time curve for the test. Letters A–E show equivalent positions on the two curves [14]

9.4.3 Unconfined Compression Relaxation Test of Cartilage

For the rapid, unconfined compression of a cylinder of cartilage modeled as a linear poroelastic cylinder of radius r filled with an inviscid fluid, the poroelastic solution for the stress is

$$\sigma(t) = \sigma_s \left[1 + \sum_{n=1}^{\infty} A_n(v) \exp\left(-\frac{\alpha_n H_a k t}{r^2} \right) \right] \tag{9.9}$$

where σ_s (=$E_s\varepsilon_s$) is the stress after complete relaxation of the tissue, and the coefficients of the series are $A_\nu(\nu) = ((1 - \nu)(1 - 2\nu)/(1 + \nu))(1/[(1 - \nu)^2\alpha_\nu{}^2 - (1\ 2\nu)])$, ν = Poisson's Ratio (between 0 and 0.5), α_ν = characteristic equation roots, $H_a = E_s((1 - \nu)/(1 + \nu)(1 - 2\nu))$, $E_s(\varepsilon_s)$ = equilibrium Young's modulus (strain) of the material after complete relaxation, and k = permeability. The characteristic equation is $\vartheta_1(\xi) - ((1 - \nu)\xi\ \vartheta_0(\xi)/(1 - 2\nu)) = 0$, where $\vartheta_0(\xi)$ and $\vartheta_1(\xi)$ are Bessel functions and α_ν are the roots of the equation. This relaxation kernel is well known as the result of solving the poroelastic constitutive model for unconfined compression. As it turns out, 1.5 is the maximum size of the formula in the square brackets of Eq. (9.9) at time $t = 0$. That the *maximum* stress ($\sigma(0)$) can only be 1.5 times the equilibrium stress (σ_s) means that one should never observe peak stresses larger than this if the KLM biphasic theory holds true; however, it was observed many years ago that the peak stresses during unconfined compression of cartilage far exceed the limit of 1.5 times the equilibrium stress, showing that the KLM biphasic theory is unable to accurately model the behavior of the cartilage when loaded at rapid physiological rates [16].

9.4.4 Additional Models for Cartilage Mechanics

In addition to the KLM biphasic theory, some have proposed additional models to understand cartilage mechanics. Using a biphasic model with a matrix that has a phenomenological (i.e., empirical) mechanical stiffness that is high in tension and low in compression can solve some of the limitations of the KLM model (i.e., its inability to predict the initial stress of an unconfined compression experiment) [17]. However, the use of this special nonlinearly elastic matrix still puts all of the tissue viscoelasticity into the fluid flow. Even though it will now be possible to reach a higher apparent stress during load, the relaxation spectrum is still too slow. The nonlinear material behavior can be modeled mechanistically (e.g., as being due to the fact that you can pull but not push on a fiber) or be simply introduced as a rule. In research, a mechanistic model helps more with understanding how a tissue functions, and therefore, the former approach is preferred.

The key to a suitable model of early rapid stress relaxation is to use a rapidly relaxing matrix that is viscoelastic [18]. The viscoelastic matrix is often introduced phenomenologically (i.e., the source of viscoelasticity is not explained) and can be combined with bi-, tri-, and other phasic fluid flow models. When a phenomenological viscoelastic material is used with a "phasic" theory, the mechanisms of fluid flow and ionic interactions are approximately mechanistic, but the viscoelasticity remains unexplained. There have been small steps in the direction of mechanistic modeling of cartilage matrix viscoelasticity using polymer dynamics models [19, 20]. Polymer dynamics is a very complex field in itself, but in theory it can account for the rapid stress relaxation caused by biopolymer:biopolymer interactions in the tissue [21]. The approach is promising and has been shown to be consistent with observations of cartilage stress relaxation in unconfined compression.

9.4.5 Modeling Tissue Charge

In addition to the mechanical properties of the collagen matrix and the effects of fluid flow, swelling effects due to the high density of fixed negative charges within the aggrecan molecule have been mathematically modeled. The electrostatic swelling pressure of the fixed charges can be accounted for by including the continuum Donnan equilibrium model to represent the effects of fixed charges [22]. In the Donnan model, the microscopic features of the location and "grain" of the molecules are absent and only the averaged responses of the tissue are accounted for. A more granular approach is to use a microstructural, molecular level solution of the electrostatic Poisson–Boltzmann equation in a unit cell of a charged glycosaminoglycan molecule and its surrounding atmosphere of mobile atoms to represent the role of charge in the tissue [23]. The micro-modeling approach accounts for the space-varying electrical potential and electrical field among the constituents of cartilage. In a head-to-head comparison, the microstructural model accurately predicted observed experimental static swelling pressure, where the Donnan approach predicted swelling pressure several times larger than observed experimentally [23]. The conclusion was that the elastostatic approach more accurately reflected the effects of GAG charge in producing mechanical swelling pressure compared to the Donnan model—at least for the static loading case.

For dynamic loading, the Donnan equation is the primary approach to modeling the effects of charge density on mechanical properties. Considering that the Donnan approach is not ideal for static cases, it would seem reasonable that a more mechanistic (i.e., non-continuum) approach to modeling dynamic behavior might be appropriate. Unfortunately, mechanistic dynamic micro-models accounting for molecular charge do not seem to exist. It is possible that polymer dynamics models accounting for free charge on the polymer molecules could be used as a starting point to model cartilage dynamics more mechanistically [24].

9.4.6 Inhomogeneity of Modulus

Schinagl and coworkers loaded excised specimens of bovine articular cartilage in compression and then measured the displacements within the tissue as a function of depth below the articular surface [25]. They allowed 40 min to elapse after applying the load to allow the displacement field within the cartilage to reach an equilibrium state. Displacements were then measured by using a microscope to record the positions of chondrocytes before and after application of the load. (The DNA in the cells was marked with a fluorescent dye to make them easier to locate.) The results showed that displacements were greater near the articular surface and diminished exponentially through the thickness of the cartilage. If all of the cartilage was fully relaxed after 40 min, the difference in strains by depth would indicate that the modulus in the upper layers was lower than that in the depths of the cartilage. Alternatively, if the surface of the tissue relaxed viscoelastically faster than the deeper parts and 40 min was not enough time to reach equilibrium, the result could mean that the tissue deformed more quickly at the top than at the bottom. Schinagl et al. experimentally found that the compressive strain at 40 min of loading was essentially the same as that for 180 min of loading, so they rejected this alternate hypothesis and concluded that the equilibrium modulus was actually different by depth [25]. They proposed that the layered moduli are functional, stating: "Upon loading, the surface layers of the articular cartilage deform, reducing pore space [decreasing permeability] and [slowing] the flow of fluid out of the tissue... maintaining hydrostatic pressurization. Deformation of the relatively soft surface layer may also lead to improved congruence between opposing cartilage-covered bones. This may have the effect of not only reducing stress concentrations in the joint, but also maintaining a fluid film between cartilage layers that lubricates loaded joints and reduces wear."

Schinagl's proposal was interesting and possibly correct, but their data were for mechanical equilibrium and the proposal was for dynamic loading of cartilage at a physiological rate, so the connection between the proposal and data was not extremely strong [25]. Silverberg et al. have demonstrated conclusively that the dynamic shear modulus of bovine cartilage (1 Hz surface shear) has depth dependence with the surface more compliant than the deeper tissue and that regularly loaded regions are more compliant at the surface than rarely loaded regions of the tissue [26]. This result is consistent with and expands the original proposal by Schinagl et al. and suggests that the depth dependence of dynamic cartilage moduli is important to the maintenance of cartilage lubrication and pressurization during physiological loading [25].

9.4.7 Effects of Strain Rate on Mechanical Properties

Figure 9.8 shows a graph of the effects of varying strain rate on the stress–strain relationship of bovine articular cartilage tested in tension [15]. As with bone, tendon, and ligament, the stiffness of cartilage increases as it is deformed more quickly.

Notice that the curve for the fastest strain rate has an inflection point, but the others do not. This phenomenon is thought to be caused by resistance to fluid flow. At slow strain rates, there is time for fluid to flow without offering much resistance to the applied deformation. At higher strain rates, fluid pressure resists the loading early in the test. This pressure also expresses fluid from the tissue, reducing its participation in the production of stress and allowing the slope of the stress–strain curve to decline (i.e., to inflect). The biphasic theory produced the following equation to describe this kind of tensile test:

$$\sigma = E_s\varepsilon + \left(h^2 d\varepsilon/dt\right)/(H_A k)\left[A_0 - \sum\left(A_n \exp\left(-4\alpha_n^2 H_A kt/h^2\right)/\alpha_n^2\right)\right] \tag{9.10}$$

where ε is the strain and $d\varepsilon/dt$ is the strain rate, E_s is the elastic modulus of the solid matrix, and A_n and α_n are constants. The summation and other symbols are as in Eqs. (9.6) and (9.7). This result describes only the low strain portion of Fig. 9.8 (i.e., $\varepsilon < 10\%$) because the biphasic theory assumes infinitesimal strains and linearity. At higher strains, the collagen fibers have been stretched to the point that their participation in resisting the applied deformation is increasingly important. It is postulated that the three curves have similar slopes at their right ends because by then the stretched-out collagen fibers are carrying virtually the entire load, and fluid effects are insignificant. To date, tensile results such as these have not been examined as polymer mechanics or electrodynamics problems. It is unlikely, however, that the viscoelasticity of cartilage in tension results solely from fluid flow.

The tensile elastic modulus of the specimens in Fig. 9.8 is in the neighborhood of 1–10 MPa, depending on the strain magnitude and rate. This value is similar to the aggregate modulus for compression of articular cartilage at low strain rates. There is evidence that the compressive modulus is at least an order of magnitude higher when dynamic loading is involved. Figure 9.9 shows how this modulus (measured at 10% strain) increases as a function of strain rate. Even higher moduli have been recorded for the dynamic modulus of cartilage when the dynamic loading is superimposed on

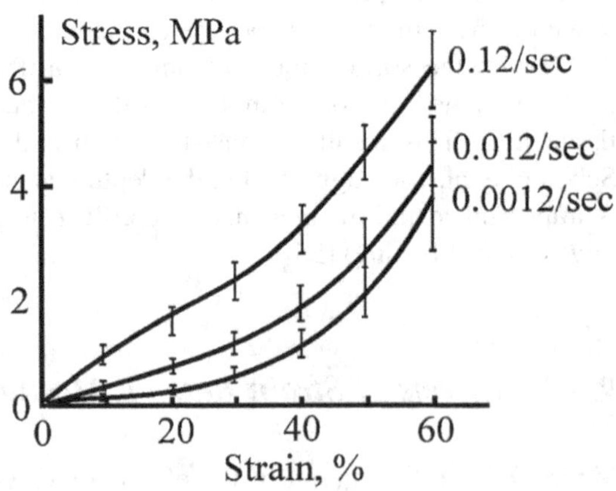

Fig. 9.8 Stress–strain curves for cartilage at three different strain rates [14]

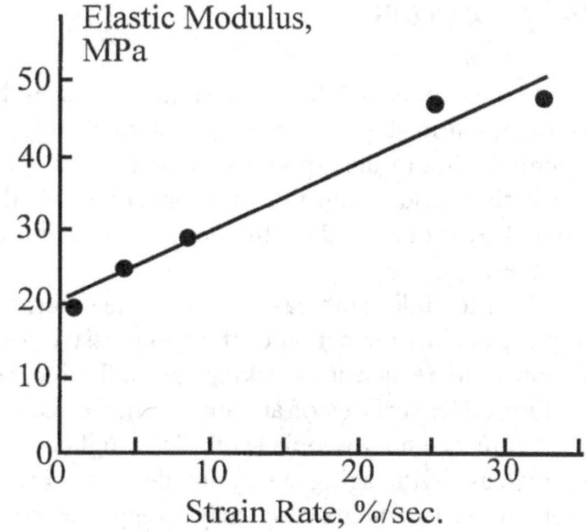

Fig. 9.9 Compressive elastic modulus vs. strain rate for bovine articular cartilage [5]

the creep strain. (Imagine the cartilage in your knees creeping into a compressed state while you stand waiting for a traffic light to change, then being dynamically loaded as you step off the curb.) Under these conditions, the cartilage behaves as a linear elastic material with a modulus of 100–500 MPa, depending on the amount of creep strain [4].

In summary, the mechanical properties of cartilage are partly determined by fluid flow through the collagen–proteoglycan matrix but models employing fluid flow through a porous elastic solid only work at very slow loading rates and fail to predict behavior at higher, more physiological rates of loading. Thus, tissue permeability is an important factor for slower loading rates but other less understood contributors to viscoelasticity, e.g., biopolymer interactions, also play an important role in the material behavior of cartilage. The relevance of viscoelasticity and fluid flow becomes clearer in the next section. Unlike most engineered bearings, synovial joints do not involve lubricants acting on rigid, inert surfaces. Joint lubricants are composed of water and hydrated polymers that are small enough to flow into and out of the cartilaginous bearing surfaces and larger molecules that are sieved and bound to joint surfaces. As a further complication, the bearing surfaces themselves deform significantly to accommodate both load and lubrication.

9.5 Mechanical Properties of the Synovial Joint Space

Synovial joints have extremely low friction coefficients, about equal to an ice skate on ice. Just how this happens is not fully understood, but much can be deduced on the basis of research on joints themselves and on the engineering lubrication theory.

9.5.1 Friction

Friction is produced between sliding surfaces because they cannot be perfectly smooth, but have projections that interlock (Fig. 9.10). Frictional forces arise from interdigitation of the asperities on the two surfaces. The compressive force between the sliding surfaces must be overcome to enable the interdigitating asperities to slide up and over one another. In addition, adhesive forces may exist between the two surfaces.

Thus, to slide more easily past one another, the two surfaces must be separated slightly against the action of the compressive force. The primary purpose of lubrication is to reduce interlocking and adhesion between the asperities on the two surfaces. The surfaces of articular cartilage seem to have ridges not unlike those in the figure, at a microscopic level. The roughness of these surfaces has been observed to increase with aging and joint disease. However, drying of the tissue during preparation for microscopy may exaggerate such features, and there is currently uncertainty about the magnitude of the surface ridges on cartilage.

The two basic laws of friction are known as Amontons' laws

- The friction force is directly proportional to the applied load.
- The friction force is independent of the apparent area of contact.

The friction force is defined as the shear force required to pull one surface across another. The friction coefficient, μ, is just the ratio of the friction force, F_F, to the applied load, or normal force, N, which pushes the surfaces together:

$$F_F = \mu N \tag{9.11}$$

Static friction coefficients usually are greater than sliding friction coefficients. Typically, friction coefficients between unlubricated surfaces are about 0.3 and those in lubricated machinery bearings are 0.05 or less (Table 9.4). Friction coefficients in synovial joints are among the smallest ever measured, about 0.001–0.005.

The reason that the friction force is independent of the bearing surface area is that when the area increases, the number of barriers to sliding increases, but the compressive force across each one decreases. Therefore, making an articular surface larger reduces the continuum level or macroscopic stresses in the tissues of the joint, but in principle it does not affect the frictional force.

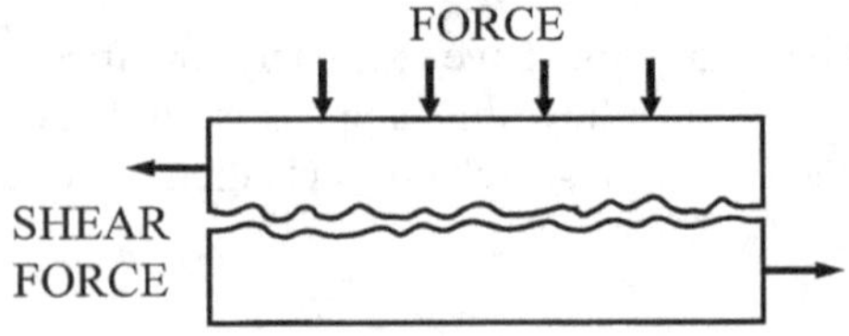

Fig. 9.10 Representation of two surfaces sliding on one another. The required shear force must exceed the frictional force resulting from interactions between the asperities of the two surfaces

Table 9.4 Typical friction coefficients

Materials and conditions	μ
Rubber on concrete, wet or dry, static	1.0
Brake material on cast iron, clean and dry, static	0.4
Brake material on cast iron, lubricated with mineral oil, static	0.1
Graphite on steel, static	0.1
Hickory on dry snow, waxed, 4 m/s, −3 °C, dynamic	0.18
Hickory on dry snow, unwaxed, 4 m/s, −3 °C, dynamic	0.08
Ice on ice, 4 m/s, 0 °C, dynamic	0.02
Articular cartilage in joints, dynamic:	
Human knee	0.005–0.02
Porcine shoulder	0.02–0.35
Canine ankle	*0.005–0.01*

Values from Refs. [27–32]

9.5.2 Wear

Wear occurs in a bearing because projections become broken by crashing into one another or tear off because of adhesion, or because foreign bodies create additional "mountains" that temporarily lodge on one surface and gouge the opposite surface. These phenomena occur in synovial joints as well, causing the cartilage surfaces to wear away over time. Foreign bodies play little part in the wear of normal joints, but can play a major role in prosthetic joints, where wear particles become foreign bodies and accelerate the wear problem. They can also migrate to the periphery of the joint, contact the synovial membrane, and cause an inflammatory response that induces chondrocytes to produce catabolic enzymes that digest the cartilage. This response may also cause bone resorption and pain. Reduced friction is associated with reduced wear (Table 9.4).

9.5.3 Types of Lubrication

There are several different mechanisms by which lubrication can be achieved. Each of the following mechanisms is thought to play a role in the lubrication of synovial joints.

Boundary Lubrication

Placing a "boundary" layer (i.e., an extremely thin layer, perhaps one molecule thick) of almost any liquid-like material between the bearing surfaces holds the surface projections slightly apart, reduces adhesion, and keeps the projections from interlocking (Fig. 9.11).

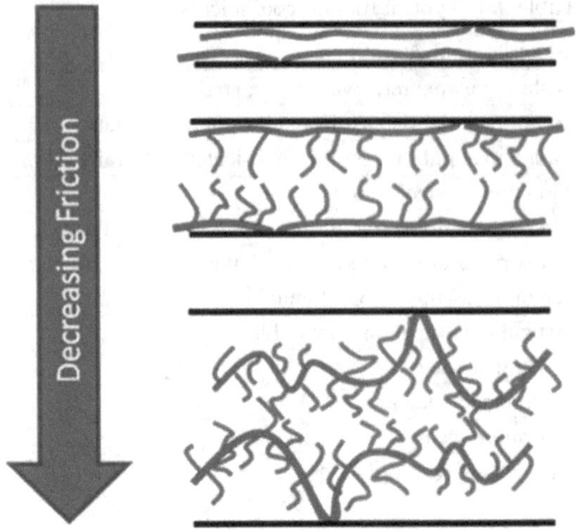

Fig. 9.11 Polymer lubricants (red and blue) between cartilage surfaces (black lines) greatly reduce friction by trapping a water film (water molecules not drawn)

Squeeze Film Lubrication

Another means to protect the joint surfaces is to separate them by the creation of a pressurized layer of fluid. One way to achieve this is as follows. When you run and your leg swings forward, its knee joint is minimally loaded, allowing the cartilage surfaces to separate and minimizing the friction. When this leg enters its stance phase, body weight pushes the femoral condyles down through the synovial fluid toward the tibial side of the joint. When contact is made, frictional forces greatly increase. The viscosity of the fluid resists the descending femur and increases the amount of time required for the two sides of the joint to meet. In the meantime, friction remains low. The bearing surface separation produced by such load-and-unload cycles is called squeeze film lubrication.

A slightly different but related mechanism for pressurization in joints that improves lubrication is the tough seal formed by the collagenous labrum that encloses articulating surfaces [33]. The labrum prevents fluid from being pushed out when the joint is loaded. If the labral seal is tight, then the joint fluid will be pressurized during loading and will reduce the friction in the joint. Squeeze film lubrication does not depend on the anatomy of the whole joint but on having the lubricant trapped between the surfaces, so it is not affected if the labrum and/or meniscus are damaged. It is not known at this time whether lack of meniscal sealing contributes to the high incidence of osteoarthritis of the knee after damage to the menisci, but it is reasonable to speculate that damage to the joint seal could play a role in increased friction and later development of osteoarthritis.

Fig. 9.12 Hydrodynamic lubrication results in complete separation of the surfaces

Hydrodynamic Lubrication

A way to achieve prolonged fluid pressure to hold up body weight and separate the joint surfaces is to use parallel rather than perpendicular relative motion between them, that is, sliding motion of one surface across the other (Fig. 9.12). The fluid next to each surface sticks to it. As one surface moves across the other, friction between adjacent fluid molecules drags the intervening fluid into the space between them, creating hydrodynamic pressure that forces them very slightly, but completely, apart. The friction force is essentially only that between the fluid molecules. For this kind of lubrication to work, the bearing or joint must be moving rapidly. Thus, it can function in synovial joints only during flexion or extension motions.

The principal argument against hydrodynamic lubrication in joints, and in favor of boundary lubrication, is the obvious fact that these bearing surfaces are not in continuous motion. Indeed, they are stationary much of the time and, it could be argued, often move slowest when carrying the highest loads. Experiments conducted so that only boundary lubrication could operate show remarkably low friction coefficients for cartilage on cartilage, supporting such a hypothesis. It has been suggested that the boundary lubricant could be (1) lipids in the cartilage (because fat solvents increase the friction coefficient), (2) a hyaluronan gel produced by synovial fibroblasts and precipitated from synovial fluid, or (3) a lubricating glycoprotein called lubricin that is found on both synovial lining and the cartilage surface [34–36]. Later, a phospholipid called dipalmitoyl phosphatidylcholine (DPPC) was put forth as the active boundary lubricant in synovial joints [37].

Elastohydrodynamic Lubrication

This term refers to situations in which the bearing surfaces are compliant enough (i.e., have a low elastic modulus) so that they deform significantly under the pressures produced in the lubricating fluid as it passes over them. It seems clear that this applies to articular cartilage, but the difficulty in mathematically modeling the situation has prevented a credible demonstration of elastohydrodynamic effects in synovial joints.

Weeping Lubrication

The source of the water that fills the pores of cartilage is the synovial fluid in the adjacent joint space. However, the hyaluronic acid molecules are too large to pass into the pore spaces. Therefore, the fluid that is contained in the pores of cartilage is actually a filtrate of synovial fluid. Lewis and McCutchen postulated that, just as this fluid is expressed from a cartilage specimen during a compression test, it is also exuded into the load-bearing region when two joint surfaces are pressed together during joint flexion [38]. This flow of fluid under pressure was hypothesized to contribute to separation of the joint surfaces. As in the case of elastohydrodynamic lubrication, it has been difficult to demonstrate convincingly that such an effect occurs.

Boosted Lubrication

This theory of joint lubrication was proposed independently by Maroudas and Walker et al. [39, 40]. The idea is quite simple. The large hyaluronic acid molecules of the synovial fluid that get left behind when synovial fluid is transported into the cartilage is hypothesized to form a gel-like layer on the articular surface. This "concentrated lubricant" is postulated to play a major role in the lubrication process. The problem is to understand where and when on the joint surface this happens. Weeping lubrication claims that fluid flows out of the loaded surface, not in. Squeeze film lubrication has the synovial fluid being squeezed out to each side of the contact region. The proponents of boosted lubrication claim, on the other hand, that because of the microscopic ridges and grooves on the surface, when the two cartilage surfaces are pressed together, pools of synovial fluid are trapped and the water is forced into rather than out of the cartilage. These conflicting ideas have generated debate.

9.5.4 Joint Lubrication Studies

In the early 1900s, simple experiments were done in which joints were set to swinging by attaching one limb segment to a pendulum, and the rate of decay of the swinging motion was studied [29, 41]. The fact that the amplitude of the motion decayed nonlinearly led to the conclusion that viscous damping was at play, providing support for the theory that a layer of fluid separated the joint surfaces, and hydrodynamic lubrication existed in the joint. Many years later, Charnley pointed out that the nonlinear decay could be for another reason: that the soft tissues about the joint resist large motions more than small motions [42]. Soon, Barnett and Cobbold did an experiment that showed that this indeed happens [43] (Fig. 9.13). A dog's ankle joint was rigged as the fulcrum of a pendulum, showing that the

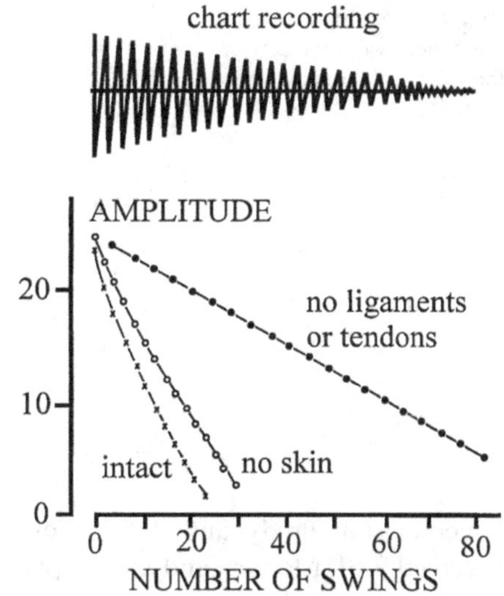

Fig. 9.13 Experimental results from an early pendulum test of synovial joint lubrication. The upper trace shows oscillation decaying over time. The lower graph shows that the nonlinearity of the decay was a function of the adjacent soft tissues [42]

resistance to motion became less, and the decay in motion more linear, when the adjacent soft tissues were removed. However, to properly interpret this experiment, it needs to be understood that the joint was rigged "upside down" so that the pendulum weight compressed the joint surfaces together continuously. Thus, this experiment did not examine the possibility of short-term hydrodynamic lubrication following an unloaded state.

Another pendulum experiment suggests that hydrodynamic lubrication does occur in joints after all if the joint load is not too great. For boundary lubrication, the friction coefficient, μ, is independent of the joint contact force, N. When the joint surfaces are separated by a layer of fluid, as in hydrodynamic lubrication, the friction coefficient will decrease as the joint load increases because of the relationship $\mu = F/N$ and the fact that, in this case, the frictional force, F, is largely independent of the normal force, N. Figure 9.14 shows how the friction coefficient for the distal interphalangeal joint of a human finger decreases with increasing force across the joint (expressed as mass), strongly suggesting that the lubrication mechanism is not simply boundary lubrication. However, the plateauing of the curve when the force exceeds 900 g indicates that sufficient joint force can reduce the fluid layer between the articular surfaces to a thin film (i.e., boundary lubrication).

Dintenfass argued that elastohydrodynamic lubrication was probably occurring because the required pressures would deform cartilage significantly relative to the fluid film thickness [44, 45]. The minimal fluid layer necessary to produce hydrodynamic lubrication should be greater than the surface roughness of cartilage, which has been measured by several investigators at 1–6 μm [44]. Several analyses of elastohydrodynamic lubrication as a mechanism for joint lubrication have yielded fluid film thickness of less than 1 μm, however [28]. To circumvent this objection, Dowson and Jin argued that the physiologic pressures produced in joints would flatten the unevenness of the cartilage surface, permitting

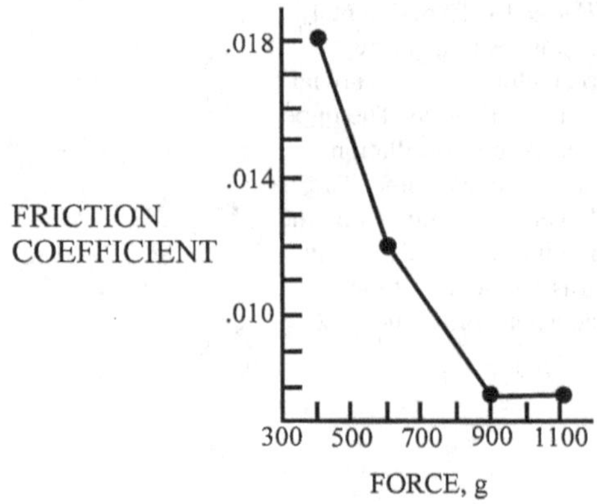

Fig. 9.14 Friction coefficient vs. force across the joint for a human finger joint [42]

"microelastohydrodynamic lubrication" to work, but this idea has not been widely accepted [46]. More recently, attempts to refine the measurement of surface roughness of cartilage have suggested that the in vivo roughness is much less than that measured on dried specimens, and that the measurement probe may "cut through" a superficial layer of smooth material and instead register the subsurface structure (e.g., proteoglycan and collagen fiber bundles). Thus, just plain hydrodynamic lubrication may be an adequate theory after all.

9.5.5 Back to Poroelastic Theory

Some of the many questions about the role played in lubrication by the movement of fluid into and out of the cartilage have been clarified by extension of the biphasic theory to the problem of lubrication, as opposed to the mechanical properties of cartilage [28]. Analysis indicates that when a parabolically shaped porous indenter is slid across a layer of cartilage at physiologic speeds, fluid is expressed from the cartilage into the leading-edge region and fluid is imbibed back into the cartilage near the trailing edge. Fluid "weeping" from the cartilage begins well in advance of the arrival of the load, and as the load increases, so does the flow. Then, elastic recoil of the cartilage allows fluid to reenter the tissue when the load is past, presumably leaving a layer of hyaluronic acid gel on the surface. Thus, the theory predicts that both the weeping and boosted lubrication theories have a role to play, as do boundary and hydrodynamic lubrication. It must be understood, however, that the values of many of the variables in the theoretical model are poorly defined, and much remains to be done in taking full advantage of the potential of the model to elucidate the etiology of various joint diseases.

9.6 Diversity of Joint Architecture

In concluding this chapter, we should take care to point out that there is great variety in the anatomy of synovial joints [47]. The implications of this with respect to lubrication are unclear, but it certainly raises the prospect that different combinations of the proposed lubrication mechanisms may predominate in different kinds of synovial joints. To make this point more tangible, we briefly describe four human synovial joints having very different anatomical conformations. Often, there seems to be a trade-off between the need for range of motion and the need for stability under load.

Temporomandibular Joint
Motions of the mandible during chewing and speaking are extremely complex, with up–down, side-to-side, and in–out motions that must be regulated with great precision. Moreover, the two temporomandibular joints are linked by the mandible and the skull, so that motion in one joint produces motion in the other joint. It is clear that a highly constrained joint would be inadequate for these needs. The articular surface of the mandible does not directly bear on the articular surface of the temporal bone; instead, elastic fibrous articular disks (menisci), 2- to 3-mm thick, lie between them. Rolling and sliding of the mandible on the disk, and the disk on the temporal bone, allow considerable freedom of motion, and it is likely that the presence of the disk is the evolutionary solution to reduce shear forces on the cartilage surface in a joint that slides. Stability is provided primarily by muscles and, to a lesser extent, by ligamentous and bony restraints.

Trapeziometacarpal Joint of the Thumb
The base of the thumb articulates with two interlocking saddle-shaped joint surfaces; that is, each surface is concave in one direction and convex in the orthogonal direction. The two surfaces are not precisely congruous. This geometry allows considerable freedom of motion when the thumb is unloaded. When the thumb is loaded (e.g., in a pinch grip), the two saddles push together and the joint geometry contributes to stability. Relative motion of the two joint surfaces is entirely by sliding because a rolling motion is not possible with this geometry. Because friction is clearly increased in sliding contact, this restriction has lubrication implications.

Hip Joint
The ball-and-socket joint of the hip consists of the spherical femoral head tethered to the socket-shaped acetabulum of the pelvis by a short ligament, seemingly a situation in which joint contact is continuous. However, the convex part is actually larger in diameter than its concave articulation, so that at low loads joint contact is confined to the periphery. As loads increase, the acetabulum undergoes elastic deformation and the two surfaces become congruent. When the joint is again unloaded, the residual force in the acetabulum pops the femoral head partially back out of the joint. This was nicely demonstrated in a cadaver study by Bay and colleagues who measured substantial periacetabular strains as a function of hip loading [48]. During gait, the femoral head is not only rotating within the acetabulum, but is moving in and out of

the acetabulum with each step. This movement may support squeeze-film lubrication in this large, heavily loaded joint. The hip joint also has a strong labrum that seals the joint when it is healthy. In fact, to remove a healthy femoral head from its acetabulum in the anatomy laboratory, it is necessary to "break the seal" of the labrum to allow air to enter the joint and permit separation of the femoral and acetabular joint faces.

Tibiofemoral Joint at the Knee

The distal femur terminates in two rounded condyles that articulate with the much flatter surface of the proximal tibia. This arrangement allows for a combination of rolling and sliding as the joint moves, but contributes little to joint stability, which depends on dynamic muscle action and on ligamentous restraints. These stability mechanisms are often inadequate, and soft tissue injuries about the knee are common. The menisci are interposed between the joint surfaces, presumably to help distribute forces across the joint, and may help provide the extraordinarily narrow fluid spaces necessary for hydrodynamic lubrication. The menisci also have a stabilizing and sealing function that is noticeable in the anatomy laboratory. This is particularly true on the medial side of the joint where the fibers of the joint capsule interdigitate with the outer rim of the meniscus when separating the femur from the tibia; the seal between the distal femur and the menisci must be broken to allow air to enter the space if the joint is to be easily separated.

These few examples give a glimpse of the diversity of joint architecture. Obviously, caution is required when applying any other theory to synovial joints. It seems likely that several different lubrication mechanisms function to varying degrees in different joints.

9.7 Summary and Further Reading

We have described the mechanics of cartilage and joint lubrication in synovial joints. Arthritic diseases involve interactions between mechanical damage and inflammatory responses within the joint and the adjacent bone. The mechanics of cartilage are dominated by its high water content, high negative charge density, and the matrix of proteoglycans and collagen molecules, which retain the water during compression and resist tensile stresses, respectively. Under static loading, electrostatic repulsion between the fixed charges explains most of the modulus of the tissue. At very slow loading rates, tissue viscoelasticity is explained by fluid flow, and permeability is an important variable. The biphasic theory has been extensively used for interpreting the results of compression tests and predicting the mechanical behavior of cartilage, but it is inaccurate for loading rates other than very slow ones. For more rapid loading, the viscoelasticity of the matrix is an essential part of the mechanical response. This has been modeled using phenomenological viscoelasticity or polymer mechanics with some success. The polymer mechanics approach has the advantage of being a mechanistic theory, but it is presently less developed than the phenomenological approach. The lubrication of synovial joints is also poorly understood, but is thought to involve both boundary and hydrodynamic lubrication mechanisms. The flow of synovial fluid in and out of the cartilage on the joint surfaces seems to be important as well.

For additional information on this subject, there are several books: *Tribology of Natural and Artificial Joints*, by Dumbleton [49], *Biomechanics of Diarthroidial Joints*, edited by Mow et al. [50], and *Mechanics of Human Joints*, edited by Wright and Radin [51]. The second edition of the book *Basic Orthopaedic Biomechanics*, edited by Mow and Hayes, devotes four excellent chapters to synovial joints [52]. For a general overview of cartilage and its relationship to bone: Bones and Cartilage: Developmental and Skeletal Developmental Biology, by Hall is a very interesting overview that takes an evolutionary standpoint [53]. A general overview of tissue mechanics that includes cartilage is: *Mechanics of Biological Tissue*, Editors: GA Holzhapel and RW Ogden, Springer-Verlag Berlin Heidelberg, 2006 [54]. This book is interesting for including modeling of tissue growth integrated with mechanical property development.

References

1. June RK, Mejia KL, Barone J, Fyhrie DP. Cartilage stress-relaxation is affected by both the charge concentration and valence of solution cations. Osteoarthr Cartil. 2009;17:669–76.
2. Nimer E, Schneiderman R, Maroudas A. Diffusion and partition of solutes in cartilage under static load. Biophys Chem. 2003;106:125–46.
3. Mow VC, Mak AF. Lubrication of diarthroidial joints. In: Skalak R, Chien S, editors. Handbook of bioengineering. New York, NY: McGraw-Hill; 1987. p. 5.1–5.34.
4. Higginson GR, Snaith JE. The mechanical stiffness of articular cartilage in confined oscillating compression. Eng Med. 1979;8:11–4.
5. Radin EL, Paul IL, Lowy M. A comparison of the dynamic force transmitting properties of subchondral bone and articular cartilage. J Bone Joint Surg. 1970;52A:444–56.
6. Unsworth A. Cartilage and synovial fluid. In: Dowson D, Wright V, editors. An introduction to the biomechanics of joints and joint replacement. London: Mechanical Engineering; 1981. p. 107–14.
7. Yamada H. Strength of biological materials. Huntington: Krieger; 1973.
8. Benninghoff A. Form und bau der gelenkknorpel in ihren beziehungen zur funktion. Erste Mitteilung Zeitschrift fu ⍰ r Anatomie und Entwicklungsgeschichte. 1925;76:43.
9. Hayes WC, Keer LM, Herrmann G, Mockros LF. A mathematical analysis for indentation tests of articular cartilage. J Biomech. 1972;5:541–51.
10. Woo SL-Y, Akeson WH, Jemmott GF. Measurements of nonhomogeneous, directional mechanical properties of articular cartilage in tension. J Biomech. 1976;9(12):785–91.
11. Williamson AK, Chen AC, Masuda K. Tensile mechanical properties of bovine articular cartilage: variations with growth and relationships to collagen network components. J Orthop Res. 2003;21:872–80.
12. Mansour JM, Mow VC. The permeability of articular cartilage under compressive strain and high pressures. J Bone Joint Surg. 1976;58A:509–16.
13. Mow VC, Kuei SC, Lai WM, Armstrong CG. Biphasic creep and stress relaxation of articular cartilage in compression. Theory and experiment. J Biomech. 1980;102:73–84.
14. Buckwalter JA, Mow VC. Cartilage repair in osteoarthritis. Philadelphia, PA: Saunders; 1992.

15. Woo SL-Y, Mow VC, Lai WM. Biomechanical properties of articular cartilage. In: Skalak R, Chien S, editors. Handbook of bioengineering. New York, NY: McGraw-Hill; 1987. p. 4.1–4.44.
16. Armstrong CG, Lai WM, Mow VC. An analysis of the unconfined compression of articular cartilage. J Biomech Eng. 1984;106:165–73. https://doi.org/10.1115/1.3138475.
17. Soltz MA, Ateshian GA. A conewise linear elasticity mixture model for the analysis of tension-compression nonlinearity in articular cartilage. Trans ASME J Biomech Eng. 2000;122:576–86.
18. Huang CY, Mow VC, Ateshian GA. The role of flow-independent viscoelasticity in the biphasic tensile and compressive responses of articular cartilage. J Biomech Eng. 2001;123:410–7.
19. Fyhrie DP, Barone J. Polymer dynamics as a mechanistic model for the flow independent viscoelasticity of cartilage. J Biomech Eng. 2003;125:578–84.
20. June RK, Neu CP, Barone J, Fyhrie DP. Polymer mechanics as a model for short term and flow-independent cartilage viscoelasticity. Mater Sci Eng. 2011;31:781–8.
21. Boyd R, Smith G. Polymer dynamics and relaxation. Cambridge: Cambridge University Press; 2007.
22. Lai WM, Hou JS, Mow VC. A triphasic theory for the swelling and deformation. Trans ASME J Biomech Eng. 1991;113:245–58.
23. Buschmann MD, Grodzinsky AJ. A molecular model of proteoglycan-associated electrostaric forces in cartilage mechanics. Trans ASME J Biomech Eng. 1995;117:179–92.
24. Boroudjerdi H, Kim Y-W, Naji A, et al. Statics and dynamics of strongly charged soft matter. Phy Rep. 2005;416:129–99.
25. Schinagl RM, Ting MK, Price JH, Sah RL. Video microscopy to quantitate the inhomogeneous equilibrium strain within articular cartilage during confined compression. Ann Biomed Eng. 1996;24:500–12.
26. Silverberg JL, Dillavou S, Bonassar L. Anatomic variation of depth-dependent mechanical properties in neonatal bovine articular cartilage. J Orthop Res. 2013;31:686–91.
27. Weast RC. CRC handbook of chemistry and physics. Boca Raton, FL: CRC Press; 1984.
28. Mow VC, Soslowsky LJ. Friction, lubrication and wear of diarthroidal joints. In: Mow V, Hayes W, editors. Basic orthopaedic biomechanics. New York, NY: Raven; 1991. p. 245–92.
29. Jones E. Joint lubrication. Lancet. 1936;230:1043–4.
30. McCutchen CW. The frictional properties of animal joints. Wear. 1962;5:1–17.
31. Linn F. Lubrication of animal joints: I the arthrotrip-someter. J Bone Joint Surg. 1967;49:1079–98.
32. Linn FC. Lubrication of animal joints: II the mechanism. J Biomech. 1968;1:193–205.
33. Haemer JM, Carter DR, Giori NJ. The low permeability of healthy meniscus and labrum limit articular cartilage consolidation and maintain fluid load support in the knee and hip. J Biomech. 2012;45:1450–6.
34. Little T, Freeman M, Swanson A. Experience on friction in the human joint. In: Wright V, editor. Lubrication and wear in joints. London: Sector Publishing; 1969. p. 110–4.
35. Swann DA, Slayter HS, Silver FH. The molecular structure of lubricating glycoprotein-I, the boundary lubricant for articular cartilage. J Biol Chem. 1981;256:5921–5.
36. Jay GD, Britt DE, Cha CJ. Lubricin is a product of megakaryocytes stimulating factor gene expression by human synovial fibroblasts. J Rheumatol. 2000;27:594–600.
37. Hills BA. Oligolamellar lubrication of joints by surface active phospholipid. J Rheumatol. 1989;16:82–91.
38. Lewis PR, McCutchen CW. Experimental evidence for weeping lubrication in animal joints. Nature. 1959;184
39. Maroudas A. Hyaluronic acid films. Proc Inst Mech Eng. 1967;181:122–4.
40. Walker PS, Dowson D, Longfield MD, Wright V. Boosted lubrication in synovial joints by fluid entrapment and enrichment. Ann Rheum Dis. 1968;27:512–20.
41. Jones E. Joint lubrication. Lancet. 1934;228:1426–7.
42. Charnley J. The lubrication of animal joints in relation to surgical reconstruction by arthroplasty. Ann Rheum Dis. 1960;19:10–9.

43. Barnett CH, Cobbold AF. Lubrication within living joints. J Bone Joint Surg. 1962;44B:662–74.
44. Dintenfass D. Lubrication in synovial joints. Nature. 1963;197:496–7.
45. Dintenfass D. Lubrication in synovial joints: a theoretical analysis. J Bone Joint Surg. 1963;45: 1241–56.
46. Dowson D, Jin ZM. Micro-elastohydrodynamic lubrication of synovial joints. Eng Med. 1986;15:63–5.
47. Adams L. The anatomy of joints related to function. In: Wright V, Radin E, editors. Mechanics of the human joints. New York, NY: Dekker; 1993. p. 27–81.
48. Bay BK, Hamel AJ, Olson SA, Sharkey NA. Statically equivalent load and support conditions produce different hip joint contact pressures and periacetabular strains. J Biomech. 1997;30: 193–6.
49. Dumbleton JH. Tribology of natural and artificial joints. Amsterdam: Elsevier; 1981.
50. Mow VC, Ratcliffe A, Woo SL-Y. Biomechanics of diarthroidal joints. New York, NY: Springer; 1990.
51. Wright V, Radin EL. Mechanics of human joints. New York, NY: Dekker; 1993.
52. Mow VC, Hayes WH. Basic orthopaedic biomechanics. Philadelphia, PA: Lippincott-Raven; 1997.
53. Hall BK. Bones and cartilage: developmental and skeletal developmental biology. San Diego: Elsevier; 2005.
54. Holzhapel G, Ogden R. Mechanics of biological tissue. Berlin: Springer; 2006.

Appendices

Appendix I

Chapter 2

2.1. Solve the second part of the hip force problem (the force equilibrium equations) by considering a free-body diagram of the weight-bearing lower extremity and using the result $P = 2W$. Remember to include the weight of the extremity itself. What is the disadvantage of using this free-body diagram to solve the moment equation?

2.2. Suppose that the distance b in women's pelves is 10% greater than men's because of their greater pelvic width. How does this change their abductor muscle force magnitude and the magnitude and direction of their hip joint reaction force, assuming the moment arm of the abductor muscles is independent of gender?

2.3. A male gymnast is performing the "iron cross" exercise on the rings, in which the arms are held straight out with a ring grasped in each hand, and the body hangs motionless and vertical between the rings. Consult an anatomy book to determine the principal adductor muscles for the arm in this position, and estimate their angle of pull and insertion point. Draw a free-body diagram and calculate the approximate magnitude of the adductor muscle force and the magnitude and direction of the reaction force in the glenohumeral (shoulder) joint during the maneuver. State the principal assumptions necessary for your calculation.

2.4. Suppose, in Exercise 2.3, that the ropes holding the rings do not rise vertically from the gymnast's hands, but pull outward or inward at an angle θ. How does this angle affect the forces in the shoulder joint?

M. R. Allen, J. Wallace, *Skeletal Tissue Mechanics*,
https://doi.org/10.1007/978-3-031-90609-1

2.5. For the equine fetlock joint, work out the magnitude and direction of the joint force when a horizontal component of the ground reaction force (i.e., Gx in Fig. 2.10) is present. Let Gx = ±0.20 Gy, and assume, as before, that $L = 0.2$ m, $\Phi = 0$, $\Theta = 40°$, and $t = 0.05$ m.

Chapter 3

3.1. Typically, a newborn infant's tibia is about 6.8 cm long. Estimate the present length of your tibia, then estimate the average rate of growth in the tibial growth plates, in millimeters/year and in micrometers/day. Again, assume that longitudinal growth of the tibia stops at age 14 in females and age 16 in males.
3.2. The rate of growth of a growth plate is equal to the daily chondrocyte production rate for each column multiplied by the height of the maximally hypertrophied cells. Explain why this is so in a steady state situation and state the necessary assumptions.
3.3. Bone can be considered a 2-phase, fiber-reenforced material. Describe the nature of this composite structure at the nano-, micro-, and macrostructural levels. Compare and contrast it with existing implant materials.
3.4. Compare the functional adaptations of bone, articular cartilage, and tendon/ligament with regard to their structure and their respective roles in skeletal support and load transmission. How do the morphological differences at each level of organization permit them to efficiently carry out their respective functions. Explain how these tissues act together to fulfill their roles and to protect each other from damage during the aging process.

Chapter 4

The following exercises are intended to start you thinking in terms that can be used to mathematically model the consequences of remodeling processes that are varying in time. To do this, it is necessary to keep track of how many BMUs are currently in each stage of remodeling. These populations depend not on the current activation frequency but on what Ac.f was at some time in the past. For example, BMUs currently in the resorbing phase were activated or "born" between the present time and RP days ago. BMUs in the reversal phase were born before that, and refilling BMUs were born even earlier than that. Also, keep in mind that we are considering the remodeling going on in a representative, two-dimensional cross-section of cortical bone, e.g., Fig. 4.7. In this context, "activation" of a BMU occurs when its leading osteoclasts first break the plane of the section, creating a new resorption cavity. It helps to imagine that you can observe such a cross section in your own femur as BMUs tunnel across it, excavating and refilling their tunnels.

4.1. Imagine a cross section through your femur. Assume that 5% of the bone in this cross section is replaced annually by osteonal remodeling. The osteons average 250 μm in diameter. What would the osteonal BMU activation rate (new osteons/mm^2 per year) have to be to produce the 5% turnover rate?

4.2. In Exercise 4.1, if each osteonal BMU produced a new Haversian canal that was 50 μm in diameter, what would be the rate of increase in bone porosity (mm^2 of canals/mm^2 of cross section/year)? Assume that the new osteons do not obliterate preexisting Haversian canals.

4.3. Draw a horizontal line representing time. Place a tick mark representing the present time at the left end of the line so that past time is measured to the right. Let the time scale be 1 cm/week. Assuming the BMU resorption period is RP = 21 days = 3 weeks, mark the region of the line containing the "birthdays" (activation days) of all BMUs that are currently resorbing bone. Assuming the reversal period is RvP = 7 days = 1 week, mark the time span during which all BMUs presently in the reversal phase were activated. Finally, assuming the refilling phase lasts FP = 56 days = 8 weeks, show the time span during which all currently refilling BMUs were "born."

4.4. Now think about the cross-sectional view of a single osteon. Assume that the mean diameter of an osteonal cement line is 200 μm and that of the mean Haversian canal is 40 μm. Using these numbers, calculate Rs.R, the mean rate at which bone is resorbed in each osteon during its resorption phase, in μm^2/day, and F.R, the mean rate at which bone is replaced during refilling of each osteon, in μm^2/day.

4.5. Calculate N.Rs and N.F assuming the activation frequency has been decreasing linearly at a rate of 5 BMU per mm^2/year and is presently 1 BMU per mm^2/year. Use the integration methods you practiced in the earlier exercises, and the same values for RP and FP as before. Then calculate something called the "bone balance":

$$\text{B.BMU} = \text{F.R(N.F)} - \text{Rs.R(N.Rs)}$$

using F.R and Rs.R from Exercise 4.4. Based on the equation, what are the units and meaning of F.R(N.F), and Rs.R(N.Rs), and B.BMU? Compare this bone balance with that which exists when the activation frequency is constant at 1 BMU per mm^2/year, assuming everything else remains the same. Why is one number positive and the other negative? Discuss the relevance of this calculation to the porosity of cortical bone in children and adults.

Chapter 5

5.1. Suppose a human bone has a cylindrical shaft whose periosteal and endosteal diameters are 2 and 1 cm, respectively. Using the data in Table 5.1, calculate the yield and ultimate load for this bone when loaded in simple compression.

(Ignore buckling as a possibility.) Then calculate the torques (twisting moments) required to yield and break the bone by twisting it. When the outer fibers of the bone have reached the ultimate shear stress, are the fibers on the endosteal surface above or below the shear yield stress?

5.2. Suppose the bone described in Exercise 5.1 has its shear strength diminished by 20% by the accumulation of Haversian canals with age. Assume also that 1 mm of bone is lost from the endosteal surface during this period of time. How much must the periosteal diameter be increased to compensate for these changes? If all these changes occurred simultaneously over a 30-year period, what would the mean periosteal apposition rate be, and how would it compare to typical values during the refilling of an osteonal BMU?

5.3. Suppose a long bone diaphysis has cleanly fractured straight across and is healing through the development of a fracture callus. Sketch a cross section through the callus, assuming circular geometry for it and the original bone. The outer diameter of the callus is 3 cm. Sketch in the original bone cross section, which had periosteal and endosteal diameters of 2 cm and 1 cm, respectively. Assume that no strength has developed between the cortices of the original bone, so that the strength of the healing fracture is entirely provided by the periosteal and medullary calluses. Calculate the bending moment required to fracture the original, intact cortex, assuming that its ultimate bending stress was 200 MPa. Now calculate the cross-sectional moment of inertia of the periosteal and endosteal calluses and compare them, individually and together, to that of the original bone. Assume the material in these calluses is homogeneous. What would the ultimate stress of the callus material have to be for it to provide the same strength as the original bone? What percentage of this strength is provided by the periosteal callus?

5.4. Consider the stress in the neck of the femur of a 70-kg person during the stance phase of gait. Use what you learned in Chap. 2 about the magnitude of the hip joint force, and assume that it is directed at an angle of 20° to the neck axis. Calculate the compressive and bending stresses produced by the joint force components parallel and perpendicular to the neck axis, respectively. Choose reasonable values for the bony dimensions. Then estimate the total axial stress on the superior and inferior surfaces of the neck. How do these stresses compare to the compressive strengths of cancellous bone shown in Table 5.4?

Chapter 6

6.1. In the Lanyon–Rubin avian model for studying the effects of mechanical loading on bone, a diaphysis is isolated as shown below. Pins extend from the metal caps over the bone ends, as shown. Assume a load is applied between the ends of the pins as shown. Calculate the longitudinal strain on the upper and lower surfaces of the bone assuming it is cylindrical with the following characteristics:

Outer diameter = 20 mm
Inner diameter = 10 mm
Load = 30 N
Moment arm (d) = 40 mm
Elastic modulus = 20 GPa

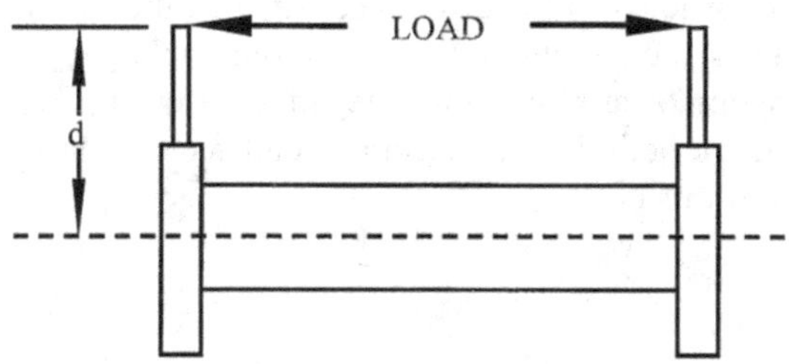

6.2. Assume that a diaphyseal cross section carries a bending moment $M = 2.5\ N\text{-}m$. Let the cross section be circular with inner and outer diameters of 30 and 20 mm, respectively, and assume that the neutral axis (NA) passes through the centroid. Draw the cross section and write an equation for the stress, $\boldsymbol{\sigma}$, as a function of angular position $\boldsymbol{\Theta}$ measured from the neutral axis. Assume that modeling adds or removes bone on the periosteal and endosteal surfaces at a rate proportional to the local longitudinal stress on the surface:

$$M_{F/R} = -K\sigma(\theta)$$

where the apposition rate $M_{F/R}$ is negative for tensile stress and positive for compressive stress and the rate constant K is the same for each kind of modeling and each surface. Sketch a line adjacent to the original periosteal surface that would represent the addition or removal of new bone at each point around the circumference after some arbitrary period of time. Do the same for the endosteal surface. (Scale your sketch so that the maximum thickness of bone added is about half the original cortical thickness.) What has happened to the bone's cross-sectional geometry? Would this change be advantageous in some way?

6.3. Reanalyze the bone in Exercise 6.2 assuming that $M_{F/R}$ is always formative and proportional to the magnitude of the longitudinal surface stress. How does the shape of this new bone distribution compare with that seen in Goodship's pig osteotomy experiment [205]? Discuss how the shape of the periosteal new bone distribution would change if M were proportional to the fourth power of the stress.

6.4. Reanalyze the bone in Exercise 6.2 assuming that MF/R is proportional to the radial gradient of stress. Assume that a positive stress gradient produces resorption ($M_{F/R} > 0$) and a negative stress gradient produces formation ($M_{F/R} > 0$). (A positive stress gradient is defined as one in which the stress becomes more tensile or less compressive, as one approaches the surface from within the bone matrix.) Let the rate constant be the same for formation and resorption and for the periosteal and endosteal surfaces. Discuss the variations of the results in Exercises 6.2, 6.3, and 6.4 and how they compare with the experimental results described in this chapter.

6.5. Consider the diaphysis of an adult's long bone subjected to a constant amount of daily torsion. Suppose that the apposition rate on the periosteal and endosteal surfaces is proportional to the difference between the average shear stress, σ_S (see Equation 6.4), and a set point or equilibrium value, $\boldsymbol{\sigma}_{sE}$. If the difference is positive, formation occurs, and if it is negative, resorption occurs. Suppose that $\boldsymbol{\sigma}_{sE}$ is the same for both the periosteal and endosteal surfaces. What would the consequences of this be?

Chapter 7

7.1. Equations 7.18 and 7.19 give the number of cycles to failure for fatigue of human cortical bone in longitudinal tension and compression, respectively. How many days would it take to achieve fatigue failure in each of these modes of loading if the normalized applied stress range, S, is selected to produce 2000 $\mu\varepsilon$ in the specimen? Assume the loading rate is 2 Hz. What experimental problems would arise from these long fatigue times? How long would the experiment take if S produced a superphysiologic 5000 $\mu\varepsilon$? Why would you not want to run the experiments at 20 Hz to reduce their duration?

7.2. Equation 7.4 states a relationship between remodeling variables and the damage burden (D_0) waiting to be repaired in an equilibrium state: $D_0 = (k_D S^q R_L)/(\pi(Cm.\ Ln.\ Rd)^2(Ac.\ F_0)F_S)$. If the loading increases, an increase in the damage burden can be avoided by increasing either the activation frequency or the diameter of the osteons, or both. Does one of these options have any advantages over the other? Think about the differences in the porosity increase that would occur during and after the remodeling, the metabolic energy required to support the cells doing the remodeling, and the effects of the altered osteonal structure on fatigue damage resistance. What are the limitations on increasing Cm.Ln. Rd? How about Ac.f_0?

7.3. Muscle fatigue may be an important cause of stress fractures. There are more muscles to control our movements than we actually need. The muscle fatigue theory of stress fracture holds that the additional muscles not only make refinements of movement possible, but also reduce the stresses in the bones when the most common movements are performed. For example, there are six principal muscles that control the foot during the toe-off phase of walking. Sharkey et al. found that the compressive strain on the dorsal surface of the human second metatarsal bone at toe-off increases from about 1700 $\mu\varepsilon$ to about 2190 $\mu\varepsilon$ when just one of these (flexor hallucis longus) stops functioning. Using Equation 7.17, find the percent decrease in fatigue life to be expected from this increase in strain.

7.4. The Pacific Crest Trail runs from Mexico to Canada through the Sierra Nevada and Cascade mountain ranges. Each year a few people succeed in walking the entire trail, starting in early spring and finishing in the fall. Many fail because of physical problems, including stress fractures. Using Equation 7.42, consider the relative effects of the distance walked and the weight of a backpack on the rate of fatigue damage formation in one of your foot or leg bones if you undertook this walk. Suppose that you ordinarily walk 5 miles/day. Plot a graph of the percent increase in damage formation rate as a function of the weight of your pack, expressed as a percent of your body weight, assuming you continued walking just 5 miles/day. Try relatively high and low values of q. Then convert this graph to a three-dimensional plot of percent change in damage formation rate vs. pack weight and distance walked per day.

Chapter 8

8.1. Equations 8.1 and 8.2 represent empirical force-strain relationships for the tendon and aponeurosis of a frog muscle-tendon unit. In these, force is expressed as percent of the maximum tetanic muscle force. Calculate the tendon and aponeurosis strains when the muscle is exerting this maximum tetanic force. Which tissue has the greater strain? In general, what is the relationship between the tendon and aponeurosis strains for this muscle-tendon unit?

8.2. You are testing anterior cruciate ligament (ACL) reconstructions. You recorded the following load-deformation data for an experimental femur–ACL–tibia reconstruction in an animal model. Failure was by bony avulsion at the tibia. The ACL cross-sectional area was 150 mm^2 and the bone-to-bone resting length of the ligament was 3.5 cm. Calculate the important structural and material properties of the reconstruction.

Displacement, mm	Load, *N*	Displacement, mm	Load, *N*	Displacement, mm	Load, *N*
0.00	0	0.12	214	0.30	1908
0.01	0	0.14	379	0.32	1915
0.02	0	0.16	583	0.34	1940
0.03	0	0.18	796	0.36	2111
0.04	0	0.20	1000	0.37	2037
0.05	0	0.22	1202	0.38	1540
0.06	10	0.24	1406	0.39	1120
0.08	37	0.26	1611	0.40	821
0.10	96	0.28	1792	0.41	698

8.3. It was noted in the main chapter that tendon temperatures in racehorses may exceed 44 °C. Suppose a typical tendon has an elastic modulus of 1 GPa and cycles between 0.5% and 2% strain during running. If the stress-strain relationship is linear in this range, what is the fluctuation in strain energy density as the tendon stress rises and falls? If that energy were expressed as heat, how much would it be in calories/ml of tendon/cycle? (The mechanical equivalent of heat is 4.19 joules/cal.) If the specific gravity and specific heat capacity of tendon tissue are similar to those of water (1 g/ml and 1 cal/g per °C, respectively), what percent of the strain energy must be converted to heat to raise the tissue temperature 1 °C in 1000 load cycles, assuming adiabatic conditions (i.e., no heat leaves the site)? Is it plausible that tendons could be heated by loading during running?

8.4. In Chap. 2 we learned that the skeleton is arranged as a system of levers that amplify the small displacements produced by contracting muscles. Instruments known as micromanipulators use levers to demagnify finger motions. In this chapter you have read that the flexor tendons of the fingers themselves have characteristics that serve to increase fine motor control; that is, they demagnify somewhat the motions of the fingers, which would otherwise be produced by the muscle contractions. Using the simple joint shown in Fig. 8.17 as a model, derive an equation to show how this can be achieved by making flexor tendons longer and more compliant. How would the tendon's length, diameter, and elastic modulus affect the resulting motion?

8.5. When a kangaroo switches from "pentapedal" (four limbs and tail) locomotion to hopping, its oxygen consumption drops, presumably because it then stores more energy in elastic tissues. One of these tissue "springs" in kangaroos and other animals is the Achilles tendon. A kangaroo's Achilles tendon was found to be 1.5 cm in diameter and 35 cm long. If both Achilles tendons had an elastic modulus of 1 GPa and were loaded to 2% strain, how much strain energy would they possess? Based strictly on energy considerations, how high could this amount of energy lift a 40-kg kangaroo?

Chapter 9

9.1. Determine the relative amounts of strain energy absorption in the cartilage and bone of the proximal half of the tibia when it is loaded uniformly over the articular surface by an arbitrary compressive force. Simplify the situation by ignoring the fibula and assuming that the tibia consists of three parts: a cylindrical diaphyseal segment of cortical bone, a metaphyseal segment of cancellous bone in the form of a truncated cone, and a disk-shaped cartilage layer. The dimensions shown in the sketch are a = 10 mm, b = 30 mm, c = 50 mm, e = 4 mm, f = 70 mm, and g = 130 mm. Assume the diaphysis is a hollow circular cylinder having inner and outer diameters as shown, and the cartilage and cancellous regions are uniformly filled with tissue. Assume also that stresses are uniformly distributed throughout each individual segment of the model and each tissue is linearly elastic with the following elastic moduli:

Cortical Bone	20,000 MPa
Cancellous Bone	200 MPa
Cartilage	20 MPa or 200 MPa

Calculate the total strain energy in each segment of the model as a percent of the total energy in the model. What is this percentage for the cartilage using the two different dynamic modulus values?

9.2. Calculate the average stress and strain values in the cartilage and cortical bone in Exercise 9.1 if the uniform compressive load is the body weight of a 70-kg person (i.e., approximately 700 N).

Appendix II

Chapter 2

2.1. Solve the second part of the hip force problem (the force equilibrium equations) by considering a free-body diagram of the weight-bearing lower extremity and using the result $P = 2W$. Remember to include the weight of the extremity itself. What is the disadvantage of using this free-body diagram to solve the moment equation?

1. Draw FBD of lower extremity:
2. From before, $\Theta = 30°$, $P = 2W$
   $\Sigma F_x = P\sin 30° - F_x = 0$
3. $F_x = P\sin 30° = 2W\,(0.5) = W$
   $\Sigma F_y = W - W/6 + P\cos 30° - F_y = 0$
4. $F_y = 5W/6 + 2W(0.866) = 2.57\ W$
5. $F = \left(F_x^2 + F_y^2\right)^{1/2} = \left(1^2 + 2.57^2\right)^{1/2}\ W$
   $F = 2.75\ W$
6. $\phi = \tan^{-1}(F_y/F_x) = \tan^{-1}(2.57/1) = 68.7°$
7. Using this FBD it would be hard to find the GRF and leg weight moment arms for $\Sigma M = 0$.

2.2. Suppose that the distance b in women's pelves is 10% greater than men's because of their greater pelvic width. How does this change their abductor muscle force magnitude and the magnitude and direction of their hip joint reaction force, assuming the moment arm of the abductor muscles is independent of gender?

1. Moments about the hip
   $\Sigma M = Pc - 5/6\,W\,b = 0$
   $P_M = 5/6\,W\,b_M/c = 2\ W$ (recall $b/c = 2.4$)

2. Let women's: $b = b_F = 1.1\ b_M$
   Then: $P_F = 1.1\ P_M$ (assuming $W_F = W_M$).
3. From $\Sigma F_x = 0$, $F_{xM} = P_M \sin 30°\ = 2W(½) = W$
   for women, $F_{xF} = 1.1$ W
4. From $\Sigma F_y = 0$, $F_{yM} = ⅚W + 2W\cos 30°$
   for women, $F_{yF} = ⅚W + 2.2W\cos 30°$
5. Then $F_F = (1.1^2 + 2.74^2)^{1/2}$ W = 2.95 W
   (Recall F = 2.75 W for men)
6. $\phi_F = \tan^{-1}(F_{yF}/F_{xF}) = \tan^{-1}(2.74/1.1) = 68.1°$
   (Recall $\phi = 68.7°$ for men)
7. Thus, a 10% increase in pelvic width would result in a 7.27% increase in hip force.
   What about the pressure on the joint surfaces?

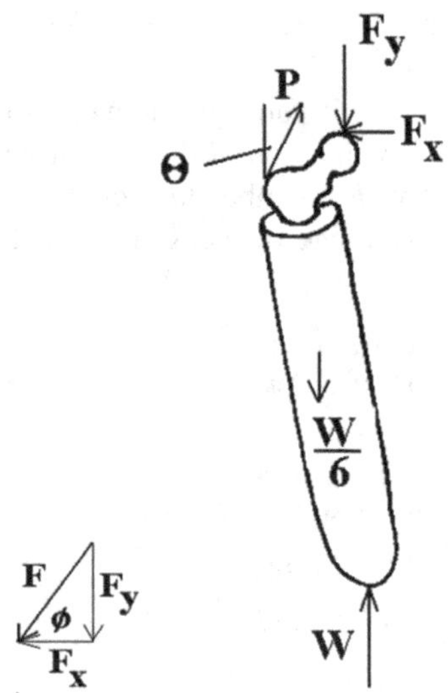

2.3. A male gymnast is performing the "iron cross" exercise on the rings, in which the arms are held straight out with a ring grasped in each hand, and the body hangs motionless and vertical between the rings. Consult an anatomy book to determine the principal adductor muscles for the arm in this position, and estimate their angle of pull and insertion point. Draw a free-body diagram and calculate the approximate magnitude of the adductor muscle force and the magnitude and direction of the reaction force in the glenohumeral (shoulder) joint during the maneuver. State the principal assumptions necessary for your calculation.

1. Neglect wt. of arm:
   $a = 0.80$ m,
   $b = 0.08$ m, $\phi = 45°$,

2. $\boldsymbol{\Sigma M} = aR - bL \sin \phi = 0$
   $L = a\mathrm{W}/2b\sin\phi = 0.8\mathrm{W}/0.16 \sin 45° = 7.07$ W
3. $\boldsymbol{\Sigma F_y} = F_y - L \sin \phi + R = 0$
   $F_y = L \sin \phi - R = a\mathrm{W}\sin\phi/2b \sin \phi - 0.5\mathrm{W}$
   $F_y = ½\mathrm{W}(a/b - 1) = ½\mathrm{W}(10 - 1) = 4.5\mathrm{W}$
4. $\boldsymbol{\Sigma F_x} = F_x - L\cos \phi = 0$
   $F_x = L\cos\phi = a\mathrm{W}/2b\tan\phi = 5\mathrm{W}$
5. $F = (5^2 + 4.5^2)^{1/2}\,\mathrm{W} = 6.73\mathrm{W}$, $\alpha = \tan^{-1} F_y/F_x = 42°$

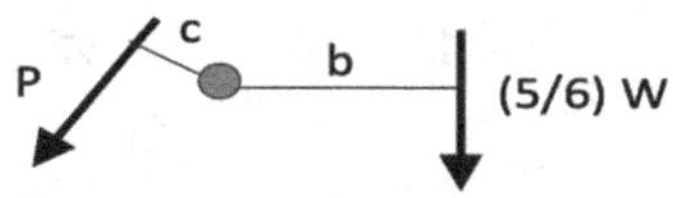

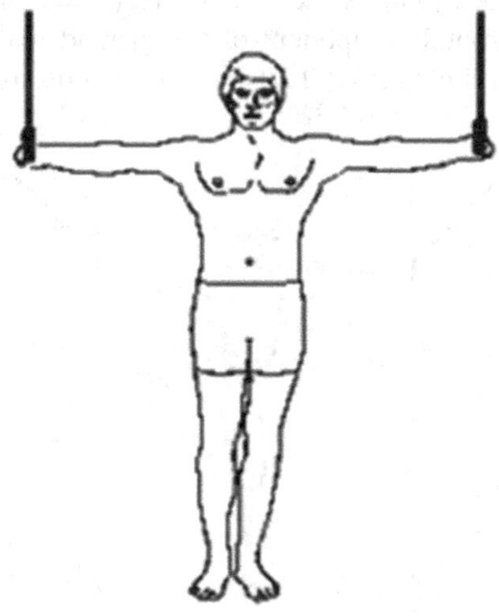

2.4. Suppose, in Exercise 2.3, that the ropes holding the rings do not rise vertically from the gymnast's hands but pull outward or inward at an angle θ. How does this angle affect the forces in the shoulder joint?

1. Assume R_x has no shoulder moment.
   $a = 0.80$ m, $b = 0.08$ m
   $\phi = 45°$, $R_y = ½\mathrm{W} = R\cos\theta$, $R = (\mathrm{W}/2) \cos\theta$
2. $\boldsymbol{\Sigma M} = aR_y - bL \sin \phi = 0$, $L = 7.07$ W as before
3. $\boldsymbol{\Sigma F_y} = F_y - L \sin \phi + R = 0$, $F_y = 4.5$ W as before
4. $\boldsymbol{\Sigma F_x} = F_x - L \cos \phi + R_x = 0$
   $F_x = L \cos \phi - R_x = ½\, a\mathrm{W}/b \tan \phi - R \sin \theta$
   $F_x = ½\mathrm{W}\ (a/b \tan \phi - \tan \theta) = ½\mathrm{W}\ (10 - \tan \theta)$

5. If $\theta > 0$, F_x is reduced. $F_x = 0$ when $\tan\theta = 10$
   $\theta = \tan^{-1}10 = 84.3°$, $R = W/2 \cos 84.3° = 5.0$ W

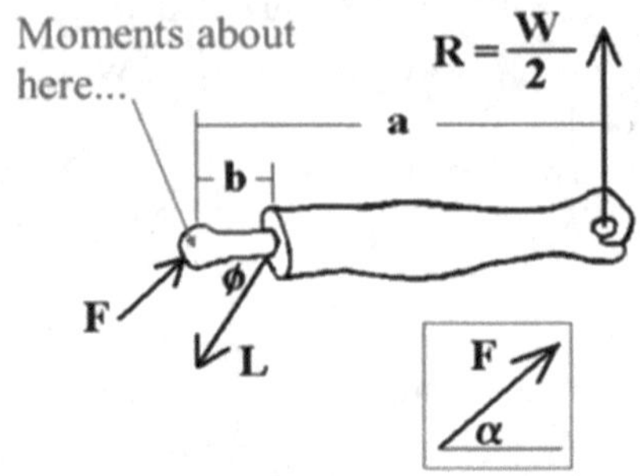

2.5. For the equine fetlock joint, work out the magnitude and direction of the joint force when a horizontal component of the ground reaction force (i.e., Gx in Fig. 2.10) is present. Let Gx = ±0.20 Gy, and assume, as before, that $L = 0.2$ m, $\Phi = 0$, $\Theta = 40°$, and $t = 0.05$ m.

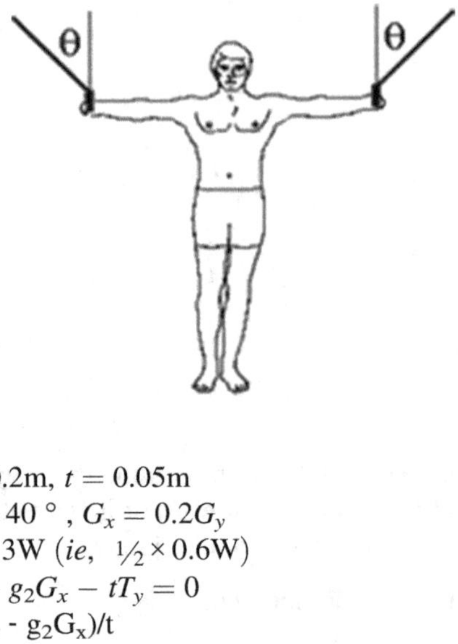

1. Assume $L = 0.2$m, $t = 0.05$m
   $\phi = 0$, $q = 40°$, $G_x = 0.2G_y$
   and $G_y = 0.3$W (*ie*, ½ × 0.6W)
2. $\boldsymbol{\Sigma M} = g_1G_y - g_2G_x - tT_y = 0$
   $T_y = (g_1G_y - g_2G_x)/t$
   Let $G_x = kG_y$, $g_1 = L\cos\theta$, $g_2 = L\sin\theta$
   $T_y = G_yL(\cos\theta - k\sin\theta)/t$
   $T_y = 0.3W(0.2)\ (\cos 40° - 0.2\sin 40°)/0.05$
   $T_y = 1.2W(0.766 - 0.129) = 0.765W$ for $G_x > 0$
   $T_y = 1.2W(0.766 + 0.129) = 1.07W$ for $G_x < 0$

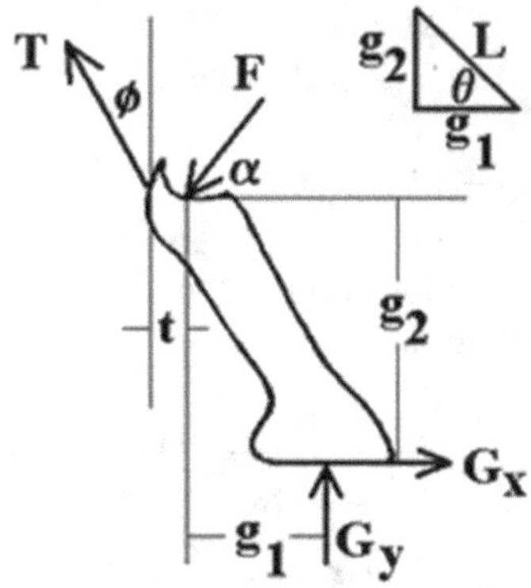

3. $\Sigma F_x = -T_x + G_x - F_x = 0$
   but $\phi = 0 = T_x$, so
   $F_x = G_x = kG_y$
   $F_x = (0.2)(0.3)\text{W} = 0.06\text{W}$ directed opposite to G_x
4. $F_y = T_y + G_y - F_y = 0$
   $F_y = T_y + G_y = 1.07\text{W} + 0.3\text{W} = 1.37\text{W}\ \ (G_x > 0)$
   $F_y = T_y + G_y = 0.765\text{W} + 0.3\text{W} = 1.07\text{W}\ (G_x < 0)$

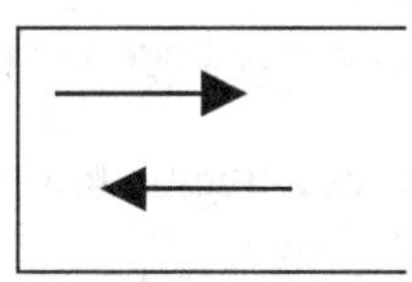

5. For $G_x > 0$, $F = (0.06^2 + 1.37^2)^{1/2}\text{W} = 1.37\text{W}$
   and $\alpha = \tan^{-1}(1.37/0.06) = 87.5°$
   For $G_x < 0$, $F = 1.07\text{W}$, $\alpha = 93°$

Chapter 3

3.1. Typically, a newborn infant's tibia is about 6.8 cm long. Estimate the present length of your tibia, then estimate the average rate of growth in the tibial growth plates, in millimeters/year and in micrometers/day. Again, assume that longitudinal growth of the tibia stops at age 14 in females and age 16 in males.

1. Adult length = L = 40 cm
   Initial length = L_0 = 6.8 cm
   $\Delta L = L - L_0 = 40 - 6.8 = 33.2$ cm
   $\Delta t = 16$ yr

2. Growth rate $= R = {}^{1}/_{2}\ \Delta L\ /\ \Delta t$

   ${}^{1}/_{2}$ because there are 2 growth plates

   Then…

   $R = \frac{1}{2} \times 33.2/16 = 1.04\ \text{cm/yr}$

   $R = 28\ \text{mm/day}$

3. $R >$ mean periosteal apposition rate,

   $R <$ BMU tunneling rate

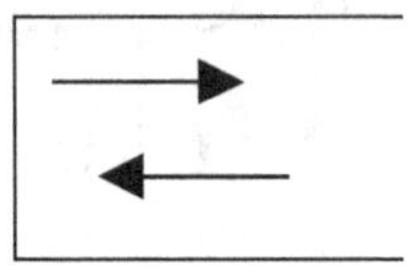

3.2. The rate of growth of a growth plate is equal to the daily chondrocyte production rate for each column multiplied by the height of the maximally hypertrophied cells. Explain why this is so in a steady state situation and state the necessary assumptions.

1. At the top of the GP, stem cells divide to produce proliferating chondrocytes.
2. These then divide several times each.
3. The number of new cells produced per day in each column is the result of all these mitoses.
4. Each chondrocyte progresses to the bottom of the column where the adjacent cartilage calcifies as the cell dies.
5. The tissue is converted to bone.
6. Because it starts as "nothing," each chondrocyte adds an element of length to the bone = to its hypertrophied height.
7. Thus, the growth rate = (#cells added/day) × (average ultimate height).
8. Assumptions:

   - All column cells originate from single stem cell.
   - All cells reach the bottom position.
   - All cells hypertrophy to average ultimate height.
   - Calculation includes the matrix above (or below) cell in the cell height.

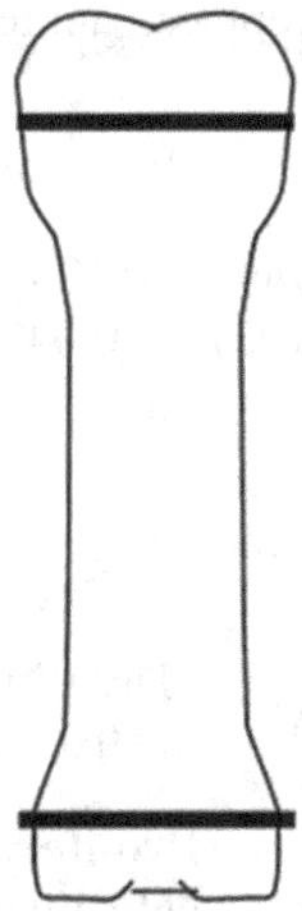

3.3. Bone can be considered a 2-phase, fiber-re-enforced material. Describe the nature of this composite structure at the nano-, micro-, and macrostructural levels. Compare and contrast it with existing implant materials.

ANSWERS MAY VARY

3.4. Compare the functional adaptations of bone, articular cartilage, and tendon/ligament with regard to their structure and their respective roles in skeletal support and load transmission. How do the morphological differences at each level of organization permit them to efficiently carry out their respective functions. Explain how these tissues act together to fulfill their roles and to protect each other from damage during the aging process.

ANSWERS MAY VARY

Chapter 4

4.1. Imagine a cross section through your femur. Assume that 5% of the bone in this cross section is replaced annually by osteonal remodeling. The osteons average 250 μm in diameter. What would the osteonal BMU activation rate (new osteons/mm^2 per year) have to be to produce the 5% turnover rate?

1. Consider a representative portion of the femoral cross-section of area A_T. Let

   osteonal cement line radius $= R_C$

   the osteon area $= A_O = \pi {R_C}^2$

   activation freq. $= f_a =$ osteons/yr/mm^2

2. # new osteons added/yr is

   $N = f_a A_T$

   The area they replace annually is

   $A_{\text{new}} = NA_O = f_a A_T A_O$

3. The fraction of the x-sect. replaced annually is
   $V_F = A_{new}/A_T = f_a A_O$
   and $f_a = V_F/A_O$
4. If $V_F = 5\ \% = 0.05$
   and $R_C = 250/2 = 125$ mm $= 0.125$ mm
   then $A_O = \pi R_C^2 = \pi(0.125)^2 = 0.0491$ mm^2
5. We have $f_a = V_F/A_O$
   $f_a = 0.05/0.0491$
   Ac.F $= f_a = 1.02$ osteons/mm^2/yr

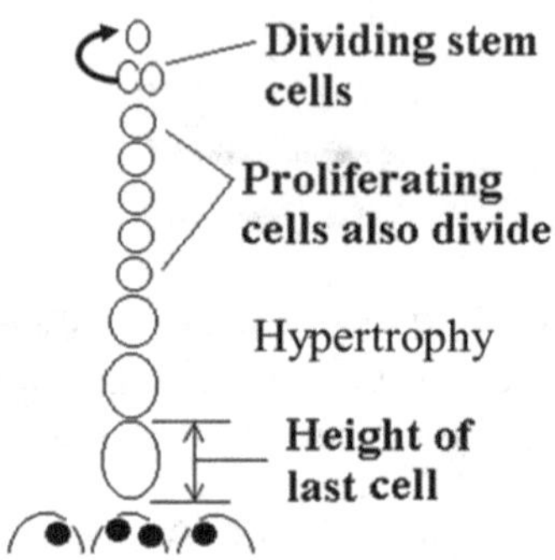

4.2. In Exercise 4.1, if each osteonal BMU produced a new Haversian canal that was 50 μm in diameter, what would be the rate of increase in bone porosity (mm^2 of canals/mm^2 of cross section/year)? Assume that the new osteons do not obliterate preexisting Haversian canals.

1. Let: A_T = Total area of a representative section
   A_H = area of Haversian canal
   dN/dt = Haversian canals added per year
   A_V = total canal (void) area in A_T
2. Then: The rate of increase in void space is
   $dA_V/dt = A_H dN/dt$
3. Because porosity $= p = A_V / A_T$,
   $dp/dt = (dA_V/dt)/A_T = A_H(dN/dt)/A_T$
   but, $f_a = (dN/dt)/A_T$ = activation frequency
   Thus, $dp/dt = A_H f_a$
4. Assume Haversian canal dia. is
   $D_H = 50$ mm $= 0.050$ mm
   Then $A_H = \pi\left(\frac{1}{2} D_H\right)^2 = \pi\left(\frac{1}{2}\, 0.05\right)^2$
5. $A_H = 0.00196$ mm^2
   From Ex. 3.3, $f_a = 1.02$ osteons/mm^2/yr

$$\text{Then}: \quad dp/dt = A_H f_a$$
$$dp/dt = 0.00196 \times 1.02$$
$$dp/dt = 0.0020 \text{ per year}$$
$$dp/dt = 0.2\% \text{per year}$$

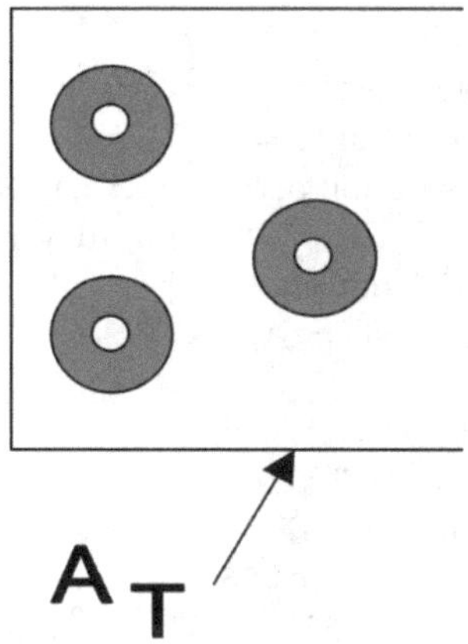

4.3. Draw a horizontal line representing time. Place a tick mark representing the present time at the left end of the line, so that past time is measured to the right. Let the time scale be 1 cm/week. Assuming the BMU resorption period is RP = 21 days = 3 weeks, mark the region of the line containing the "birthdays" (activation days) of all BMUs that are currently resorbing bone. Assuming the reversal period is RvP = 7 days = 1 week, mark the time span during which all BMUs presently in the reversal phase were activated. Finally, assuming the refilling phase lasts FP = 56 days = 8 weeks shows the time span during which all currently refilling BMUs were "born."

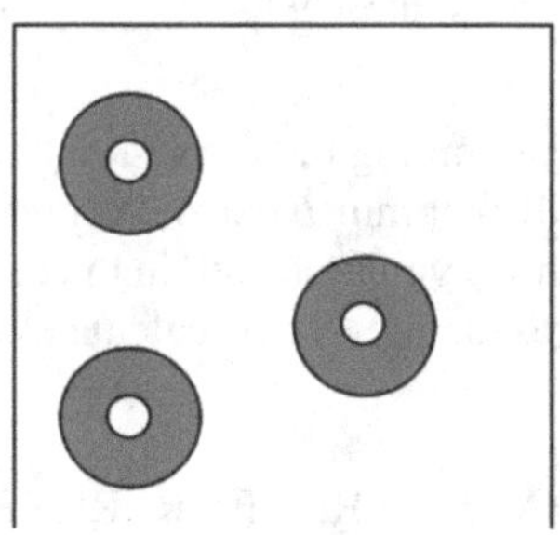

4.4. Now think about the cross-sectional view of a single osteon. Assume that the mean diameter of an osteonal cement line is 200 μm and that of the mean Haversian canal is 40 μm. Using these numbers, calculate Rs.R, the mean rate at which bone is resorbed in each osteon during its resorption phase, in μm^2/day, and F.R, the mean rate at which bone is replaced during refilling of each osteon, in μm^2/day.

1. Cement line radius = R_C = 100 mm
2. Area resorbed by osteoclasts = $A_C = \pi R_C^2$
   Rate of resorption = Rs. R = $A_R/R.\ P = \pi R_C^2/\sigma_R$
   Assume: RP = resorption period = 21 days
   Then: Rs. R = $\pi 100^2/21$ = 1496 mm^2/day
   Rs. R = 1496 × 10^{-6}mm^2/day
3. Area refilled by osteoblasts = A_F
   $A_F = \pi(R_C^2 - R_H^2)$
   Rate of refilling = $Q_B = A_F\ /\ FP$
   F. R = $\pi(R_C^2 - R_H^2)/FP$
   Assume: FP = refilling period = 56 days
   Then: F. R = $\pi(100^2 - 20^2)/56$
   $F.\ R$ = 539 mm^2/day = 539 × 10^{-6}mm^2/day
   Note that Rs.R/ F.R = 1496/539 = 2.8

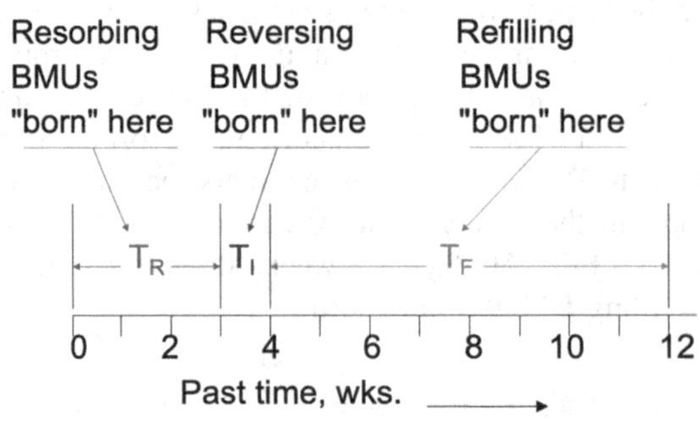

4.5. Calculate N.Rs and N.F assuming the activation frequency has been decreasing linearly at a rate of 5 BMU per mm^2/year and is presently 1 BMU per mm^2/year. Use the integration methods you practiced in the earlier exercises, and the same values for RP and FP as before. Then calculate something called the "bone balance":

$$B.BMU = F.R(N.F) - Rs.R(N.Rs)$$

using F.R and Rs.R from Exercise 4.4. Based on the equation, what are the units and meaning of F.R(N.F), and Rs.R(N.Rs), and B.BMU? Compare this bone balance with that which exists when the activation frequency is constant at

1 BMU per mm^2/year, assuming everything else remains the same. Why is one number positive and the other negative? Discuss the relevance of this calculation to the porosity of cortical bone in children and adults.

1. The plot shows how Ac.f is linearly decreasing at a rate of 5 BMU/mm^2/yr/yr or 2.5 BMU/mm^2/yr every 6 mo. Therefore, since the current value is 1.0, the value 6 mo. ago was 3.5 BMU/mm^2/yr and: Ac.f = 1 + 5t
2. For N.Rs=N$_R$, $N_R = \int_0^{T_R} Ac.f\,dt$

   where T_R = RP= 21 days = 0.058 yr.

   $N_R = \int_0^{0.058} (1 + 5t)\,dt = \left[t + \frac{5}{2}t^2\right]_0^{0.58}$

   N_R = 0.058 + 2.5(0.058)^2

   N_R = 0.0664 BMU/mm^2
3. For N.F = N_F, $N_F = \int_{T_R+T_I}^{T_R+T_I+T_F} Ac.f\,dt$

   where T_R+ T_I = 4 wks = 0.077 yr, T_R + T_I + T_F = 12 wks = 0.231 yr with T_F = FP:

   $N_F = \int_{0.077}^{0.231} (1 + 5t)\,dt = \left[t + \frac{5}{2}t^2\right]_{0.077}^{0.231}$

   $N_F = \left[0.231 + 2.5(0.231)^2\right] - \left[0.077\text{-}2.5(0.077)^2\right]$

   $N_F = 0.364 - 0.092 = 0.272\ \mathrm{BMU/mm^2}$
4. F.R(N.F) = mm^2/mm^2/day of bone being formed in refilling BMUs

   Rs.R(N.Rs) = mm^2/mm^2/day of bone being resorbed in resorbing BMUs

   B.BMU = mm^2/mm^2/day net formation rate of bone due to all active BMUs

   B.BMU = F.R(N.F) – Rs.R(N.Rs)

   B.BMU = [(539 × 0.272) – (1496 × 0.0664)] × 10^{-6}

   B.BMU = [146.6 – 99.3] × 10^{-6}

   B.BMU = 47.3 × 10^{-6} mm^2/mm^2/day
5. If Ac.f = 1 BMU/mm^2/yr, then

   $N_R = \mathrm{N.Rs} = (\mathrm{Ac.f})(\mathrm{RP}) = 1 \times (3/52) = 0.058\ \mathrm{BMU/mm^2}$

   $N_F = \mathrm{N.F} = (\mathrm{Ac.f})(\mathrm{FP}) = 1 \times (8/52) = 0.154\ \mathrm{BMU/mm^2}$

   $\mathrm{B.BMU} = \mathrm{F.R(N.F)–Rs.R(N.Rs)}$

   $\mathrm{B.BMU} = (539 \times 0.154) - (1496 \times 0.058) \times 10^{-6}$

   $\mathrm{B.BMU} = [83.0 - 86.8] \times 10^{-6}$

   $\mathrm{B.BMU} = -3.8 \times 10^{-6}\ \mathrm{mm^2/mm^2/day}$
6. When Ac.f = constant, B.BMU < 0 because each BMU removes more bone than it forms due to the Haversian canal. When Ac.f is decreasing, B.BMU may be positive because the number of new, resorbing BMUs is constantly declining, and the backlog of refilling work exceeds the current resorption work.
7. In children, d(Ac.f)/dt ≪ 0, and cortical bone porosity is decreasing because B.BMU > 0. In adults, Ac.f is approximately constant, and B.BMU tends to be slightly negative.

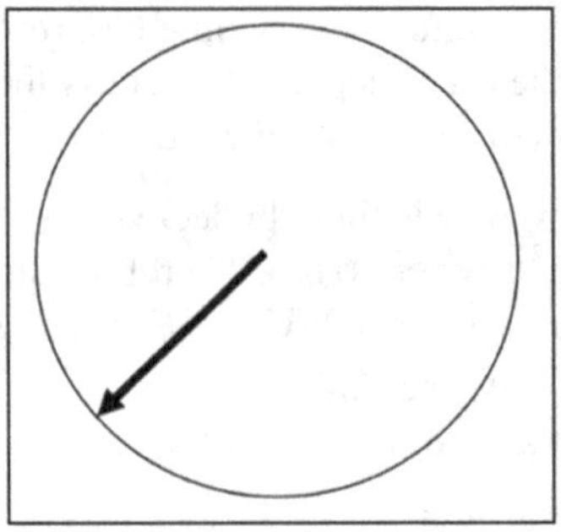

Chapter 5

5.1. Suppose a human bone has a cylindrical shaft whose periosteal and endosteal diameters are 2 and 1 cm, respectively. Using the data in Table 5.1, calculate the yield and ultimate load for this bone when loaded in simple compression. (Ignore buckling as a possibility.) Then calculate the torques (twisting moments) required to yield and break the bone by twisting it. When the outer fibers of the bone have reached the ultimate shear stress, are the fibers on the endosteal surface above or below the shear yield stress?

1. Loading in compression:

$R_p = 1.0 \text{ cm} = 0.01 \text{ m}, \quad R_e = 0.5 \text{ cm} = 0.005 \text{ m}$

$\sigma_{\text{yld}} = 182 \text{ MPa}, \quad \sigma_{\text{ult}} = 195 \text{ MPa}$

$\text{xsect.area} = \text{A} = \pi\left(\text{R}_\text{p}{}^2 - \text{R}_\text{e}{}^2\right)$

$A = \pi\left(.01^2 - .005^2\right) = 2.36 \times 10^{-4}\text{m}^2$

$F_{\text{yld}} = \sigma_{\text{yld}} A = 182 \times 10^6 \times 2.36 \times 10^{-4}$

$F_{\text{yld}} = 42,883 \text{ N}$

$F_{\text{ult}} = \sigma_{\text{ult}} A = 195 \times 10^6 \times 2.36 \times 10^{-4}$

$F_{\text{ult}} = 45,946 \text{ N}$

2. Loading in Torsion:

Polar moment of inertia $= J = \pi\left(R_p{}^4 - R_e{}^4\right)/2$

$J = \pi\left(0.01^4 - 0\,.005^4\right)/2 = 1.47 \times 10^{-8}\text{m}^4$

$\tau_{\text{yld}} = 54 \text{ MPa}, \quad \tau_{\text{ult}} = 69 \text{ MPa}$

$T_{\text{yld}} = \tau_{\text{yld}} J/R_p = 54 \times 10^6 \times 1.47 \times 10^{-8}/0.01$

$T_{\text{yld}} = 79.4 \text{ N-m}$

$T_{\text{ult}} = \tau_{\text{ult}} J/R_p = 69 \times 10^6 \times 1.47 \times 10^{-8}/0.01$

$T_{\text{ult}} = 101 \text{ N-m}$ at periosteal failure

The endosteal shear stress at failure is

$\tau_e = T_{ult} R_e / J = 101 \times 0.005 / 1.47 \times 10^{-8}$

$\tau_e = 34.5$ MPa

This is less than the yield shear stress, 54 MPa. That is, when the bone fails at the periosteal surface, the endosteal surface has not yet yielded.

5.2. Suppose the bone described in Exercise 5.1 has its shear strength diminished by 20% by the accumulation of Haversian canals with age. Assume also that 1 mm of bone is lost from the endosteal surface during this period of time. How much must the periosteal diameter be increased to compensate for these changes? If all these changes occurred simultaneously over a 30-year period, what would the mean periosteal apposition rate be, and how would it compare to typical values during the refilling of an osteonal BMU?

1. Assume ultimate shear strength is reduced 20%, so that after 30 years,
   $\tau_{ult} = 69 \times 0.8 = 55$ MPa
   The endosteal radius is increased from
   $R_e = 0.005$ m to 0.006 m.
   The failure torque would be
   $T_{ult} = \tau_{ult}\ J/R_p = \tau_{ult}\ \pi(R_p^4 - R_e^4)/2R_p$
   This cannot be solved for R_p, so solve by trial and error, get
   $R_p = 0.01089$ m for $T_{ult} = 101$ N - m
2. R_p is increased from 0.01000 to 0.01089 m, i.e., 0.89 mm = 890 μm are added over 30 years. The required apposition rate is

$$M_F = 890/(30 \times 365) = 0.081\ \text{mm/day}$$

which is $\ll$ 1mm/day, the typical apposition rate in a BMU.

5.3. Suppose a long bone diaphysis has cleanly fractured straight across and is healing through the development of a fracture callus. Sketch a cross section through the callus assuming circular geometry for it and the original bone. The outer diameter of the callus is 3 cm. Sketch in the original bone cross section, which had periosteal and endosteal diameters of 2 cm and 1 cm, respectively. Assume that no strength has developed between the cortices of the original bone, so that the strength of the healing fracture is entirely provided by the periosteal and medullary calluses. Calculate the bending moment required to fracture the original, intact cortex, assuming that its ultimate bending stress was 200 MPa. Now calculate the cross-sectional moment of inertia of the periosteal and endosteal calluses and compare them, individually and together, to that of the original bone. Assume the material in these calluses is homogeneous. What would the ultimate stress of the callus material have to be for it to provide the same strength as the original bone? What percentage of this strength is provided by the periosteal callus?

1. Original bending strength

$$I_o = \tfrac{1}{4}\pi\left(R_p{}^4 - R_e{}^4\right)$$
$$I_o = \frac{1}{4}\pi\left(0.01^4 - 0\ .005^4\right)$$
$$I_o = 7.36 \times 10^{-9}\text{m}^4$$
$$M = \sigma_b I/R_p$$
$$M = 200 \times 10^6 \times 7.36 \times 10^{-9}/0.01$$
$$M = 147\ \text{N-m}$$

2. CSMI of callus

$$I_p = \tfrac{1}{4}\pi\left(R_c{}^4 - R_p{}^4\right)$$
$$I_p = \frac{1}{4}\pi\left(0.015^4 - 0.01^4\right)$$
$$I_p = 3.19 \times 10^{-8}\text{m}^4 \ldots 4 \times I_o$$
$$I_e = \tfrac{1}{4}\pi R_e{}^4 I_e = \tfrac{1}{4}\pi \times 0.005^4 I_e = 4.91 \times 10^{-10}\text{m}^4 \ldots \tfrac{1}{15} \times I_o$$

Total callus CSMI:

$$I_c = I_p + I_e = 3.24 \times 10^{-8}\text{m}^4$$

3. Strength of callus material

$$\sigma_c = M\ R_c/I$$
$$\sigma_c = 147 \times 0.015/3.24 \times 10^{-8}$$
$$\sigma_c = 68.2\ \text{MPa}$$

…This is about 1/3 of s_b

4. Periosteal callus strength

$$M_p = \sigma_c I_p/R_c$$
$$M_p = 68.2 \times 10^6 \times 3.19 \times 10^{-8}/.015$$
$$M_p = 145\ \text{N-m} \ldots \text{This is 98\%of the total}$$

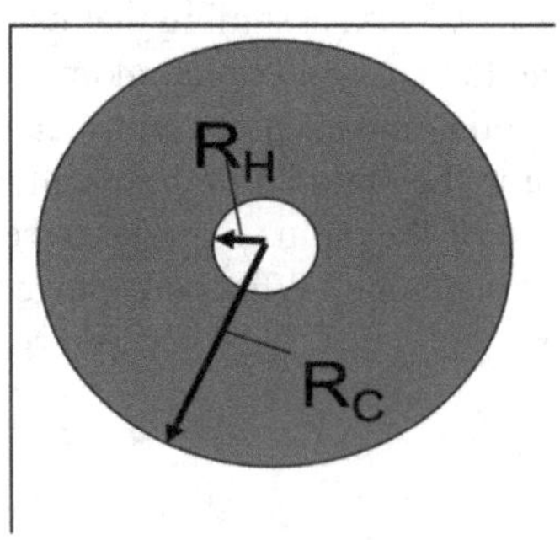

5.4. Consider the stress in the neck of the femur of a 70-kg person during the stance phase of gait. Use what you learned in Chap. 2 about the magnitude of the hip joint force, and assume that it is directed at an angle of 20° to the neck axis. Calculate the compressive and bending stresses produced by the joint force components parallel and perpendicular to the neck axis, respectively. Choose reasonable values for the bony dimensions. Then estimate the total axial stress on the superior and inferior surfaces of the neck. How do these stresses compare to the compressive strengths of cancellous bone shown in Table 5.4?

1. Assume the force acting on the head of the femur in a 70 kg person is
   $F = 2.75 \times \mathrm{W} = 2.75 \times 70 \times 9.8 = 1887\ \mathrm{N}$
   Neck radius:
   $R = 0.015\ \mathrm{m}$
   Compressive stress:
   $\sigma_C = \frac{F \cos 20^\circ}{\pi R^2}$
   $\sigma_C = \frac{1887 \cos 20^o}{\pi 0.015^2} = 2.51\mathrm{MPa}$
2. The bending stress is:
   $\sigma_B = \frac{Mc}{I} = \frac{F \sin 20^o LR}{\frac{\pi}{4} R^4}$
   $\sigma_B = \frac{1887 \cdot \sin 20^\circ \cdot .03 \cdot .015}{\frac{\pi}{4} 0.015^4}$
   $\sigma_B = 7.30 MPa$
3. Superior neck surface: $\sigma_S = + \sigma_B - \sigma_C$
   $\sigma_S = 7.30 - 2.51 = + 4.79\ \mathrm{MPa}$
   Inferior neck surface: $\sigma_S = - \sigma_B - \sigma_C$
   $\sigma_I = -7.30 - 2.51 = -9.81\ \mathrm{MPa}$

   $|\sigma_{\mathrm{I}}| > |\sigma_S|$
4. Ultimate compressive stress values for cancellous bone range from less than 1 MPa to more than 10 MPa. $\sigma_{\mathrm{I}} = 9.81$ MPa is near the upper end of this range.
5. To account for a cortical shell, the neck could be modeled as an inner cylinder of cancellous bone and an outer cylinder of cortical bone. Composite beam theory would need to be used to apportion the stresses between the two kinds of bone. An alternative would be to develop a finite element model for the proximal femur.

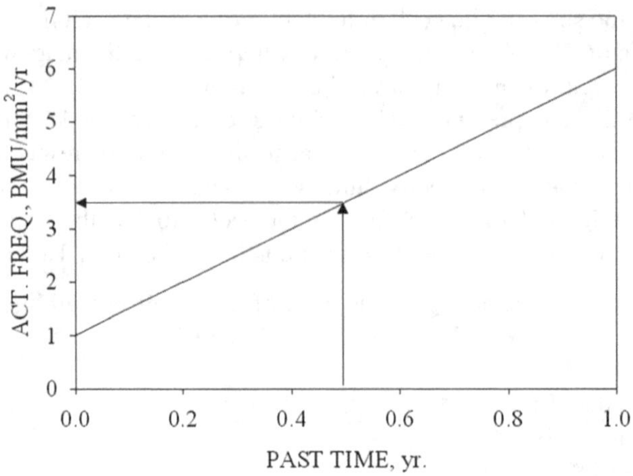

Chapter 6

6.1. In the Lanyon–Rubin avian model for studying the effects of mechanical loading on bone, a diaphysis is isolated as shown in Fig. 6.9. Pins extend from the metal caps over the bone ends, as shown. Assume a load is applied between the ends of the pins as shown. Calculate the longitudinal strain on the upper and lower surfaces of the bone, assuming it is cylindrical with the following characteristics:
Outer diameter = 20 mm
Inner diameter = 10 mm
Load = 30 N
Moment arm (d) = 40 mm
Elastic modulus = 20 GPa

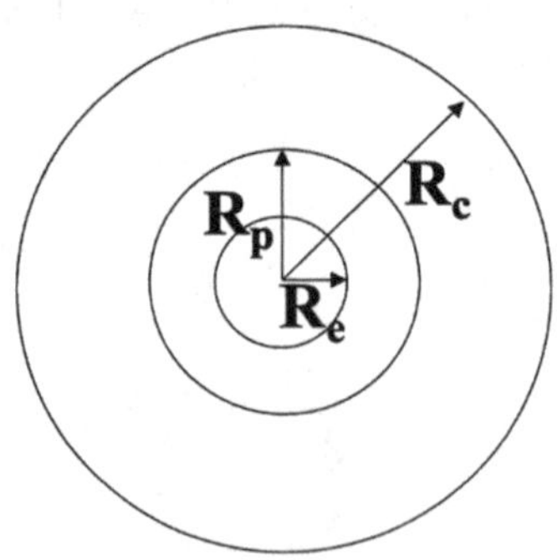

R_c=0.015 m
R_p=0.010 m
R_e=0.005 m
σ_b=200MPa

1. $R_p = 10$ mm, $R_e = 5$ mm
   $F = 30$ N, $d = 40$ mm
   $E = 20$ GPa
   Tensile load effect:
   X-sect.Area $= A = \pi\left(R_p{}^2 - R_e{}^2\right)$
   $A = \pi\left(10^2 - 5^2\right) = 236\ \text{mm}^2$
   $\sigma_T = F/A = 30/235.6 = 0.127$ MPa
   Bending load effect:
   $I = \pi\left(R_p{}^4 - R_e{}^4\right)/4 = \pi\left(10^4 - 5^4\right) = 7363\ \text{mm}^4$
   $\sigma_B = Mc/I = (F \times d)\ R_p/I$
   $\sigma_B = (30 \times 40) \times 10)/7363 = 1.63$ MPa $\quad (\pm)$
   On the upper surface, the bending stress is tensile:
   $\sigma_U = \sigma_T + \sigma_B = 0.127 + 1.63 = 1.76$ MPa
   On the lower surface, the bending stress is compressive:
   $\sigma_L = \sigma_T - \sigma_B = 0.127 - 1.63 = -1.50$ MPa
2. In terms of strain:
   $\varepsilon_U = \sigma_U/E = 1.76 \times 10^6/20 \times 10^9 = 88\ \mu\varepsilon$
   $\varepsilon_L = \sigma_L/E = -1.50 \times 10^6/20 \times 10^9 = -75\ \mu\varepsilon$

6.2. Assume that a diaphyseal cross section carries a bending moment $M = 2.5\ N\text{-}m$. Let the cross section be circular with inner and outer diameters of 30 and 20 mm, respectively, and assume that the neutral axis (NA) passes through the centroid. Draw the cross section and write an equation for the stress, $\boldsymbol{\sigma}$, as a function of angular position $\boldsymbol{\Theta}$ measured from the neutral axis. Assume that modeling adds or removes bone on the periosteal and endosteal surfaces at a rate proportional to the local longitudinal stress on the surface:

$$M_{F/R} = -K\sigma(\theta)$$

where the apposition rate $M_{F/R}$ is negative for tensile stress and positive for compressive stress, and the rate constant K is the same for each kind of modeling and each surface. Sketch a line adjacent to the original periosteal surface that would represent the addition or removal of new bone at each point around the circumference after some arbitrary period of time. Do the same for the endosteal surface. (Scale your sketch so that the maximum thickness of bone added is about half the original cortical thickness.) What has happened to the bone's cross-sectional geometry? Would this change be advantageous in some way?

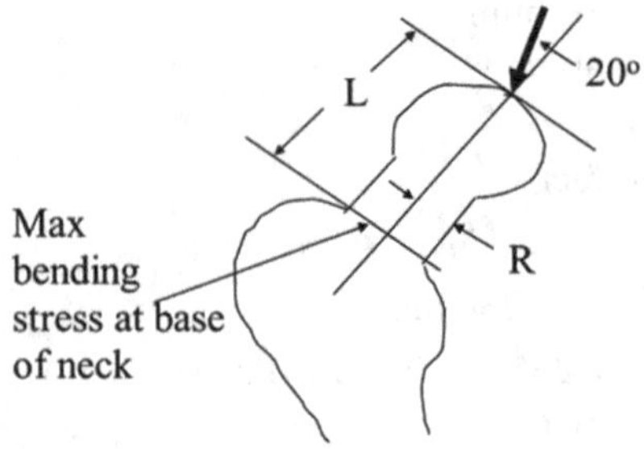

- The stress is σ = Mc/I
  On periosteal surface,
  $c = R_p \sin \Theta$
  $\sigma(\Theta) = -M R_p \sin \Theta / I$
  $I = \pi(R_p^{\ 4} - R_e^{\ 4})/4$
  $I = \pi(15^4 - 10^4)/4 = 31,900 \text{ mm}^4$
  $M = 2.5\ N - m = 2500\ N - \text{mm}$
  Then $\sigma(\Theta) = -(2500 \times 15/31,900) \sin \Theta$
  $\sigma(\Theta) = -1.176 \sin \Theta$
- The periosteal apposition/resorption rate is
  $M_{F/R} = -K\,\sigma(\Theta) = 1.176 \text{ K} \sin \Theta$
  Choose K so that $M_{F/R} = 2.5$ mm/yr at $\Theta = 90^o$
  $K = 2.5/1.175 = 2.13$ mm/yr/MPa
  Then $M_{F/R} = 2.5 \sin \Theta$
- For the endosteal surface $c = R_e = 10$ mm and
  $\sigma(\Theta) = -M R_e \sin \Theta/\text{I}$
  Then $\sigma(\Theta) = -(2500 \times 10/31,900) \sin \Theta$
  $\sigma(\Theta) = -0.784 \sin \Theta$
- The endosteal apposition/resorption rate is
  $M_{F/R} = -K\,\sigma(\Theta) = 0.784 \text{ K} \sin \Theta$
  Keeping the same K = 2.13 mm/yr/MPa
  Then $M_{F/R} = 0.784 \times 2.13 \sin \Theta$
  $M_{F/R} = 1.67 \sin \Theta$
- Note that, for both the periosteal and endosteal surfaces, $M_{F/R}$ is formative (+) on the compression side ($\sin \Theta > 0$) and resorptive (−) on the tensile side of the bone ($\sin \Theta < 0$).
- The results after 1 year are shown in this plot of half the bone:

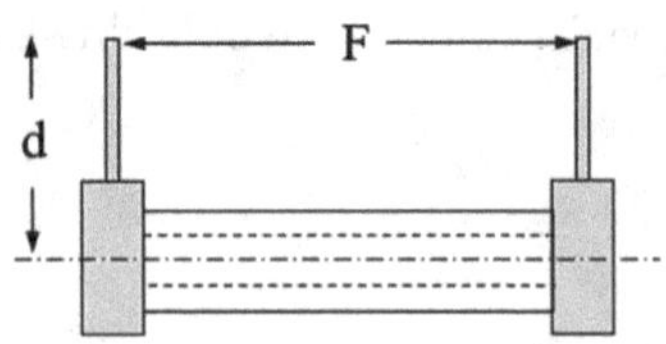

6.3. Reanalyze the bone in Exercise 6.2 assuming that $M_{F/R}$ is always formative and proportional to the magnitude of the longitudinal surface stress. How does the shape of this new bone distribution compare with that seen in Goodship's pig osteotomy experiment [205]? Discuss how the shape of the periosteal new bone distribution would change if M were proportional to the fourth power of the stress.

1. If $M_{F/R}$ is always positive and proportional to the magnitude of the surface stress, the periosteal diameter would be elongated in the direction of the bending load, and the endosteal diameter would become shortened in this direction. The cross-sectional moment of inertia of the bone would increase and the bending stresses would decrease.
2. The addition of bone in this model is not congruent with the result seen in the FIGURE TO THE RIGHT. There was no addition of bone on the lower periosteal surface, and the addition to the endosteal surface was asymmetric.
3. If $M_{F/R}$ were proportional to the 4th power of stress, the result would be an even more elongated cross-section.

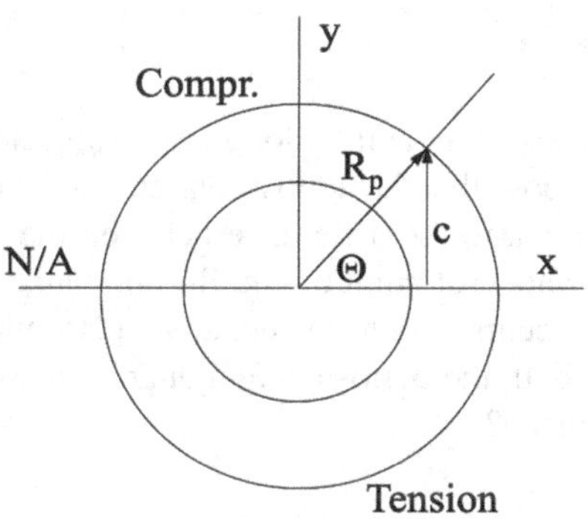

6.4. Reanalyze the bone in Exercise 6.2 assuming that MF/R is proportional to the radial gradient of stress. Assume that a positive stress gradient produces resorption ($M_{F/R} > 0$) and a negative stress gradient produces formation ($M_{F/R} > 0$). (A positive stress gradient is defined as one in which the stress becomes more tensile or less compressive, as one approaches the surface from within the bone matrix.) Let the rate constant be the same for formation and resorption and for the periosteal and endosteal surfaces. Discuss the variations of the results in Exercises 6.2, 6.3, and 6.4 and how they compare with the experimental results described in this chapter.

- Let, $M_{F/R} = \text{constant} \times d\sigma/dr$

  $\sigma = Mc/I = M\ r\ \sin\Theta/I$

  $d\sigma/dr = (M/I)\ \sin\Theta$

  $M_{F/R} = \text{const.} \times (M/I)\ \sin\Theta$

- On the tension side, the stress gradient is positive (resorption) at the periosteal surface and negative (formation) at the endosteal surface; vice versa for the compression side. The section drifts in the direction of the bending load. If this rule existed, a curved bone would straighten itself.

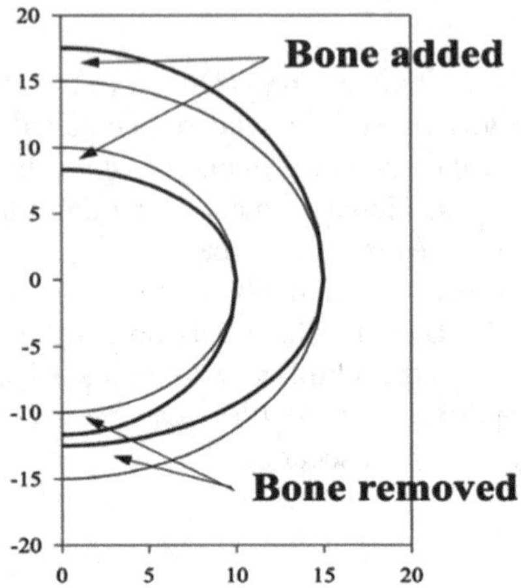

6.5. Consider the diaphysis of an adult's long bone subjected to a constant amount of daily torsion. Suppose that the apposition rate on the periosteal and endosteal surfaces is proportional to the difference between the average shear stress, σs (see Equation 6.4), and a set point or equilibrium value, σ_{sE}. If the difference is positive, formation occurs, and if it is negative, resorption occurs. Suppose that σ_{sE} is the same for both the periosteal and endosteal surfaces. What would the consequences of this be?

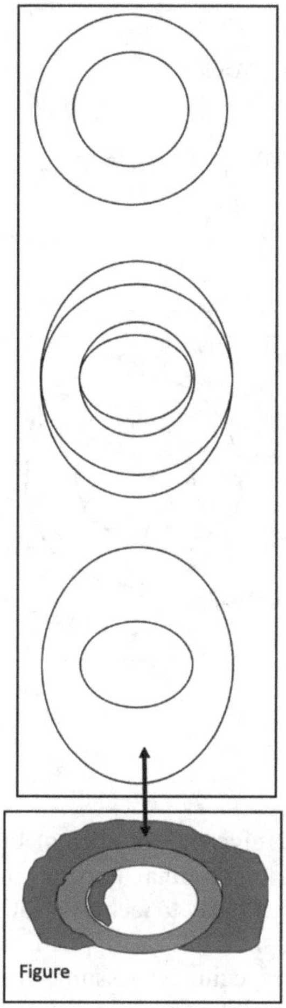

1. We have for the apposition rate
   $M_{F/R} = k\ (\sigma_s - \sigma_{sE})$
   The periosteal stress is:
   $\sigma_{sp} = Tr/J = 2TR_p/\pi(R_p^4 - R_e^4)$
   and the endosteal stress is:
   $\sigma_{se} = Tr/J = 2TR_e/\pi(R_p^4 - R_e^4)$
2. Because $\sigma_{sp} > \sigma_{se}$, if the set point or equilibrium stress, σ_{sE}, is the same for both surfaces, it is impossible for both surfaces to be in equilibrium simultaneously.
3. If $\sigma_{se} = \sigma_{sE}$, then $\sigma_{sp} > \sigma_{sE}$, and periosteal formation occurs. If $\sigma_{sp} = \sigma_{sE}$, then $\sigma_{se} < \sigma_{sE}$, and endosteal resorption occurs. Perhaps constant switching between these two conditions leads to radial expansion with aging.

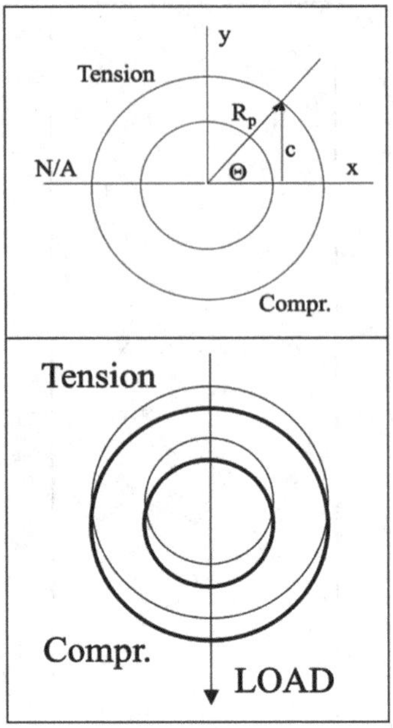

Chapter 7

7.1. Equations 7.18 and 7.19 give the number of cycles to failure for fatigue of human cortical bone in longitudinal tension and compression, respectively. How many days would it take to achieve fatigue failure in each of these modes of loading if the normalized applied stress range, *S*, is selected to produce 2000 **με** in the specimen? Assume the loading rate is 2 Hz. What experimental problems would arise from these long fatigue times? How long would the experiment take if *S* produced a superphysiologic 5000 με? Why would you not want to run the experiments at 20 Hz to reduce their duration?

Tension: $N_F = 1.445 \times 10^{53} \mathrm{S}^{-14.1}$

$S = 2000$ με, $N_F = 4,124,249$ cycles

There are 60 × 60 × 24 s/day, so the number of days required for failure would be

$t_F = 4,124,249/60 \times 60 \times 24 \times 2\mathrm{Hz} = 23.8$ days

Compression: $N_F = 9.333 \times 10^{40}\, S^{-10.3}$

S = 2000 με, $\mathrm{N_F} = 9,320,000$ cycles

and, $t_F = 9,320,000/60 \times 60 \times 24 \times 2\mathrm{Hz} = 53.9$ days

It is difficult to keep the testing equipment operating for this long, and the specimen will decay biologically, producing enzymes that will cause mechanical deterioration.

Tension: $N_F = 1.445 \times 10^{53} S^{-14.1}$
$S = 5000\ \mu\varepsilon, \quad N_F = 10$ cycles

$t_F = 10/2$ cycles/ sec $= 5$ secs
Compression: $N_F = 9.333 \times 10^{40} S^{-10.3}$
$S = 5000\ \mu\varepsilon, \quad N_F = 742$ cycles

$t_F = 742/120$ cycles/ min $= 6.2$ min
You would not want to load at 20 Hz because fatigue life is strain rate and frequency-dependent.

7.2. Equation 7.4 states a relationship between remodeling variables and the damage burden (D_0) waiting to be repaired in an equilibrium state: $D_0 = (k_D S^q R_L)/(\pi(Cm.\ Ln.\ Rd)^2(Ac.\ F_0)F_S)$. If the loading increases, an increase in the damage burden can be avoided by increasing either the activation frequency or the diameter of the osteons, or both. Does one of these options have any advantages over the other? Think about the differences in the porosity increase that would occur during and after the remodeling, the metabolic energy required to support the cells doing the remodeling, and the effects of the altered osteonal structure on fatigue damage resistance. What are the limitations on increasing Cm.Ln. Rd? How about Ac.f_0?

The damage removal (repair) rate is proportional to the rate at which bone is removed and replaced. This rate of bone replacement depends linearly on activation frequency and the cross-sectional areas of the osteons. Increasing either would require more osteoclast and osteoblast activity, and it is not clear that either choice has an advantage in terms of the metabolic cost to support the additional cells.

Increasing Ac.f_0 would produce more osteons, and insofar as osteons make the bone tougher and control crack propagation, this could be advantageous. However, the porosity of the bone would increase due to the remodeling space and the additional Haversian canals produced.

Increasing Cm.Ln.Rd would increase remodeling space, too. It would also increase the maximum distance that nutrients would have to travel to reach osteocytes at the periphery of each osteon. This could conceivably be detrimental.

Data suggests that bone toughness is greatest for many small osteons *or* fewer numbers of large osteons. It is not clear what this means in the context of this problem!

There are data showing that damage increases Ac.f_0. There are no data showing an effect on Cm.Ln.Rd, but the experiment has apparently not been performed.

7.3. Muscle fatigue may be an important cause of stress fractures. There are more muscles to control our movements than we actually need. The muscle fatigue theory of stress fracture holds that the additional muscles not only make refinements of movement possible but also reduce the stresses in the bones when the most common movements are performed. For example, there are six principal muscles that control the foot during the toe-off phase of walking. Sharkey et al. found that the compressive strain on the dorsal surface of the human second metatarsal bone at toe-off increases from about 1700 **με** to about 2190 **με** when just one of these (flexor hallucis longus) stops functioning. Using

Equation 7.17, find the percent decrease in fatigue life to be expected from this increase in strain.

From Eq. 7.17, $N_F = c/S^q$

Let * denote the situation in which muscle fatigue has increased strain values. Then:

$N*_F/N_F = S^q/S*^q$

If $q = 5$, $N*_F/N_F = 1700^5/2190^5$

$N*_F/N_F = 0.28$

If $q = 15$, $N*_F/N_F = 1700^{15}/2190^{15}$

$N*_F/N_F = 0.022$

Thus, the fatigue life might be reduced to a small percentage of the normal value.

7.4. The Pacific Crest Trail runs from Mexico to Canada through the Sierra Nevada and Cascade mountain ranges. Each year a few people succeed in walking the entire trail, starting in early spring and finishing in the fall. Many fail because of physical problems, including stress fractures. Using Equation 7.42, consider the relative effects of the distance walked and the weight of a backpack on the rate of fatigue damage formation in one of your foot or leg bones if you undertook this walk. Suppose that you ordinarily walk 5 miles/day. Plot a graph of the percent increase in damage formation rate as a function of the weight of your pack, expressed as a percent of your body weight, assuming you continued walking just 5 miles/day. Try relatively high and low values of q. Then convert this graph to a three-dimensional plot of percent change in damage formation rate vs. pack weight and distance walked per day.

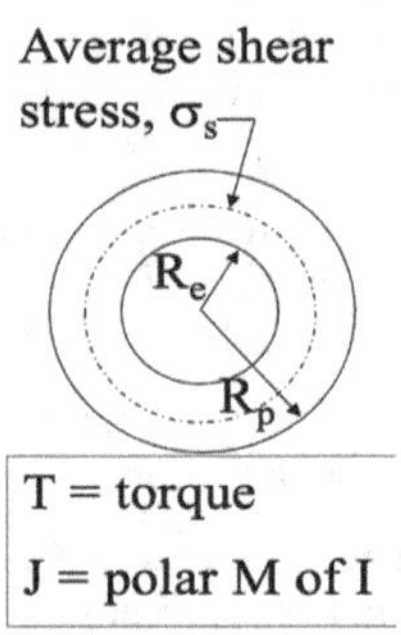

1. Assume the fatigue damage formation rate is (Eq. 7.42)

   $\dot{D}_F = k_D\, S^q\, R_L$

   Let: $W_T = W_B + W_P$

   $W_T = (1 + p)W_B$

   $R_L = R_M d$

2. Assume $S = C\, W_T$

   Then if: $\lambda = k_D C^q R_M$

   Equation 7.42: $\dot{D}_F = \lambda(1 + p)^q W^q{}_B\, d$

3. We have $\dot{D}_F = \lambda(1+p)^q W_B{}^q\, d$

   Normal walking without a pack,

   $p = 0,\ d = d_0$

   and, $\dot{D}_{F0} = \lambda W_B{}^q\, d_0$

4. Hiking with a pack, relative to normal walking,

   $$\dot{D}_F/\dot{D}_{F0} = \frac{\lambda(1+p)^q W_B^q d_0}{\lambda W_B^q d_0}$$

   $$\dot{D}_F/\dot{D}_{F0} = (1+p)^q d/d_0$$

   The second equation is the normalized equation.

5. Normalized damage form. rate = $(1+p)^q\ d/d_o$

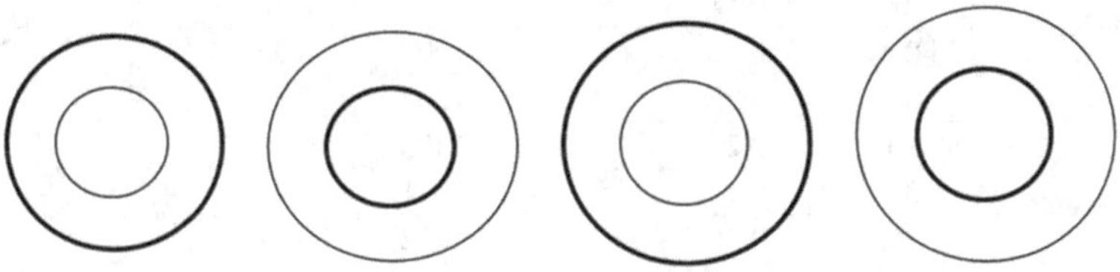

Chapter 8

8.1. Equations 8.1 and 8.2 represent empirical force-strain relationships for the tendon and aponeurosis of a frog muscle-tendon unit. In these, force is expressed as percent of the maximum tetanic muscle force. Calculate the tendon and aponeurosis strains when the muscle is exerting this maximum tetanic force. Which tissue has the greater strain? In general, what is the relationship between the tendon and aponeurosis strains for this muscle-tendon unit?

1. Equation 8.1 Tendon $F_T = 10^{(\varepsilon_T+0.633)/1.35}$

   Equation 8.2 Aponeurosis $F_A = 10^{(\varepsilon_A+3.63)/5.66}$

2. Assume that $F_T = F_A = 100\%$. Then:

   $$F_T = 100 = 10^{(\varepsilon_T+6.33)/1.35}$$
   $$\log_{10}(100) = \log_{10}\left[10^{(\varepsilon_T+0.633)/1.35}\right]$$
   $$2 = (\varepsilon_T + 0.633)/1.35$$
   $$\varepsilon_T = 2 \times 1.35 - 0.633 = 2.07\%$$

   $$F_A = 100 = 10^{(\varepsilon_A+3.63)/5.66}$$
   $$\log_{10}(100) = \log_{10}\left[10^{(\varepsilon_A+3.63)/5.66}\right]$$
   $$2 = (\varepsilon_A + 3.63)/5.66$$
   $$\varepsilon_A = 2 \times 5.66 - 3.63 = 7.69\%$$

   The aponeurosis has nearly 4× as much strain as the tendon. ($\varepsilon_A/\varepsilon_T$ = 7.69/2.07 = 3.7)

8.2. You are testing anterior cruciate ligament (ACL) reconstructions. You recorded the following load-deformation data for an experimental femur–ACL–tibia reconstruction in an animal model. Failure was by bony avulsion at the tibia. The ACL cross-sectional area was 150 mm^2 and the bone-to-bone resting length of the ligament was 3.5 cm. Calculate the important structural and material properties of the reconstruction.

Displacement, mm	Load, *N*	Displacement, mm	Load, *N*	Displacement, mm	Load, *N*
0.00	0	0.12	214	0.30	1908
0.01	0	0.14	379	0.32	1915
0.02	0	0.16	583	0.34	1940
0.03	0	0.18	796	0.36	2111
0.04	0	0.20	1000	0.37	2037
0.05	0	0.22	1202	0.38	1540
0.06	10	0.24	1406	0.39	1120
0.08	37	0.26	1611	0.40	821
0.10	96	0.28	1792	0.41	698

1. Plot the data:
   yield load = 1900 N
   ultimate load = 2100 N
   slope = 1400 N/0.14 mm
   = 10^4 N/mm
2. Cross-sectional Area
   $A = 150\ mm^2$
   Initial Length
   $L_0 = 3.5$ cm = 35 mm
   Elastic Modulus (ACL)

$$\boxed{E = \frac{1400/150}{0.14/35} = 9.33/0.004 = 2330 MPa}$$

   $E = 2.33$ GPa
   Yield stress (bone?) = 1900/150 = 12.7 MPa
   Ultimate stress (bone) = 2100/150 = 14.0 MPa

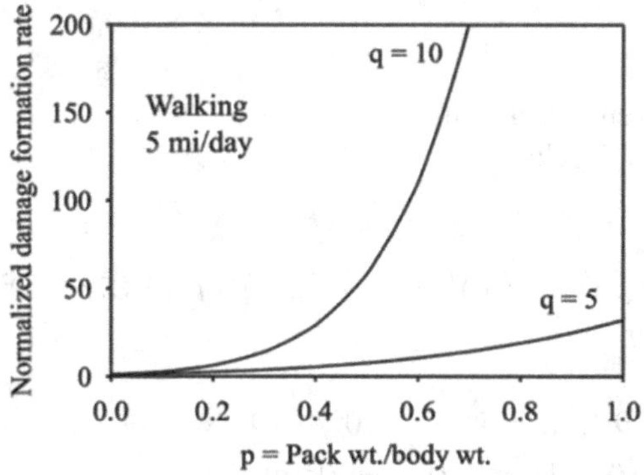

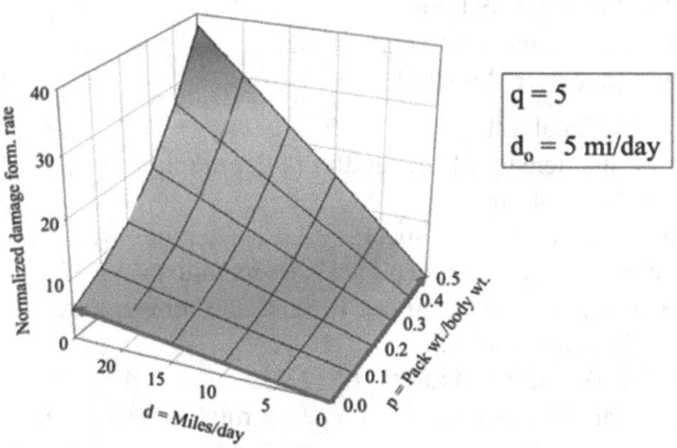

8.3. It was noted in the main chapter that tendon temperatures in racehorses may exceed 44°C. Suppose a typical tendon has an elastic modulus of 1 GPa and cycles between 0.5% and 2% strain during running. If the stress-strain relationship is linear in this range, what is the fluctuation in strain energy density as the tendon stress rises and falls? If that energy were expressed as heat, how much would it be in calories/ml of tendon/cycle? (The mechanical equivalent of heat is 4.19 joules/cal.) If the specific gravity and specific heat capacity of tendon tissue are similar to those of water (1 g/ml and 1 cal/g per °C, respectively), what percent of the strain energy must be converted to heat to raise the tissue temperature 1°C in 1000 load cycles, assuming adiabatic conditions (i.e., no heat leaves the site)? Is it plausible that tendons could be heated by loading during running?

Exercise 4.7:

1. E = 1 GPa
   SED = strain energy density
   Change in SED/cycle:

$$\Delta u = ½E{\varepsilon_2}^2 - \frac{1}{2}E{\varepsilon_1}^2$$

$$\Delta u = ½E\left({\varepsilon_2}^2 - {\varepsilon_1}^2\right) = ½ \times 10^9\left(.02^2 - .005^2\right)\Delta u = 1.875 \times 10^5 J/m^3$$

2. Heat:

$$\Delta Q = \Delta u/4.19\ \mathrm{J/cal} = 1.875 \times 10^5/4.19$$

$$\Delta Q = 44.8 \times 10^3\ \mathrm{cal/m^3} \quad 1\mathrm{m^3} = 10^6\ \mathrm{ml}$$

$$\Delta Q = 44.8 \times 10^{-3}\ \mathrm{cal/ml}$$

3. For 1000 cycles, the total heat will be

$$\Delta Q = 44.8 \times 10^{-3} \times 1000$$

$$\Delta Q = 44.8\ \mathrm{cal/ml}$$

   Letting the density of the tendon be 1 gm/ml,
   ΔQ = 44.8 cal/gm
   If tendon has the same specific heat capacity as water, i.e., 1 cal/gm/°C, this heat would raise the specimen's temperature 44.8°C.
4. The percent hysteresis required to raise the temperature 1°C during 1000 cycles of loading would be
   $\%u$ = 100 × (1 ° C/44.8 ° C) = 2.23%
5. It seems plausible that the strain energy might be sufficient to significantly raise the tendon's temperature several degrees during a long run, even accounting for heat transport away from the tendon.

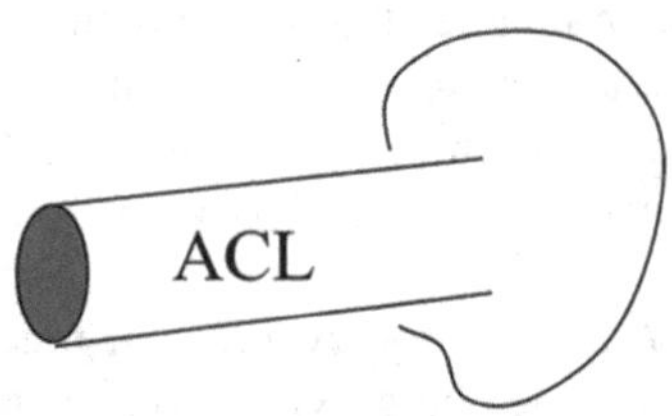

8.4. In Chap. 2 we learned that the skeleton is arranged as a system of levers that amplify the small displacements produced by contracting muscles. Instruments known as micromanipulators use levers to demagnify finger motions. In this chapter you have read that the flexor tendons of the fingers themselves have characteristics that serve to increase fine motor control; that is, they demagnify somewhat the motions of the fingers, which would otherwise be produced by the muscle contractions. Using the simple joint shown in Fig. 8.17 as a model, derive an equation to show how this can be achieved by making flexor tendons longer and more compliant. How would the tendon's length, diameter, and elastic modulus affect the resulting motion?

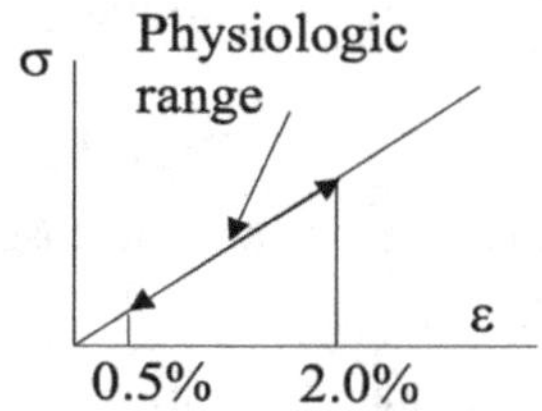

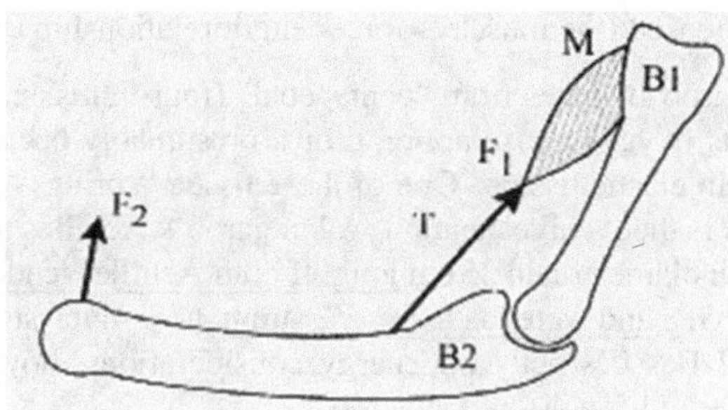

1. Suppose the muscle (M) contracts an amount ΔL_m, a fraction γ of its resting length, L_{m0}. Then the change in length of the muscle is
   $\Delta L_m = \gamma\ L_{m0}$
   Let the force produced by this contraction be F. Then
   $\boxed{\frac{F}{A_T} = E\frac{\Delta L_T}{L_{T0}}}$

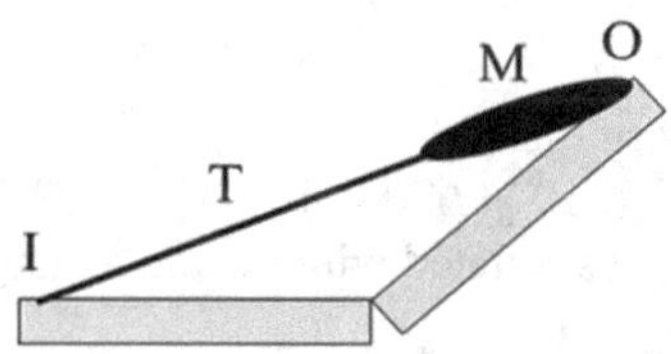

2. Solving for ΔL_T,
   $\boxed{\Delta L_T = \frac{L_{T0}}{E}\frac{F}{A_T}}$
   The net movement of point I toward point O is $\boxed{\Delta L_I = \Delta L_m - \Delta L_T}$ because the tendon stretches as the muscle contracts. We have
   $\boxed{\Delta L_I = \gamma \Delta L_{m0} - \frac{L_{T0}}{E}\frac{F}{A_T}}$
3. Assuming that the amount the muscle contracts is proportional to the force it produces ($\gamma = \beta\ F$), then the movement at I is
   $\boxed{\Delta L_I = \gamma \Delta L_{m0} - \frac{L_{T0}}{E}\frac{F}{A_T} = \beta F \Delta L_{m0} - \frac{L_{T0}}{E}\frac{F}{A_T}}$

$$\boxed{\Delta L_I = \left(\beta \Delta L_{m0} - \frac{L_{T0}}{EA_T}\right)F}$$

Consider the things that would reduce this movement.

$$\boxed{\Delta L_I = \left(\beta \Delta L_{m0} - \frac{L_{T0}}{EA_T}\right)F}$$

The things that would reduce ΔL_I and increase fine motor control are:

- L_{T0} is increased (longer tendon)
- E is decreased (more compliant tendon)
- A is decreased (thinner tendon)
- or β is decreased (the muscle's force-length relationship changes)

8.5. When a kangaroo switches from "pentapedal" (four limbs and tail) locomotion to hopping, its oxygen consumption drops, presumably because it then stores more energy in elastic tissues. One of these tissue "springs" in kangaroos and other animals is the Achilles tendon. A kangaroo's Achilles tendon was found to be 1.5 cm in diameter and 35 cm long. If both Achilles tendons had an elastic modulus of 1 GPa and were loaded to 2% strain, how much strain energy would they possess? Based strictly on energy considerations, how high could this amount of energy lift a 40-kg kangaroo?

The strain energy density in each Achilles tendon is

$u = \frac{1}{2}\,\sigma\varepsilon = \frac{1}{2}\mathrm{E}\,\varepsilon^2$

$E = 1$ GPa, $\varepsilon = 0.02$, so

$u = \frac{1}{2} \times 10^9 \times (.02)^2 = 2 \times 10^5\ \mathrm{J/m^3}$

1. The volume of each tendon is

   $V = \pi R^2 L = \pi\left(.75 \times 10^{-2}\right)^2 \times .35$

   $V = 0.620 \times 10^{-4}\mathrm{m}^3$
2. The total stored energy is

   $U = 2\ \mathrm{uV} = 2(2 \times 10^5)(.620 \times 10^{-4}) = 24.8$ J
3. The potential energy associated with raising a kangaroo's center of mass to a height H is

   $P = \mathrm{mgH}$

   Then $H = P/\mathrm{mg}$

   Substituting the strain energy, U, for the potential energy, P, we have,

   $H = U/\mathrm{mg}$

   $m = 40$ kg,

   $H = 24.8/(40 \times 9.8) = 0.0632$ m

   $H = 6.32$ cm

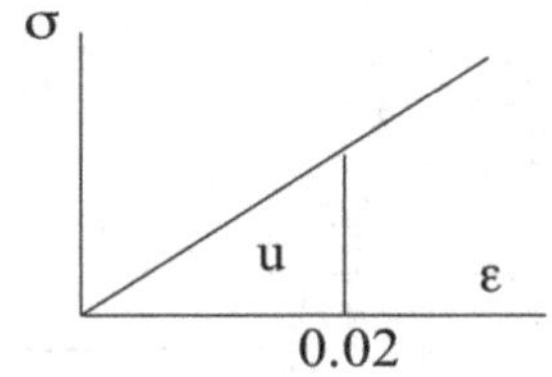

Chapter 9

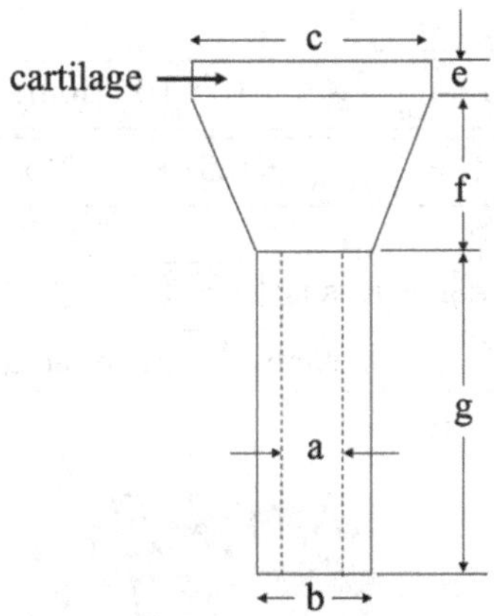

9.1. Determine the relative amounts of strain energy absorption in the cartilage and bone of the proximal half of the tibia when it is loaded uniformly over the articular surface by an arbitrary compressive force. Simplify the situation by ignoring the fibula and assuming that the tibia consists of three parts: a cylindrical diaphyseal segment of cortical bone, a metaphyseal segment of cancellous bone in the form of a truncated cone, and a disk-shaped cartilage layer. The dimensions shown in the sketch are $a = 10\,\text{mm}$, $b = 30\,\text{mm}$, $c = 50\,\text{mm}$, $e = 4\,\text{mm}$, $f = 70\,\text{mm}$, and $g = 130\,\text{mm}$. Assume the diaphysis is a hollow circular cylinder having inner and outer diameters as shown, and the cartilage and cancellous regions are uniformly filled with tissue. Assume also that stresses are uniformly distributed throughout each individual segment of the model and each tissue is linearly elastic with the following elastic moduli:

Cortical Bone	20,000 MPa
Cancellous Bone	200 MPa
Cartilage	20 MPa or 200 MPa

Calculate the total strain energy in each segment of the model as a percent of the total energy in the model. What is this percentage for the cartilage using the two different dynamic modulus values?

1. Cartilage

Stress: $\sigma = \frac{P}{A} = \frac{P}{\frac{1}{4}\pi c^2} = \frac{4P}{\pi c^2}$

SED: $u = \frac{\sigma^2}{2E} = \frac{16P^2}{2E\pi^2 c^4} = \frac{8P^2}{\pi^2 E c^4}$

Total strain energy: $U = u \times V$

Where volume of cartilage is: $V = \pi(c/2)^2 e$

$$U = \frac{\pi c^2 e}{4} \times \frac{8P^2}{\pi^2 E c^4} = \frac{2P^2 e}{\pi \ E c^2}$$

Let $P = 1$ Newton (unit force). Then

$$U = \frac{2 \times 1^2 \times .004}{\pi \times .05^2 E} = \frac{1.02}{E}$$

$E = 20\text{MPa}$: $U = \frac{1.02}{20 \times 10^6} = 5.10 \times 10^{-8}\text{Joule}$

$E = 200\text{MPa}$: $U = \frac{1.02}{200 \times 10^6} = 0.51 \times 10^{-8}\text{Joule}$

2. Trabecular bone

Assume the average stress is: $\overline{\sigma} = \frac{P}{\overline{A}}$

$\overline{A}$ = area in middle of metaphysis where radius is $r = (c + b)/4$. Then

$$\overline{\sigma} = P/\overline{A} = 16P/\pi(c+b)^2$$

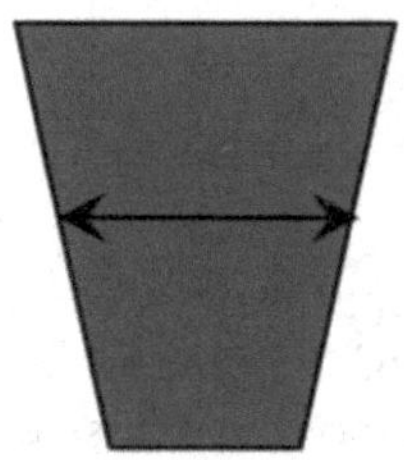

The SED in this average section is

$$u = \frac{\sigma^2}{2E} = \frac{16^2 P^2}{2E\pi^2 (c+b)^4} = \frac{128P^2}{\pi^2 E (c+b)^4}$$

The volume of the trabecular region is

$$V = \frac{1}{3}\pi f\left[\frac{1}{4}\left(c^2 + cb + b^2\right)\right] = \frac{1}{12}\pi f\left(c^2 + cb + b^2\right)$$

$$V = \frac{1}{12}\pi \times 0.07\left(0.05^2 + 0.05 \times 0.03 + 0.03^2\right)$$

$$V = 8.97 \times 10^{-5} m^3$$

The total strain energy is $U = V \times u$: $U = 8.97 \times 10^{-5}\left[\frac{128P^2}{\pi^2 E (c+b)^4}\right]$

$$U = 8.97 \times 10^{-5}\left[\frac{128 \times 1^2}{\pi^2 E (0.05+0.03)^4}\right] = 28.4/E$$

For the trabecular bone, $E = 200$ MPa, so

$$U = 28.4/200 \times 10^6 = 1.42 \times 10^{-7} \text{ Joule}$$

3. Cortical Bone

Stress: $\boxed{\sigma = \frac{P}{A} = \frac{4P}{\pi\left(b^2 - a^2\right)}}$

SED: $\boxed{u = \frac{\sigma^2}{2E} = \frac{16P^2}{2\pi^2 E\left(b^2 - a^2\right)^2} = \frac{8P^2}{\pi^2 E\left(b^2 - a^2\right)^2}}$

Volume: $\boxed{V = Ag = \frac{1}{4}\pi\left(b^2 - a^2\right)g}$

Total Strain Energy:

$$\boxed{U = uV = \left[\frac{8P^2}{\pi^2 E\left(b^2 - a^2\right)^2}\right]\left[\frac{1}{4}\pi\left(b^2 - a^2\right)g\right]}$$

$$\boxed{U = uV = \frac{2P^2 g}{\pi E\left(b^2 - a^2\right)}}$$

$$\boxed{U = \frac{2 \times 1^2 \times 0.130}{\pi E\left(0.03^2 - 0.01^2\right)} = \frac{103.5}{E}}$$

$E = 20$ GPa, $\boxed{U = \frac{103.5}{20 \times 10^9} = 5.17 \times 10^{-9}\text{Joule}}$

4. For cartilage, E = 20 MPa, total strain energy:

$U_{\text{total}} = U_{\text{cart}} + U_{\text{trab}} + U_{\text{cort}}$

$U_{\text{total}} = 5.10 \times 10^{-8} + 1.42 \times 10^{-7} + 5.17 \times 10^{-9}$

$U_{\text{total}} = 1.98 \times 10^{-7}$ Joule

Percentages:

$$\boxed{\%U_{\text{cart}} = \frac{2.04 \times 10^{-8}}{1.98 \times 10^{-7}} \times 100 = 26}$$

$$\boxed{\%U_{\text{trab}} = \frac{1.42 \times 10^{-7}}{1.98 \times 10^{-7}} \times 100 = 72}$$

$$\boxed{\%U_{\text{cort}} = \frac{5.17 \times 10^{-9}}{1.98 \times 10^{-7}} \times 100 = 2}$$

9.2. Calculate the average stress and strain values in the cartilage and cortical bone in Exercise 9.1 if the uniform compressive load is the body weight of a 70-kg person (i.e., approximately 700 N).

Exercise 5.2:

1. Cartilage, $P = 700$ N $\boxed{\sigma = \frac{P}{A} = \frac{4P}{\pi c^2} = \frac{4 \times 700}{\pi 0.05^2} = 3.56 \times 10^5\text{Pa}}$

E=20MPa $\boxed{\varepsilon = \frac{\sigma}{E} = \frac{0.356\text{MPa}}{20\text{MPa}} = 0.0183}$

E= 50MPa $\boxed{\varepsilon = \frac{\sigma}{E} = \frac{0.356\text{MPa}}{50\text{MPa}} = 0.00712}$

2. Cortical Bone, P = 700 N

$$\boxed{\sigma = \frac{P}{A} = \frac{4P}{\pi\left(b^2 - a^2\right)}}$$

$$\boxed{\sigma = \frac{4 \times 700}{\pi\left(0.03^2 - 0.01^2\right)} = 1.11 \times 10^6 \text{Pa}}$$

$E = 20$ GPa, $\boxed{\begin{array}{l} \varepsilon = \dfrac{\sigma}{E} = \dfrac{1.11\text{MPa}}{20,000\text{MPa}} = 55.7 \times 10^{-6} \\ \varepsilon = 55.7\ \mu\varepsilon \end{array}}$

Index

M. R. Allen, J. Wallace, *Skeletal Tissue Mechanics*,
https://doi.org/10.1007/978-3-031-90609-1

B

C

T